The Launchin

Modern American S

1846–1876

THE LAUNCHING OF MODERN AMERICAN SCIENCE

1846–1876

---

by Robert V. Bruce

Cornell University Press
Ithaca, New York

Published by arrangement with Alfred A. Knopf, Inc.

Cornell Paperbacks edition first published 1988 by Cornell University Press.

Library of Congress Cataloging-in-Publication Data

Bruce, Robert V.
The launching of modern American science, 1846-1876 / by Robert V. Bruce.—Cornell paperbacks ed.
p. cm.
Bibliography: p.
Includes index.
ISBN 0-8014-9496-6 (pbk. : alk, paper)
1.Science—United States—History. I.Title.
[Q127.U6B78 1988]
509.73—dc19 87-34010
CIP

Printed in the United States of America

The paper in this book is acid-free and meets the guidelines for permanence and durability of the Committee on Production Guidelines for Book Longevity of the Council on Library Resources.

To Marilyn

"The true spirit of science is of no country it may be said, but take the world as it is, and science is still essentially national."

—ALEXANDER DALLAS BACHE, 1844

"If we do not hitch onto the moon and quarry our granite there, it won't be the fault of the Yankees."

—CLEVELAND ABBE, 1866

Contents

Part One

SCIENCE IN A NEW WORLD, 1846–1861

Chapter 1

Prologue:

The Lay of the Land

Science and technology are the prime instruments of irreversible change in the thought and life of mankind, and for much of our own century the United States has led in wielding them. This is the story of how American scientists and engineers designed and launched their momentous enterprise.

They did it in the years from 1846 to 1876. Of course the roots and branches of so complex a story cannot be lopped off clean at those edges. But neither is the choice of bounds altogether arbitrary. Each of those two years witnessed crucial and symbolic events. To the nation, 1846 brought the Mexican War, and with it a last great surge of westward expansion. The expansion in turn touched off a sectional quarrel that flamed into civil war and ended only with the election of 1876. To American science, 1846 brought the Smithsonian Institution, the Yale Scientific School, and the arrival of Louis Agassiz; and 1876 brought the American Chemical Society, the Johns Hopkins University, and J. Willard Gibbs's epochal monograph "On the Equilibrium of Heterogeneous Substances." To American technology, 1846 brought the Hoe printing press, the Howe sewing machine, and Orson Munn's *Scientific American;* and 1876 brought Bell's telephone, Edison's Menlo Park "invention factory," and an apotheosis of technology in the great Centennial Exhibition. But these tallies merely bear out the fundamental point: between those two years Americans established national patterns and institutions in science and technology that still prevail. By 1876 modern American science and technology were fully formed and rising to their cloud-wrapped destiny.

In science as in other matters, the nineteenth century was a time for organizing. Until then, the scientific pursuit had for the most part been a small-scale, spare-time indulgence of individual curiosity. But science and technology had lately begun getting together to offer mankind a new range of possibilities for ease or adventure, pleasure or pain, increase or extinction. With such vistas opening, men began pursuing science more urgently, and the farther they pursued it, the more it ramified. The proliferation of its branches meant that individual scientists had to become more specialized. The growing complexity of science demanded formal scientific education and full-time professional work, not the casual, intermittent attention of self-taught amateurs. The spread of scientific investigation in so many directions called for the recruiting of scientists, the systematizing of communication among them, and the reliable evaluation of their work. And all this required money, even beyond the rising cost of more elaborate materials and apparatus.

In short, the pursuit of science had to become a collective enterprise, like those in business. Modern science needed labor, capital, and management. Its character at a given time and place was influenced by the kinds of raw materials available to it. It also had to consider markets, in the sense of offering grounds for financial support.

During the mid-nineteenth century, American science responded to all these new requisites. Science came to see itself, and society came to see it, as an established profession. Scientific education was lengthened and specialized by means of scientific schools, graduate education, and the modern university. These institutions in turn gave scientists a livelihood. Scientists learned to proselytize the public, and support by both private philanthropy and public agencies not only assumed a new scale but also developed new institutional patterns. Government agencies became founts of employment and scientific leadership. A national scientific society evolved to recruit manpower and public support for science, to give clearer direction and professional status to the work, to quicken communication, and to grapple with the peculiarly American dilemma of intellectual elitism in a social and political democracy. Congress chartered a national academy to stimulate scientific zeal through the prospect of official honor. Thus were the blueprints of modern American science drawn and the foundations laid.

Technology was so closely intertwined with industry and agriculture that its corresponding development was half-hidden. Government agencies (aside from the Patent Office) were less vital to it than they were to science, though federal subsidies for technological education loomed large. But the parallels to science were there: in the rise of professionalism through societies, higher education, and specialization; in the passage from the lone hand to the team, from Howe to Edison; in the appearance of professional journals and other media of communication; and even in the conscious adoption of scientific methods, quite apart from the application of scientific discoveries. The greatest new technological institution was the business corporation; and the years from 1846 to 1876 were the formative years of the corporate age.

Not only American science and technology, but also American life in general had by 1846 entered what has been called the Age of Enterprise.

In that year the nation's population stood at about twenty million, its area at well under two million square miles, its total wealth (in mid-twentieth-century terms) at less than five billion dollars. Thirty years later, its population had increased to forty-six million, its area to more than three million square miles, and its wealth to nearly twenty-five billion —all this despite one of the bloodiest civil wars in history. The nation's phenomenal economic growth, derived partly from land and partly from technology, accounts for much of what happened in American science during those years and most of what distinguished it from the science of other nations. Indeed, that growth accounts for most of what was distinctive in American life generally, then and later.

The New World's material promise influenced American life in two divergent ways, one coming first but overlapped by the other. The earlier influence, felt from the beginning, was that of abundant land that could be exploited by individual action in relative security. This tended toward the diffusion of capital, labor, and control; toward local and individual nonspecialization; and toward local and individual self-sufficiency. But commercial and (especially after 1790) industrial technology challenged these tendencies with the promise of still greater wealth to be won through collective, organized action. The new influence tended toward the concentration of capital, labor, and control; toward local and individual specialization; and toward interdependence of both individuals and localities. By 1846 the influence of technology was proving to be a match for that of abundant land, and by 1876 the way of collective, organized enterprise clearly dominated American life.

The tensions between these two ways to wealth were mirrored in American science during the nineteenth century, and not by coincidence. The same vastness of unexploited territory that shaped the life of early America had also offered immediate and direct, though limited, rewards for individual action in the earth and life sciences. So to some degree the scientists had shared the general tendencies toward diffusion, nonspecialization, and self-sufficiency. But, as with the nation at large, the diminishing of returns (only temporarily offset by territorial expansion) began to discourage those American scientists who took the individualist path. Meanwhile the very nature of scientific development called more and more urgently for larger-scale, collective effort, hence concentrated, specialized, and interdependent.

The relationship between science and society in mid-nineteenth-century America involved more than parallels. Developments in American life during that period directly affected American science. The national pride that supported wars to win or keep territory also spurred scientists and gave them a claim to public support. Their ambitions for American science had overtones of Manifest Destiny and the War for the Union. Territorial expansion drew them back for a time toward descriptive natural history and gave them government employment. The strengthening of democratic ideals and representative government forced American scientists to cultivate the general public. Technology made possible a new range of scientific communication through the telegraph, the modern postal system, the high-speed press, the extension of railroads, and the acceleration of city growth, a key to scientific vitality. And while the Civil War impoverished, distracted, or killed scientists, it also opened some institutional opportunities.

The history of the scientific enterprise in nineteenth-century America requires some account of its output. But measured against what Europeans were doing in those years, that output was modest. The emphasis of this book will be on the process more than the product, on the internal sociology, economics, and politics of science and on its interaction with the larger society. In these lies the real significance of the period.

Certain sciences will not be covered. Medicine is one of them, since medical science in this period cannot be easily separated from medical intuition and skill. Also omitted are sciences concerned largely with human ways, such as anthropology, philology, geography, ethnology, and psychology, because of their different character and also because of limits on time and space.

Technology will be treated with emphasis on its relationship to science. At the practical level, the rise of higher education in technology created teaching jobs for scientists, who in turn introduced future technologists to scientific norms and techniques. Furthermore, technology engendered new wealth to which scientists could appeal for financial support, and the triumphs of technology were themselves artfully seized upon by the scientists as justification for that support. During most of the nineteenth century, technology actually owed much less to scientific knowledge than the public was led to believe. Yet the scientists' claims were not so much unfounded, as they were premature. The boundary between technology and science was indeed becoming more flexible and permeable. And as technology increasingly adopted scientific methods and institutions, it even began to qualify as something of a science in itself. On that score too, technology has a place in this story.

Chapter 2

The European Model

"*American* science: is there such a thing?" asked the *New York Times* science reporter in 1860. He had no doubt of the answer. Though scientific principles are presumed to be universal, "yet the mode in which the human intellect formulates these phenomena will follow the structure of the nation's mind." Most scientists and historians, then and later, have concurred in this. Scientific emphasis, style, and institutions bear the stamp of a nation's culture and circumstances.

Nevertheless, scientists of one nation build on the work of scientists in all nations. In so doing, they pick up intimations of how that work was done, in what circumstances, with what facilities and encouragement, under what auspices; and they are sometimes led to follow suit or try to. To some extent, therefore, the manner as well as the matter of science tends to jump national boundaries. So a study of nineteenth-century American science must begin with European science—its structure, its characteristics, and its lines of transatlantic propagation.

By mid-century Europe had already shown what had to be done to advance science as the times demanded. Some nations had learned faster than others, and the German states fastest of all. The German response may have owed something to the rivalry of small states, within each of which a sovereign could spend money without consulting the taxpayers. A lot still went for royal frills. But the armies of Napoleonic France had knocked it into the crowned heads of Germany, especially that of Prussia, that to survive they needed the power of science, or at least of the national pride it helped foster.

Most of the princely sums spent on German science went for labor and housing rather than apparatus, since so much raw material of science still lay within reach of naked hand and eye. Yet science's growing scale of operations made even that kind of cost increasingly heavy. An

American scientist marvelled in 1846 at the Royal Museum in Berlin. "Everything here is on a grand scale," he wrote home wistfully. "The comparative anatomy alone occupies rooms more than 300 feet in length and 50 wide. . . . But let me talk of the insects—yes, the *insects.* [He was an entomologist.] Just think of it, they have over 40,000 species of coleoptera alone. . . . All the other orders are equally well represented. There are 4 or 5 men constantly at work on them—cleaning, arranging, classifying, describing, &c. &c."

The great vehicle of German science, however, was the university. For that also, the state paid the fare. "Oh! that we had such a university," wrote the envious entomologist in Berlin. "Such a corps of Professors—such philosophical [i.e., scientific] apparatus—such buildings—such a library of 600,000 volumes and," he added, "such patronage."

The German university—there were a score of them by 1846—was a community dedicated above all else to research. It met all the basic requirements of scientific advance. It lengthened education by four or five years, often to the student's middle or late twenties. Its decentralized structure, its freedoms of faculty to teach and students to elect what they chose, and its seminar method, all encouraged specialization. The easy and frequent transfer of students and scholars from one university to another aided scientific communication. Desire for university rank helped keep up professional standards and responsibility. And the university often guided its students to areas needing investigation, thus helping to coordinate scientific development.

In recruiting to science, Germany's political fragmentation gave universities all the more scope by limiting the competition of politics, colonial empire, and world trade. The university offered scientists a livelihood. To the better ones it also gave prestige. By word and example it conveyed to some students (all too few) the zest of scientific seeking and the exultation of finding.

In the year 1846 nothing like a modern university nor even serious graduate training existed in the United States.

There were other organs of science in Germany. Having no national capital as a mart of scientific knowledge, the Germans had gone far ahead of other peoples in publishing scientific periodicals. As elsewhere, local scientific societies had been formed in the cities. And in 1822 a score of German scientists had organized the Gesellschaft Deutscher Naturforscher und Ärzte, a scientific association for all the German states, meeting annually in one city after another. Though only actively publishing scientists could belong, membership rose to nearly five hundred in a decade.

Money, manpower, and organization had thus by 1846 given German science a lead in quantity of research. In quality, too, it stood high. Germany in 1846 could muster a company of giants: Helmholtz, Ohm, and Mayer in physics; Gauss in mathematics; Liebig, Wöhler, Bunsen, and Mitscherlich in chemistry; Bessel and Encke in astronomy; Von Baer,

Müller, Schleiden, and Schwann in biology; Humboldt in geography.

But France could call the roll of Foucault and Arago in physics, Cauchy in mathematics, Gay-Lussac and Dumas in chemistry, Leverrier in astronomy, Bernard and Magendie in physiology. The high noon of French science was already past for that era, but not long past; and the morning had been glorious. The Revolution had killed Lavoisier but brought a new birth of science as a source of light for humanity, of glory for France, of strength for industry, and of expertise for the armies. Old, cramping forms were smashed and swept away. In their place arose and grew the Ecole Polytechnique, meant to be useful, proving to be great. The Ecole saw alliance, not conflict, between research and teaching; and its brilliant faculty, including Laplace, Lagrange, Ampère, Sadi Carnot, Cauchy, Gay-Lussac, and Arago, drew visitors from abroad who carried its spirit back to their own schools.

Napoleon, the centralizer, the maker of rules, the stalker of the main chance, communicated too much of his own spirit to the Ecole Polytechnique and French scientific education generally. The central government harnessed the Ecole to military ends, and so its scientific pace and spirit slackened. The Ecole Normale Supérieure had its scientific section, the Sorbonne its Faculty of Sciences, the provinces their schools and colleges. But the central government on which they all depended was heavy-handed and closefisted. French savants followed the German example in 1833 with the Congrès Scientifique de France, meeting yearly in one city or another, with membership drawn from local scientific societies. Nevertheless, the dominance of Paris continued to shade the growth of provincial science. Journals were published, museums maintained, and yet the Germans seemed in most things institutional to be abreast or ahead of France.

In one institution, however, the French outdid the Germans and caught the attention of American scientific organizers: the national society called the Académie des Sciences of the Institut de France. For generations government money had helped support its researches and its published reports, the *Comptes Rendus.* The rare distinction of election, possible only when death made a vacancy, whetted individual ambition in France as the rivalry of universities did that of groups in Germany.

The English also had giants among them in 1846: Faraday and Joule in physics, Babbage and Sylvester in mathematics, Adams and John Herschel in astronomy, Hooker and Darwin in biology, Lyell and Murchison in geology. But the giants stood more nearly alone, not so firmly supported as in Germany by pyramids of pygmies, the rank and file workers who gathered data and applied conclusions. New local societies had grown up and flourished since the late eighteenth century, but amateurs predominated on their rolls. And otherwise England lagged sadly in organized collective effort.

The Prussian universities rested on a broad base of free, universal, compulsory education. The British had nothing of the sort. Prussia had forty-four free public libraries, France had more than a hundred, England had one. The work of bringing science before the people of England depended largely on private lecturers, cheap books, subscription libraries, local societies, and the "mechanics' institutes," which, for lack of literate mechanics, had slipped into middle-class entertainment.

The Royal Society, unsupported by government, had come under the control of nonscientific members and was sinking to little more than a London club for titled dilettantes. It had scarcely strength to tremble at the threat of the more vigorous national societies being organized around specialties like zoology, geology, and chemistry. The Royal Institution supported only one or two resident professors (though one of them was Michael Faraday).

Saddest of all was the scientific stupor of Oxford and Cambridge Universities. Scientific chairs often went to clergymen, who regarded them as sidelines or stopgaps. Their classrooms, as well as those of genuine scientists, were nearly or entirely deserted by the mid-1840s.

In 1830 the mathematician Charles Babbage published his *Reflections on the Decline of Science in England.* Only with organization, state aid, and paid professionalism, he asserted, could science keep up its rate of advance. Babbage's book struck sparks especially at Edinburgh, whose scientific circle came nearest to the spirit of continental science. From Edinburgh came a push that in 1831 helped create the British Association for the Advancement of Science, modelled on the German Gesellschaft, with similar annual migratory meetings, but less demanding membership requirements. The expressed aims included encouragement and coordination of research, better contacts between scientists, and better relations with both the public and the government.

During the forties, German university ways began to show up in London's Royal College of Chemistry, University College, and King's College. By the fifties, even Oxford and Cambridge had begun admitting the sciences as peers to the classics. All this had the backing of Prince Albert, Victoria's German consort. But continental science moved faster. Smugness, insularity, laissez-faire, class and sectarian cross-hauling left England farther behind when Albert died in 1861 than it had been a decade earlier.

Among the Russians, no doubt, the incidence of genius was as great as in any other nation, but not yet the revelation of it. The tyranny of Nicholas I blanketed the intellect of his people. In 1839 the great observatory at Pulkovo opened its interstellar eye with the guidance of the German-born Friedrich von Struve and the help of government largesse. But the czar considered higher education too unsettling for the common people. In 1846 the universities were forbidden to admit any student without a certificate of his high social origin. The aristocracy cared little for science; other classes were barred from it; the bureaucracy cramped the exceptions. Not until after 1855, when the death of Nicholas brought the thaw of Alexander II, could Russian science begin its flowering.

National characterizations were often ventured by contemporary observers. French science was seen as classical in form, mathematical in method, elegant in finish. By 1846 its clarifying influence had supposedly helped dispel the fog of metaphysics from German science. To French precision and logic, the Germans added thoroughness and massive organized effort, a sort of scientific strip mining carried down to basic principles. The English were regarded as self-financed individualists. The originality of genius in England thus had freer rein but less fodder, so that it often started insights to which other nations gave chase. For the great majority of English scientists, however, individualism merely reinforced the Baconian approach, the feeling that only the pressure of myriad discrete facts could warrant a hypothesis, just as English economists felt that only the pressure of innumerable individual decisions could properly govern the economy. The most typical English scientist was thought to be a fact-hunting amateur, buoyed up by a Baconian faith in the importance of his gamebag. So, at any rate, ran the consensus.

. . .

How was the European experience transmitted to America? The published proceedings of a few American societies listed their library acquisitions, and these included European books and journals. Such lists of titles gave Americans some hints of what Europe was doing. But in 1846 the only periodic general account published in America appeared in the *American Journal of Science,* which regularly imported European publications for review, excerpting, and summary.

American publishers reprinted or (there being no international copyright agreement) pirated some European scientific books in toto. But the more advanced a European work was in its field, the fewer were those Americans who could understand it, let alone buy it. Benjamin Peirce, a leading American mathematician, planned in the early forties to translate and publish the recent memoirs of Cauchy and others, but got no encouragement. No serious chemist could well ignore the works of Berzelius, yet American publishers balked at anything heavier than his short treatise on the use of the blowpipe. And at that, special type had to be cut for his chemical formulas. "There was a mess of trouble and an expensive job too," wrote the translator, "so I have not been near the printer for a week and think it very likely that he would like to back out of the whole scrape if he could."

So American scientists depended heavily on imported books and journals. Getting them took time and money. A chemist ordered the works of Berzelius and Rose one February; he was still waiting in November. He figured the cost at thirty-seven dollars, probably as much as he earned in a week or more. A few booksellers in the large port cities made a specialty of importing books, sending their own agents to Europe, even opening branches there. English books predominated, however, for lack of polyglot customers. Even the large and well-connected house of

Little & Brown in Boston, which had a book-importing contract with Harvard College, did not satisfy the college as to French material. German books, more and more important to scientists, were especially hard to get, at least until the failure of the 1848 revolutions drove some German intellectuals into American exile and thus expanded the American market.

Colleges therefore sometimes commissioned footloose faculty or even sent them especially to buy books and apparatus in Europe. Any scientist who somehow got to Europe would buy books for himself and usually for colleagues at their request. Some Americans subscribed to European journals or got society publications as corresponding members. Such men, their friends and neighbors, and those who lived near well-stocked college or society libraries might hope to scramble up onto the shoulders of European giants. Most of the rest could only play about dispiritedly in the shadow of the great ones, eventually getting some filtered light through the passage of European findings into general surveys or American writings.

Europe, for its part, took little interest in American scientific books before the mid-forties. Of 382 American books reprinted in England, 1833–43, only 9 were scientific. Nor did Americans push their wares. In 1845 the astronomer royal of Great Britain remarked to an American with "some surprise" that not one copy of the transactions of an American scientific institution had ever been sent him or the Royal Observatory. The British charged full domestic rates for American mail passing through England to the Continent. Postal charges on books going to and from Europe, complained an American scientist in 1848, often cost ten times the price of the books themselves. American scientists and societies had little money to spare for that. Besides, he said, the American societies gave their officials no standing instructions to exchange publications with European societies; and so little exchanging was done, and that tardily.

In 1848 an Anglo-American postal treaty lowered rates. Yet in 1852 the Royal Library of Berlin still got only one American periodical (the *American Journal of Science*) and the transactions of only one American scientific society (the American Academy of Arts and Sciences)—and those four or five years late. Four years later an American in London heard "many lamentations" about the difficulty of getting American scientific papers and books. British libraries had few state or federal scientific publications, though they wanted all they could get; and in general American science was "very badly represented in Great Britain through books."

For information and example, American science needed easier access to European publications. For encouragement and criticism, it needed more European comment on its own publications. But it needed something more. The European stimulus to American science, the call to competition, would have been much weaker without transatlantic communication man to man, both in writing and in person.

Even with letter rates at twenty-one cents a half-ounce to most European ports in the 1850s, those Americans who had (or hoped they had) some claim to European notice, and a number of humbler workers who could at least offer data or specimens from the American natural environment, corresponded directly with European counterparts. And fortunately for American self-confidence, the initiative was not always American. Europeans thought little of American ideas, but they needed American facts. From such transatlantic correspondence was to come one of the most famous private letters in the annals of science, Charles Darwin's 1857 letter to the American botanist Asa Gray, on the strength of which Darwin later established priority in his theory of biological evolution.

Yet even more important to American science than letters were the travels of Americans to Europe and of Europeans to America.

Chapter 3

A Procession of Pilgrims

Since the mid-eighteenth century, Americans studying medicine at Edinburgh University had also been getting heavy doses of other sciences. But they had not gone there for that reason. The European pilgrimage that heralded a new era of purposeful American scientific study abroad was made by a young Connecticut Yankee named Benjamin Silliman.

One hot July morning in 1801, in the shade of Yale's great elms, President Timothy Dwight stopped to chat with Silliman, back as a college tutor five years after graduating at sixteen. Dwight was seeking a man to fill a newly created professorship of natural history and chemistry, and though Silliman had a law career in mind, Dwight saw in him the prime requisites of love for God and Yale and the intelligence to pick up the necessary science. The law might pay better, Dwight conceded, but in science "the field will be all your own." The new nation was rich in unexplored materials for science, he pointed out, and by helping develop them the young man would be serving his country and winning reputation for himself. After consulting his conscience and a few elementary books on chemistry, Silliman accepted, and in 1802 his appointment became official.

Not content with two preparatory years of studying chemistry at the medical school in Philadelphia, Silliman persuaded Yale in 1804 to send him to Europe with ten thousand dollars for the purchase of books and apparatus. Napoleon's war-making kept Silliman from most of the Continent, but in London he frequented the learned societies and visited Humphrey Davy in his laboratory. The great revelation was Edinburgh. Half a century later Silliman wrote of the lectures he attended at Edinburgh University: "Upon that scale I endeavored to form my professional character, to imitate what I saw and heard, and afterwards to introduce such improvements as I might be able to hit upon." Silliman, moreover, made friends easily, and those he made abroad helped him keep Yale and American science in touch with the wider world.

Some thought Silliman overbearing and self-important. Occasionally he affected to know more than he did, and his own researches never came to much. Nevertheless, he did more than any other man in the first half of the nineteenth century to establish science in America as a profession. Tall, dignified, handsome, with a melodious voice, he was unsurpassed as a recruiter. By the end of his long life, no other American college teacher could count so many former pupils and assistants in the upper ranks of science. His public lectures also drew young men to science and helped persuade the general public that science was worth supporting.

Won over by Silliman's enthusiasm, a wealthy amateur named George Gibbs sold Yale his collection of minerals, the finest in America at that time. The Gibbs collection enabled Silliman to offer the first fully illustrated American college course in mineralogy and geology. In 1818 Gibbs prodded Silliman into founding the *American Journal of Science and Arts,* commonly known as "Silliman's *Journal.*" In 1846 it was still the nation's main channel of scientific communication. And Silliman, by then the unchallenged patriarch of American science, still edited it.

Meanwhile the booming American economy in the 1820s and early 1830s generated more of the wherewithal for European study, whether from family resources or the kind of college subsidy that had underwritten Silliman's own pilgrimage. And so a small wave of Americans broke on the European scientific community in the middle and late 1830s, among them such future luminaries as the botanist Asa Gray, the astronomer Elias Loomis, the naturalist Lewis Gibbes, the chemist James Booth, and the geologist Charles Jackson. They sat in on scientific meetings, hobnobbed with European professional scientists (some of whom they had previously corresponded with), toured and utilized European scientific facilities. The four last-named travellers even prefigured the decades to come by studying for a time under leading European scientists.

Of that decade's pilgrims, two in particular—Joseph Henry and Alexander Dallas Bache—were to stand far above the others as leaders in the shaping of modern American science. And so they merit a fuller introduction here.

Joseph Henry was born in Albany, New York, in 1797, the son of an obscure "cartman" who died when the boy was sixteen. Poverty had sadly limited young Henry's schooling, but something in him responded to a popular book on science that fell into his hands about then. Once that spark had been struck, the Albany environment kept it aglow and fed it. Albany was not only the state capital but also the ninth largest city in the nation, a center of trade and industry. It supported libraries, bookstores, public lectures, and two educational institutions of good repute, the Albany Institute and the Albany Academy. In his early twenties Henry spent some time studying at the academy, though he always considered himself "principally self educated." He also got help and encouragement from several local scientists, including one of Benjamin Silliman's former pupils, Amos Eaton, the founder in 1824 of the Rensselaer Institute in nearby Troy. After spells of studying anatomy and physiology, assisting the academy's lecturer on chemistry, and running a state road survey, Henry settled down in 1826 as Professor of Mathematics and Natural Philosophy at the Albany Academy. At one time he had seriously considered becoming a professional actor, but with this appointment he was confirmed as a professional scientist.

At Albany during the next six years Henry carried out the most notable series of physical experiments undertaken in America to that time. He chose to investigate electricity. Using multilayered windings of insulated wire, he made the first electromagnets strong enough for practical use. Independently of Ohm's theory, he worked out properly proportioned resistances in electric circuits. He made the first electric motor to use electromagnets and a commutator, and the first electromagnetic telegraph with a polarized armature, embracing the basic principle of the commercial telegraph. Most fundamentally he discovered mutual induction independently of Michael Faraday and self-induction prior to Faraday, whence the term "henry" for the standard unit of electrical inductance.

Meanwhile Henry had to make a living by seven hours a day of routine instruction in elementary science, along with added hours of paperwork and discipline. Only the vacation month of August left his time free and a schoolroom vacant for sustained experiment. Simple equipment and supplies—zinc, for example—were often unobtainable in Albany or too costly for Henry's thin purse. With few books or periodicals available, Henry all too seldom felt the spur and guidance of European work.

In his cramped circumstances Henry was wise or lucky to choose the field he did. The heights reached by electrical science in the 1820s were not yet too demanding for an Albany schoolteacher with little time and few books. Anyway, Henry was a limited theorist, though an outstanding experimenter.

Henry got a taste of what Europeans feasted on when Princeton gave him its chair of natural philosophy in 1832: a higher salary, beginning at a thousand dollars and rising to fifteen hundred; a rent-free home; more time for research; a little money for apparatus; and nearness to libraries and intellectual companionship in New York and Philadelphia. In the latter city he came to know not only Robert Hare, the chemist with whom Silliman had first studied, but also Hare's young colleague at the University of Pennsylvania, Alexander Dallas Bache.

Henry succeeded as a teacher at Princeton. Like Silliman, he cut a prepossessing figure, spoke fluently and impressively, and induced in his students at least a transient current of enthusiasm for science. He had classroom showmanship. And more decidedly than Silliman, he had a sense of humor.

At Princeton his research yielded more riches: the electromagnetic relay, noninductive windings, the concept (basic to the modern transformer) of stepping voltage up or down by properly proportioned coils, the relations of high-order induced currents, the action of induction at a distance, and other discoveries. Telegraphy, telephony, radio, the electric light and power industry, all depend on one or more of these ideas. As early as 1836, long before the consequences of his discoveries had unfolded, the Princeton trustees rewarded Henry with a year off at full pay. Thus in February 1837 he was at last able to sail for England.

Henry's young friend Alexander Dallas Bache had already gone abroad. Bache was the great-grandson of Benjamin Franklin, who had been the most notable American scientist before Joseph Henry. More than that, Bache's grandfather, his uncle by marriage, and his brother-in-law all served as secretaries of the treasury; and his mother's brother, George M. Dallas, became vice president. Young Bache seemed in a hurry to extend the family's distinctions. After graduating first in his West Point class at not quite nineteen, he spent three years teaching at West Point and on engineering duty. At twenty-two he became professor of natural history at the University of Pennsylvania. On his thirtieth birthday in July 1836 he was chosen first president of the projected Girard College and assigned to study European schools at first hand.

Some of his great-grandfather's traits seemed to reappear in Alexander Dallas Bache. Hearty, jovial, physically vigorous, broad-mouthed and blunt-nosed, he drew people to him at first sight. A firm handshake, a smile, a quick pleasantry in a gentle voice, and Bache at once seemed a longstanding confidant. He had a winningly evident capacity for affection. Though not often provoked there lurked in him an equally formidable capacity for combat. Kindly, tactful, and suave though he almost always was, his solid self-confidence and sheathed force had already imbued him with the habit of command.

As a scientist, however, Bache fell far short of both his famous ancestor and his friend Professor Henry. Unlike Henry's, his mind ran to quantitative observation and measurement. Aside from an article on the relation of color to heat radiation, he produced little of striking insight or originality. An English translation of a French translation of Berzelius, an analysis of Pennsylvania coals, the pressure-temperature variations of mercury vapor, the rates of salt water corrosion of various metals, improvements in steam boiler safety valves—these suggest the nature of his work. For some years he observed, recorded, and tried to fathom meteorological phenomena. His longest sustained scientific effort was, characteristically, a series of painstaking measurements of variations in terrestrial magnetism at Philadelphia from 1830 to 1845.

Bache left for Europe in September 1836, some months before Henry. By the time he returned in October 1838 he had piled up exhaustive data on nearly three hundred secondary and technical schools throughout western Europe, including Great Britain, France, and the German states —enough to make a report of 666 printed pages. Most of all he admired German education, from elementary school to Ph.D. The purpose of the report seemed to exclude data on universities, but "from the character of my associations, before leaving home," wrote Bache therein, ". . . I felt highly interested in this class of institutions. . . ." Apart from his formal mission, he fed hungrily on news of European science and the gossip of scientists, treasuring every personal contact with those of note. Ten years later he would write: "I do not know when I have been more gratified by a small matter than to find from my brother that as he was going thro' the Berlin Observatory, Encke recognized his name & asked for me."

Henry arrived in London on St. Patrick's Day, 1837, and Bache joined him there a week later. Together they bearded what Henry called the "lions of the city," attended lectures, demonstrations, and society meetings, and bought apparatus for their respective colleges. Henry and Faraday met at last and got on well. To Charles Wheatstone, then developing an electric telegraph, Henry explained his electromagnetic relay, a great help to Wheatstone's system. He met other scientific notables—Daniell, Sabine, Babbage—in England. Later he went with Bache to Edinburgh, which Henry, son of Scottish parents, loved. (Bache, a bon vivant like his great-grandfather, thought city and people cold and much preferred the freer-flowing wit and whiskey of Dublin.) In Paris, Henry sat in on a meeting of the Académie des Sciences and was annoyed at newspaper reading by members during the proceedings. En route home he made some remarks at the annual meeting of the British Association for the Advancement of Science, whereupon a boorish Briton derided him for presuming that an American should be taken seriously. For all that, Henry came home feeling up to the mark in the latest findings, experimental techniques, and institutions of European science.

Thus for both Bache and Henry, as for Silliman before them, a European pilgrimage helped shape a vision of what American science should be. Ten years later each would be in command of an institution which would do much to make the vision real: Bache as superintendent of the United States Coast Survey and Henry as secretary (i.e., director) of the Smithsonian Institution. Moreover, Bache and Henry would form the nucleus of a half-dozen or so scientific organizers and policymakers who called themselves the "Lazzaroni," and Bache would be the group's acknowledged "Chief." The Lazzaroni, singly and collectively, will dominate Part Three of this book. Indeed, as specimens, spokesmen, and leaders, they might be called the soul of it.

. . .

Though no other pilgrims of their generation did more to lead American science along the European path, Bache and Henry were not typical of those who made the grand tour in the years that followed. Bache and Henry came as mature observers. They did not study the works of European science so much as its workings—its structure, tone, mores, and polity. The true prototype of subsequent scientific Americans in Europe was a young Philadelphian, James C. Booth, who in 1832, "deficient in manipulative skill and not well grounded," as his European mentor put it, enrolled at the technical school of Hesse-Cassel as the first American laboratory student of Friedrich Wöhler, a founder of organic chemistry. Back home in 1836 Booth himself opened a private teaching laboratory that for forty years was a spawning ground for American chemists.

The main body of this new breed began to arrive in Europe during the 1840s, and in the van was a short, tough-fibered New Englander named Josiah Dwight Whitney, already hardened by childhood discipline into the brusque, self-sufficient man he would always be. First drawn to science as a boy by Silliman's popular lectures, Whitney graduated from Yale, studied chemistry with Robert Hare, and came under the wing of Charles T. Jackson, by then a leading geologist and chemist in Boston. It was Jackson who rescued Whitney from the unwelcome prospect of Harvard Law School by persuading the young man's father that science was now a tenable profession and a wide-open field with plenty of work for "practical intelligent men"—provided (as Whitney relayed the argument) that one devoted his "whole energies to that subject and that alone." Whitney sailed for Europe in May 1842.

For three years Whitney revelled in study and sights. Notwithstanding his crusty nature, he loved music and felt its spell from Russian churches to Liszt's recitals in Paris. But libraries and bookshops enchanted him too. His formal studies began at the Ecole des Mines in Paris, where he also attended free, government-supported lectures by leading French scientists. In Stockholm the great Berzelius himself gave Whitney a letter of recommendation to the noted chemist Heinrich Rose in Berlin; and when Whitney's funds ran short, Jackson came to the rescue by getting him a commission to translate one of Berzelius's works. (Jackson later referred to this as "the first & best service [Whitney] can render the Science of his Country.") And Jackson sent Whitney later commissions to hire German workers for Jackson's mining clients and to gather information on German processes for Jackson's cotton textile clients. So Whitney was able to follow in Booth's footsteps as a serious student of chemistry under the German masters.

Compared with Paris, Berlin seemed a dull city, unfrequented by Americans (after a year Whitney had not a single American acquaintance there). But that very dullness was at least conducive to study. And there was much to learn. The University of Berlin, Whitney wrote his father, led all other German universities in numbers of students and reputations of professors, especially in science. One might borrow books without limit from the Royal Library, always open. Professor Karl Rammelsberg, young but highly regarded, showed Whitney in a summer laboratory course what rigorous accuracy could be maintained in chemical analysis. Heinrich Rose, who accepted only four students at a time from applicants throughout the world, presently took Whitney into his laboratory, and Whitney began to speculate that a Berlin Ph.D. in chemistry might pay well at home.

In 1845 Whitney was joined by another young New Englander, who would become even more influential than Whitney in helping to shape the American way of science—in part because, unlike the fiercely independent Whitney, he would be admitted into Bache's exclusive coterie, the Lazzaroni. The newcomer's name was Oliver Wolcott Gibbs.

In embarking on a scientific career, Wolcott Gibbs (as he always styled himself) had no parental doubts to allay, his father being the George Gibbs who had lent support to Benjamin Silliman. At Columbia College in New York, moreover, the younger Gibbs studied under James Renwick, who, though more an engineer than a scientist, gave his students laboratory work in chemistry and electricity, a rare practice in American colleges of that day. After graduating, Gibbs, like Whitney, gained further experience in the laboratory of Robert Hare. From there, like most who sought advanced chemical study in America during those years, Gibbs went on to medical school. But his M.D. degree in 1845 did not satisfy him, and so he turned up in Berlin that spring.

Like Whitney, Gibbs had his prickly side. As a city, Berlin displeased him. The wind blew sand in his face, the stagnant gutters stank, the sidewalks were never swept, the beds were too short, and the cooking was "detestable," though the Berliners seemed to be always either discussing or ingesting it. Still, his fourth-story room rose above street smells, and on summer evenings in the beer halls he liked the local brew, the concerts, the air of jovial ease. At twenty-three (two years younger than Whitney), Gibbs was tall and well-proportioned, bore himself gracefully, and sported a new set of glossy black whiskers, hyphenated by a mustache, which led the Germans to think him French. Coming from a wealthy line of Connecticut governors, he had social poise and could make himself agreeable when he chose. Professor Rose invited him to dinner now and then, and he was a favorite at the home of Johann Poggendorff, editor of one of Germany's leading scientific journals. Besides, he considered the Germans ahead of other nations "in every branch of science" and Berlin "the very centre of circulation of philosophy & science."

Their bent for belittlement probably helped sustain Gibbs and Whitney. "This Europe," Whitney had once written home, "is a sad place to take down one's ideas of self, one's pride and vanity." But Whitney soon got over his humility, and Gibbs may never have been afflicted. Gibbs sized up Professor Rammelsberg as "a little bow-legged fellow of no great refinement of manners but withal a very good & careful analyst. . . . we have sometimes fierce disputes about nomenclature [but] on doctrinal points I believe we agree very well." Professor Rose, on the other hand, impressed Gibbs as "a fine looking man," and Rose in turn seemed quite taken with the cocky young New Englander. In Rose's lab Gibbs put in eight hours a day at inorganic analysis. His first analysis "succeeded very well," whereupon Rose rewarded him by expressing "a very high opinion of American science" and the belief that "for a young nation they had done much."

In 1846 Gibbs met another compatriot, the twenty-one-year-old Bostonian Benjamin A. Gould, Jr., who like Gibbs would eventually be accepted as one of Bache's Lazzaroni circle. At Harvard, Gould had followed the example of his scholarly father in studying ancient languages, until he fell under the spell of the mathematician Benjamin Peirce. Soon after graduating, Gould resolved to devote his life to astronomy and to the advancement of science in America. Though Peirce had not been to Europe, he and his disciple were highly conscious of it, Peirce having revised Bowditch's translation of Laplace, and young Gould having written a senior thesis on the publications of the British Association for the Advancement of Science. So Gould sailed for Europe in July 1845, visiting the astronomer royal, George Airy, at Greenwich, then Arago and Biot at Paris, before fixing at last on Germany for his astronomical studies.

Gould was brilliant, witty, articulate, and temperamental. In youth he made friends easily, though in later life he became famous among American scientists for making enemies. At Berlin he won the friendship of the renowned geographer and naturalist Alexander von Humboldt, who in turn commended him to Carl Friedrich Gauss, one of the greatest mathematicians of all time; later, while studying at Göttingen, Gould lived in Gauss's home.

Gould had a knack for finding or stirring up excitement. In Europe he joined liberal societies at some risk to his life. At Berlin in October 1846 he happened to be an assistant in Encke's observatory when the young astronomer Galle, at Leverrier's request, turned a telescope to the spot where, from disturbances in Uranus's orbit, the Frenchman had calculated an unknown planet would be found. There Galle indeed saw the faint disk of the planet later called Neptune. This dramatic evidence that the force of gravitation spanned the universe and that the mind of man could follow it can be seen in retrospect, and was felt then, as one of the great moments of science.

It was no wonder that the two high-spirited Yankees, Gibbs and Gould, became fast friends in Berlin and presently, over their plates of sauerbraten at the Cafe Belvedere, began mapping great futures for themselves and for science in America.

. . .

Neither John P. Norton nor Eben N. Horsford made Berlin his Mecca, nor did either of them join the Bache coterie in later years. Nevertheless, each of those young scientific pilgrims of the 1840s brought significant aspects of European science back to America.

John P. Norton had considered various careers, none with enthusiasm. It was his father, an Albany businessman and part-time farmer, who suggested agricultural chemistry. The young man studied chemistry for a time under Benjamin Silliman and his son at Yale. In 1844, when the twenty-one-year-old Norton decided to prepare in Europe for teaching agricultural chemistry in America, the elder Silliman steered him to Edinburgh and the laboratory of a leader in the new field, Professor James Johnston of the Highland Agricultural Society. There Norton spent eighteen months on a chemical analysis of the oat plant, a rather narrow subject, however dear to Scottish hearts. But he also analyzed soils and fertilizers for the farmer members of the society, visited farms and advised their operators, and gave public lectures in various towns. All this readied him to introduce the same sort of work later into Connecticut, work he had done much to create a demand for with his letters from Europe to American farm journals. Meanwhile, Norton's father was persuaded, probably by Benjamin Silliman, Jr., to donate five thousand dollars toward establishing a Yale chair in agricultural chemistry for his son. This led to the organizing of a scientific school at Yale in 1846, which in turn set Yale on its way toward becoming a European-style university.

Eben N. Horsford, son of a progressive farmer in upstate New York, came to science not through the one-year engineering course he took at Rensselaer Institute but through assisting New York State geologist James Hall for a time in collecting fossils. Hall's friend Professor John W. Webster of Harvard amounted to little as a scientist, but he happened to be one of the first in America to recognize the work of the German chemist Justus von Liebig, proposing a course on it in 1841, only a year after the publication of Liebig's *Organic Chemistry in Its Application to Agriculture and Physiology.* Thanks to Webster, Horsford, like Norton, chose agricultural chemistry as a field; but at Webster's urging, Horsford, unlike Norton, sought out that field's acknowledged founder at the University of Giessen in 1844. By then Horsford was a bland, balding young man of twenty-six, keenly alive to opportunity, though inclined to scatter his abundant energies in too many directions.

Horsford was not Liebig's first American student. John Lawrence Smith, a South Carolinian with an M.D. degree, had studied awhile at Giessen a year or two earlier. But Horsford's letters from Giessen in a widely read farm journal, the *Albany Cultivator,* made Liebig and his work known to many American farmers, as Norton's published letters had done for Johnston. Moreover, the Horsford letters were perhaps the first comprehensive account of a German scientific education published in the United States. As a symbol and model of that education for a generation of American students and teachers, Liebig's establishment deserves a close look here.

Horsford found the master in his private lab, seemingly absorbed in thought, curt and businesslike in manner. As one of a hundred students at the first of that season's lectures, however, Horsford was won over by the dexterity of Liebig's demonstrations, the clarity of his explanations, the store of knowledge he drew upon. "He is all mind," wrote the new disciple. The elder Silliman, in a second visit to Europe some years later, dropped in on a lecture and confirmed the description. He found a hundred students attentively taking notes while Liebig, tall and suave, lectured in a quiet, musical voice and a natural manner without French-style histrionics.

As a Rensselaer graduate Horsford had known a teaching lab before, but Liebig's reached a new level and scale. Seventeen men had graduated with Horsford at Rensselaer; that had been considered a big class. Liebig had thirty or forty advanced students in his lab and nearly as many beginners in that of an assistant. The two teaching labs were big, yet crowded with apparatus, reagents, and hard-working students (Horsford put in fourteen hours a day). The beginners and those not in organic chemistry got an occasional gracious nod from Liebig as he strolled around with his hands in his pockets and a "smoking cap" on his head. At least twice a day each of the others, working on a problem assigned him by Liebig, would find the master watching him, perhaps looking him in the face for ten disconcerting minutes without a word, more likely showing the student a better way to do something, asking questions, raising doubts, offering insights. Twice a week Liebig would formally review their work, his own, and that of chemists elsewhere. Everyone worked from dawn to dusk. Chemistry was their chief dissipation and for most the only one. They seemed chronically a little drunk with the excitement and comradeship of a great united quest. The janitor kept complaining that he could not get them out of the lab at night so that he could clean it.

One measure of Liebig's impact on students was the conversion of Whitney and Gibbs. That hardheaded pair wanted to go on to organic chemistry but felt sure that Liebig had been overrated. Whitney doubted that "his genius can communicate itself to his pupils by his merely looking at them once a day," and Gibbs noted that Liebig's fees were four times Rammelsberg's. But Professor Rose and Eben Horsford both urged them to give Liebig a try, and they yielded, though with misgivings. "I suppose," wrote Whitney rather sourly, "I should be run after for a Professorship [at Boston] if I had studied at Giessen, as it seems to be a settled point that no young man can be expected to know anything of chemistry, unless he has studied with Liebig."

The two Yankees were quick to find fault. Giessen was "a wretched hole" with streets "narrow & filthy beyond all description." Liebig seemed aloof, though polite. Whitney disliked him at once, and Gibbs resented his meddling in lab work. Whitney likened the lab to a malodorous blend of "the apothecary's shop, the barn-yard, and the dissecting-room." If it had not been outside the town, he thought, it would have been driven out. He and Gibbs were working together on sheep's bile and a fusion of caustic potash with old horse meat, while their neighbors explored the inwardness of rotten eggs, dried blood, a collection of animals' eyes, and so on ad nauseam.

Yet even Whitney began to find Liebig tolerable, and Gibbs admitted that the master "improves somewhat on acquaintance and can be agreeable when he chooses." Most of all, wrote Gibbs, "it is a great advantage to be where so many are at work. One's mind gets well stirred up. One converses continually on scientific subjects."

Of the academic pilgrims we have followed, only the astronomer Benjamin Gould stayed on long enough to bring back his doctorate, Göttingen '48. All it got him for a start was a precarious living coaching Harvard students. Whitney did not even remain faithful to chemistry. He went home in 1847 to become Charles Jackson's assistant on a geological survey of the Lake Superior copper region, and this led him into a distinguished though sometimes stormy career as a geologist.

Even without the Ph.D. degree, Horsford made capital of his Liebig connection. At Harvard the Rumford Professorship of the Application of Science to the Useful Arts had fallen vacant. Joseph Henry was sounded out but declined. The geologist Henry D. Rogers had weighty credentials and lobbied for the job, but Benjamin Peirce thought his work too sloppy and speculative. Meanwhile John Webster alerted Horsford and his friends, and recommendations for Horsford began to come in. Horsford had ardently cultivated Liebig's friendship and had published a couple of solid papers at Liebig's instance (better papers than he ever did thereafter). Liebig's warm recommendation to President Edward Everett clinched the case. What was more, at Liebig's urging Everett promised Horsford a well-equipped laboratory and time for research. In apprising Liebig of all this, Everett congratulated him on exerting "a powerful though indirect influence over American science." The congratulations were in order. Though Horsford contributed little to science thereafter, he worked with missionary fervor to replicate Liebig's laboratory teaching in America, and the terms of his appointment quickened a nascent movement toward research as a legitimate function of American higher education.

In that last respect Gibbs was less fortunate than Horsford and Norton. He felt that a private teaching laboratory in New York, like Booth's in Philadelphia or Jackson's in Boston, would be too risky for a heavy investment in chemicals and equipment. Yet his heart sank at the prospect of college teaching as it went in that day. Most college teachers in America, he wrote, "are worked to death and many never in the course of their whole lives publish one single original paper or contribute one single new fact to science." So in 1847 he settled for a teaching position at the New York College of Physicians and Surgeons, where a course of four months a year would at least leave him time for research.

Among the returned pilgrims, specific European influences can be traced. Some repatriates, like Horsford and Norton, consciously modelled themselves on particular Europeans. Bache looked to Arago, Gauss, and, in developing the Coast Survey, Sir Edward Sabine of the British Ordnance Survey. Gould adopted the astronomical methods of Gauss, Bessel, and Struve, as did well-posted stay-at-homes like Benjamin Peirce, William Chauvenet, and the men of the Naval Observatory, thus making a reading knowledge of German indispensable to American astronomers by the 1850s. Gould furthermore kept up a steady correspondence with Gauss, Encke, and other Europeans, confiding his projects and aspirations to them and getting sympathy and counsel in return. But more fundamental effects were evident in the temper and vision of the pilgrims.

Even in the colonial period, Americans had fretted about the dependence of American science on Europe for everything but native talent and raw data. With political independence came national consciousness and pride. American scientists tried harder to work up and publish their own data and to win some measure of respect from their European counterparts. Occasional modest successes spurred them on. In vertebrate paleontology, Americans quickly applied Cuvier's new theoretical framework to their fossils and took pains to transmit their findings to Europe, which accorded them gratifying respect after 1830. European scientists showed esteem also for the botanist Asa Gray, the meteorologist James Espy, the zoologist and philologist Samuel Haldeman, the geologist James Hall, the entomologist John L. LeConte, and others. Bache's Coast Survey won praise from European geophysicists. Benjamin Gould could boast of having refused a European university position (at Göttingen), and Henry D. Rogers, though not deemed good enough for Harvard's Rumford Chair, actually came to hold a chair at Glasgow—though no other American scientist seems to have got a European chair before the Civil War. Americans snatched at the crumbs of European approval that fell to them. Liebig liked to tell his American students ("whether from flattery or from ignorance of us I don't know," remarked one of them) that America would soon surpass Europe in science. In 1843 Charles Jackson proudly published a noted French geologist's rather faint praise of the Americans' "numerous and minute explorations . . . careful and continuous observations" in geology, and the elder Silliman reported a flattering allusion to American scientific progress by Humboldt in 1851.

But occasional tepid compliments like these were not enough for most Americans of the 1840s. By the end of that decade the American eagle was full-fledged and screaming, in cultural and intellectual matters as well as political and diplomatic. Bache and Henry had been stung by the ill-concealed contempt of some European scientists in the 1830s, and Bache saw little improvement in the next decade. In 1844 he complained that French scientists were not only chauvinistic in general but also indifferent to American work especially and, like the Germans, quite content with such dilutions of it as seeped into English journals and books. Even the English, he thought, were uninterested or patronizing.

Looking back, Benjamin Gould saw American awe of European science as having bred "not self-reliance, but self-distrust and intellectual timidity." Now, however, the unblinkable fact of European scientific superiority inspired not humility and resignation but appeals to national honor and calls to action. In 1846 an American congressman pronounced it "a disgrace to the most enlightened people on earth" that Americans got their astronomical data chiefly from Europe. Bache and Henry deplored the tendency to base American scientific reputations on European approval, and a Philadelphian viewed it as a despicable "want of moral courage." In 1849 the geologist James Dana mourned that "our country is the only one that seems to take no pride in its scientific men," and Henry thought it high time for Americans to show "a scientific esprit de corps."

Americans began to follow the French in scientific chauvinism. The Scottish chemist James Johnston protested the tendency in his American visit of 1849–50. The naturalists called for an American nomenclature, he reported, the geologists disdained to label their formations with the names of European localities, the mineralogists insisted on an independent analysis of American minerals, and a New York writer on agricultural science felt bound to analyze from scratch all species grown in his state, though most had already been thoroughly done in Europe. A German entomologist complained that Americans, to forestall European descriptions of American insects, rushed prematurely into print. In 1849, as one twenty-six-year-old American to another, Spencer Baird congratulated Joseph Leidy on a recent contribution dealing with the internal plant life of animals. "It is a great discovery," Baird wrote, "and we are a great people."

The scientific pilgrims had seen America in better perspective, but they had also seen Europe up close. They saw how far they had to go, but they also saw that mere mortals could make it. To Whitney and Gibbs, Rammelsberg was "a little bow-legged fellow of no great refinement," and Liebig was no god (except perhaps to Horsford). "Your hint about 'foreign lions' losing by a nearer view much of the enchantment lent by distance is wonderfully true," Gould wrote a Boston friend. "Day after day my respect & admiration for what we have at home is increasing."

So they dared to make more sweeping and farsighted plans than did most of their flag-waving countrymen. In 1838, soon after their own grand tour, Henry had written Bache: "I am now more than ever of your opinion that the real working men in the way of science in this country should make common cause . . . to raise their own scientific character.

To make science more respected at home [in order] to increase the facilities of scientific investigators and the inducements to scientific labours." In similar vein we find Gibbs writing from Berlin in 1846: "I intend to ... work my way up till I can command a professorship and do something solid for science.... I want to see some good big things done for science in America and not such dribbling as we have so long had to tolerate."

It would take longer than they probably supposed for America to overtake Europe in science. None of them lived to see the day, though Gibbs came fairly close. Nevertheless, as we shall see, they would do much to point the way and clear the path.

Returning pilgrims were not the only carriers of European scientific culture to America. A number of able European scientists came to America, some to visit, others to settle. European sojourners, such as the distinguished geologists Charles Lyell and Edouard de Verneuil in the mid-1840s, were well received. And permanent recruits from Europe found as cordial a reception among American scientists as did other immigrants among the population at large. Tradition, humanity, and the interests of American science demanded hospitality especially for European scientists coming as refugees from the failed revolutions of 1848. A Jesuit astronomer, after being exiled from Italy, taught for twenty years at Georgetown University. Germany, as the leading producer of both European science and political refugees, was the main source of those refugee scientists who came to stay. One cursory survey of notable refugee scientists lists seven German Forty-eighters; and there were others, including some before and after 1848, who were not exiled outright but voluntarily sought a politically freer society.

Of 475 leading American scientists active at any time between 1846 and 1876, about 12 per cent were born in Europe—about the same proportion as in the general population. But among the scientists, the proportion of foreign-born was significantly lower among those active in 1876 than among those active in 1846. Since scientists like others usually came to America for material advancement, this falling-off implies a decline in the economic attractiveness of America for European scientists. Perhaps a growing number of Americans competed with the immigrants. Perhaps demand for scientists rose faster in Europe than in America. The Civil War blighted American science and so may have deterred scientific immigrants as it did others. Probably all of those factors played a part. Nevertheless the figures in themselves, trends aside, show how significant the transplanting of European scientists must have been in the antebellum development of American science.

Numbers do not tell the whole story. One of those immigrant scientists had an effect far beyond statistical measurement. In the perspective of seventy-two years an eminent zoologist would write: "It was not until the fortunate circumstances which brought the Swiss naturalist, Louis

Agassiz, to our country in 1846 that the modern conceptions of biological science were established in America." Earlier, thirty years after the event, an equally eminent scientist would recall that "with his presence a gradual but entire change took place," while the younger Silliman would equate Agassiz's role in "the great scientific awakening about 1845" with his own father's contribution twenty years before that.

And near the close of Agassiz's own lifetime, a quarter century after his westering, a leading geologist would write him directly and simply: "The new era in America dates precisely from the day of your arrival in Boston."

Chapter 4

Agassiz's Boston: The City in Science

The best way to survey American science at the dawn of its "new era" is through the eyes of the newcomer who would do so much to transform and lead it. No other European observer left so comprehensive and knowledgeable a commentary as did Louis Agassiz, and no American observer had so rich a background of European experience as a basis for comparison.

Born in French Switzerland in 1807, the son of a Protestant minister, Agassiz grew up in an unshakable religious faith that would constrict and finally stultify his scientific thought. Though fascinated by the living nature around him, the boy showed little interest in science as a possible career until his mid-teens, when he chose it as a way out of the middle class. A local museum not only confirmed his bent for natural history but also made him a lifelong believer in museum teaching. He went on to Ph.D. and M.D. degrees at Munich, meanwhile publishing a solid, precise study of Brazilian fishes. While studying further in Paris, he won the affection of Georges Cuvier, the great master of comparative anatomy, who virtually adopted him as a professional heir. Humboldt, likewise charmed, got the young man a college and museum job at Neuchâtel; and there, over a ten-year period, Agassiz produced a masterful, five-volume work, *Researches on the Fossil Fishes.* It established him as a leading naturalist.

The *Fossil Fishes* set forth Agassiz's lifelong view of the biological cosmos, a view that coincided with Cuvier's. The characteristics of a species never changed, Agassiz insisted. God made each species separately out of nothing, discontinued it when he saw fit, and replaced it with a new line. "A Thought of a Supreme Intelligence manifested in material reality; that is the view I take of the animal kingdom," Agassiz told his American colleagues in 1849. Though he talked of development, and though evolutionists were to draw upon his paleontology, the relationship Agassiz saw among species was theological, not biological, the plan of God, not the mindless interplay of genetics and mortality. It simply pleased God to give a beautifully developed sequence to the characteristics of successive species, and what pleased God delighted Louis Agassiz. As a museum high priest he glowed with reverence in tracing the logical progression of species toward the ultimate ideal. Agassiz was sure he knew God's final intent. "The creation of man," he said confidently, "was the aim of the plan from the beginning."

Louis Agassiz's God was a busy, imaginative, enthusiastic, benevolent God with a love of form and order, a concern for detail, an appetite for variety, a passion for progress, an eternal readiness to start fresh. He was, in short, a God in whose image Louis Agassiz had been made with extraordinary fidelity.

But the tide of scientific thought was turning against so simple and comforting a cosmology. More and more scientists, as well as economists and historians, preferred ascribing change not to the will of hero, king, or God, but to the impersonal weight of gradual cumulations and aggregations. Thus the older school of geologists, the "catastrophists," who like Agassiz believed in a past of abrupt, cataclysmic, God-ordained changes, were yielding to the "uniformitarian" ideas of Charles Lyell, the belief that the ceaseless repetition of still-continuing processes had gradually made the earth what it was. So also in biology the intimations of evolution, the gradual development of species through long ages of small variations, were growing louder.

Agassiz's most widely known European work could be accommodated to either world-view. Others had seen traces of long-vanished glaciers in certain aspects of Alpine terrain, but Agassiz perceived that ancient glaciers had left their tracks throughout much of Europe. He hypothesized an epoch of all-encompassing glaciation, which a friend of his christened the Ice Age, and he gathered convincing evidence for it through summer work over eight years. Lyell himself promptly accepted the theory. It may have strained the definition of "uniformitarianism," but the glaciers had certainly worked gradually and were still to be seen in action. Agassiz, however, insisted that the freeze had been so sudden and widespread as to destroy whole species and so require at least one special creation after the first.

Not only with his paleontology and geology, but also with his ichthyology, zoology, and marine biology, Agassiz won a reputation as an almost universal naturalist. His output was astounding—eleven published works in fourteen years at Neuchâtel. He all but took over a local print shop for the purpose, and by 1839 had a dozen assistants at work on scientific projects. But in 1845 his wife, feeling neglected, went home to her family with the children, the print shop collapsed under a load of debt, Agassiz's egoism and his demands on his associates began turning some friendships sour, and his multifarious projects overburdened even his own phenomenal vigor and will. So he decided to start fresh, to perform for his personal world an act of special creation, by going to America, where he had taken care to cultivate leading naturalists by correspondence. The trip was supposed to be a visit, but he wound up his European affairs with suggestive thoroughness. His destination was Boston, where his friend Lyell had got him a lucrative engagement for a Lowell Institute lecture series.

Agassiz's spirits ran boyishly high as he put Europe behind him. At thirty-nine he was a tall, strong, florid man with a large, well-shaped head and flowing black hair, dark, alert eyes, and a full, mobile, gently humorous mouth. He conveyed an instant impression of intelligence, sensitivity, and boundless good humor. A rich voice, spiced with a French accent, added to his charm, as it would also to his success as a popular lecturer. Smoking cigars incessantly, practicing his English on the captain, he thought already of studying fishes, zoology, and glacial action in the areas of the Great Lakes, the Ohio and Mississippi rivers, and the Rocky Mountains—a life's work, though the visit was ostensibly for only two years. When the ship touched at Halifax, he "sprang on shore" and headed for the heights, where signs of glacial action delighted him. Then came the southward leg of the voyage, the first sighting of Cape Ann light, the long line of coast growing more distinct until Salem, Nahant, Boston Harbor, and at last the dome of Bulfinch's State House on Beacon Hill came into view. The day was October 3, 1846. John A. Lowell, sole trustee of the Lowell Institute, met him at the wharf, and they walked to Lowell's Beacon Hill home through crisp October air. The streets and buildings reminded Agassiz of London and Paris, but the October sky seemed as pure as that of Italy. The trees along the way blazed with fall colors, and Agassiz saw five different species of oaks. His American years were off to a brisk and brilliant start.

It was fitting that Agassiz should first encounter American science in a thriving city. The city was the natural habitat of the scientist. In 1840 only 11 per cent of the American people lived in cities, yet 44 per cent of the most productive American scientists between 1800 and 1860 were born in cities, and most of the rest made for the cities when they could. Even though the earth and life scientists found their subject matter elsewhere, they, like their brethren in laboratories, operated from urban bases.

One reason for this is easily surmised, and the scientists themselves made it plain. Like Wolcott Gibbs in Liebig's lab, they welcomed the electricity generated by the rubbing together of minds. "I find it of great importance in experimenting," wrote Joseph Henry in 1838, "to have some person with whom I can talk over the subject in the way of sharpening myself." And in 1846 he wrote of his preference for "the conveniences and sympathies of a city life." An amateur scientist from Mississippi, visiting New York City in 1854, marvelled at the "endless sources of information and gratification" that surrounded him there. "How tamely and unconsciously we vegetate in the remote little villages," he wrote wistfully. "I have crowded today a year of enjoyment and knowledge in a few hours."

As the Mississippian attested, scientists drooped in isolation. Fifteen years spent away from books and co-workers weighed heavily on Herman Haupt, struggling in rural Pennsylvania to make a science of bridge design. A young entomologist in Maine had to write a Philadelphian for help in naming specimens, being "acquainted with no one in this vicinity and indeed in this state who has paid any attention to this branch of natural history." A teacher at a small Georgia college longed for "the scientific sympathy of which I am here utterly destitute." "We have no botanists in the up country," complained a South Carolinian, "& consequently I have always been alone, or nearly so, & for many, many years I have almost neglected [botany]."

Yet the scientist could be lonely in a crowd. From only six miles out of Cleveland, a conchologist wrote in 1850: "I am insulated from the rest of the world, so far as natural science is concerned . . . [and so] my investigations are either superficial or are left unfinished." From Montreal a scientist wrote: "It is a great loss to me, there being no entomologist here. . . . It is always pleasant to have some one with whom you can converse." In the booming San Francisco of 1860 a meteorologist found "hardly a single individual in town to whom I can talk upon such matters, and not one from whom I can gain information or receive instruction."

It follows that other cities supported scientific activity out of proportion to their populations. Boston, New York, and Philadelphia together had less than a third of American city-dwellers, yet produced more than half of the city-born scientists. Four out of five leading scientists in the first half of the nineteenth century worked in those three cities plus Washington, New Haven, Albany, and urban Ohio. Of the rest, a large part could be found in a few minor centers: West Point, Princeton, Amherst, Charleston, South Carolina, and (before the 1840s) New Harmony, Indiana.

If size alone did not account for a city's scientific vitality, what else did? Since Boston happened to have taken the lead in American science just before Agassiz arrived, a look at what he found there may turn up some clues.

In the spring of 1846 Josiah Whitney called Boston "the only city in America where anything of any account is done for science." He meant this as hyperbole, of course, but it reflected a sense among scientists that Boston had lately forged ahead. One modern statistical study plots the geographical distribution of leading scientists at that time and finds that whereas Philadelphia led in 1840, with Boston running second, the order was reversed in 1845. A similar count of 145 leading scientists on the basis of states shows Massachusetts edging out Pennsylvania in 1846, with New York third. The scientific fertility of Massachusetts is especially striking in view of the fact that Pennsylvania had more than twice as many people and New York more than three times as many. With less than 5 per cent of the nation's population, Massachusetts produced more than 20 per cent of the leading scientists, and nearly that many of them worked there. What was more, the Bay State would hold that per capita lead undiminished into the last quarter of the century.

For an American city, Boston was old. That fact had something to do with its scientific strength. The very presence of a scientific community draws and holds more scientists. So the first American nuclei of science, the bridgeheads of European culture, tended to become the perennial centers. Still, Philadelphia and New York City were old, too; and yet with a population of 115,000, less than half the size of Philadelphia's and less than a third of New York's, Boston had outstripped both.

Quite apart from its scientific attractions, Boston appealed to some scientists as a desirable place to live. Having veered in good season from fishing and shipping to finance and industry, Boston's spirit of enterprise remained strong. Once almost an island, the city was now being linked to its busy hinterland by several railroad embankments. Yet its physical appearance and scale seemed more civilized and humane than those of most cities on the make. "This is a beautiful and pleasant place," wrote one young scientific student. "There is so much life and activity here that I don't know how anyone that is used to it could ever live in such a dull place as North Carolina is now." In 1846 the English visitor Charles Lyell recalled the "feelings of hope and pleasure" he had "more than once experienced . . . when I returned to Boston" and which he now felt more than ever after travelling in the South and West.

Boston's natural environment did not explain its scientific eminence. To be sure, more than one young New Englander went on to conchology from the shell collections, local and exotic, he saw in some Yankee seaport. But by the late forties New Bedford whalers had turned from the South Pacific and Indian Oceans to the cold seas off Siberia and Alaska, and the more eye-catching shell species had got scarce. Anyway, though a third of all leading life scientists active 1846–76 had grown up in New England, so had even larger proportions of American mathematicians and astronomers.

It was evidently the human factor, the spirit, values, and resources of the people, that chiefly nurtured Boston science.

To begin with, there was the traditional commitment of Boston to learning. New England Puritanism had set much store by it, including science as witness to God's handiwork. More than two centuries before Agassiz arrived, the Puritan Commonwealth had accordingly led the British colonies in establishing both public schools and higher education, and those seedlings had taken firm root. Far from blighting that aspect of the Puritan legacy, the recent rise of Yankee industry had nourished it. With no natural endowment except waterpower, industry in New England, more than elsewhere, needed whatever advantage a literate and adaptable work force could give it. Besides, an educated work force would presumably be more stable, responsible, and moral at a time of dangerously unsettling social change. There seemed to be grounds for this view. At a meeting of three thousand Boston workmen and clerks for the purpose of forming a library, Agassiz was amazed to see them "listening attentively in perfect quiet for two hours to an address on the advantages of education, of reading, and the means of employing usefully the leisure moments of a workman's life." The broader the base of education, the more recruits and popular support there would be for science. Whether consciously or not, amateur scientists were indirectly serving the field they loved when they took a strong and active part in the Massachusetts school reforms of the 1830s and 1840s.

Boston's elite gave science strong support. Puritanism had bequeathed them an ideal of stewardship, of *richesse oblige,* and science was regarded as a worthy cause. As an avocation it attracted well-born amateurs, who were in turn part of a network of public-spirited elite groups. Wealthy lovers of science for its own sake could enlist other patrons through ties of college, marriage, friendship, business, politics, and shared membership in philanthropic organizations. Science-minded clergymen had influence with members of their congregations, educators with their former pupils, doctors with their rich patients. Some professional scientists were themselves born into the elite, and reputable newcomers were readily admitted. Agassiz, of course, was lionized at once, and after the death of his first wife he married into the Boston aristocracy. He worked his social connections to great effect in the interest of science.

Wealthy patrons of Boston science had other motives besides. Local pride, reinforced increasingly by national pride, saw scientific leadership as supporting a claim to the world's respect. Even those whose allegiance was to other fields of learning believed that progress in one field would stimulate all, and so they seconded the appeals of science. The economic motive also entered. Maritime enterprise, still far from being displaced by industry, drew upon mathematics and astronomy. Industry and commerce were easily persuaded that their technology, or at least some elements of it, sprang from science.

It may even be speculated, though not demonstrated, that scientists derived more than financial and moral support from the Boston milieu. The Puritan ethic of grace through work and the Yankee spirit of enterprise may have communicated themselves to Yankee scientists. A South Carolina mineralogist found his Yankee counterparts to be "regular mineral traders, and . . . a Southerner stands a bad chance at 'driving a trade' with a Yankee."

But if the culture, ideals, and resources of the people laid the foundation for Boston science, certain local institutions formed the pillars of the structure: the societies, the libraries, and the college.

An urban institution important in more ways then than now was the local scientific society. Americans in the 1840s felt a strong urge to organize, and science seemed as good an excuse as any. Any village might, and some did, organize a self-styled "scientific society" of a half-dozen amateurs almost as readily as a sewing circle. But a scientific society of any consequence needed the numbers, wealth, and cognate institutions of a city.

An effective scientific society could multiply points of intellectual exchange among its members and with scientists elsewhere. It has been described in modern terms as an "information system," much more important in that day because alternatives were far less numerous, efficient, and wide-ranging than now. It could assemble, house, and care for books, periodicals, specimens, and apparatus, collections beyond the purse of a private individual. Its regular meetings for the presenting and discussion of papers could encourage sound research and put down quackery. Its published *Proceedings* or *Transactions* or *Memoirs* or *Journal* could do the same before a wider, more expert audience over a longer period and could more permanently enlarge scientific knowledge. In such ways it could set standards, evaluate quality, minimize duplication of effort, and raise questions for further study.

Such a society could also give science a firmer footing in the world at large. It could tap the affluence and muster the influence of amateur members for the financing of scientific projects; in this sense its typically lenient admission requirements could pay off. With more weight and dignity than an individual, a society could lobby for government support of science. And by winning public notice, stirring young minds with lectures and exhibits, and leading those minds into serious study and research, it could help recruit for the scientific calling. Indeed, as government, universities, national organizations, and independent journals outdid local societies in most other functions, that of mediating between science and the public became increasingly important in justifying the society's continued existence.

With all its usefulness, however, a local scientific society had both weaknesses and dangers. Even the largest ones gave scientists no livelihood beyond occasional meager stipends as curators. Rarely if ever did an American scientific society make research grants to individuals like the grants Agassiz had got from European societies, though it might help finance its own or a joint project or expedition.

Another limitation was the strong amateur element in a local society's membership, especially in democratic America as compared to elitist Europe. An 1876 report on the Academy of Natural Sciences in Philadelphia described it well:

> Though the register of its members contains the names of very many true scientists, a large majority of the members of the society always has been and is now composed of gentlemen who, to use an expression of the founders, are "friendly to science" and its cultivation. Many of them pursue science only as a recreation during leisure hours, some are pleased to observe and know what others do, and others are content to encourage those who work.

Useful as they may have been in fund raising and public relations, amateurs seemed increasingly out of keeping with the specialized expertise and full-time commitment of "true scientists." The professionals feared adulteration, the amateurs resented elitism, and the division strained the bonds of the society.

In 1846 New England had fourteen of the nation's thirty-two existing scientific societies, as compared to eleven for New York, New Jersey, and Pennsylvania combined. Boston's American Academy of Arts and Sciences had been organized in 1780 at the urging of John Adams, anxious to outdo Philadelphia's temporarily comatose American Philosophical Society. Nodding in its own turn, the American Academy awoke in 1830 to a newer and nearer rival in the Boston Society of Natural History. The Academy remained moderately active and respected. In 1840 it made a small grant to the new Harvard Observatory, and later in the decade it brought out two able volumes of its *Memoirs* and began publishing its *Proceedings.* But not until the fifties did it fully recover from the effect of the new local competition.

Unlike some other scientific societies, the young Boston Society of Natural History had managed to struggle through the depression of the early 1840s, and though short of room in 1846, it stood on firm financial ground. Already it had begun publishing its *Proceedings* and its *Journal.* The Society's *Journal,* wrote Benjamin Gould from Europe in 1846, was in the libraries and highly valued, though he had never seen any of the Academy's *Transactions* there. As for the *Proceedings,* James Dana of Yale found "much more in them of general interest—that is . . . not confined so exclusively to brief description of species in natural history & catalogues of books—as those of Philadelphia." By 1850 the Society had bought, remodelled, and occupied debt free a thirty-thousand-dollar building.

Boston men dominated the Society of Natural History, while Cambridge scientists leaned toward the American Academy. Still, most leading scientists in the area belonged to both. The Society of Natural History, looking less to Harvard, depended on amateurs, especially practicing physicians, but there were able men among them, such as the botanist Jacob Bigelow, the conchologist Augustus Gould, and the zoologist Amos Binney. Several taught at the Harvard Medical School, which despite its name was situated in Boston and ran its own affairs. But the mainstay of the Society throughout its first forty years was the skewed yet formidably talented Charles T. Jackson. Through his last three strife-ridden decades, until rage and despair drove him mad, the Society stood by that erratic genius.

Jackson's turbulent career illustrates several aspects of American science in that era. Born in 1805 of an old Massachusetts family, he took an M.D. degree at Harvard in 1829. Geology had already captured his interest, however, and he studied it along with medicine in Europe during the next three years. Back home in Boston he practiced medicine until 1837, when he was appointed state geologist and chemist of Maine. In that year he also opened a private teaching laboratory in chemistry at Boston. Jackson had a brilliant mind, fertile in ideas, but possessed by a craving for fame. On his voyage home from Europe in 1832 he had talked freely with a fellow passenger, Samuel F. B. Morse, about recent developments in electricity. Jackson had already thought of using it to send messages and had made a working model of such a device. Six years later, after Morse developed his telegraph, Jackson claimed credit for it. When the German chemist Christian Schönbein announced his discovery of guncotton in 1846, Jackson claimed credit for that too. An ominous pattern was taking shape in Jackson's life. Then as now, fame through priority was a powerful spur to scientific achievement; but scientific ambition may also goad its captive to destruction.

Jackson's career took its fatal turn after one of his former pupils, a Boston dentist named William T. G. Morton, persuaded an eminent Boston surgeon to operate on a patient whom Morton had anesthetized with ether. This was the first such operation made known to the medical profession. An astonished audience of physicians and medical students in the operating theater of the Massachusetts General Hospital witnessed the historic experiment on October 16, 1846. At its conclusion the surgeon pronounced the death sentence of surgical pain: "Gentlemen, this is no humbug." But there was no anesthetic for what Morton was soon to suffer. Like other chemists, Jackson had on occasion accidentally inhaled chlorine and then used ether to ease the pain. In the early 1840s he had even talked of ether's possible usefulness in surgery. Though Morton claimed to have conceived that idea independently, he admitted that Jackson had suggested ether to him as a local anesthetic in dentistry and had given him essential advice about preparing it. But Jackson had left it at that. It was Morton who tried ether on animals and himself, devised an apparatus and had it made, arranged for a surgical trial, and administered the ether.

If Jackson had indeed thought of ether anesthesia before Morton, he had been so desperate for success that he could not bring himself to risk failure. Not until the operation had succeeded did Jackson claim credit and insist that Morton had been merely his agent. By the spring of 1847 the savage and protracted "Ether Controversy" had broken out in earnest. Jackson's scientific reputation, the advice Morton acknowledged receiving from him, and Jackson's pathological obsession with the struggle kept the war hot until the tragic deaths of both men.

The affair was, of course, a classic instance of the lust for scientific priority run mad. But it has another significance more to the present point. The famous event could not have occurred where and when it did without the basic concept, the suitable anesthetic, the workable apparatus, the anesthetist, the consenting patient, the open-minded surgeon, and the will to act. To these ends Boston provided the expert chemist, the skilled instrument maker, the layman with confidence in scientific progress, the skillful and self-assured physician, the spirit of innovation, and the scientific standing that could catch the world's attention. Whatever share of credit one may allot to Morton or Jackson, some ought to go also to the scientific, medical, and technological communities of Boston.

Like an effective scientific society, an adequate scientific library needed money, more money than a village or a small college could scrape together, and so it too had to be the product of a city or a large university. "In this place," wrote John Torrey from Princeton, "I labour under many disadvantages—chiefly from want of books, that I might command in a large city." Setting out to study the Infusoria in 1843, the West Point microscopist Jacob Bailey found it "almost impossible to procure any works relating to them." A Cincinnati conchologist wrote a Philadelphia friend that "my opportunities for determining species are not near as good as yours, for I have no large public collection like the Academy of Natural Sciences to refer to, nor can I consult so valuable and extensive a library as they possess." Something could be done with time and travel. In 1845 the meteorologist James Coffin managed to write a report on the winds of the Northern Hemisphere by exhausting the libraries of New York, New Haven, Philadelphia, and Washington. But a dozen years later another scientist still could not find "in this country any single library at all equal to the entire demands of our whole body of investigators."

Sharper still were the pangs of those who had once used such libraries in Europe. One American remembered with envy how the library of the Pulkovo Observatory in Russia had in fifteen minutes yielded "a great mass of evidence on the subject" he was pursuing. "No one," wrote Louis Agassiz in 1855, "has felt more keenly the want of an extensive scientific library than I have since I have been in the United States." Certain German works, he complained in 1853, "are hardly to be seen in any American library, though they constitute now the basis upon which modern physiology has been renovated." And one of his students that fall dutifully recorded in his lecture notes that "our libraries are deficient in scientific works."

A Harvard alumnus in 1816 reported that Göttingen, with forty professors to Harvard's twenty, had more than two hundred thousand volumes to Harvard's less than twenty thousand. Though Harvard more than tripled its holdings by the late 1840s, Göttingen still outstripped it six to one. Yet Harvard by then had the largest library in the nation, surpassing the Library of Congress. Yale, the runner-up among colleges, had less than half as many.

By the late fifties Harvard College holdings had passed seventy thousand. In those days the college library was freely open to Boston scientists and scholars. But in any case Boston had equivalents. In 1858 the Boston Public Library opened with seventy-two thousand volumes. The Boston Athenaeum had seventy thousand, and it was strong in science. The specialized scientific libraries of the American Academy and the Natural History Society contained ten thousand and six thousand volumes respectively, while Nathaniel Bowditch's private mathematical library of three thousand volumes had been opened by his heirs to public use.

By 1850 Boston in particular and Massachusetts in general led the nation in scholarly library resources. Aside from public school libraries, Massachusetts surpassed all other states in total library volumes, more than half a million, and more significantly in the categories of college and learned society libraries. Nevertheless, Philadelphia's two leading scientific society libraries were more than twice as large as their Boston counterparts, while single libraries in Munich and Paris each had more books than the Massachusetts total. As Agassiz soon learned, even Massachusetts had far to go.

In the scientific firmament of 1846, Boston proper constituted one twin of a binary star. Its small partner Cambridge, linked to it by a couple of toll bridges across the Charles, had grown in population by fifty per cent to thirteen thousand in the previous five years, and so had just swapped its town meeting for a city charter. There were still farms all over town, however, and woods full of wildflowers.

Harvard supported twenty instructors, of whom four taught science (the Rumford Chair being vacant just then). In addition the astronomers William and George Bond, father and son, worked at the Harvard Observatory. And like Boston, Cambridge had some serious amateur scientists, such as the entomologist Thaddeus Harris, who worked as the College librarian. Of the four faculty members in science, Joseph Lovering, the physics teacher, is remembered not for any significant research but for effective lecturing and a fifty-year record of devotion to routine duty, 1838–88; and the chemist John W. Webster chiefly for fiscal fecklessness, a taste for the macabre, and a sudden impulse that got him hanged for murder in 1850.

The remaining pair, Asa Gray in botany and Benjamin Peirce in mathematics, led the nation in their fields. Gray had no appetite for scientific politics and so would exercise his considerable influence in American science by sheer weight of individual accomplishment. Peirce, on the other hand, would, with his future colleague Agassiz, round out the six-man inner circle of American science, Bache's Lazzaroni, along with Bache, Henry, Gibbs, and Gould.

Gray met Agassiz on the evening of October 8, 1846. "He is a fine, pleasant fellow," wrote Gray to a friend next day. "We shall take good care of him here." So they would, in the literal sense, over the next quarter century. And in an ironic sense, so would Gray.

Asa Gray, at thirty-five still boyish in his short, slight, agile body and clean-cut, beardless face, would remain lively in spirit to the end. For through his long life, he would be spared sickness, tragedy, even serious trouble, while blessed with a wife he loved, a home he loved, and unbroken success in a profession he loved. And in one of the great intellectual contests of the century, a struggle which enlisted Agassiz in the opposition, Gray would be a leader on the winning side. Born of Massachusetts stock in upstate New York, he came to his lifelong passion for botany through an encyclopedia article on it. Though like so many of his generation he prefaced his scientific career with an M.D. degree, botany soon claimed him. Enthusiastic, keen-eyed, with a quick, retentive mind and a pleasant though sometimes rather doggedly self-assured manner, he became the protégé of John Torrey, the leader of American botany. The eminent Torrey even welcomed young Gray as collaborator on a *Flora of North America,* until waves of specimens from the West swamped the venture. Gray soon supplanted his mentor as leader in the field, though without losing Torrey's friendship.

Gray's swift rise signified more than his personal ability. He had resolved to be a specialist just as science was coming to demand that commitment and American society was beginning to support it. All his life the botanist Torrey had to make his living as a chemist. "Although chemistry is my profession," he wrote wistfully in 1852, "botany occupies much of my time." Botany occupied all of Gray's time. Gray in 1842 had fallen easily into an endowed chair just created at Harvard, one which President Quincy unprecedentedly agreed should not only be confined to botany but also should emphasize research at least as much as teaching. With cool realism, Gray jettisoned field work also and specialized in "closet botany," developing his herbarium into the nation's main clearinghouse for the study, classification, and exchange of specimens sent by fieldworkers. And though he gave the Lowell Institute lectures a dutiful go, his awkward and unhappy platform manner and slight impediment of speech saved Gray from squandering time and energy on popular lecturing.

Gray's most eminent Harvard colleague in science, the mathematician Benjamin Peirce (pronounced "purse"), had the same safeguard. "I am not gifted with the qualities of speech, which make the popular orator," he conceded. He was in fact doubly secured by his field. "Tell me not in mournful numbers," the average Boston lecture-goer might have said, quoting Peirce's colleague Professor Longfellow and speaking also for Peirce's students.

Broad-chested and erect, with long, dark brown hair, regular features, a high forehead, and deep-set eyes, Peirce at thirty-seven had already been teaching mathematics at his alma mater for fifteen years. Without question he had great mathematical ability. Bowditch had seen it and used young Peirce's help in his famous version of Laplace's *Mécanique Céleste.* More than thirty years later Peirce in turn dedicated his most notable work to "my master in science, Nathaniel Bowditch." After Bowditch died in 1838, Peirce was easily the foremost mathematical astronomer in the United States. His crowning work, *Linear Associative Algebra* (1870), has been characterized with authority in our own time as "the first major original contribution to mathematics produced in the United States."

Like John Farrar, his predecessor in the chair of mathematics, Peirce made high, esoteric drama of his lectures. Farrar had customarily begun in an easy-going tone, then let his voice rise in pitch and volume, shifting into the rhythm of a temple chant, exalting mathematics as God's grammar, and ringing out over the dismissal bell in a final soprano climax. Peirce too would be carried away in a mystical frenzy. His mind would outrace his chalk. He would skip over recondite deductions as self-evident, and even so would be brought up short with a sigh of awakening at the furthest corner of his crammed blackboard. He had great personal presence; he had once mounted a chair and averted a crowd panic during the crush at one of Jenny Lind's Boston concerts. His own student audience would sit uncomprehending but fascinated, as if watching a wizard at work (and he was indeed Salem-born, though fortunately for him in a skeptical age). "The less we knew of what was going on," remembered one, "the more attractive was the enthusiasm of the man." They respected him as a genius, admired him as a spectacle, and liked him as a man, calling him "Benny" among themselves. With rare exceptions, they shunned him as a teacher. In the late 1830s, mathematics was made elective, except for freshmen; by 1847 only six upperclassmen took it. After 1850 sophomores were once again conscripted. The sophomores regretted this keenly. So did Peirce.

But the ways and means of higher scientific education in mid-nineteenth-century America will be dealt with at length in a later chapter. This one is meant as a group portrait of the Boston-Cambridge scientific community, in which Harvard faculty, soon to include Agassiz, were prominent figures.

No institution demonstrated the hold of science on the Boston mind more convincingly than the Lowell Institute. Its namesake and benefactor, John Lowell, of the remarkable Massachusetts family, a manufacturer of textiles whose wide reading included considerable science, had died young but rich in 1836, not long after his wife and children had perished from scarlet fever. His will left half of his estate to endow free public lectures in Boston, these to be principally on evidence in support of Christianity and on various scientific subjects. The endowment, wisely earmarked for lectures and not for buildings, yielded twenty-five thousand dollars a year to start, enough to pay speakers royally by the standards of that day.

Science, not theology, proved to be the biggest drawing card and accounted for the largest share of lectures. When the series was inaugurated with twelve lectures on geology by Benjamin Silliman, Sr., in 1839, the rush for free tickets broke windows at the place of distribution, and thousands were turned away. Silliman was paid two thousand dollars, well above his annual salary at Yale. Popular enthusiasm continued through the 1840s, though the lecturers, including such noted scientists as Charles Lyell, Jeffries Wyman, Asa Gray, and Ormsby Mitchel, were serious and sometimes rather technical. Not only did the Lowell Institute encourage popular interest in science and inspire young men to scientific careers, but it also gave Boston scientists added income and attracted others to the city, the most notable being Louis Agassiz.

Agassiz's Lowell Institute lectures would be sensationally popular, a curtain raiser to years of triumph on American platforms, warming him with public adulation that helped induce him to live out his life in Boston. But the lectures were not to begin until December 1846, two months after his arrival in America. Meanwhile, on the well-founded hypothesis that for American science even in 1846 there was life after Boston, Agassiz set out with characteristic enthusiasm to inspect it on its native ground.

His tour is worth following.

Chapter 5

Agassiz's America: The Geography of Science

On October 16, 1846, the day that introduced surgical anesthesia to the world, Louis Agassiz's thoughts were happily occupied with the ancient moraines and polished rocks flashing past his train window —unmistakable evidence of long-past glaciers. He also mused on the newer and speedier force that bore him along. "There is something infernal in the irresistible power of steam, carrying such heavy masses along with the swiftness of lightning," he wrote home. American haste and impatience must have been fostered by the railroad. "If here and there something goes to pieces, no one is astonished; never mind! we go fast—we gain time—that is the essential thing."

Hustled and exhilarated, Agassiz arrived that same day in New Haven, the City of Elms, whose trees had turned golden at the touch of fall. Yale's outward aspect did not glitter. Its buildings were plain, poor, and inconvenient. It enrolled more undergraduates but fewer professional students than did Harvard. A row of brick dormitories faced the public green, and near them rose the red sandstone library building, opened less than a year before—"books badly chosen, ninety per cent mere lumber," wrote one caustic visitor that spring. Under the new president Theodore Woolsey, more rigid discipline, new courses, and able recruits to the faculty would do more than buildings to maintain Yale's standing, but Woolsey did not formally assume office until October 21.

Since August the School of Science had existed in prospect and on paper, but half its faculty, namely Benjamin Silliman, Jr., the professor of "Practical [i. e., industrial] Chemistry," seemed low on pioneer spirit. "All goes on in the old plodding way here," he had written that summer. He and John P. Norton were "University Professors," he commented sarcastically, "without Salaries & dependent on our own exertions for the quid pro quo."

"Young Ben," as students nicknamed the twenty-nine-year-old huntsman of the quid pro quo, lived with his family next door to his father, referred to as "Uncle Ben" by the students and as "the patriarch of American science" by Agassiz. Here Agassiz probably met Professor Denison Olmsted, astronomer and meteorologist, who had made significant contributions to knowledge of the origins of meteors and the dynamics of hailstorms, but dedicated himself to teaching rather than research. Agassiz was most struck, however, by the senior Silliman's son-in-law, thirty-three-year-old James Dwight Dana, whom Agassiz immediately judged as "likely to be the most distinguished naturalist of the United States."

Like Asa Gray, Dana was a small, slender man, full of nervous energy, quick in thought and movement, a man simple, direct, forceful, yet kindly in manner and winning in his smile. Like Gray, he shrank from public speaking, though when put to it he did better than Gray. His alert eyes were blue, his shock of hair a light brown, and his features were given distinction by a large, aquiline nose. He disliked the demands of classroom discipline and academic routine, but did his duty conscientiously. Love of science, not mere duty, drove him to a pace of scientific work that sapped his strength and brought him at last to years of invalidism, in that respect unlike Gray.

Also unlike Gray, Dana had spent the years 1838–42 as geologist of the United States South Seas Exploring Expedition under the navy's Captain Charles Wilkes. Gray had intended to go as botanist, but withdrew before the Wilkes Expedition sailed. That was a loss to the expedition, which had to settle for a mediocre political appointee as botanist. It was a loss to Gray, whose "closet botany" would have been vitalized by the experience, and who came to know as well as anyone what Darwin had gained from his voyage with the *Beagle.* And meanwhile Dana had gone on to an immensely enriching experience, his observations yielding him sixteen hundred pages of notable geological and zoological reports over a period of a dozen years.

The Gray-Dana parallel resumes, however, with Gray's *Elements of Botany* (1836) taking an instant lead in the botany textbook field and Dana's *System of Mineralogy* (1837) doing likewise in mineralogy. And most fundamentally, Dana became a great coordinator and collator of American geology, just as Gray was of American botany.

For many years Dana's presence and his power to attract able young faculty would help Yale keep its leadership in geology and its front-rank position in science generally. Another buttress was the *American Journal of Science*—Silliman's *Journal*—which one commentator in 1857 thought had done more than anything else to develop an American scientific community and win European respect for it.

Sheer longevity worked for the *Journal.* The life expectancy of a scientific journal in that era was five or six years. Scientists preferred to publish in one that libraries would not discard as ephemeral; and by appealing to philanthropy and reaching into his own pocket, the elder Silliman had somehow kept his quarterly issues coming. But age slowed him, the times began to pass him by, old subscribers died off and were not replaced. By 1843 death and hard times had so pared the list that Young Ben spent two months on the road drumming up new subscribers. The year 1846 brought a new day. In January appeared the first issue of the *Journal*'s second series, which published six issues a year rather than four. Also in that year, James Dana joined the two Sillimans as an active editor. Without Dana the *Journal* could scarcely have survived.

From New Haven Agassiz went on by steamboat to New York City, "your insular city of the Manhattoes, belted round by wharves as Indian isles by coral reefs—commerce surrounds it with her surf," as Herman Melville wrote soon after. Seen from the city's magnificent, ship-studded harbor, its skyline, fringed with masts, had not yet reared up to swallow its church spires. The newest, that of Trinity Church, commanded Wall Street, at least architecturally. Charles Lyell, revisiting the city after five years' absence, was impressed by the new church, "seen from great distances in this atmosphere, so beautifully clear."

Aside from Trinity, the changes that struck Lyell most forcefully were works of technology: the fountains, supplied by the new Croton aqueduct, and the telegraph poles along Broadway, streaking the sky with their wires. The surf of commerce did not stop at the wharves; it roared through Broadway and on to the West. More than a third of a million New Yorkers now felt its power, and by 1850 it would draw in enough people to make half a million. Boom times were back again. Lyell saw lavish furnishings in private mansions—satin and velvet draperies, rich carpets, marble and inlaid tables, huge mirrors—but far fewer private libraries or even lone bookcases than in Boston.

Like learning generally, science in New York City had not kept pace with population and prosperity. The New York Lyceum of Natural History had lost its heavily mortgaged building during the depression of the early forties and after some months of homelessness had come to roost temporarily in three rooms on Broadway. James Dana could not count more than a half-dozen working members in it. "The rest of the members," he commented, "have a few shells or a little smattering in some kind of science or perhaps hardly that." "The general spirit of the city is not in their favor," wrote Agassiz of New York scientists generally. "The objection is everywhere the same, poor collections and poor libraries."

Still, however weak New York science may have been in proportion to the city's size and wealth, in absolute terms it stood third in the nation, after Boston and Philadelphia. One measure of scientific activity pointed to its future: in 1840, New York published half as many scientific journals (including society proceedings) as Boston; by 1850 it had come up even;

in 1870 it would outpublish Boston by three to one, and in 1880 by seven to one. As scientific societies and periodicals became more truly national, New York City's size and commercial leadership attracted their headquarters.

So also the city's rate of growth held great promise for higher education. On Park Place in a single building (a long, narrow, three-story structure with a squat little observatory tower), Columbia College still slumbered as it had in Wolcott Gibbs's student days. It had fewer than 150 students. James Renwick in chemistry and one lackluster mathematics professor still constituted its entire scientific faculty. But like the Astors, Columbia had land in Manhattan, and so every passing year brought an unearned increment to its endowment.

Columbia's chief rival, New York University, with about the same number of students, surveyed the grassy sweep of Washington Square from a grand, crenellated, marble building in English Gothic style. But its expensive home was the tomb of its pioneer graduate school, a victim, like the Lyceum, of boom-time overspending and depression reckoning. Columbia's library was small; New York University's was negligible.

In science, however, the newcomer led, having the glory of Professor Morse's telegraph and the continuing prestige of John W. Draper, its English-born professor of chemistry. Draper's substantial scientific contributions dealt mainly with radiant energy and light: its chemical effect, the interpretation of its spectrum, the application of photochemical knowledge to photography. With his sister as subject, he took the world's first portrait photograph. More significant to science was his work in applying photography to microscopy and spectrum analysis. And in the winter of 1839–40 he opened the era of astronomical photography with the first known photograph of the moon showing detail.

Elias Loomis, a Connecticut Yankee in his mid-thirties, further strengthened the college in astronomy. As a student and then a junior colleague of Denison Olmsted at Yale he had done good work on meteor showers and with Olmsted had rediscovered Halley's Comet on its 1835 return. Later, at Western Reserve College, he began a distinguished career as a meteorologist, going beyond Olmsted in quantitative methods and introducing isothermal and isobar graphic representation. In 1846, two years after coming to New York University, he published a classic paper on two storms. A tight-lipped, sharp-featured, intense type of the rural Yankee, quick and abrupt in speech, he went to Princeton in 1848, but was driven back to New York after a year by that strongly Southern-tinged school's prejudice against his Yankee origin and ways. Not until 1860 did he leave New York University again for his final quarter-century of teaching as Olmsted's successor at Yale.

On his way to Philadelphia Agassiz stopped at Princeton, meeting John Torrey and Joseph Henry, admiring the physical and electrical apparatus Henry had collected for the college, and tactfully passing over the deficiencies of the library (which had not yet reached ten thousand volumes).

Philadelphia made a quiet contrast to New York. Along its broad, straight, tree-lined streets ran brick pavements and rows of brick houses with green latticed shutters, polished brass door mountings, and white marble trim. Long before New York's Croton aqueduct opened, Philadelphia's Fairmount waterworks had given the city cleanliness next to its godliness. When summer heat reverberated from its brick walls, the city's people could look to gardens and shady, green squares. Yet sobriety did not mean stagnation. With a population nearing three hundred thousand, Philadelphia remained second only to New York. The Pennsylvania Railroad, chartered in 1846, guaranteed it a hearty helping of the nation's commerce. And it was shifting more rapidly to an industrial base than were New York and Boston; by 1860 18 per cent of Philadelphia's people would be engaged in manufacturing, as against 10 per cent in each of the other two cities (Boston's environs were another matter).

Though Boston had just nosed out Philadelphia for the scientific lead, the gap between them was not great; and Europe still regarded Philadelphia as the headquarters of American science. The Philosophical Society's *Proceedings* were in most respectable European libraries. Unlike Boston, however, Philadelphia drew its scientific strength more from societies than from colleges, which fact ranged Boston with the future and Philadelphia with the past. At the two classically simple downtown buildings of the University of Pennsylvania, one for the college and one for the medical school, complacency, rigidity, and lack of imagination kept the university out of the first rank. A library of less than five thousand volumes did little to help. To the university's further shame, Philadelphia's Central High School, for a time under the energetic leadership of Alexander Dallas Bache, showed considerably more scientific spirit. In the early forties the high school even had the best-equipped astronomical observatory in the nation, winning a European reputation with it.

Philadelphia's eminence in medicine had long helped to sustain it in nonmedical sciences also, especially the life sciences. Here again, as both science and medicine specialized, Philadelphia science leaned on a failing support. In one respect, however, it had a sturdy prop. That was in chemistry. The world's first chemical society had been founded in Philadelphia in 1792; one of its committee studies had led Robert Hare to invent his widely used oxyhydrogen blowpipe. Philadelphia medical leadership, with its demand for pharmaceuticals, had nurtured the city's chemical industry. From these the industry had expanded to paint, sulfuric acid, and other industrial products. By the twenties Philadelphia was the undisputed center of American chemical activity, and its industrial growth both reflected and maintained that lead. At an industrial exhibition during his four days in town, Agassiz was especially impressed with the chemical section.

Most of Agassiz's time and attention, however, went to "the magnificent collections of the Academy of Science and of the Philosophical Society." The American Philosophical Society, established in 1743 by Bache's great-grandfather Benjamin Franklin, had lately ended its first century with fanfare and entered its second in near despair. After the sheriff attached the property in Philosophical Hall, an appeal to members for contributions had finally rescued the Society in the summer of 1846. It still faced the competition of the Academy of Natural Sciences, as it had since the Academy's establishment in 1812; but an accommodation evolved, and three years after Agassiz's visit the Society sent its fossils to the Academy.

Fortune smiled more steadily on the Academy of Natural Sciences, which had lately moved into a new building and begun publishing its *Proceedings.* It met faithfully every Tuesday evening through the rest of the century. In June 1846 Dr. Thomas B. Wilson's gift of a ten-thousand-specimen European bird collection made the Academy's ornithological holdings "the greatest & most magnificent collection in the world," as one young member, Joseph Leidy, jubilantly wrote. It led to an immediate enlargement of the building. And Wilson later added many thousands of paleontological and bird specimens, as well as money and books. "Our library," wrote Leidy in 1850, "is destined to be the best scientific one in the U.S." It probably was. By 1858 the Academy of Natural Sciences had twenty-five thousand volumes to the Philosophical Society's twenty thousand, the American Academy's ten thousand, and the Boston Natural History Society's six thousand. In 1854 a budding Albany paleontologist took it for granted that Philadelphia was the best place in America to study natural history.

Without realizing it then or later (for he died within two years), Amos Binney of Boston became another of the Academy's great benefactors by buying a life membership for the twenty-one-year-old Joseph Leidy in the summer of 1845. Born in Philadelphia of German ancestry, Leidy took his M.D. at the University of Pennsylvania in 1844, but soon abandoned practice for teaching. Later he joined the university faculty and from 1853 until his death in 1891 held its chair of anatomy. He was recognized as one of the foremost American anatomists of his time. But he was more: the last great all-around zoologist, making permanent contributions in every branch from protozoa to man. For more than twenty years, until pushed aside by better-heeled and smaller-souled rivals, he led in vertebrate paleontology. His Academy paper of 1847 "On the Fossil Horse of America" showed the appearance and extinction of the horse in America before the coming of Europeans. It opened a line of investigation that eventually produced, from other hands, one of the most telling evidences in support of Darwinism.

Leidy's own tireless labor went toward the collection of facts, not the spinning of theories. "I am too busy to theorize or make money," he was quoted as saying. Nevertheless, his facts sometimes forced theories upon him in true Baconian style. In that eventful month of October 1846 appeared his Academy paper identifying a minute parasitic worm in pork as the one associated with human trichinosis, and from it he went on to still another line of brilliant work in parasitology. In a paper of 1851 he even outlined the role of environment in developing, eliminating, and replacing species. Yet he said little about Darwinism after it appeared in 1859.

Leidy was a man of simple tastes and simple goodness. He liked black bread, Swiss cheese, lager beer, and, it seems, everyone he met. Long after Leidy died, a veteran staff member of the Academy recalled his own first encounter with him:

> A heavily built man, slightly round-shouldered, but with a face described by one of his friends as of "the Christ type," crossed the floor and said "Are you the new librarian?" "Yes, Sir." " 'Um, 'um—well if I can be of any help to you, let me know, I am Dr. Leidy." And that was the *leit-motif* of his association with everyone in the Academy during the entire period of his membership.

Leidy was too unassuming and kindly—and perhaps too busy—to figure in the scientific infighting of the time. Thus his eventual abandonment of paleontology to Edward Cope and Othniel Marsh, both of whom exhibited the spirit as well as the bones of *Tyrannosaurus rex.* Leidy's quiet self-containment had its drawbacks, however, and Philadelphia's science mirrored them. So, at any rate, implied a friend who urged Leidy to attend a scientific convention and added: "It is a shame you Philadelphians (beliers of the name) don't mingle more with your brethren."

Leidy's friend, the Pennsylvanian zoologist Spencer Baird, spoke with authority. He had been a skilled collector of distinguished friendships since the age of seventeen, when with a question about a flycatcher he initiated a seven-year correspondence with John J. Audubon. In 1846, having at twenty-three just been appointed professor of natural history at his alma mater Dickinson College, Baird was a tall, energetic young man with an insatiable passion for the slaughter, stuffing, and storing of specimens. As early as May 1846 he had begun preparing to add a prize specimen to his cabinet of cordial correspondents: "I am about giving my birds a complete overhauling, new labelling &c; and putting into new cases, so that if, when Agassiz comes, he wishes to see birds, I shall be able to show him the largest collection of American birds in this, or perhaps in any country." Agassiz did not get out to Carlisle on this trip, but Baird came to Philadelphia and pleased Agassiz with an offer of duplicate specimens.

If Leidy met Agassiz, he made no recorded impression, unfortunately. Still more unfortunately, a tall, pale, cadaverous, somewhat stooping physician and scientist named Samuel G. Morton did make a deep impression on Agassiz, the deepest of any zoologist after Cuvier. Morton, a forty-seven-year-old anatomist, a pioneer in American invertebrate paleontology, and (as Agassiz put it) "the pillar of the Philadelphia Academy," was best-known for his unequalled collection of six hundred human skulls, assembled over many years from among different races of men throughout the world and especially strong in American Indian crania. On the latter, he had in 1839 published *Crania Americana,* which concluded on the basis of seventy-one skulls that differences in cranial capacity clearly distinguished the races of mankind. Without publicly and categorically saying so at that point, Morton headed toward the doctrine that the races were distinct species, separately created, and endowed by their Creator with certain unalterably different mental capacities, of which the "Caucasian" had the greatest and the "Ethiopian" the least. Agassiz thought Morton's collection, known in scientific circles as "the American Golgotha," in itself worth coming to America to see, and Morton "a man of science in the best sense; admirable both as regards his knowledge and his activity."

On his first American swing, Agassiz managed to touch nearly all the elements in the later development of American science. He visited the major centers and met most of the men who, along with the young pilgrims then in Europe, would dominate American science in his time. He also encountered the agents and institutions of development: the college, the well-disposed layman, the local society, the professional journal, even, in the Philadelphia exhibition, the science-powered industry. When he left Philadelphia there still remained the national society, the engineering school, and the scientific agencies of state and federal government. Before he got back to Boston, he would see specimens of those too.

In Washington he dutifully visited the White House and Capitol and quoted the phrase "City of Magnificent Distances." Though the city's population had not reached forty thousand, clumps of buildings in its muddy expanse suggested what would come, like patches of embroidery on an unfinished tapestry. So too with Washington science. In the 1830s only about one American scientist in forty had lived there. Even among the scientists of 1846, only 9 per cent would spend any part of their working lives in Washington. When a congressman that year asked James Dana to live there while working on the Wilkes Expedition collections, Dana refused. He laid the request to "some promptings of ill-nature" and thought it "perfectly absurd" to be asked to work "in a city where there are no books."

But times had already begun to change in Washington. Ordered to make astronomical observations there in conjunction with the Wilkes Expedition, Lieutenant James Gilliss of the navy had seized the occasion to get a Congressional appropriation for a permanent Naval Observatory in 1842. By 1846 observations were well begun at the new building near the Potomac mud flats. Though Agassiz (no astronomer) passed up the observatory, he did visit the Capitol Hill headquarters of the Coast Survey, which in 1843 had entered its great years under Alexander D. Bache as the leading federal scientific agency. Bache happened to be away, but Agassiz was impressed by the Survey's charts.

At the Coast Survey, Agassiz met Bache's cousin, Colonel John J. Abert, chief of the Army Corps of Topographical Engineers. In the era of Manifest Destiny, the corps acted as a clearinghouse and general staff for the great scientific reconnaissance of the trans-Mississippi West. Abert gave Agassiz "important information" and printed reports on the West, as well as a collection of freshwater shells from that region. Thus, Agassiz felt a Western wind that would influence the course of American science in his time.

At the Patent Office's grandiose Greek temple on F Street, the commissioner was, like Gilliss, Abert, and Bache, bringing scientific investigation within his agency's purview—in his case primarily for the benefit of agriculture. It was neither that work nor the display of patent models that drew Agassiz there, however. The rich collections of the Wilkes Expedition, arriving ahead of the expedition itself, had been entrusted to the new, federally chartered National Institute for the Promotion of Science, disdained by most professional scientists for its heavy taint of amateurism and politics. Through ignorance and stupidity, the National Institute's agent, a political appointee, had shockingly mishandled and damaged some of the collections, until Wilkes himself persuaded the Patent Office to house them under his supervision. Not all had been lost, however. Agassiz was "agreeably surprised by the richness of the zoological and geological collections." He thought the results as good as those of any European expedition, and the collections generally the only ones outside of the Philadelphia Academy equal in interest to those of Europe.

Early in the somewhat furtive liaison of government and science, Agassiz had thus caught a glimpse of its progress and first fruits. On his way back to Boston, he saw still other aspects of the alliance.

In the United States Military Academy at West Point he encountered what amounted to the first American engineering school, founded in 1802. "*Original investigations* of physical truth," wrote a scientist in 1838, "are not the objects contemplated or mainly pursued in that establishment." But the teaching of science in engineering schools could provide a living for original investigators. West Point's only scientist outside of mathematics and astronomy, the professor of chemistry, mineralogy, and geology, Jacob W. Bailey, was able to pursue his true passion of microscopic botany. Early in 1846 he had at last begun working in "a beautiful and well furnished laboratory," with plenty of light, air, and space, many closed cases, and a separate room for delicate apparatus. It was, he wrote happily, "in striking contrast to the vile hole where I have lost health, time and patience for 6 years past." After three "delightful days" with microscopy and local glacier traces, Agassiz went on up the Hudson with Bailey to Albany.

Albany, with its medical school, preparatory school, agricultural society, and museums, and nearby Troy, with its Rensselaer Institute, formed another binary star of American science, though fainter than Boston-Cambridge. Albany produced Joseph Henry; Troy's Amos Eaton kindled a number of lesser lights. But Albany's science was not too strong to be dominated for half a century by one man, James Hall, and one patron, the state government.

Son of an English-born mill superintendent in Hingham, Massachusetts, Hall as a boy came under the spell of nature along the seashore and in the woods. A Boston Natural History Society man gave him some guidance. He found his greatest inspiration, however, as a Rensselaer student under Amos Eaton, whose influence helped get him a place on the New York State Geological Survey when it began in 1836. Luck made him chief geologist for the westernmost of the four divisions, a ten-thousand-square-mile stretch which turned out to offer the clearest and most complete known development of the Devonian System, a region overflowing with fossils and laid open to the geologist by the cuts of the Genesee River and the Finger Lakes. Hall made the most of it. His final report in 1843 became a classic of geological literature. It brought order to American stratigraphy from the Upper Cambrian to the top of the Devonian. Its introduction ventured to suggest the continuity of species lifelines, the unlikelihood of special creations. This apparently disturbed Agassiz, who must have been relieved when for some reason—perhaps the abundance of data to collect, classify, and describe—Hall drew back from that grand overview and, with one exception, from other sweeping interpretations for the rest of his life.

Hall's first report by itself made Albany a required stop for visiting European geologists. Yet it merely began his career. The richness of his district in fossils brought him to life as a paleontologist, to become the recognized leader on the invertebrate side of the field as Leidy was to be in the vertebrate. He wangled a state assignment in 1843 to prepare a one-volume report in one year on New York paleontology. He finished it in thirteen massive and masterly volumes, the last of which he saw through the press in 1894. This project stands as the most ambitious single scientific enterprise undertaken by a state government in its time. Indeed, it lasted beyond its time, into the period when most significant geological work done for government was under federal rather than state auspices.

Beginning in 1845 with his first request for more time and money, Hall became an expert at bullying legislative committees into support of his researches. He would stare down or blast aside reminders that the original deadline was now twenty or thirty or forty years past. If appropriations fell short, he cast his own savings into the work. The time and money already sunk in it, the long list of previous extensions, struck Hall not as matters for apology but rather as arguments for more of the same. His foibles became sources of strength. He was egotistical; so he overawed legislators with unblushing (but not altogether unwarranted) praise of his own international reputation. He was a hypochondriac, forecasting his imminent demise at intervals over a period of sixty years; with complete sincerity, therefore, he could appeal to the pity of legislators, limping in sciatically with one arm in a sling to protect a bruised finger. He had a fierce temper, being capable of pummelling his employees, clinching an argument by snatching down a shotgun from a rack over his table, living apart from his wife and family for many years in another house on the same property; and accordingly, crossed in his piteous debility, he could pull his arm from the sling, crash his fist on the table, and stamp out of the room with a firm and even gait. His profanity approached art; so did his politicking. He was in fact a supremely healthy and vigorous man, stocky, hard-muscled, with thick black hair and keen, suspicious, pale blue eyes, when Agassiz met him at the age of thirty-five, and still in perfect shape more than fifty years later when he crossed the Caucasus Mountains on horseback.

Hall's influence went far beyond New York. At various times he served as state geologist of Iowa and Wisconsin, though always keeping his Albany base. He corresponded widely for data and specimens. His temper sped the flow of brilliant young assistants through his laboratory, men who became leaders themselves. Service with Hall was the equivalent of graduate education in paleontology; it also suggested a ritual ordeal. But the upbraidings that constituted Hall's standard commencement address came to be discounted by the scientific community; their hastily departing subjects lost no standing outside Hall's laboratory.

Bailey's introduction of Hall to Agassiz began a long and, for Hall, a remarkably untroubled friendship. Hall found Agassiz his "beau ideal of a Naturalist." Agassiz was delighted with Hall's fossil collection, whipping out a jackknife without leave and exposing bone after bone in a prized chunk of Pennsylvania stone. Hall (wrote Bailey) "had no reason to regret the results." From that moment until his dying day twenty-seven years later, Agassiz never ceased to covet and angle for the Hall collection. But he had met his match in acquisitiveness, shrewdness, and tenacity. Hall was a fine specimen of Yankee "mineral trader," tireless and insatiable in buying, selling, and swapping. Once he even formed a small joint-stock company to mine the canal banks at Lockport for trilobites. Like Agassiz he was quick to borrow specimens and infuriatingly slow to return them, even under pressure, hoping perhaps that he might outlive the lender; he kept one batch forty-six years.

And so, having taken the first measure of his new colleagues, Agassiz turned again toward Boston. They had taken his measure also and found it heroic. They had all looked forward long and anxiously to his coming, had expected much, and like Hall had not been disappointed. "Wonderful! Inspired!" wrote Augustus Gould. Samuel Morton, the racist skull collector, was delighted with Agassiz's "astonishing memory, quick perceptions, encyclopedic knowledge of Natural History and most pleasing manner." Agassiz's charm and good humor won the liking of all, as his intellect and learning won their respect. In more than one sense, the scientific community of the United States had become Agassiz's America.

On his first tour Agassiz missed the West and South but not much in the way of scientific enterprise. In both regions the profusion of nature had stirred the curiosity of a lonely few. The richly fossil-bearing and scarcely disturbed strata of the Ohio and Mississippi valleys inspired some. A German naturalist in 1844 came upon an Indiana cooper who turned out to be a self-taught fossil collector. Two years later, in the same state, Charles Lyell met a rifle-toting shoemaker who had a collection of more than 150 well-mounted local birds.

In Wisconsin an educated young Swedish frontier farmer named Thure Kumlien was also trying heroically to pursue ornithology with little leisure and no books on American birds. After nearly a decade he somehow got a correspondence started in 1851 with a member of the Boston Society of Natural History. "I have not seen anybody that takes an interest in anything else but wheat, potatoes and corn," wrote Kumlien. "You do not wonder if I am greedy for books." Kumlien's Yankee correspondent helped him reach Easterners like Spencer Baird, who determined species for him, ordered insect pins and glass eyes for his specimens, and found buyers for his finished work. By the mid-1860s Kumlien was selling specimens to museums not only in Washington and Cambridge, but also in Stockholm, Leyden, Berlin, and London; and he could put the plow aside for full-time professional collecting.

Yet for every Kumlien who broke out of isolation there must have been many unfulfilled Midwesterners whose scientific curiosity finally died of frustration and neglect. Except for Audubon's work, journals in the East and in Europe ignored Midwestern naturalists during the first half of the century. The able amateur botanist William Sullivant of Columbus, Ohio, had money and so could get books as well as guidance from Torrey and Gray. But light came only from the East. In Columbus he could look to no college, no scientific society, no botanical comrade; and the nearest publisher suitable for his work was eight hundred miles away.

Midwestern scientific societies were formed in emulation of the East and in the belief that science would stimulate economic as well as cultural progress. Thirteen had appeared by 1860. Their newly settled environs turned them toward natural history. All tried to develop museums, and most managed some sort of publication program. But only those in the larger cities survived for long. In 1851 a Louisville amateur botanist declined the presidency of a proposed Louisville Natural History Society because of his "experience and observation of all such efforts in different portions of the Western Country." He counted it a lost cause from "the very small number of persons in this quarter who pay any attention whatever to any of the departments of Nat. Science." As forecast, the infant society did not live long. It lost its collections in a fire and in 1854 finally collapsed when several of its leading members moved away.

One scientific society flourished briefly in Cincinnati during the early 1820s only to go down in hard times along with the struggling college that had given it shelter. A successor in the same city tried in 1846 to revise the taxonomic system of local fossils, but lacked the experience, resources, and standing to make a success of the project, and faded away in the 1850s. Another at St. Louis, 1837–43, failed because, as its president said, "people are too occupied with the wants of life." Not until 1856, after St. Louis University acquired an observatory and Washington Institute (later Washington University) was founded, did the Academy of Science of St. Louis come to lasting life.

The Cleveland Academy of Natural Sciences had meanwhile appeared in 1845, chiefly the creation of professors at the Western Reserve College's medical school. Cleveland's strong New England heritage ought to have helped; two of the new Academy's leaders had studied under the elder Silliman at Yale. Yet the ninety members of 1846 could not produce a quorum in 1849, as the president pointed out to the other member present. General prosperity and the Cleveland meeting of a national society revived the Academy in the mid-1850s. But the city's rich men took little interest. Too poor for publications of its own, offering little more than a meeting place for a handful of local enthusiasts, the Academy died before the Civil War.

In the boom year 1856 and the boom city of Chicago, one Midwestern scientific society did take permanent root: the Chicago Academy of Sciences. Its organizer, young Robert Kennicott (soon to become a notable professional collector), had managed to enlist a number of wealthy Chicagoans in support. But the Illinois State Natural History Society, started in the depression year 1858 by enthusiastic amateurs, struggled on for only a few years, dauntless but penniless, until absorbed by a state agency.

In 1860 a St. Louis mathematician could still write, "What I miss most is scientific association." And in 1862 a Cincinnati conchologist could still complain that "I am lost here amid such scientific isolation. . . . We have . . . in this great city literally not a public collection of science of any value & nobody that cares whether we have or not." A chemist in Indiana summed up his impressions of the whole Midwest in 1861:

> I find the West even more deficient in scientific vigor than I was led to expect. Scientific men we have & of the best; but they are few & far between. The country is given to agriculture of vast proportions in mechanical contrivances, to trade & to the building up (in consequence) of new cities.... I like the Western people and the West but feel here like one of those "square pegs in a round hole."

In 1846, with about a fifth of the nation's population, the north central states had only about a twelfth of the leading scientists. In 1861 its share of scientists remained the same, though it now had a quarter of the total population. In both years Massachusetts alone had twice as many leading scientists.

Urbanization accounted for some of the disparity. In the Massachusetts of 1850, Worcester was bigger than Cleveland or Columbus, Lowell bigger than Chicago or Detroit, Nantucket bigger than Indianapolis. Yet Cincinnati's population approached that of Boston. The other factors have been implied in the scientists' comments. With so much in material resources still waiting to be exploited, Midwesterners chased after more tangible rewards than science offered. In their hot pursuit of happiness, furthermore, they moved around so often that local societies lacked continuity in leadership and membership. And Midwestern scientists also lacked the public interest and support that tradition and a long head start had given Boston science. So the Midwest, as a new country, chiefly produced scientific raw materials for the East and Europe to process. Easterners like Jackson, Whitney, and Hall even ran the geological surveys of some Midwestern states and took their data back home to work up and report on.

Two years after Agassiz's initial reconnaissance of the United States, the nation's jurisdiction leaped to the Pacific Coast, promptly followed by the gold rush. Naturalists of several nations, including the scientists of the Wilkes Expedition, had already been there, and they kept coming. By the summer of 1850 the New York entomologist John L. LeConte was chasing butterflies near San Diego and wondering if he should invest in what looked like coal deposits nearby. Otherwise his views were decided: "the climate is heavenly ... San Francisco is damnable.... A scientific man is completely lost & useless in this country." Others agreed. At Sacramento in 1854 an amateur meteorologist complained of having not one scientist nor even an educated man to talk to. As with the Midwest, population growth did not mean scientific growth in proportion. Burning with "California fever," young George Davidson of the Coast Survey got Bache to assign him to Pacific Coast duty in 1850. He became one of the leading lights of West Coast science for more than half a century. Yet after forty years there, Davidson complained that his work had been "sadly hampered" for want of frequent contact with scientific men and associations.

A start at supplying the want had been made at San Francisco in 1853, when seven men (five of them physicians) organized the California Academy of Natural Sciences, proclaiming California a "virgin field" that had not been "subjected to critical scientific examination." Compared with Eastern societies, the California Academy took its time in the deflowering. But an early sign of cultivation and urbanity could be seen by 1856 when, on the day the vigilantes hanged a man for shooting its first president, the Academy calmly met to discuss recent additions to its library. The Academy survived those turbulent times to see its second century.

In 1853, when Louis Agassiz finally toured the South, he found it far richer in specimens than in scientists. The South's lag may be measured statistically. In 1830 (the median birth date of leading American scientists 1846–76) the slave states south of Maryland had 41 per cent of the nation's population. Yet less than 9 per cent of the leading scientists had been born in those states, and less than 12 per cent worked in them. One reason for the disparity was that more than a third of their people were enslaved and thereby forcibly prevented from even learning to read, let alone pursuing science. Yet the free people of those states constituted about 30 per cent of the nation's total. Other factors besides enslavement obviously entered.

Lack of sizable cities surely hampered Southern science. If we count St. Louis and Louisville as Midwestern and Baltimore as Middle Atlantic, the South in 1850 was left with New Orleans (116,000), Charleston (43,000), and Richmond (28,000). Richmond being of little consequence in science, we need look only at the other two cities before turning to the colleges and the planter aristocracy.

The languidly elegant city of Charleston, South Carolina, had age, tradition, and location on its side. As a seaport it had been part of the transatlantic community in colonial times. This fact and an early leaven of Scottish intellectuals gave it a tradition of cosmopolitan learning. Its museum, though musty and moribund when Agassiz first came in 1847, antedated all others in the nation; and Agassiz stirred the museum and its supporters to develop excellent collections in paleontology and natural history during the fifties. The inspiration of Agassiz also called to life a small natural history society, which managed to publish a couple of volumes of papers before the war came. But the society got little notice and less support, either from men of wealth or from the public at large. Agassiz returned for two winter seasons of lecturing at the Charleston Medical College in the early 1850s, thus keeping the spark of science alive. "He is a very remarkable man," wrote a Charlestonian in 1850. "He was much delighted with my collecting. . . . This stimulates me to try to make it more valuable." The city could boast of Dr. John E. Holbrook, the Massachusetts-raised herpetologist, and the Reverend John Bachman, the New York State-born Lutheran minister, distinguished ornithologist, and collaborator with Audubon. Two or three planters, several physicians, and a couple of professors at the College of Charleston finished out the scientific colony. Such was the leading center of Southern science.

"New Orleans is a great place," wrote the expatriate Charlestonian James D. B. De Bow in 1846, "but has scarcely yet one scientific idea." He had just pounded out the first number of that resounding though ineffectual tom-tom for Southern enterprise, the *Commercial Review of the South and Southwest,* commonly known as *De Bow's Review,* but this remark was private. His fellow citizens corroborated it in detail. "I know of no individual in this city who gives any attention to conchology," wrote one. Another knew of no one in the entire Southwest likely to have made observations of the recent transit of Mercury; and the death of the geologist Michael Tuomey in 1857, according to a Mississippi scientist, took away "almost the only man of mark as a naturalist in all the South-West."

New Orleans science, such as it was, concerned itself mostly with natural history and medicine, because of the region's abundance of flora, fauna, and fever. Most New Orleans scientists were M.D.'s, and the leaders taught at the Medical College of Louisiana. Of the three foremost, two were from Massachusetts and one from Pennsylvania. The most original was the Bay Stater John L. Riddell, professor of chemistry at the Medical College. In 1847 he gave a popular lecture on the practical problems of eventual moon flights, such as oxygen supply, meteorite hazards, and instrumentation. Of more immediate use was his pioneering binocular microscope. A hearty extrovert, he collected botany and bullion with equal zeal. Some thought him unscrupulous and whispered that he had lost a job at the New Orleans Mint "upon suspicion that its yellow dust adhered to his fingers."

For public consumption, De Bow in 1854 asserted that New Orleans had as many scientists and scientific facilities as any city of its size in the Union. But the New Orleans Academy of Sciences, founded by five physicians a year earlier, enrolled only about seventy nominal members before the war, issued only one seventeen-page volume of trivial proceedings, and heard little but extemporized remarks from those who showed up at meetings. In science as in other ways, New Orleans seems to have been to Charleston as New York was to Boston.

In the cities, and still more markedly outside them, Southern science, like that of the Midwest, lacked the public appreciation and support that sustained science in Boston. Science had no effective way to reach the Southern people generally. Illiteracy was not only enforced by law among a third of them, the slaves, but also by state and local neglect among many of the remainder. Only North Carolina and Kentucky had respectable public school systems before the Civil War, and those were recent. Popular lecturers, who did much to rally the public behind science in the North, found travel difficult in the South, towns small and scarce, and crowds hard to gather for anything less visceral than horse races, cockfights, and revival meetings. So their "scientific exhibitions" amounted to magic shows—though they might profess, as did one advertisement, to "render the entertainment instructive as well as amusing," and audiences might be "particularly asked not to keep time with the music." Some good lecturers came, notably Agassiz, but they kept to the few cities and large towns. And though nearly every Southern town had a little hodgepodge of curiosities it called its "museum," only Charleston had one worthy of the name.

Even literate Southerners encountered little science. Scientific books were hard to get in the South. In the capital of Mississippi, no bookstore had anything on geology but a small elementary textbook, and so the state geologist botched his report for lack of reference works. In 1850 the slave states as a whole, including the border states, had one public library volume for every twenty-one free inhabitants, whereas Massachusetts had one for every four of its people. And such public libraries as the South had were poor in science, even in Charleston.

Periodicals did not fill the gap. In 1850 more than five times as many copies of scientific periodicals were printed in Massachusetts as in all the slave states combined. The latter accounted for only a thirteenth of the national total, though they had more than a fourth of the nation's nonslave population. Scientific publication in the South had to depend largely on the hospitality of Southern medical journals, which tried consciously to make up for the paucity of more appropriate outlets, and on the *Southern Quarterly Review* and *De Bow's Review,* which published a scattering of scientific reviews and articles.

Higher education in the South did little to mend matters. Southern political orators often glorified science and urged their listeners to study and advance it. Jefferson Davis in 1852 exhorted students at the University of Mississippi to study science for the honor of the state, and for the good of the cotton crop through geology, chemistry, and entomology. Some Southern colleges, such as the University of Georgia, were generous enough with funds for scientific apparatus and supplies. But, for lack of previous exposure to science, most Southern college students knew little about it and cared less. The classics took precedence in their minds as the hallmark of gentlemanly culture. One professor remarked in 1844 that he had known some competent students during his long service at the University of Nashville, but never one with an enthusiasm for science. In the mid-fifties the University of Georgia faced facts and ruled that failure in mathematics and sciences (a third of all courses) should not keep a student from rising with his class.

The few young Southerners of means who somehow acquired a taste for science tended to seek further helpings in Northern colleges, and with good reason. Southern colleges were likely to have small faculties, usually with only two scientific members available to offer an absurdly wide range of superficial courses. Heavy teaching loads, committee work, low pay, and the never-ending duty of disciplining notoriously unruly student bodies disheartened and handicapped even the ablest of

Southern academic scientists. They suffered also from isolation and indifference. Frederick A. P. Barnard of the University of Mississippi considered the geologist Eugene Hilgard to be "the only man here, or (so far as I know) in the state, in whom I can find any sympathy as a lover of science for its own sake." Nevertheless he advised young Hilgard to move to Washington, because "to detain him here longer would only be to do our part to bury him."

In the end Hilgard remained staunchly loyal to the South, but Barnard himself went north as soon as he got the chance, there to become president of Columbia University. A still more notable figure in American science, William Barton Rogers, likewise made the break. Rogers, forty-one when Agassiz made his first American tour, held the chair of natural philosophy in the University of Virginia. He was the ablest and most influential of four brother scientists, sons of Patrick Rogers, a Scotch-Irish immigrant who had taught science at William and Mary College. William and his brother Henry became noted geologists; James and Robert served as professors of chemistry at various colleges. The "Brothers Rogers," as they were known in scientific circles (or the "Brodgers," as their adversaries privately called them), worked as a formidable team in scientific affairs, but it was William who (as we shall see) eventually led the principled opposition to Bache's Lazzaroni clique and became founder and leader of the Massachusetts Institute of Technology. Though William had the lean, craggy, Yankee face of an Emerson, nothing in his ancestry and upbringing turned him northward, only his discontent with the working conditions and intellectual milieu of Charlottesville, which he likened to a stagnant pond. Henry moved from the University of Pennsylvania to Boston in 1845. William likewise jumped to that bigger pond in 1853, though he secured no academic position there until he created his own.

Producing few scientists and losing some of those, Southern colleges needed all they could draw from outside. Yet as the South's upper-class mind closed against an outside world hostile to slavery, imported savants met with more suspicion than gratitude. After bringing James J. Sylvester, one of the greatest mathematicians of his time, from England in 1841, the University of Virginia lost him three months later, partly because of local anti-Semitism. Hostility became mutual. In turning down a Louisiana position in 1837, Asa Gray wrote bluntly: "I do not like the Southern States." In 1846 Wolcott Gibbs wrote, "I have no objection to go anywhere except South." A higher salary at the University of Georgia in 1854 could not induce a Delaware College astronomer to go that way; he chose Indiana instead. And in 1857 one of the nation's two leading mathematicians, William Chauvenet of the Naval Academy, wrote the other, Benjamin Peirce, as to a bid from South Carolina College:

> I will not go *south* any further . . . if I can help it. I certainly need more salary to educate my children but I will not sacrifice them to

> money—for I should regard it as a sacrifice of the best part of their characters to bring them up under S. Carolina influences. I do not talk in this way to my good friends in Columbia, because it is useless to stir up their prejudices.

Weak in the collective support that cities, societies, and colleges gave science in the Northeast, Southern science depended more heavily on individual amateurs—planters, politicians, lawyers, and especially physicians. But more than elsewhere, the Southern amateur scientist stood alone. "With the exception of three or four individuals," wrote one man from Little Rock, "science is not desired by this community, and *black-legs* are much more welcomed than gentlemen of science." A University of Georgia professor complained in 1854 that "no one could see the *cui bono* of exposing myself to sun for 3 hours, merely to watch the progress of the late eclipse." In South Carolina, shell gathering was put down as "complete tomfoolery." In Mississippi a clergyman and amateur naturalist wrote pathetically to a new-found friend:

> The interview I had with you was the first I have ever had with any one in this part of the country on those subjects. . . . My collections . . . look like toys after seeing yours. . . . There is so little pleasure in displaying them to those "who view them with brute unconscious gaze" that . . . I have many things of interest to me, which I have never exhibited to any one.

It was a rare lover of science for its own sake who could long shrug off the "brute unconscious gaze" or outright mockery of those around him.

In that respect, at least, the institution of slavery bolstered some amateur scientists. Planter aristocrats, proud monarchs of their own small worlds, could dismiss popular scorn. A back-country planter in South Carolina, tired at last of hearing his botanizing called "child's play," quit the field and gave his plants away—but only after thirty years of collecting. "I do not know a single individual along the coast of Georgia who devotes any attention to natural history," wrote a lordly sea-island planter. Nevertheless he went serenely on with it.

The habit of command had other consequences, however. It inhibited some planters who had the stamina, will, and intellect to make worthy contributions. "The idea of publication never even entered their minds," wrote a Georgia scientist. "What right had anyone to publish unless it was something of the greatest importance, something that would revolutionize thought?"—the work, in short, of an unchallenged master.

The South's peculiar institution stunted the life of the mind among masters as well as slaves. The system's culture, values, and compulsive self-justification fostered a pseudomedieval romanticism at odds with the scientific temperament. Pride, not free of guilt and fear, bred conservatism that became rigidity, an insistence on conformity that became thought control, and defensiveness that hardened into xenophobia. Absorption in economic gain and proslavery polemics displaced science in the minds of some planters. Not a product of slavery, but consistent with its intellectual tendencies, was religious fundamentalism, hostile toward science that cast doubt on Biblical chronology and sometimes acting to suppress that science.

The proslavery ideology rejected logic when it simultaneously embraced the Biblical doctrine that blacks and whites alike belonged to the family of man, the Jeffersonian doctrine that all men had the right to liberty, and the Calhoun doctrine that blacks had no such right. One might suppose that no one able to defy logic in that way could be a true scientist. Yet some reputable Southern scientists did so, and our own experience shows that scientific reason can coexist with political and philosophical unreason.

Philosophy aside, slavery corrupted the temperament and mental habits of the masters. It gave them leisure to do science, but also invited them to langour and self-indulgence, mental as well as physical. "The study of science," says a modern authority on the mind of the Old South, "requires concentration, perseverance, and great curiosity—qualities in which Southerners as a whole seem to have been deficient in comparison with Northerners." Whether or not the work ethic spurred Boston scientists, it seldom moved Southern aristocrats.

Some gentlemen scientists used slaves as research assistants, even in fields of collecting that were more sport than toil. From one comfortable plantation villa, set among the magnificent evergreens and moss-hung live oaks of St. Simon's Island in Georgia, Charles Lyell strolled pleasantly along a beach with his host and picked up twenty-nine species of marine shells. Nevertheless, wrote a conchologist, "in our Southern States generally we have few who care any thing about the study of Conchology and fewer still who are willing to stretch out a hand or wet a foot in personal collection. They depend on their servants for such labor and Negroes are far from being the best collectors." A Mississippi gentleman scientist commented sarcastically on the state geologist's use of a convict to keep apparatus in order: "This is one mode of patronizing science and keeping it respectable." The same scornful gentleman put a slave to work mounting specimens for the museum at the University of Mississippi.

In justice to Southern scientists, however, it must be said that they faced greater physical hardships than did their Northern counterparts. A transplanted Massachusetts man, Moses Curtis, wrote Asa Gray from North Carolina:

> You northern fellows know nothing of what we southerners endure in the collection of objects of Nat. Hist. Ask Harris if he knows what Red Bugs [also known as "chiggers"] are. . . . The creatures are barely

> perceptible to the naked eye, & your first knowledge is from the ten thousand bites all over your body, blotching it like the measles, & setting the whole surface on fire. These, with millions of Ticks, Musquitoes, Gnats, Yellow Flies, & Gauze wing Flies, render Botanizing here anything but agreeable.

By the mid-fifties he confessed to Gray that he rarely went into the woods anymore: "Ticks & Red Bugs have fairly beaten me from the field." Another Carolinian preferred red bugs to mosquitoes; he paid tribute to both, but thought he would rather be ambushed than charged. "The vermin are intolerable," wrote Curtis to John Torrey. "It is only a fresh and untried passion that can resist them for a few seasons. . . . The working botanists in the Southern States . . . all seem to give out in a few years." In South Carolina, heat and mosquitoes discouraged Lewis Gibbes from making astronomical observations. Even indoor work, if delicate or tedious, contended with the hot, sticky climate. One Southerner reminded a friend about the latter's prediction that the South's "ennervating" climate would soon blight "the intensity of first love" for science. "I would not admit for a moment then that my zeal could ever flag," he wrote, "but tempera mutanta," and so he had given up working with the microscope. When in 1837 James Dana rejected the idea of teaching in a Southern college, he gave two reasons. One was unruly students. The other was the climate.

The South was so rich and varied in natural history, and that field was so open to the lone amateur, that Southern scientists pursued it despite physical hardships. But a greater proportion of them than was the case elsewhere preferred less physically arduous fields. Meteorology, of practical concern to planters anyway, had many nonslave data gatherers. The weather came to their verandahs and called for nothing more than checking of thermometers, barometers, and rain gauges. And the field of science in which the South most nearly approached its population quota was mathematics. Of all sciences, mathematics is freest of physical effort.

In several respects, slavery thus left its peculiar brand on Southern science, as did physical environment and population distribution. The South was a distinct scientific region. But as compared with the Northeast, the South and Midwest both showed the effects of ruralism and westward expansion. And as compared with Europe, so did all regions of the United States. Without denying regional differences, these and other characteristics common to all the states, but not to Europe, justified Agassiz and his circle in talking of "American science" and the "American scientist."

Chapter 6

Science, American Style: Targets of Opportunity

"In a world quite new . . . so full of interesting objects," wrote Louis Agassiz in April 1847, "I have been carried away by the objects immediately around me . . . and brought into such a state of excitement that I at last was taken sick." What happened to Agassiz also happened to American science—and in some respects to the American people generally.

American science in 1846 was growing up outdoors. Mathematicians, chemists, and physicists together comprised only about a quarter of the leading scientists active in that year. The remainder, including the biologists, spent little time in experimental laboratories. "Sir," said a Vermont congressman in an 1846 debate, "a laboratory is a charnel house, chemical decomposition begins with death, and experiments are but the dry bones of science." Outdoors there waited an open frontier of earth, sea, and sky, a bonanza of undescribed natural objects for every man.

One observer saw the "lonely American entomologists" as being

> like so many scientific Robinson Crusoes, each with the insect-wealth of a new island at his disposal. They are monarchs of all they survey. With what affluence they exhibit their dozens of undescribed species; with what autocratic power they divide and recombine genera!

"Something new is arising at every turn," wrote a newly enlisted collector of lichens and fungi in 1849, "nor need I move a half mile from my door." From time to time most came back to city bases, where books and museums and expert colleagues helped them put their finds in order. Some, like Joseph Leidy in paleontology or Asa Gray in botany, let others do the outside work. But they dealt with outdoor things. To observe, to describe, to classify, perchance to name; then to observe again. That was the happy and simple round of most American scientists.

Within limits, it was a natural and proper course. European science had run a similar course—indeed was to a considerable degree still running it. In 1844 Alexander Dallas Bache saw the logical sequence in the history of the Royal Society "from its early day of descriptive natural history to the latest times of the highest flights of mathematics and physics." A reviewer in Silliman's *Journal* also granted that "in a country so unexplored as ours . . . for a time, general and abstract researches will, and legitimately may, give place to the labors of the literal historiographer of nature." But, he added, "this discrimination ought on no account to survive the occasion for it."

In certain fields and places it looked as though the occasion were beginning to pass. In 1844 an entomologist, touting his own hobbyhorse, remarked that "other departments of natural history have been so thoroughly explored as to leave little room for new discoveries." The gloating fungologist of 1849 wrote three years later that "this region is all worked to death." "I do not find anything new," a New Jersey botanist wrote John Torrey. "I have pretty thoroughly explored this region." The pioneer bonanza seemed to be running out, and that augured a move indoors.

But a long renewal was at hand. For along with so much else of import to American science, the year 1846 had brought the Mexican War. Out of it came the Mexican Cession, a million square miles of rich new material for American geologists and naturalists.

In their passion to exploit this windfall, American scientists, especially young Spencer Baird of the Smithsonian Institution, moved with vigor and finesse to enlist the United States Army as it explored and surveyed the new domain in what has been called the "Great Reconnaissance." Meanwhile the navy sent out its officers and ships as pathfinders of the American commercial frontier: to Japan and the North Pacific, to Panama and Chile, to the Río de la Plata and the Amazon, to West Africa and even the Dead Sea. And science sailed with them. The scientists' military and naval alliances, their tactics, strategy, and campaigns, will be detailed in a later chapter. What concerns us in analyzing the American style of science is the fact that during the 1840s and 1850s these rovers by land and sea flooded American science with data and collections.

As the flood of data rose, willy-nilly it bore along Torrey, Gray, Baird, Hall, Leidy, and lesser men. "No end to the accession of rich treasures," wrote Baird. "Oh, for time enough to develop them." The flood drowned the Torrey and Gray *Flora of North America.* "Last Saturday was my birthday," wrote Gray. "44! Oh! Oh!" Gray could scarcely keep up with what came in from day to day, let alone develop strategy for years ahead as his friend Charles Darwin was doing. "The forces of American expansion in the 1850's," writes his modern biographer, had made him "their prisoner."

No doubt Gray was, like most, a prisoner of his times. He and his peers were also prisoners of their desires, which is to say eager prisoners. He need not have taken over the botany of the Wilkes Expedition collections, yet he did so, alleging "a sense of duty to the science I cultivate & to the country." He did not finish it. Baird's lamentation for more time was followed immediately by the words: "By all means send me lots of *Salamandrosus.*" "Collect everything you can find," he wrote one correspondent, and to another, "I need not say that I want *everything.*" With several projects unfinished, including work for the Wilkes collections, Agassiz quite deliberately sent out more than six thousand circulars soliciting specimens for a grandly ambitious study of all United States freshwater fishes, and he got his fishes by the thousands. He did not finish that project either. And from his sepulchre of extinct vertebrates, Joseph Leidy wrote: "This new pliocene collection only makes me thirst for more." They all seemed happily drunk on data.

Not content with the government bonanza, they hired professional collectors, of whom the greatest was Charles Wright, Gray's touchy and eccentric, but nonetheless hardy, resourceful, and intelligent legman. Dr. George Engelmann, the St. Louis botanist, worked with Gray to recruit, equip, and instruct transients passing through for the West, while Gray processed the collections and sold duplicate specimens to help defray expenses. Gray even resorted to professional collectors in Central and South America. And still unsatisfied, Hall, Baird, Gray, Torrey, and Agassiz all tolerated, patiently answered, even encouraged inquiries and offers of specimens from amateurs of widely varying expertness. With his Washington vantage, Baird made use of Americans abroad, including the footloose author of *Home Sweet Home,* and of science-dabbling army officers at various posts, especially surgeons.

By the spring of 1861, when the last topographical engineer to leave the West came back with his dustpoke of data from the Northwest Boundary Survey, an enormous accumulation of facts had been labelled and filed by American scientists. The government land expeditions alone had made reports to match their mountains. Dozens of articles besides had been quarried, shaped, and set in scientific journals. "Great things are done when men and mountains meet," or so wrote William Blake the poet (not to be confused with William P. Blake, the geologist of Lieutenant Williamson's Pacific Railroad survey). What great things, then, came out of this mountainous labor?

More than those in other fields, the geologists managed to draw conclusions from their facts. In the Southwest especially the evidence of past geological events was written large. Geologists did not need to jump at conclusions; conclusions jumped at them. The Great Basin had obviously once been a sea. The Grand Canyon region made stunningly evident the immense cumulative effect of age-long gradual erosion. In the Grand Canyon one of James Hall's protégés, John S. Newberry, saw the mountain-slicing possibilities of a great slow uplift of land beneath a river which eroded it at about the same rate. William P. Blake saw the erosive power of wind-blown sand. The unaccustomed intoxication of large ideas led some to venture too far and too fast, in theories of mountain-building, for example. Yet even some of these daring theories approached the truth. And the demonstrated power of erosion, by accounting for what had been taken as evidence for catastrophism, cast doubt on theories of wholesale species destruction followed by special new creations. Thus it helped prepare the way for Darwinism.

Even though "too busy to theorize or make money," Joseph Leidy also placed a stone in the evolutionary edifice. His reports and articles on the vertebrate paleontology of the West, his study of *The Ancient Fauna of Nebraska,* all stretched conceptions of life's antiquity on earth and so made room for the slow work of evolution, as well as storing up facts for the eventual testing and particularizing of the theory.

More conformably to the classifying, theory-dodging tendency of American science in other fields, the geologists succeeded in determining a typical Western stratigraphic column, or sequence of strata, with which strata over a large part of the West could be correlated.

In the other major scientific fields, only the last sort of triumph in classification could be claimed. In botany Torrey and Gray completed the main outlines of Western flora, leaving out no major Western region. Torrey looked for no more. He had as little inclination to generalize as Leidy; and since he did not deal with sequences in time as Leidy did, theories of causation were not thrust upon him by implication. (Darwin's *Origin of Species* had been in print three years before Torrey promised Gray in 1862 that he would get himself a copy.) Gray's livelier mind now and then sniffed curiously at such matters as plant chemistry and physiology, questions involving change and therefore causation, but the unceasing pressure of empirical data kept pushing him back into classification or taxonomy.

In a narrowly Baconian way, the botanical reports did include efforts to correlate plant distributions with altitude, geological formations, and other environmental characteristics. These, however, led to no insights into past development or causes. The zoological reports on animals, birds, and fishes, awesomely comprehensive as they were, did not come even that close to generalizing, though Baird, their chief contributor and organizer, had once talked of searching out "the mutual connection and causes of many natural phenomena." All the scientists had taken meticulously clear taxonomic snapshots; but only those of the geologists and paleontologists had motion, direction, the semblance of life.

. . .

"Facts," Agassiz was quoted as having said one day, "facts are stupid things, until brought in connection with some general law." This remark might seem strange coming from Agassiz. By its definition, not only most of the facts out of the Great Reconnaissance but also most of those dredged up (figuratively and literally) by Agassiz himself were "stupid things." But Agassiz connected his facts with his general law that all things and events are made by God. This proposition makes a search for anything but superficial causes unnecessary, if not impertinent or impious. Still, in closing the question of cause, Agassiz's view opened the question of purpose; and thus Agassiz could claim for science a goal beyond the immediate facts: to fathom God's plan. Assuming, however, that God is never aimless in his work, it follows that every fact, large or small, has significance for his grand plan. Agassiz, the Great Reconnoiterers, indeed most American scientists of the mid-century seemed to act as if that were in fact the case. As God's eye was on the sparrow, so Agassiz's was on the minnow. All this, of course, merely holds aimlessness at one remove; for if all things are alike significant, then choosing things to examine becomes itself a random act. But to Americans, anything at one remove was as good as out of sight.

There was another way of hallowing the gathering of facts, and American scientists seized upon it. It was called Baconianism or the Baconian philosophy or the system of inductive reasoning. Reacting against an excess of theories over facts, Sir Francis Bacon in the seventeenth century prescribed that all possible facts on a given subject should be collected, all possible experiments performed, and all the results classified; and that only then should a general law be framed to describe their relationship. By then the law would appear almost of itself, he had assumed. If moderated to allow a hypothesis from something less than all possible facts, discarding it if new facts require, Baconianism works well enough. In Bacon's own day it was an altogether healthy influence. During the eighteenth century France made more of the doctrine than did England, but the nineteenth century brought a new emphasis on conscious Baconianism in England and the United States.

The more explicit American preachers of Baconianism (few of whom were active scientists) often seemed more grudging of hypotheses than Bacon himself had been. With American facts accumulating in quantities undreamed of by Bacon and no end in sight, they despaired of final generalizations and spoke as though classification alone should be the goal. Not surprisingly, the simplest and baldest statement of this view was by a popular journal, the *Scientific American,* in 1852: "Science is but a collection of well-arranged facts."

But preaching was not necessarily practice. Joseph Henry could complain about "anti-Baconian" remarks by a colleague, and yet tell his own students that Bacon's rules were not worth much and that a real scientist needed some imagination. The geologist of the Grand Canyon region, John S. Newberry, not only contributed to the properly Baconian correlation of Western stratigraphy, but also came near the truth in quite un-Baconian hypotheses about the history of the Grand Canyon and of the whole continent. More than that he pointed out the usefulness of studying a single mountain chain or peak to derive the principles of its creation, without waiting for all the facts about all the mountains. Ten years later a presidential address before the American Association for the Advancement of Science seemed almost a direct reproof to such doings. The key to modern progress, said the speaker, was in "the universal adoption of the comparatively new system of inductive reasoning . . . the Baconian philosophy." Not proposing to enlarge on that proposition, he went on, "I must . . . trust to its acceptance without argument." The speaker was John S. Newberry.

Baconianism, in short, was for most American scientists an accessory after the facts, a rationalization more than a reason. If like Asa Gray they were prisoners, it was of their times and temperaments, not of a dead philosopher.

As remarked earlier, the first mainspring of American development, in society as well as science, was an unprecedented abundance of potential wealth in land open to immediate taking by most individuals. Ralph Waldo Emerson celebrated one result of the open-land influence in his tributes to self-reliance. In 1847 another aspect struck him less happily: "Alas for America, as I must so often say, the ungirt, the diffuse . . . one wide ground juniper, out of which no cedar, no oak will rear up a mast to the clouds! It all runs to leaves, to suckers, to tendrils, to miscellany."

The westward scattering of energies and people had its counterpart in science. "I feel that I am now stretching over too many departments," wrote James Dana in 1848. "It compels me to throw aside entirely one branch after I have got up a deep interest in it, and bend all my force & memory to another." Many American scientists took this condition for granted. One declined the presidency of a society because the president "ought to have some general acquaintance with all the different branches of Natural Science, whereas I make no pretensions to any other than Botany." Another naturalist, when asked what field he was in, replied: "I am studying Wisconsin." (He never got around to studying Darwin.) But European scientists in America, and Americans in Europe, saw and deplored the contrast between European specialization and American diffuseness, superficiality, and overextension. Bache even hinted at the economic parallel, quoting "an intelligent foreigner" as saying of American science: "You have so much ground to clear . . . that you cannot give all your time to one garden spot."

By analyzing the four chief ingredients of his own scientific success, Charles Darwin clarified another facet of the American way in science. He believed his most significant qualities to have been: "the love of science—unbounded patience in long reflecting over any subject—industry in observing and collecting facts—and a fair share of invention as well as of common sense." Many American scientists matched Darwin in industry and love of science; or as Agassiz put it in a letter to Europe, "in ardor and activity they even surpass most of our savans." Many had a fair share of invention, though perhaps inhibited by Baconianism or submerged in data. Their crucial lack was of "unbounded patience in long reflecting over any subject."

The tendency was national. With so much prospective wealth to go around, logic should have lessened the urgency of taking it. On the contrary, its very accessibility and visibility incited haste. Americans rushed for gold, rushed for land, rushed for business, even rushed for fossils in the inexhaustible Dakota Bad Lands. Bache admitted to having felt in his early Coast Survey work "that same nervous impatience which belongs, I suppose, to all Americans." A Cincinnati astronomer, rushing to Paris with a popular subscription fund for an observatory object glass, quailed before the prospect of slow, meticulous, special grinding. "I knew," he wrote, "from the character of my countrymen," that a four-year delay "was equivalent to an abandonment of the whole enterprise." American naturalists developed a European reputation for grabbing at priorities through premature, slapdash species descriptions, sometimes unwittingly of species already named. "I will not run a race with those," Agassiz wrote Baird, pointedly urging him to "take your stand like a man" against the tendency. (Baird denied having followed or encouraged it.)

The habit of haste was enduring. Early in 1847 Agassiz was quoted as saying he "can't learn the American fashion of doing up science *running—must walk.*" A leading American astronomer much later felt that "prevalent among us during all the middle part of the century . . . [was a] lack of the power of continuous work necessary to bring important researches to a completion." A veteran conchologist admitted in 1869 that "we are all of us too hastily inclined & jump at conclusions too rapidly." In presidential addresses the American Association for the Advancement of Science was told in 1873 that "we are in fact a fast people, . . . not content to devote patient and laborious study" to science; and in 1885 that "possessed by the demon of scientific haste, we continually spoil our own performances." And even as late as 1898 a noted scientist commented that "the European pioneer puts stepping-stones where the American lays a bridge."

Another facet of the American outlook, closely related to diffuseness and impatience, is nevertheless distinguishable from them. And so influential was this quality in American science that the distinction, though subtle, should be drawn.

Again an English expression of it helps make it clear. In 1833, John Stuart Mill compared the French and English intellect:

> The English public think nobody worth listening to, except in so far as he tells them of something to be *done,* and not only that, but of something which can be done *immediately.* What is more, the only

> *reasons* they will generally attend to, are those founded on the specific good consequences to be expected from the adoption of the specific proposition.

Forty years later, another Englishman, John Morley, similarly found the political spirit "which is incessantly thinking of present consequences and the immediately feasible" to be "the strongest element in our national life." This sounds like simple impatience. Yet one may proceed patiently toward an immediate objective and impatiently toward a remote goal. The difference between impatience and the cult of the specific goal or line-of-sight target is that between procedure and agenda.

Perhaps English and Americans shared the characteristic because both had long seen and hotly pursued great potential wealth open to the enterprising individual, wealth from technological resources in the English case and natural resources in the American. At any rate the quality fitted their mutual Baconianism, which in theory looked a jump or two ahead to generalizations, but in practice bagged the nearest visible fact, counted it a goal achieved, then went on to bag the next. Even if the fact hunter rose to brilliance in experimentation, as Americans sometimes did, it was a case of tactics without strategy. "Few, indeed," complained an astronomer in 1876, "are the American investigators who have followed up their subject during their whole lives, working it out from step to step, as Europeans have done." They made no effort "to look beyond the necessities of the immediate present."

The annual soul-searchings of the American Association for the Advancement of Science made the point clearly. In 1869, referring to "our national fault—want of thoroughness," Benjamin Gould added that "thoroughness will never flourish while only those pursuits are encouraged which promise immediate recompense of the most tangible sort." In 1873 a speaker charged American universities with "making short cuts in going over the fields of science." In 1875 an astronomer complained that "livings and prizes have enticed men to work where practical results are directly in view, in the applied rather than in the pure mathematics." And in 1892 a speaker noted that "that line of work which promises the quickest returns, in the proper form for publication, is most attractive to the young student of Physics and Astronomy."

As the latter comments imply, both economic utilitarianism and the fact-gathering, priority-seeking spirit were corollaries to the line-of-sight approach. The work of men like Asa Gray or Spencer Baird has often and quite properly been cited to show that American science was not merely utilitarian in an economic sense; nevertheless, it illustrates the second motive, which might be called professional utilitarianism. Being first in print with a fact gratified Baird and Gray as surely as being first at the Patent Office with a caveat gratified an inventor. The reward, moreover, was direct, prompt, and often repeated—a good way to engrain a habit.

Less innocent, and in paradoxical contrast to the Baconian approach, was another corollary of the quick payoff: charlatanism. Coming back from Europe in 1838, Henry wrote Bache that he had been struck more than ever by its prevalence in American science. "I thought," he added, "of your frequent remark 'that we must put down quackery or quackery will put down science.'" In place of steppingstones, Americans sometimes laid bridges of smoke and moonbeams. "In this country," wrote Henry to a European friend in 1846, "our newspapers are filled with the puffs of quackery and every man who can burn phosphorous in oxygen and exhibit a few experiments to a class of young ladies is called a man of science." As late as 1868 Henry referred scornfully to "the humbugging system which prevails in this country to so great an extent." Much conflict in the organizing of American science, as will be seen hereafter, would arise from its leaders' determination to "put down quackery." And even short of quackery, Americans too often jumped to conclusions, as in mountain-building theories, overleaping the middle ground of patient thought and testing.

Whether or not nineteenth-century Americans, lay or professional, were peculiarly biased in favor of scientific research with a practical goal has been debated by historians. Europeans were more successful in putting science to practical use, but they happened to be stronger than Americans in sciences that lent themselves to immediate application. There are no manometers to measure *pressure* for a material payoff in America, but both visiting and American scientists seemed to think it stronger than in Europe. "Neither [Henry D.] Rogers nor any other American have a mind for purely scientific searches," wrote an immigrant Swiss naturalist in 1852; they look for "a practical result." Joseph Henry was less sweeping but just as positive. "In this country," he wrote for public consumption in 1847, "many excel in the application of science to the practical arts of life, few devote themselves to the continued labor and patient thought necessary to the discovery and development of new truths." He put it less politely in private that same year: "in a country like ours . . . since it will probably be found necessary to make a few oblations to Buncombe, practical science must have a share." He never wavered in his estimate, and as late as 1870 testified that "the great demand in the United States is for applied science, not theoretical science."

Young American scientists sometimes chafed in the harness of mere economic utility. "To study *pure* science has always been my cherished wish," wrote twenty-five-year-old William H. Brewer in 1854, "yet I have done most in applied science. My daily bread has been earned by my labors, and I have been *obliged* to cultivate such fields as yielded *that.*" He blamed his case on "the demands of American society," the "American atmosphere." Yet he was to hold back none of his intellect and energies from forty years of labor to improve American agriculture. Others, like Eben Horsford, had no misgivings at all. Near the century's end it was said in praise of one dead geologist that his "most abstract work was consciously for the benefit of the community," and of another that no "theoretical or speculative science ever diverted him from the single aim of making his scientific knowledge and discoveries useful to his fellow men."

Anyway, the demands of American society were not as confining as Brewer implied. Alexis de Tocqueville, that uncannily perceptive French visitor of the Jackson era, commented that a democracy favored not only applied science but also "that portion of theoretical science which is necessary to those who make such applications." Scientists then as now knew how to use their would-be users. As one scientist put it in 1865: "We must make the application of science pay our way in abstract studies." Thus James Hall could get large sums over a long period from a democratic state government for unabashedly nonutilitarian researches, once he had got the work started on some plausibly utilitarian grounds. But the tactic was easier to use in some sciences than in others.

Besides economic and professional utility, scientific utility—the goal of maximum scientific gain from a given investment—influenced the direction of effort. In 1860 the doubts of a geographer about a proposed Arctic expedition anticipated the debates of a century later over allocation of funds for space exploration:

> While I feel an intense desire to know more than we have thus far learned of . . . these mysterious & unconquered regions, I have in my mind the fresh & sad remembrance of so many noble lives, of so much devotion to scientific inquiry, of so vast resources sacrificed for so slender results; a richness of means, indeed, which applied to the investigation of any other portion of our globe would have sufficed for solving geographical & natural problems of far greater scientific & practical importance.

Near the end of the century a physicist put the case more generally. Scientific progress, he said,

> may be expressed by a curve approaching truth asymptotically. . . . So long as investigators find that they are working upon the steep part of the curve where it approaches truth rapidly, there is no lack of interest; this, however, seems to die out quickly when much labor and great patience are required to extend experimentally the curve now more slowly approaching complete knowledge. . . . As soon as a startlingly new or curious line of investigation is suggested every one pounces upon it and older problems are left far from completion. That we in America are especially inclined to this weakness in physical investigations I believe to be the case.

In the time span of this study there came small rushes to new fields like electricity and organic chemistry, though superimposed on the continuing strong pull of descriptive natural history and the geophysical sciences.

In short, the effect of the American itch for the immediate return, for the target of opportunity, was not as simple as might be first supposed. It led Americans toward relatively unexploited fields, whether Western botany or organic chemistry, until pickings got too slim or competition too keen. It led toward fields requiring the least previous training and expense, like meteorology, especially those not requiring higher mathematics. It put a premium also on fields that could draw financial support by raising hopes of economic return. And within a given field, it lured them to the description and classification of data, upon which they could also invoke the blessing of Bacon.

It was all in tune with that most American of formal philosophies, pragmatism. And while pragmatism, like the general attitude of American scientists, had its roots in the whole American experience, it was natural that the formal expression of pragmatism should have come not from American political thinkers, as a casual observer might have expected, but from the American scientific community, specifically that of mid-nineteenth-century Cambridge: from Charles S. Peirce, mathematician, chemist, geodesist, and son of Benjamin Peirce; from William James, biologist, physiologist, psychologist, and one-time Agassiz student; and with some assistance from Chauncey Wright, mathematician, computer for the Nautical Almanac office in Cambridge, and recording secretary of the American Academy of Arts and Sciences in Boston.

Chapter 7

Becoming a Scientist

After external influences and mere chance are taken into account, there often remains a touch of mystery about one's choice of paths in the maze of life. For example, a banker's son named George Searle seemed set on his course as a computer and observer at the Harvard Observatory. Yet in 1868 he wrote:

> I mean to quit Astronomy. . . . I take some pleasure in working out problems that present themselves in computing, because my head is somewhat mathematical; but . . . the composition of the sun & stars, the nature of the nebulae, the meteor & comet question &c are absolutely indifferent to me. And the same is true of all the physical sciences. Nothing except the honor and credit of discovery would induce me to spend a moment's thought upon them, and even that motive has a very slight influence.

So George Searle became a Catholic priest, wondering why he had waited so long to be happy, and eventually rose to a high place in the Paulist order. His brother Arthur, who had tried farming, statistics, brokerage, sheep raising, and the teaching of English, filled in temporarily—he supposed—at the observatory. Thus began Arthur Searle's brilliant career of more than forty years as a professional astronomer.

The difference in the way the brothers went was clearly not in their immediate circumstances, for each moved counter to them. They shared the same heredity, early environment, and nurture. Evidently Arthur had an inborn passion for science, needing only arousal and opportunity, whereas George had none. Why some should harbor it and others should not remains unfathomable.

However derived, whether innate or acquired, the distinctive traits of scientists were much the same in nineteenth-century America as those described in nineteenth-century England or twentieth-century America. Scientists tended to be self-sufficient, self-controlled, independent-minded, assertive, even domineering. Intellectually they were precise, methodical, yet inquisitive and observant. Mathematical and mechanical aptitudes were common among them. They liked order and were given to making lists and tables. As boys they usually collected and arranged specimens of natural history, perhaps especially in mid-nineteenth-century America, where so many grown-up scientists did likewise ("often prompted by the same spirit that one sees among children," as an American zoologist remarked in 1876, apropos of hasty misnaming of species).

Those drawn to science were sometimes hard put to it to analyze the impulse themselves. "I cannot express in words the happiness I have experienced from my experiments and investigations," wrote a meteorologist. "To me," wrote a geologist, "there is great pleasure in research. . . . It is a delight." "Natural sciences," wrote a botanist, "beautify and elevate every pursuit of life and . . . rarely fail to enchain the whole being if they have once fairly begun." "I love science and feel like devoting my life to it, that's all I can say for myself," wrote a young paleontologist.

More graphically eloquent was a botanist in 1849: "The worst of these things is that they haunt me at night. I dream of them—there stands a stellate folliaceus Parmelia covered with Apothecia, & turn as I will, there it is stamped upon my brain." Or a young paleontologist in 1853: "I could endure cheerfully any amount of toil, hardship, and self denial . . . to labor in the field as a naturalist. . . . I could live as the wild Indian lives . . . without a murmur." Or the young man in rural Maine, making a voltaic pile in 1816 with no guide but an elementary book on chemistry, who could not feel the shock on his skin but stuck pins under it and so got a good effect. A conchologist in 1862 thought the chief use of his science was as an "innocent recreation to the mind." Many a scientist would have scorned a phrase so pallid.

Scientists and their admirers usually mentioned as motives "the love of truth and the desire to be useful to their fellow men," as Joseph Henry put it. Another word for "love of truth" was curiosity, or in the phrase of an English biologist, "a mere desire to unravel the marvelous." "It is, perhaps," wrote an American chemist in 1859, "the love of the mysterious, the longing to become acquainted with the unknown, that has, at least in the beginning, chiefly influenced" those who became scientists.

Modern scientists have been found to have "high ego strength." So had those of a century ago. An ego-gratifying sense of power comes with driving a salient of light into the unknown. "Never think of trade my boy," wrote a young scientist in 1864, "when the world is under your feet and you can weigh the stars in your brain." Less grandiose but also tonic to the ego was the hope of leaving a remembrance, like a fossil, to unending posterity. Science is presumed to be irreversible and cumulative; so a name attached to a priority in discovery or embalmed in the naming of a species would presumably endure through all recorded time. "Facts that seem to be neglected," wrote James Dana hopefully, "will some time or other find their place; and it is sufficient for me to have studied them out, and given them to the winds." One might suppose that faith in spiritual immortality would serve as well, but the Reverend John Bachman, whose faith burned high, nevertheless took special pains to have a biographer do his zoological contributions posthumous justice. (The biographer never got around to it.)

Social and religious motives often strengthened the inner drive to be a scientist. These were especially evident among a half-dozen Göttingen-trained agricultural scientists of the 1850s. They had all been raised in the spirit of evangelical pietism, which stressed selfless dedication to a calling. They disdained the selfish materialism and the grubby trifling with ethics that already seemed necessary for business success. They took it for granted that science, in whatever form, would inevitably advance the welfare of mankind, and that agricultural science in particular would benefit the American people, still largely rural. Science, in short, opened a path to distinction consistent with all their precepts—morality, patriotism, democracy, altruism, and service to God.

The starting nudge to obey the inner drive and to act on one or more of the aforesaid conscious motives might come from any of several directions. Of the leading scientists of that era (those later enshrined in the *Dictionary of American Biography*), about a quarter were first stimulated by family influence, and about a tenth were led to science by their jobs. The family influence was more common among life scientists than among earth scientists, perhaps because natural history was somewhat more likely to be the hobby of a parent. On the other hand, none of the life scientists found their calling through their jobs, whereas more than a fifth of the earth scientists did, doubtless because there were more untrained jobs associated with earth sciences. But in every major scientific field, about half the recruits took their cue from school or college.

Some scientists got their first remembered impulse from a book—from a popular treatise as in the case of Joseph Henry, from an encyclopedia volume as with Asa Gray, perhaps from Silliman's account of his European travels, or from more serious books by the English anatomist Richard Owen or the English evolutionist Erasmus Darwin or the Scottish anti-evolutionary geologist Hugh Miller, sound in detail and felicitous in style. Other boys caught their scientific fever from good popular lecturers like Silliman or even from the flashy showmen of chemical and electrical demonstrations at lyceums and cattle fairs. Still others responded to local environment, like young Spencer Baird in a countryside of plentiful bird life, fish, and fossils. A Boston teenager and frequent maker of tabulations, who could buy minerals, chemicals, and apparatus cheap and in variety from nearby specialty shops, went on eventually to become a great compiler of chemical and physical constants. The impulse to be an astronomer could come out of a clear sky, especially one with a comet in it.

The inborn proclivity, the temperament, the awakening, and the rationale were not enough to insure a career in science. Opportunity and means were also essential. Yet American society in the antebellum era denied that opportunity to millions of its members, whatever their talents and yearnings, on the irrational grounds of race and sex.

Enslavement absolutely barred most American blacks from science. Among the free black minority, most were excluded from public schools and almost all from higher education—by then more important to a scientific career than it had been for the eighteenth-century black mathematician and astronomer Benjamin Banneker. Even if a black had somehow acquired a sound scientific education, race prejudice—not only among Southerners but also among Northerners such as Henry, Bache, Peirce, and Agassiz—would surely have blocked him or her from a career or even a hearing in science. A few blacks managed to break away from slavery and rise above prejudice, but the extraordinarily gifted among them, like Frederick Douglass, quite rightly gave themselves to the growing fight against slavery. So no black American in that era made a name for himself in science.

Less oppressive, but in some ways analogous, was the lot of American women. As in the case of whites over blacks, men rationalized the denial of opportunity by stereotyping the victims as inherently unfit for it. God and nature made women to raise families and adorn society (or "man" the factories), most men liked to believe. Intellectual pursuits were manifestly outside women's sphere. For a woman to aspire to serious scientific work was deemed especially grotesque, unseemly, hopeless, and impermissible.

Yet there were rents in the crinoline curtain. Mothers, after all, raised sons, who might benefit from a home atmosphere of polite learning and informed reverence for God's works in nature. So no outcry arose when, in the second quarter of the nineteenth century, genteel girls and women took up science as a polite study rather than as preparation for a career. Female academies and seminaries, like their male counterparts in those years, added a variety of science courses and even invested in expensive telescopes, physical and chemical apparatus, and natural history collections. Many sought to equal the quality of courses at men's schools, though no women's school achieved a full range of college-level science courses until 1865. A few such schools could draw on nearby male professional scientists of some eminence for occasional lectures.

Science reached women outside the classroom also. Many women attended the public lectures on science that were especially popular in the 1830s and 1840s. Both men and women wrote popular books and magazine articles on science for the growing women's market. Women wrote ten such books on botany alone, 1819–59. From the late 1820s to the early 1870s, Mrs. Almira Phelps sold more than a quarter million copies of her *Familiar Lectures on Botany,* the most popular of her books on science for children, drawing mostly on a few serious books by others. Several women, notwithstanding stereotypes, liked insects well enough to write popular yet knowledgeable articles on entomology.

Thus drawn in, many upper-class women began taking a more active though still discreetly modest part in the scientific enterprise. They organized informal clubs or study groups for self-improvement and recreation, botany being a favorite study as somehow emblematic of womanhood, though entomology was popular without recourse to metaphor. Such women, however, eschewed serious scientific publication as unladylike. Other women began to find a livelihood in science, though peripheral and humble, as teachers of science in girls' schools and as librarians and curators for all-male natural history societies. The womenfolk of some male professional scientists made unsung contributions as research assistants. The elder Benjamin Silliman, James Hall, Spencer Baird, William Sullivant, Ormsby Mitchel, Alexander Dallas Bache, and Asa Gray all got help from their wives in arranging, collecting, observing, and especially drawing. The astronomer Sears Walker's sister Susan had mathematical ability and helped her brother until his untimely death. Afterward she thought of carrying on in his profession, but apparently did not.

On the frontier, where less was made of gentility and where the raw materials were plentiful, a few women could make more direct and unabashed contributions. A pioneer wife in Colorado, unusual in being a college graduate, learned taxidermy from a claim-jumping German in the sixties, hunted, shot, and mounted her own specimens, and in the seventies opened a small museum that pioneered in the idea of reproducing natural settings. In northeastern California, three women (two of them mother and daughter) studied the botany of the area and provided useful data to Eastern botanists. Asa Gray named a couple of species after the mother and daughter—using, of course, the surnames bestowed by their husbands.

But the male-imposed limits on women scientists remained narrow. Most scientific societies excluded women on principle, fearing the taint of dilettantism and constraints on male give-and-take. The American Association for the Advancement of Science admitted them, but only three joined before the Civil War. The wartime organizational activity of women made them bolder in seeking admission to scientific associations afterward. More and more joined the AAAS. In 1874 the AAAS introduced a higher category of "fellows" to which few women were chosen, and this male countertactic spread to other scientific organizations. But however slow, grudging, and qualified the movement to admit women, it went forward. In Boston, known for its strong-minded women of learning, the

American Academy of Arts and Sciences admitted its first woman member in 1848, and the Society of Natural History at last followed suit in 1876, though officially confining women to an inferior status.

The postwar years brought forth independent women's colleges that not only offered good scientific courses but also hired women to teach them. But graduate education for women came more slowly; and when it did arrive late in the century, women found that degrees and achievements together still did not guarantee fair recognition.

One stellar exception to the low rank of women in antebellum science was the astronomer Maria Mitchell, born and raised on Nantucket Island off the Massachusetts coast. In studying astronomy she had help from her father, who was a banker and amateur astronomer, and his scientific friends in Boston. In 1847, at the age of twenty-eight, she won a Danish royal medal for discovering a comet, calculating its position, and tracking it across the sky. After that, the sheer novelty of a woman's being honored as a scientist made her widely known. But she had to wait another fifteen years for a respectable job in science, that of professor in the newly established Vassar College. Even then, young Simon Newcomb probably expressed the opinion of other male savants when he wrote privately that "Miss M. is only a female astronomer after all."

The Royal Society catalogue of scientific articles lists only three American women as writers, 1800–64. The *Dictionary of American Biography* includes only three women among nearly five hundred men on the basis of their activity in science, 1846–76; one, of course, was Mitchell, and the other two were popular writers. Where this book uses the masculine pronoun in referring to the typical *DAB* scientist of those years, it does so in recognition of historical reality. The striving of women in science during this period doubtless prepared the public mind for real progress after 1876, but that is another story.

Even for a white male, becoming a professional scientist in antebellum America took nerve and determination—and if he happened to have been born in the wrong class or place, a good deal of luck besides.

Although Joseph Henry believed in 1841 that there were "more interested in popular science among us than in any other part of the world," most Americans considered science a poor choice of callings. The profession had only just been christened. Though Benjamin Gould later claimed credit for the word "scientist" as a substitute for "savant," "philosopher," or "scientific man," the Englishman William Whewell seems to have coined it in 1840. During the great exhibition of 1851 in London, the word "scientist" was heard only once, and then from Whewell, who added parenthetically, "if I may use the word." The professional scientist himself was a new species. In the United States of 1800 hardly a score of full-time scientific jobs, all of them academic, had existed outside medical practice. As late as 1847 the treasurer of Harvard remarked that science as a profession had "been only recently known or required in this country," and a year later Joseph Henry could see few scientific positions in America that encouraged "high scientific attainment."

The professional scientist was looked down on by most Americans. Since the beginning of the century, the public had viewed him as lacking interest and ability in practical affairs, and in the 1850s scientists complained that polite society accorded high status to material success but not to intellectual attainments. A physicist in 1859 laid American backwardness in science to the people's absorption in earning a living and more lately in "the rapid accumulation of wealth." An English scientist in America in 1849 was struck by the low pay and status of professional men generally. He blamed this on political egalitarianism, which undermined intellectual meritocracy, and on the wider diffusion of education, which lessened its distinction. A few scientist-explorers, like Humboldt or Frémont or the Arctic explorer Elisha Kent Kane, caught the imaginations of some young men, but the days of glamorous derring-do in science were passing.

Louis Agassiz in Switzerland had turned to science as the best or only ladder to a higher class. In America, on the contrary, many young men who might have been scientists set out on more promising roads to fortune. "There is a good deal of scientific talent and taste in my present College Class," wrote Asa Gray sadly in 1854, "but they will all be drawn off into active & more lucrative pursuits;—and I cannot wonder at it, nor indeed wish it otherwise." At Göttingen in the fifties, young John Pierpont Morgan turned down an assistantship in mathematics with the prospect of an eventual professor's chair, came home, and went into banking. He does not seem to have ever regretted his decision. Some men even spent years as professional scientists and then gave up. "I have turned land speculator," wrote one in 1855, *"and* I am going to make some money. Good bye to science and good taste and hail to the almighty Dollar & hard work." In 1846 a University of South Carolina science professor quit the classroom to manage an ironworks; in 1860 a Washington University science professor became a merchant in Indiana.

Parents often tried to dissuade their offspring from so untried, unhonored, and unrewarding a career as science. James Dana's father opposed the idea in the 1830s. So did Josiah Whitney's and Spencer Baird's. "My love for Natural History is so great," wrote young Ferdinand V. Hayden in 1854, "that I hardly feel any disposition for anything else. But . . . my relatives and friends (who can appreciate nothing which does not bring in an immediate return of cash) strongly oppose my having anything more to do with it." In 1859 a friend congratulated a young Baltimorean, later a distinguished scientist, "that the old governor relented, and enabled you to become a man of pure science, one of the highest aims of humanity! . . . That scene with your Mammon-worshipping northern relatives amused me a good deal."

In the 1850s the young chemist Samuel W. Johnson looked to the lives of great scientists for courage to go on in science, "though alone and exposed to the sneers of the vulgar and ignorant." It was small wonder that in the United States of 1846, with a population of about twenty million, there lived only seven or eight hundred men who in their lifetimes had published or ever would publish a scientific article or book.

Those with the dedication and self-confidence to forswear "the rapid accumulation of wealth," brave the opposition of parents, and face down the "sneers of the vulgar" were not all equal in opportunity and means to become scientists. As is obvious from what has been said about the geography of antebellum science, a Massachusetts boy was more likely to become a scientist than one of identical talent and temperament born in Mississippi. And geography aside, other accidents of birth strongly affected career choices. This can be inferred from the social and economic origins of leading scientists, those in the *Dictionary of American Biography* who were active at any time between 1846 and 1876 and whose fathers' occupations are stated.

Of those 249 *DAB* scientists, only one—Joseph Henry—was the son of an unskilled laborer, and only nine were sons of skilled workers. Only 15 per cent were sons of small farmers, although that category comprised 77 per cent of all Americans with listed occupations in the 1840 census. Seventeen per cent of the scientists were sons of entrepreneurs. But a remarkable 59 per cent were sons of professional men, though that class amounted to less than 2 per cent of those with listed occupations in the 1840 census. Of all the scientists, 17 per cent were sons of clergymen, 12 per cent of physicians, and 10 per cent of educators.

Only minor changes occurred in the pattern during the period 1846–76. The relatively small proportion of farmers' sons declined further after the Civil War, but so did the proportion of farmers in the whole population. The proportion of scientists who were sons of government workers fell off steadily, for whatever reason. The proportion of clergymen's sons declined until the Civil War, then rose again sharply. Perhaps the coming of Darwinism had something to do with that, or perhaps the urge to enter a more potent priesthood than was any longer supported by theology. In any case, the extraordinarily high proportion of professional men's sons was getting higher still at the end of the period; of those who entered science between 1861 and 1876, 65 per cent fell in that category.

It is true that among *DAB* subjects in all periods and fields (including science) whose fathers' occupations are listed, about 50 per cent were sons of professional men. But the tendency was even more marked in the case of the scientists, and it implies something fundamental about what affected their choice of a career. No doubt family income in the professional classes was somewhat higher than the national average, and this may have eased the way. But, as will be seen, poor boys could go to college even in those days if they really wanted to. The key seems to have been in wanting to, in deliberately choosing a scientific career over a variety of more accessible, immediate, and lucrative possibilities in an explosively developing nation. The very thought of a scientific career may not have occurred to most farm boys or many sons of businessmen. If it did, their esteem for a life of intellect rather than of action and material reward would presumably have been far lower than among boys whose fathers exemplified professional pride and commitment. In particular, a clergyman would have been most likely to impress his son with dedication to a calling and indifference to the laying up of earthly treasures.

Given the resolve to become a scientist, it remained to be carried out. By mid-century that seemed to mean going to college. Eleven out of twelve *DAB* scientists active in 1846 had attended college, and four out of five had acquired bachelor's or other degrees, such as M.D.'s. Among younger scientists, the proportion was still higher. A survey of Americans who published scientific articles shows that 96 per cent of those born in the 1820s acquired some college education. There still were scientists without formal higher education, but they were increasingly rare.

Since only one in a hundred young American men went to college, the scientists obviously belonged to an elite group. But it was not exclusively an aristocracy of wealth. In the mid-1840s tuition and living expenses together cost undergraduates $195 a year at Yale (of which only $33 went for tuition), $245 at Harvard, and $332 at the University of Virginia. The newer country colleges of New England competed for students by cutting tuition still more, making charitable grants, offering cheap living, adapting the academic calendar to local work seasons, and acquiescing in long absences by schoolteaching students. At Amherst in the early 1850s one student held his total costs for four years to a little less than $1,000, which he managed to raise through loans, grants, and outside work. Still, average costs amounted to more than half of a skilled laborer's income in 1860, and they were rising. Schools that sank money into scientific facilities might charge two or three times the average tuition. Scientifically inclined students whose families could not help out needed unusual stamina, resourcefulness, and motivation.

By later standards, they got little for their pains. One fact above all limited what could be offered in scientific instruction: the pitiful smallness of the colleges. The largest undergraduate enrollment in 1846 was Yale's 424; the largest total enrollment, including professional schools, was Harvard's 628. Columbia and New York University together enrolled only 247. Lafayette's trustees outnumbered its students. In 1854 the nation's colleges had enrollments averaging 100 and faculties averaging less than 9. In the Midwest and South, most college enrollments ranged from 25 to 80.

The colleges were provincial. In 1854 only three Columbia students came from more than fifty miles away, and four-fifths of Harvard's students were from New England. Most colleges were rural and isolated. They were also poor, with annual incomes averaging only about ten thousand dollars, and so their libraries had painfully few volumes.

Nevertheless, antebellum colleges strove valiantly to give science its due. Science had always been part of the American college curriculum, first as mathematics and "natural philosophy" (physics, chemistry, and astronomy), then joined in the late eighteenth century by "natural history" (earth and life sciences), then in the early Republic splitting into branches like botany, chemistry, and mineralogy. As scientific knowledge proliferated further, it seemed logical to add still more science courses to the general curriculum. It also seemed conducive to good public relations, since the public (egged on by the scientists) expected science to yield great material benefits. Furthermore, a show of relevance and modernity might attract students, a hope given urgency by a widespread, though temporary, drop in enrollment during the middle and late 1840s, especially in New England.

So in the antebellum years, science came to bulk even larger in the undergraduate curriculum. Religion was not an obstacle, except sporadically in the South. Even the so-called denominational colleges usually offered chemistry and some other sciences, and left theology to the seminaries. Including mathematics, science accounted for a third or more of all undergraduate studies at Yale, Princeton, and Lafayette, a third of upperclass studies at Mississippi, and a quarter of all studies at Alabama in 1859, though Harvard settled for a little less. By 1850 the typical college had four faculty members in math and science out of a total of nine.

Science was taught by lectures, recitations, and, here and there, laboratory work. Lectures included the usual demonstrations and blackboard diagrams by the professor and note-takings by the student. In recitations, the instructor systematically questioned students on the textbook or previous lecture, recording a grade for each student at each session. The effectiveness of this procedure depended on the teacher's ability and attitude. Joseph Lovering at Harvard was a good lecturer, but in his recitation classes he, like almost all other Harvard faculty, confined himself to extracting rather than imparting knowledge. At Yale in the 1860s it was regarded as improper for a student to ask a question during recitations. In most recitations students were called on at random, but some teachers, like Asa Gray at Harvard, followed a well-known order, so that each student knew in advance what he would be asked about. In twenty years of teaching before the war, Gray produced no notable botanists from his classes, except for one who had come already committed to the field. Still, by the 1840s some teachers were bringing more life and light into the recitation ritual; and to the extent that it gave attention to individual students, it was useful in a time of widely varying student ages and backgrounds.

Professors customarily used scientific apparatus in lecture demonstrations, but laboratory work for students was less common. Cost was a factor. Chemicals and apparatus often had to be imported from Europe. Southern colleges spent more than Northern on scientific apparatus, though they had little success in producing scientists. Between 1850 and 1864, on the other hand, Lafayette College in Pennsylvania allotted only 2 per cent of its expenditures for library books, chemical supplies, and laboratory equipment combined. Even Harvard and Yale gave inadequate support to the chemistry laboratories of their scientific schools. Microscopes were seldom seen in student labs before 1870.

The growing importance of experimental science and the example of European schools could not be wholly ignored. By the 1840s colleges were spending more on physics apparatus, and some of them expected students to use it for experiments in their spare time. Agassiz introduced American naturalists to laboratory methods of study in comparative anatomy and embryology. Two Wöhler students with Göttingen Ph.D.'s brought undergraduate chemistry lab instruction to Amherst in 1855. Union College, the most heavily endowed in the nation at that time, opened the best undergraduate chemistry lab of its day under still another Wöhler man in 1857. And in the following year Harvard College began chemistry lab instruction for undergraduates. But the practice did not become common among colleges until the 1870s.

For science textbooks, colleges relied mainly on European works, especially since there was no international copyright system. Joseph Henry complained in 1846 that college textbooks were "generally reprints of English books, with the name of some third rate man attached as editor." In mathematics, French texts in translation predominated. This dependence galled some American scientists, like Spencer Baird, who in 1847 assigned his Dickinson College freshmen a popular account of zoology by the writer of children's books "Peter Parley." It was "very poor," he confessed, but had "the merit of being more American than others." A year later Louis Agassiz and Augustus Gould brought out their *Principles of Zoology,* which put an end to Parley in that field. Asa Gray's texts in botany and James Dana's in mineralogy and (after 1862) geology dominated in those fields. But except for texts by the younger Silliman and by John W. Draper, chemistry continued to depend largely on Europe until the 1870s. And so did mathematics.

Of all college courses in science, mathematics had the longest tradition. A Princeton graduate of 1845 recalled that "piety and mathematics rated extravagantly high." By then most colleges had a separate professor of mathematics and had supplanted the classical Greek authors with more lucid, straightforward, and advanced French methods and texts.

All this notwithstanding, American students regarded the subject with what one student of the 1860s later called a "deep-seated aversion . . . an inward disgust and hatred . . . not capable of adequate expression in any ordinary way," though it found some vent in the "Burial of Euclid," a mock-savage, nocturnal student ritual. Looking back after twenty years of mathematics teaching, he asserted that 90 per cent of American college students had no hope of ever grasping mathematics and thought it dull, obscure, and pointless, except on the dubious grounds of mental exercise or the sort of spiritual benefit to be derived from a hair shirt. In 1857 a Columbia trustee's diary summed up a report on the mathematical vacuity of the juniors: any proposition relating to "time, space, quantity or proportion would be considered by them equally mysterious, incomprehensible, and disgusting."

Hopeless students meant hopeless teachers. A Southern professor in 1838 complained that the teaching of math was a stultifying drill in mere elements—"the repetition of the alphabet forever"—the only advantage being that the teacher needed no preparation for the classroom. A Harvard student, offered a teaching job in 1852, remarked on "how discouraging it is to force Math. into those who have neither ability or inclination to receive it." Most math professors taught astronomy or other subjects also, and probably with relief.

College entrance requirements in mathematics were minimal. Harvard, where Benjamin Peirce gave more advanced courses than were offered anywhere else, required only arithmetic, elementary algebra, and a smattering of geometry for admission. Most antebellum college courses carried math no further than what would now be the secondary school level, though a few made a stab (usually lethal) at calculus for their seniors. After the mid-1840s students were more and more commonly given the option of substituting other courses.

The rationale for mathematics in college was shifting from mental discipline to its necessity in science. Yet even students who wanted to be scientists regarded mathematics with mixed feelings. One in the early 1850s found it "a very great advantage in chemistry," but another of the same vintage thought calculus "devilish queer stuff at the outset anyway" and had yet to see its usefulness. Not many chose mathematics as a field, and more than a quarter of those who did combined it with astronomy, like Benjamin Peirce. Europeans looked down on American mathematicians, and the Americans knew it. Not until after 1876 did mathematics win some respectability as a professional field in America.

Worst of all, aversion to mathematics handicapped Americans in other sciences. Columbia's concern for its juniors arose because their mathematical infantilism barred them from certain aspects of physics. As the trend toward mathematical sophistication continued in physics and other fields, it forced American scientists to beg for mathematical help, to limit their researches, to abandon certain of them in mid-course, or to drop out of research altogether.

In physics by mid-century separate texts were appearing on mechanics and optics as well as general physics. Physics lectures now included electricity, magnetism, meteorology, and the steam engine. More sophisticated apparatus came into use. The new texts took a mathematical approach, and for the first time professors were well trained enough to go beyond the texts when necessary.

In astronomy, Yale had led in the 1830s as the first college with an adequate telescope. Denison Olmsted drew students to his course when he spotted Halley's Comet through the five-inch lens in 1835 weeks before news came of European sightings. But the instrument was placed in a steeple with low windows that blocked observations more than thirty degrees above the horizon, President Woolsey resisted giving astronomy more time in the curriculum, and Olmsted failed to adopt the new, more generalized and mathematically rigorous German methods. So in the 1840s Harvard took the lead. Peirce's course emphasized mathematical astronomy; and the Harvard College Observatory with its fifteen-inch German lens, one of the two largest in the world, offered experience in observational astronomy. Twenty per cent of all active American astronomers, 1825–75, got some or all of their training at Harvard. In 1854 the German astronomer Franz Brünnow came to the University of Michigan and offered the only course in the United States that taught the German method fully. It lasted only two years but produced an able astronomer and outstanding teacher in James C. Watson, who carried on and strengthened Michigan prestige in the field. By 1915 Michigan graduates made up a quarter of the leading American astronomers. Brünnow's significance in transplanting European methods and standards in astronomy may be likened to that of Agassiz in natural history.

In the 1840s, colleges began scrambling for telescopes as signs of intellectual grace. Even the Central Masonic Institute of Selma, Alabama, in 1851 had a telescope as big as Yale's. By 1850 most colleges had observatories or were building them. Besides Yale, Harvard, and Michigan, Dartmouth produced able astronomers. But only a couple of college observatories had endowments for salaries and publication; and in most cases, as public enthusiasm waned, the nominal "director," given no relief from his teaching duties, soon let the stars shift for themselves. Most college observatories were used not for research but for instruction. The usual requirement was a full course and some mathematical astronomy in a math course, amounting to perhaps eighty hours in class.

By 1840 most colleges had cabinets of geology and natural history, and geology had a course to itself. During the forties, course expansion in science was largely in zoology and courses combining biology and organic chemistry, in which Liebig's work had stirred interest. Progress in college science during the 1850s came with better texts rather than additional courses. The multiplication of science courses in the general curriculum had reached its limit. At the end of the decade, mathematics, physics, astronomy, chemistry, geology, and biology were being taught in the general undergraduate curriculum, usually by three or four professors.

. . .

Most antebellum college students thus encountered a greater variety of scientific fields than do most nonscience majors today. Some may have been recruited to science thereby who would not have been otherwise. But those who acquired a taste for science got little more than a taste.

College catalogues, like seed catalogues, expressed an ideal. Some courses listed were never actually given. Most that were given were necessarily superficial, especially since new nonscience courses were also being shoehorned in. Consider, for example, the science offerings of the College of Charleston in 1850 as advertised in the *Charleston* (S.C.) *Courier.* Lewis R. Gibbes, M.D., was professor of mathematics, chemistry, mineralogy, and natural philosophy, which consisted of "Dynamics, Mechanics, Hydrostatics, Hydraulics, Aerostatics, Pneumatics, Astronomy, Optics, Electricity and Magnetism." The Reverend John Bachman, D.D., concurrently with his church duties served as professor of natural history, "embracing Zoology in all its branches, Physiology and Comparative Anatomy, Mammalogy, Ornithology, Technology, Herpetology, Entomology, Botany, and the Elements of Geology." And Assistant Professor William P. Miles, former lawyer and future congressman and planter, encompassed mathematics, geometry, algebra, plane and spherical trigonometry, mensuration, and navigation. As Oliver Wendell Holmes later remarked, the professors of that day occupied not chairs but settees.

The inadequacy of that thin-spread melange led students at several colleges to organize their own chemical, mathematical, natural history, and other scientific societies, some with libraries richer in the field than those of the colleges themselves. One such at Williams heard many papers by its members, built up an outstanding natural history collection, and even sent its own expeditions as far afield as Greenland and South America. But the colleges could not possibly meet the implied challenge by further expanding science coverage within the common curriculum. As it was, many students, overwhelmed, simply shirked their studies. The only solution was to let nascent scientists specialize. The scientific students might be allowed to elect more science courses and fewer nonscience courses. A prescribed parallel curriculum might be set up within the college. They might be enrolled in a separate scientific school, either independent or affiliated with a liberal arts college. Or they might be offered graduate work in science. In the antebellum period, all these expedients were tried.

Here and there since the beginning of the century, one course had occasionally been designated as an alternative to another. Jefferson's new University of Virginia in the 1820s had gone further by allowing students to attend any one or a combination of eight "schools," including those of mathematics, natural philosophy, and natural history. But its enrollment, faculty, and funds could not sustain so lavish an assortment, and so by the 1830s Virginia had fallen back on what amounted to the standard college curriculum. In the late thirties Harvard made several courses optional, and by 1843 almost all courses after the freshman year were elective. But Harvard also lacked the faculty to cover so many choices. Reaction set in, and by the 1850s the options of Harvard students were being increasingly curtailed.

In Schenectady during the late 1820s Union College, already science-oriented, instituted a parallel curriculum, the Scientific Course, of which a third was in mathematics and 30 per cent in other sciences, and which was especially notable for leading to a B.A. degree equal to that of the college's classical curriculum. The new curriculum, while allowing some options near the end, prescribed most of its courses. Moreover, its only significant difference from the classical curriculum was in substituting two science courses for advanced classical languages. Nevertheless, it seemed more in tune than the conventional curriculum with the path-breaking spirit and practical needs of the time, and it certainly succeeded in the practical sense. A third of all Union College students were soon enrolled in the Scientific Course, overall enrollment soared, and from 1830 to 1860 Union ranked consistently among the top three American colleges in the size of its graduating classes. Yet its example did not become general.

Union's triumph may have owed something to its having been first with its wares in a limited market. One of Union's alumni, President Francis Wayland of Brown, later tried and failed to outdo his alma mater. Science already took up a third of all class time at Brown, and the college's seven-man faculty included four scientists. Responding to the enrollment drop among colleges generally in the late forties, Wayland's report of 1850 called upon Brown to give the customer even more of what Wayland supposed he wanted: a freely elective system, a flexible program ranging from two to six years according to each student's needs, and new programs in applied science, agriculture, law, and teaching. All this was aimed at preparing students for the new age of technology-based industry, agriculture, and commerce. The experiment got bravely under way in 1851. Besides the new curriculum, it lured students with easier entrance requirements, low charges, good faculty, and a three-year bachelor's degree. Yet after an initial rush to enroll, students began dropping out at an early stage, Wayland resigned in 1855, and Brown relapsed into conventionality.

Wayland's defeat cannot be blamed on curricular changes alone. Because Wayland lorded it over the faculty, John A. Porter, professor of applied chemistry, and William A. Norton, professor of civil engineering, jumped from Brown to Yale, taking many students with them. The self-discipline of students did not rise to the new freedom Wayland permitted them, and their quality declined with the lowering of requirements. But the new curriculum gave trouble also. Brown lacked the funds to support it, and enrollment did not rise enough to provide them. The cutback in degree requirements lowered the prestige of the degree. The fact that enrollments in other colleges recovered while Brown's began to fall suggested that Wayland had misdiagnosed the problem. So did the ineffectiveness of his therapy. There was evidently not as much pent-up demand for scientific education as Wayland had supposed.

Nevertheless, the demand would eventually come, and then Wayland's ideas, or some of them, would make more sense. Meanwhile the elective system, and its less permissive competitor the parallel curriculum, made a few gingerly converts among colleges of the 1850s. Between 1846 and the Civil War, a number of colleges also established scientific schools with their own B.S. or Ph.B. degrees, the story of which will be told with that of engineering education.

The remaining alternative, graduate study, would find its ultimate niche in universities on the European model. An account of that crusade must be reserved for a later chapter, after the organization and leadership of American science have been analyzed. But antebellum substitutes in several forms may be considered at this point.

In a loose sense, learning on the job under a senior scientist might be called a sort of graduate education. The elder Benjamin Silliman at Yale in the 1830s hired and instructed good students as laboratory assistants. Government observatories, expeditions, and scientific bureaus provided training and experience for young college graduates. But in the 1840s no American college or self-styled university offered formal graduate work in the modern sense. New York University had tried to create a master's degree program in the 1830s, but the bud had been nipped by the Panic of 1837 before fully opening. Courses could be taken at Harvard and elsewhere by "resident graduates," but they got neither a degree nor much attention from college authorities.

Some students, like Joseph LeConte, went on to medical training "as the best preparation for science." Of the leading scientists (those in the *Dictionary of American Biography*) who were active at any time between 1846 and 1876, one in seven had an M.D. degree, the proportion being one in four among life scientists and one in six among chemists; and many of these, like LeConte, had taken the degree solely for science, with no intention of practicing medicine. But at the beginning of the period, in 1846, the proportion of M.D.'s among all *DAB* scientists then active had been not one in seven but one in four. As specialization widened the gap between scientists and physicians, medical training was rapidly losing its utility as preparation for a scientific career. This inference is further supported by a study of geologists with M.D.'s. A graph of the numbers born in successive years shows a strikingly smooth downward curve, especially steep after the 1820s, from the beginning of the nineteenth century to near bottom at its midpoint.

While fewer would-be scientists resorted to medical training, more chose advanced study in Europe. Of the *DAB* scientists active at any time 1846–60, a quarter had some European training. They were almost evenly divided between those born and educated in Europe and those who went there to study.

Even when America offered equivalent training, American students found Europe cheaper. In 1849 one chemistry student went to Heidelberg because it cost less than the $600 a year charged by James Booth's private teaching laboratory in Philadelphia. In the mid-fifties all academic and living expenses at both Munich and Heidelberg came to less than $500 a year, and living expenses at Leipzig were about $250. But in any case, American equivalents usually could not be found at any price.

For the chemists in particular, European laboratory facilities counted heavily. In 1847 the Smithsonian Institution trustees urged that the Institution's projected chemistry laboratory be equipped like Liebig's and opened to students, who would otherwise have to go to Europe. But it fell far short of that. The heavy investment could not be justified by the number of American students. The few private teaching labs in America were mostly for beginners. In 1849 Booth's remained the only one teaching analytical chemistry, and two attempts to compete with Booth in Philadelphia during the fifties failed for lack of students. Even Booth's students felt it worthwhile to study in Germany afterwards. In 1857 the German-style lab set up at Union College was still what one observer called "the only really complete Laboratory in this country." So of the thirty-five American-born *DAB* scientists who studied in Europe, 1840–60, twenty-one were chemists.

Of those thirty-five, eight went to Europe in the years 1844–47, mostly to Liebig at Giessen and to Paris. A greater surge, thirteen, went in the years 1853–55, mostly to Freiberg, Göttingen, Heidelberg, Paris, and Munich, to which Liebig had lately moved. The slackening after 1847 doubtless reflected the decline in Liebig's reputation. The upsurge of the 1850s is not so readily explained. A rush to Wöhler at Göttingen by students coming out of Eben Horsford's Harvard laboratory has been ascribed to good preparation at Harvard and to a visit there by one of Wöhler's American Ph.D.'s. But half of the students at the Yale Scientific School made the same move at the same time. A rebound from earlier disillusionment with agricultural chemistry and a growing interest in industrial chemistry may have been factors. As for the falling off after 1855, the supply of chemists in America seems to have exceeded the demand as early as 1856, and word may have got around fast.

Like European students, the Americans were *Wandervögel,* birds of passage. Almost all on the *DAB* list visited several schools, and more than half of them formally attended two or more, seeking out the best ones for particular courses. Those who did not shop around encountered the others who had, and they all compared notes. The Americans often passed through Great Britain, but seldom paused for formal study. Some resented the rudeness and arrogance of British students toward Americans. Most went on because French and German universities had more to offer. More Americans studied science in Paris than anywhere else, but most of them for only a few months, doing the bulk of their work elsewhere. Paris was cheaper to live in than London, but not as cheap as Germany. The Jardin des Plantes drew students in natural history, and the Ecole des Mines drew them in geology and mineralogy. Paris also led in the mathematically oriented physical sciences. But though it had good private teaching labs, and though both Jean-Baptiste Dumas and Adolph Würtz were at the Sorbonne, Paris could not match Germany in chemistry, for which most Americans had come to Europe.

Berlin had Eilhard Mitscherlich, the discoverer of isomorphism, Heinrich Rose, judged by a knowledgeable American student to be "the greatest analytical chemist in the world," and A. W. Hofmann, a leader in organic chemistry. Rose took only four students a year into his laboratory, but like other German professors he seemed to have a special fondness for Americans. At Heidelberg, physical chemistry with Robert Bunsen was the great attraction. Like Rose he took only four lab students a year, but the chosen few found him inspiring. With Liebig's departure, Giessen had dropped out of the running; but in his new position at Munich, Liebig lacked his old drawing power, and not only because the inadequacies of his soil theories were becoming apparent. He was tired in body and spirit, perhaps in part because of his long, furious controversies with fellow chemists. At Munich he charged double the usual fees and looked in on his lab students only once a month or so.

The new lodestone for American chemistry students was Friedrich Wöhler, one of the founders of organic chemistry, at Göttingen. One American student described him as "a thin-faced careworn looking man" with an untrimmed and unruly head of curly hair; but he was a great teacher in lecture and lab, full of easy-going charm and offhand humor. He had inexhaustible patience with and interest in his students. Chided for spending so much time on the American students, he replied, "I know, I know, but they are all such nice boys." For chemistry in general and organic chemistry in particular, Göttingen ranked first with students in the 1850s. The American influx to Göttingen crested in the mid-fifties, when a dozen or more were on hand at the same time, then fell off until Wöhler in 1858 and 1860 remarked plaintively that he had only three Americans working with him.

American scientific students in Germany were reputedly held in such high esteem at the leading universities that their German classmates were jealous of them. All the Americans at Göttingen were respected as hard workers. The chemists put in fifteen or sixteen hours a day in lecture, lab, and study. Two American Ph.D.'s from Heidelberg considered the doctoral examination "a farce" and the degree "a humbug," yet another at Heidelberg a year later boasted that German and English students usually failed the examination or settled for the lowest of the three grades, while no American had taken less than the first.

Like the pilgrims of 1844–47, those of the 1850s helped to convey the methods and mores of European science to America. Their example supported the drive in America for universities on the European model, a campaign that, as will be seen, gathered force during the fifties. But most immediately they brought to American science a cadre whose advanced training could not well have been secured otherwise.

Becoming a scientist in the antebellum years thus involved many factors: race, sex, geographical location, social and economic background, the attitudes of parents and peers, the availability of training, the innate temperament, ability, and craving, and the element of chance or luck. It also involved the tides and currents of social and economic change, or in other words, history.

The influence of historical events can be surmised from statistics of scientific recruitment. In the period 1846–50, nearly twice as many *DAB* scientists began their careers as in the preceding half-decade. The economic boom of the early and middle forties may have made college attendance easier and job prospects brighter. During the years 1851–55, the rate of entry levelled off, but in the last half of the decade it rose again by about 25 per cent. As before, the boom years had come in mid-decade. There is little mystery about the decline of more than 30 per cent in the numbers entering during the Civil War years, 1861–65. As will be spelled out hereafter, the war dealt harshly with American science in several ways. The slowness of postwar recovery in scientific recruitment—in the early 1870s still not up to the level of the late 1850s, despite population growth and economic boom—can be explained in large part by the delayed effects of the war.

Chapter 8

Being a Scientist

What were the antebellum scientists' working ways, their tools, their achievements? The answers must be sought field by field, for by 1846 two out of three scientists confined themselves to a single major field.

Surprisingly, despite contemporary impressions of increasing specialization and historians' later assumptions of it, that ratio held steady until at least 1876. But the paradox can be resolved by closer analysis. Americans tended toward descriptive science, which minimized the need of theoretical grounding. So a chemist, for example, might on occasion analyze and describe geological specimens without much training in geology, a physicist might dabble in terrestrial magnetism, a mathematician might help out in geodesy, an astronomer might take a turn with wind and cloud movements. Such limited descriptive forays would statistically obscure a fundamental shift toward greater commitment to a specialty. Furthermore, statistics based on major fields are blind to the intensifying of specialization within those fields. The botanist William Sullivant, for example, began shedding divisions of his field after 1840—first grasses and sedges, then lichens and fungi, then all flowering and seed-bearing plants, and so on until by 1860, no longer a rolling stone, he confined himself to gathering mosses. A study of leading American chemists, 1820–80, shows a steady decline in the average number of chemical subfields in which each man published, that average dropping by half from the 1840s to the 1880s. And the chemists who stuck closest to a single subfield turned out to be the most productive.

So on the whole it is more realistic to follow the scientists into their working lives not as a single class, but separately as mathematicians, astronomers, physicists, chemists, life scientists, and earth scientists.

. . .

In the last half of the 1840s, chemistry recruited more scientists who eventually won places in the *Dictionary of American Biography* than did any of the other five major fields—30 per cent of the total, though only 15 per cent of the *DAB* scientists already active in 1846 were chemists. Liebig's sudden vogue after 1840 must have contributed to the swing toward chemistry, and the partial disenchantment with his methods by 1850 may account for some of the decline in chemistry recruits to about 20 per cent of the total in the fifties. Still, over the whole period 1846–76 chemistry won 24 per cent of the entering *DAB* scientists.

Besides the growing economic promise of chemistry, chemical reactions made an appealing visual impact on the young. Hence the frequent chemical demonstrations in popular lectures. But by the same token, chemistry had its hazards. Too many young heroes of its early days, men like Scheele, Davy, or the American James Woodhouse, never grew old, because of their lighthearted readiness to taste or smell new compounds; Woodhouse almost certainly died of carbon monoxide poisoning. Survivors learned more caution and passed it along, so that by mid-century chemistry had such hale veterans as Robert Hare and the elder Silliman. Even the cautious chemist nevertheless ran a risk. The German-born Frederick A. Genth, one of the most painstaking, wrote in 1857 that "the gas NO has made me sick and I fear it more than anything else." Silliman in his time had been sickened by fumes and temporarily blinded by an explosion, and explosions during lecture demonstrations often startled students and sometimes injured them. At Harvard in Professor John W. Webster's class a copper fragment once buried itself in the back of a seat, vacant that day only because of a student's absence. "The President sent for me and told me I must be more careful," Webster later remarked to his class. "He said I should feel very badly indeed if I had killed one of the students. And I should."

Until the forties, American chemists had most of their apparatus made to order on their own design. Then came the era of ready-made equipment. By 1846 New York City had a well-stocked dealer, who purveyed a variety of electric batteries, portable furnaces, air-pumps, blowpipes, retorts, crucibles, tubes, mortars, lamps, weights, balances, tongs, and so on. Europe had skilled workmen at low wages, and educational equipment entered the United States duty free. The spread of laboratory instruction in Germany gave German makers of chemical apparatus a head start. Americans trained in German labs got used to German apparatus and brought home as much as they could. So through the antebellum period, American chemists turned first to German catalogues and agents. Even after American apparatus became cheaper or better or both, some American colleges still sent off one big order to Europe every year or so, usually to Berlin, although some favored France for fine chemicals. As late as 1864 Harvard bought heavily from Europe.

In 1848 the first American wholesale manufacturer of laboratory metalware, a German immigrant, began work in Philadelphia. In the 1850s James Renwick of Columbia got Christopher Becker, a Dutch immigrant, to design and make an analytical balance that established Becker & Sons as the leading American supplier of weights and balances. Boston dealers emphasized electrical and physical rather than chemical apparatus, and as late as 1861 no accurate weights were for sale there. But a Boston chemical glassware factory, established in the mid-fifties, led the nation in that specialty by then. Not until late in the century, however, did American demand support a well-developed domestic production of chemical laboratory apparatus.

Because the basic types of chemical apparatus had been worked out in Europe before chemistry made much headway in America, and because ready-made European equipment came cheap, antebellum Americans did not contribute as much to laboratory technology as might have been expected. Charles Jackson's widely used alcohol air-blast lamp was displaced in the mid-fifties by coal gas and the Bunsen burner. But during the sixties, Wolcott Gibbs made a notable contribution with his application of electrolysis to quantitative analysis.

American contributions to chemistry in the antebellum period were mostly practical, as in multiple-effect evaporation, or else descriptive and analytical. Highly refined measurement fascinated Americans, and Josiah Cooke's work during the fifties on atomic weights eventually set his Harvard pupil Theodore Richards on the road to the 1914 Nobel Prize in chemistry for the precision and ingenuity of his atomic weight determinations. Wolcott Gibbs made classic analyses of the complex cobalt compounds in collaboration with Frederick Genth during the fifties. J. Lawrence Smith developed ingenious and useful methods for determining alkalis in minerals and for other analytical purposes. So it went—measurement, technique, ingenuity, analytical data gathering, but little depth or long-range strategy.

If one counts among life scientists those styling themselves naturalists, as well as the zoologists and botanists and the more specialized entomologists, ornithologists, conchologists, and paleontologists, the life sciences claimed a third of all *DAB* scientists active during the period 1846–76, the largest share among major fields. The field held up well in recruiting, being nosed out by chemistry in the late 1840s and equalled by it during the Civil War, but in other years and over the whole period enlisting more new scientists than any other.

An astronomer once wrote another urging "better physical education" for the sake of scientists' mental powers. Fieldworkers in the life sciences had no need to worry about exercise. "My researches upon the banks of streams and in the water," wrote an entomologist, "my exertions in climbing trees, ascending hills, beating bushes, sweeping the grass with the insect net, turning stones and logs," all contributed to "the exercise of the muscles, the expansion of the chest, and to mental and bodily recreation." He died at the age of ninety-one.

But like the chemists in their labs, the naturalists in the field put up with discomfort and danger, including an occasional threat of Indian attack in the Far West. John Bachman noted on a specimen of poison oak that it "was once near putting an end to my botanical studies." As we have seen, insects tormented Southern naturalists beyond endurance, and even in the Blue Hills of Milton, Massachusetts, Thaddeus Harris found that "often, just as I was about to throw my handkerchief on a Cicindela, my hand was arrested & almost involuntarily carried to my face to brush off the flies." The insects' counterattack extended indoors, especially in the South, where they devoured botanical and entomological specimens. One year the cockroaches ate all the labels in Bachman's collection. But the North too had its counterparts in the Anthreni, whose ravages appeared in Ohio and Pennsylvania and even in Boston, where the Natural History Society waged a thirty years' war against them with much loss to its collections.

Almost no Americans undertook experimental biology in those years. Little was done in cytology before 1880, though cell theory was established in Europe by the early forties. Yet even Americans needed good microscopes for such tiny specimens as fungi, insects, and Jacob Bailey's beloved Infusoria. "I rejoice," wrote Bailey in 1847, "at the spread of the taste for microscopic research." The achromatic microscope, giving an image free from extraneous or false colors, was just coming into use in America. But microscopes were costly and hard to get. Agassiz waited eighteen months for one, and a North Carolina botanist shunned certain lines of work because he could not afford one.

In 1838 Charles A. Spencer of Canastota in upstate New York opened an optical shop and at once began experimenting with optical glass mixtures and achromatic lens combinations. By 1847 he felt ready to report success to Bailey, the leading advocate of microscopy. Delighted with the first instrument, Bailey grew still more enthusiastic as he tried successive products of Spencer's shop. "We ought to take pride in him as an American, and do what we can to make him known," Bailey wrote John Torrey. In 1848 a Bailey article in Silliman's *Journal* proclaimed Spencer's microscopes as good as any made in England or France; and James Dana, as editor, continued to give Spencer's work frequent notice. A committee of the American Association for the Advancement of Science reported in 1851 that "Spencer's objectives are now the best in the world." In the early 1850s German immigrants entered the field in New Haven and Philadelphia, but Spencer held his lead and remained unsurpassed by any others in the world during most of the remaining years of the century. A scientist in 1859 credited Spencer's microscopes with drawing more Americans into laboratory research at last.

Ten per cent of the *DAB* life scientists bore the old-fashioned label of naturalist. Of the rest, 62 per cent were zoologists or in more specialized branches of zoology. Finding American zoology almost exclusively given to description and classifying, Louis Agassiz was leading it into greater concern with species relationships (as God willed them), structure, and embryology. His philosophy left no room for evolution. Yet the mid-1840s opened a new era in paleontology, which would lay the foundations for Darwinism. In 1846 Hiram Prout of St. Louis delivered a paper on a titanothere jaw fragment from the Bad Lands of South Dakota. Published in 1847, it started a rush of paleontologists to the West in general and the immensely rich fossil deposits of the Bad Lands in particular. For twenty years, beginning with his first paper on the American fossil horse and continuing with his brilliant *Ancient Fauna of Nebraska* in 1853, Joseph Leidy dominated American vertebrate paleontology and did much to win European respect for it. In the spirit of American science generally, the paleontologists shied away from fundamental questions, fixed their attention on immediate, visible goals, and excelled as technicians, notable for their skill, daring, and ingenuity in preserving, restoring, and mounting specimens.

Excluding "naturalists," 38 per cent of the *DAB* life scientists, 1846–76, were botanists. Torrey, Gray, Sullivant, and Engelmann led during the antebellum period. Gray's *Manual of the Botany of the Northern United States* in 1848 made an organized science out of American systematic botany. With Torrey, Gray drew the outlines of North American flora generally. Gray's international exchanges and correspondence, and his work on the collections of naval exploring expeditions, enabled him to grasp the dynamics of plant distribution throughout the Northern Hemisphere. In particular, his study of Japanese flora in comparison with North American gave him a worldwide reputation. Nevertheless, taxonomy dominated the field through the antebellum period, and along with it, despite Gray's unwilling drift toward evolution, a belief in the immutability of species. The leaders of American botany expressed concern during the early fifties about the scarcity of new faces in the field. Perhaps what really troubled them was a sense of intellectual sterility.

A quarter of all *DAB* scientists active in the period 1846–76 were earth scientists, including those in geophysical fields such as meteorology, geodesy, and oceanography. Of 116 scientists so classed, 60 were geologists and 8 were mineralogists. Agassiz in 1847 lumped the latter two fields together as the most numerous class of American scientists, and in the sixties Benjamin Gould asserted that geologists comprised a larger proportion of scientists in America than in any other country. Neither offered statistical evidence, but both may well have been right, depending on the definition of fields.

Like other outdoor sciences, geology had its hardships, graphically set forth in a letter by Josiah Whitney in 1859:

> Yesterday . . . came one of those very severe days which make the geologist's life not so easy a one. The whole forenoon I spent in floundering, wriggling, writhing, creeping and crawling through the mud holes of an extensive and interesting, but intolerably uncomfortable mine, whereof my bones are all aching and my flesh battered. Coming out a mass of wet mud with a human being encased in it, the said human was further inhumanly treated by being obliged to ride 20 miles over a road, which was for all the world the mine with its top cut off, and a driving northeast storm of rain beating into his face all the way, and washing the mud off him into his boots. Wasn't that comfortable?

Whitney accepted this in the field, along with his bedroll and fried pork, but at home insisted on clean sheets every day, a fresh towel every time he wiped his hands, and an epicurean table. Nevertheless, his wife fretted about his frequent long absences, another drawback of life as a geologist—"its worst feature," as one geologist wrote. In the early seventies a young geologist, defending his departure for Brazil, assured his wife that "I do love my science but I do love *you* better," perhaps spoiling the effect by adding that "the six best years of my life have been devoted to the study of South America." Three years later he was writing: "Oh how I wish . . . that I could be relieved of the distracting thought that I am making you miserable by my insane persistence." He was relieved of it by his death soon after. James Hall, on the other hand, lived apart from his wife on the same property, which practice apparently did not make *her* miserable.

In 1844 Alexander Bache rejoiced that American geology had gone beyond mere description to "bold generalizations." He may have had in mind Henry Rogers's 1842 theory of mountain building by subterranean waves, which turned out to be much too bold. Nevertheless, Bache was not far off the mark. In America, with the Far West thrown open, the natural record was too rich, emphatic, and suggestive for geologists to stop short with data gathering. After 1840 serious generalizations came often enough from American geologists to put them ahead of the Germans and on a par with the British and French. The Far West, as has been pointed out, inspired John S. Newberry's recognition of river erosion and William P. Blake's of wind-blown sand as shaping forces. The tall, quiet, scholarly Fielding Meek worked in perfect complement with the short, exuberant, daring Ferdinand Hayden to explore, map, and describe the geology and paleontology of the northern Plains and northeastern Rockies, establishing Western strata and matching them with European, as well as reconstructing the environment and fauna of the Cretaceous sea.

More than anyone else, however, it was James Dana who gave American geology its intellectual stature and a theoretical framework that lasted into the early twentieth century. In 1846 he set forth the doctrine of the permanence of continents and ocean basins, widely accepted for many years though now overridden by plate tectonics. In 1847 he took up and elaborated on the theory that mountains and even continents were formed by unequal contraction of the earth's crust—a sort of wrinkling—as the earth cooled. James Hall challenged that idea in 1857 with the hypothesis that unequal surface loading, due to erosion and deposition, resulted in underlying pressure and temperature changes that in turn made mountains. Dana held to his contractionist views, but from Hall's theory he picked up the concept of a sediment-filled trough in the warped crust which he eventually christened a "geosyncline." That concept has also been radically modified by plate tectonics, but it figured significantly in a century of geological searching and thinking. Dana furthermore integrated his geosynclinal and mountain-building ideas in a complete theory of continental and ocean-basin development, the first such comprehensive theory put forward by an American.

Mountain-building theories, the permanence of continents, the implicitly evolutionary findings of Western paleontology, the Ice Age theory (imported with Agassiz, but enhanced by American data and strongly supported by Dana), and the power of erosion by wind, rain, and rivers—these were large ideas especially associated with American geology.

In mineralogy also Dana took a broader view than his American contemporaries. Earlier mineralogists had chiefly considered the external characteristics of minerals. Then, with the rise of analytical chemistry in the thirties and forties, they swung to the other extreme, making elaborate chemical analyses. Dana almost alone kept crystal forms and other physical characteristics in view along with chemistry, until a better balance at last developed. Even Dana, however, failed to make full use of the microscope in studying rock sections, a technique pioneered by the English and developed by the Germans into the science of petrology.

The roving Americans, not yet having forsaken the high seas, and possessing a greater range of terrain and climate than most peoples, emphasized the geophysical side of the earth sciences—geodesy, terrestrial magnetism, meteorology, oceanography, and hydrography—more than European scientists did.

Bache in 1844 hailed the transition of meteorology from a branch of natural history to "a physical science." Even in meteorology Americans lagged behind Europeans in theoretical contributions. But, as earlier mentioned, Elias Loomis introduced isobaric and isothermal lines into weather mapping, and William Redfield discovered the circular movement of wind around a storm center. Americans could—and did—boast of James P. Espy, whom they called the "Storm King." Though Espy was wrong about the mechanics of storms, he pointed out the role of convection in forming precipitation through the upward rise, expansion, and consequent cooling of moist air. Espy also pioneered in synoptic weather charts and in the use of telegraphic weather reports over a large area, wherein the size of the nation gave it an advantage. And William Ferrel, a Tennessee schoolmaster, made still another notable contribution in analyzing the effect of the earth's rotation on winds and currents.

American meteorologists might have profited by paying more attention to the balloonist John Wise, whom a sympathetic chemist praised as "devoting his life to exalt aeronautics from quackery to a science." The chemist urged Bache to use aerial photography in Coast Survey work, but Wise's own chief scientific interest was meteorology. In the 1840s he correctly described the early stages of hail as seen during a balloon ascent and reported the strong air current that moved steadily eastward at high elevations. But scientists regarded him with amusement, like a University of Pennsylvania scientist who joked about the notion of "a Prof. of Aeronautics in the Dept."

Of the six major fields as defined here, mathematics depended least on instrumentation and physical activity. It enrolled less than 9 per cent of *DAB* scientists, 1846–76. And its antebellum achievements were few. Benjamin Peirce's work having already been noted, there is little more to be said about the field. But American astronomy calls for something more, though its share of *DAB* scientists was only 10 per cent.

From its high estate in colonial times, American astronomy sank low in public esteem during the republic's first half-century. It concerned itself almost exclusively with what was called "practical astronomy"—measuring the positions of celestial bodies, extrapolating them into the past and future, and determining local time and geographical coordinates. Yet it bore the character of an aristocratic hobby, and so the new democracy howled down President John Quincy Adams's proposal of a national observatory in 1825. But Adams lived to see and help along a dramatic turnabout in the thirties and forties.

An American edition of John Herschel's popular *Treatise on Astronomy* in 1834 opened the eyes of the public to new and more exciting dimensions of astronomy, made possible by advances in telescopes and other instruments. "Theoretical astronomy" probed the dynamics of the universe, working out orbits on the basis of gravitational forces. Still more appealing to the American taste for the concrete and tangible was "physical astronomy," which regarded celestial objects not as mere generators of mathematical figures or gravitational attraction, but as real bodies with physical characteristics to be discerned or deduced. The cosmos cooperated. Spectacular meteors showered down in 1832 and 1833, Halley's Comet came back brilliantly in 1835, and early in 1843 another great comet drew its luminous train across a fifty-degree arc of the night sky. The taint of aristocracy faded as professionals began to supersede rich amateurs. In Congressional debates on the Smithsonian bequest, Adams, no longer the Jacksonians' scapegoat but a battler for human freedom, kept astronomy before the public. In Cincinnati an observatory became the goal of a people's movement, and Adams laid its cornerstone in 1843. Three years later, the prediction and sighting of Neptune captured the public's imagination.

Fixed observatories began to give American astronomers employment and dignity along with the tools of discovery. The National (or Naval) Observatory at Washington—"smuggled" into existence, to Adams's glee, by navy lieutenant James Gilliss in 1842—the Cincinnati Observatory, half a dozen built by wealthy amateurs, and those of the colleges, altogether numbered at least twenty-five by the mid-1850s. Astronomers at the two or three best-endowed observatories deplored the other establishments as duplicating effort and diluting funds. But those others provided training and jobs, and they helped keep the public interested. Bache's Coast Survey eased the situation by farming out astronomical work to sundry individuals and observatories. So did the navy's Nautical Almanac in Cambridge.

In calculating, computing, and theorizing, the astronomer could keep conventional hours. But in observing, his schedule was ruled by the stars. Like the chemist, he was further set apart by his instruments. Benjamin Gould remarked in 1856 that "it is the mechanical artist to whom . . . we owe the chief advances of modern astronomy."

During the thirties and early forties, European instruments had no American competitors except the small reflector telescopes of Amasa Holcomb in Massachusetts. In the mid-forties, as observatories multiplied, other Americans entered the field, and after 1850 no more foreign telescopes were ordered for serious work. Henry Fitz of New York City led in the forties and early fifties with refractors up to sixteen inches. Then the Cambridge firm of Alvan Clark & Sons, formed in 1852, pulled ahead with an eighteen-and-a-half-inch refractor. Though Clark had been a portrait painter until his forties, he lived to lead the world in the field, make a thirty-inch refractor for the principal Russian observatory, and see his firm complete a thirty-six-inch lens for the Lick Observatory in the 1880s.

Americans were quick to apply new technologies to astronomy. As soon as commercial telegraphy appeared in 1844, Commander Charles Wilkes used it in determining longitude by simultaneous star observations in widely separated places. Four years later several Americans independently contrived electromagnetic devices to record the observation times more precisely. The English-born John W. Draper of New York University inaugurated celestial photography with his 1840 daguerreotype of the moon. The Harvard College Observatory carried photography to the stars during the fifties, and in the following decade Lewis Rutherfurd of New York used it in pioneering stellar spectroscopy.

Until 1839 Harvard lacked funds for an observatory or even an instrument capable of any worthwhile observation. Then a fifty-year-old Dorchester clockmaker and self-taught astronomer named William C. Bond was hired by the navy to make observations in conjunction with the Wilkes Expedition. Harvard got Bond to come with his equipment to

Cambridge in the unpaid position of "Astronomical Observer to the University." When a wave of inquiries about the comet of 1843 revealed the Cambridge observatory's instrumental deficiencies, a citizens' meeting presided over by the textile tycoon Abbott Lawrence launched a drive that provided a new observatory in 1846 and a $20,000 refractor telescope in 1847. The fifteen-inch German object glass alone cost $12,000 and equalled the big lens at the Pulkovo Observatory in Russia, till then the largest in the world. Harvard put Bond and his son George on modest salaries. In 1848 a close friend of George Bond killed himself after bequeathing a $100,000 endowment for the observatory and for a professorship of astronomy, given to the elder Bond. By the mid-fifties, gifts to the observatory, including the initial bequest, had exceeded $150,000.

William Bond, a kindly gentleman with silvery hair and a rosy complexion, and his son George, tall, thin, pale, blue-eyed, and dark-haired, worked so closely together that their announcements of discoveries usually began "We noticed. . . ." Most notable were those of the seventh satellite of Saturn and a previously undetermined ring of that planet. The Bonds emphasized observation and technique more than theory. William Bond designed a chronograph controlled by a spring governor for electrical recording of star transit times. Widely adopted, it became known as the "American method." In 1862 George published the century's most complete and careful study of a particular comet, Donati's Comet of 1858. Using the fifteen-inch telescope under William Bond's direction, a professional photographer in 1850 took the first photograph of a star. By the late fifties, with improved photographic processes and a better control clock for the telescope, George Bond had brought stellar photography from second to sixth magnitude stars and was demonstrating the use of timed exposures as measures of stellar brightness. Though photography did not replace visual observations in many areas before the eighties, George, who died at thirty-nine in 1865, foresaw its future and became an early proponent of mountain-top observatories for clearer and steadier air.

As the nation's best equipped observatory outside government, the Harvard College Observatory worked with nearly all the great government surveys of the forties and fifties. In 1859 George Bond verified a long-standing Harvard assumption by announcing that "Cambridge is now the central geographical point of this continent." Its longitude having "undoubtedly been investigated with more care than that of any other spot on the globe," it had become a basic reference point throughout the hemisphere and even on British Admiralty charts.

In astronomy as in other fields of science, Americans were strongest in observation, description, cataloguing, and technique. By mid-century they were doing good work in practical astronomy, though one commented in retrospect that some of them had not understood correct methods of reduction and had shown "evidence of hasty and ill-considered plans." Benjamin Gould and others catalogued the positions of thousands of stars. By 1876 Americans had discovered ten comets, several hundred double or binary stars, and forty-nine asteroids, the latter count being greater than any other country's. In theoretical astronomy, Benjamin Peirce and Sears Walker contributed to the understanding of Neptune's orbit, and William Ferrel showed that the tides produced by the moon must retard the earth's rotation. In physical astronomy, where skill in observation counted for more than mathematical sophistication, Americans made their most important contributions. Denison Olmsted and others perceived the extraterrestrial origin of meteors, and James Dana studied the possibility of lunar volcanoes. The Bonds, with Harvard's telescope, discovered many new physical features in planets, comets, and nebulae.

Physics in the modern sense, as distinct from physical astronomy and geophysics, was last and least among the major fields of antebellum science, embracing about 4 per cent of *DAB* scientists active at any time during the years 1846–60. As "natural philosophy" it might be called the oldest of the sciences. In terms of its coming transformation and significance it might be called the newest. But that transformation, with its demands on mathematical expertise, had not yet touched the United States. Handicapped by weakness in mathematics and swayed by the data-gathering bent of American science generally, the handful of American physicists preferred experiment to theory. Indeed, the technological excitement of those years tended to draw them away from basic science to practical application or invention. Joseph Henry was the only American physicist with a European reputation.

Few as antebellum physicists were, they exceeded the demand. The brilliant young astronomer Simon Newcomb tried all through the spring of 1860 for an appointment in physics at Washington University, writing letters, travelling to St. Louis, getting recommendations from Benjamin Peirce and other scientific notables. Yet the appointment went to Lieutenant John M. Schofield of the West Point faculty, who took it, he remembered, simply "because there was no captaincy in sight for me during the ordinary lifetime of man." A hint from Jefferson Davis as to coming military opportunities soon led Schofield to quit his professorship and reclaim his commission, eventually rising to lieutenant general.

But a change was at hand in the market for physicists as well. By 1863 an academic recruiter from the University of Michigan was complaining that while astronomers and mathematicians were plentiful, physicists were not to be had. After the war students seem to have responded to the new demand. In the early seventies physicists accounted for 11.5 per cent of new *DAB* scientists, as compared to about 5 per cent in the preceding quarter century.

With hindsight we can detect portents even in the antebellum years. Electricity had once been the province of chemists, but Joseph Henry and others had claimed it for physics by the 1840s. Like chemistry, its visual effects had made it a staple of popular lecture demonstrations, and boys of the fifties made a toy of it. A Yankee boy in 1858 mentioned that his ten-year-old friend's "Brother Henry has got home and is going to fix his Electric Machine." "Brother Henry" was young Henry Adams, who would eventually make the dynamo the mystical symbol of a new age and would voice what seem like uncanny premonitions of nuclear war. In 1855 seventeen-year-old Edward Morley of Attleboro, Massachusetts, wrote: "I have finished the Electrical Machine at last and hope and expect to electrify all the neighborhood." Eventually he and Albert Michelson would electrify more than the neighborhood with an American triumph of precise measurement showing the observed speed of light to be constant whatever the observer's relative motion. It took Einstein to explain that. Today it would take another seer-historian like Henry Adams to perceive the ultimate consequences of the explanation.

James Dana was proud of American geology. It was, he remarked in 1854, "almost the only science in which progress has been made in the country except some departments of zoology." He himself enjoyed an international reputation. When he visited Paris in 1859, it was reported, "the savants ran after and feasted him, and kept him up late at nights. . . . No scientific man from America ever before created such a sensation in Paris." Yet even Dana could not shake off an oppressive awareness of European superiority in science; he conceded ruefully in 1858 that the progress of American geology was largely due to Lyell of England and De Verneuil of France. American scientists still worried about European opinion of them. Alexander Bache in 1856 thought that while the English had grown somewhat more respectful of American science in recent years, the French were even more contemptuous of it than before; and a scientific American in Paris in 1864 saw no change. But on two occasions and in two fields, antebellum scientists took the measure of Europe's best.

The first encounter was in astronomy. The astronomers might have been expected to be above national rivalries. But in finding that Antares was a triple star, Sears Walker exulted: "This is an important discovery, and is truly American." Three years after arriving from Germany, the astronomer Christian Peters saw every American discovery as helping "to teach old-fogy Europe to look a little more around towards this side of the Atlantic." Rivalry with the British played a part in establishing the navy's Nautical Almanac in 1849. Popular support was rallied for the Cincinnati Observatory in the early 1840s by challenging American democracy to match the astronomical achievements of Russian autocracy at Pulkovo, and George Bond in 1860 sounded the same note in behalf of an American astronomical expedition competing with a Russian counterpart. The space race with Russia remained friendly, Pulkovo gently suggesting to Harvard that since their similar telescopes led both to the same objects, cooperation made more sense than competition. But France and America, in the persons of Urbain Leverrier and Benjamin Peirce, had at each other more hotly.

In 1846 Leverrier postulated the existence and location of a previously unknown planet on the basis of perturbations in the orbit of Uranus. When Johann Galle sighted the planet (later named Neptune) at the point indicated, the uniformity of natural law throughout the universe, and the power of science to reveal it, were borne upon the minds of the laity as never before. But Sears Walker of the Naval Observatory pored over earlier records and discovered a 1795 catalogue recording Neptune as a fixed star. From this he calculated Neptune's orbit as being quite different from Leverrier's postulation, except at the point of first sighting.

Walker's bumptious correction of Leverrier created a sensation not only in America but also in Europe, and the uproar grew when Benjamin Peirce published a paper in mathematical support of Walker, calling the discovery of Neptune a mere "happy accident." Joseph Henry thought Peirce had been "too hasty," and President Everett of Harvard fretted for weeks lest Peirce bring discredit on the college. When Europe failed to accept the Walker-Peirce calculations, Everett could not help suspecting some fallacy in them. European astronomers were furious at the Yankee impudence. Four years later George Bond, on a visit to France, found Leverrier still "irritated to the last degree." James Dana in 1848 privately regretted "as a national calamity" that Peirce had been so "quick upon the trigger." But as it eventually turned out, though Peirce had overstated his case, he and Walker were right, especially in calling Leverrier to account for failing to see that several different orbits could have met the requirements. Peirce and Bond were also right in their calculation of Neptune's mass, though their figure was set aside for thirty years in favor of an erroneous value found by Otto Struve. The whole episode tends to support American charges of European arrogance.

A dozen years after the Leverrier episode, another scientific controversy, the greatest of the century, gave Europe occasion to raise its estimate of American scientists. In this case the alignment was quite different, not one of America against Europe, but one of Americans on both sides as acknowledged peers of the Europeans. And in the course of this struggle, paradoxically, American science showed its growing self-respect most plainly in deflating the very man whose arrival in 1846 had done so much to hearten it.

Though Louis Agassiz's coming had swelled American pride, the adulation that greeted the "foreign lion" had revealed how modest that pride was. Agassiz himself, for all his public touting of American science, privately remarked in 1855 that he had stayed because he could influence science so much more easily in the New World than in the Old. Only Agassiz's charm and ebullience kept such assurance from looking like arrogance. But that same buoyant charm supported a growing addiction to public acclaim. And the more he played to the gallery, the less time and need he had for sustained reflection and intellectual growth, for exploring or even recognizing new directions in scientific thought.

In 1851, remarking on the eyeless creatures of the Mammoth Cave, Agassiz raised the issue of whether "physical circumstances ever modified organized beings" and held it up as "a great aim for the young American naturalist who would not shrink from the idea of devoting his life to the solution of one great question." Evidently his hypothetical naturalist had a short life expectancy, for in 1857 Agassiz announced the final solution of this and all other great problems in natural history. In that year, the "Essay on Classification" in the first volume of his *Contributions to the Natural History of the United States* explained for all time (as Agassiz saw it) the workings and meaning of natural history. His mentor Cuvier had, wrote Agassiz, been proved right in his system of four great divisions of the animal kingdom, and right also in his savage rejection of Lamarckism or any other hypothesis of variability in species. Species were "thoughts of the Creator," had been created from time to time and place to place in final form and full numbers, had periodically been wiped out by divinely willed catastrophes, and had been replaced by new and improved species in accord with God's Cuvierian master plan. Henceforward (it seemed to follow) questions of cause and process could be answered with as little thought and trouble as they had been before the birth of modern science, to wit, with the simple words, "God willed it so."

But James Dana's review in the *American Journal of Science,* while praising the useful aspects of the volume, found the "Essay" intellectually static, unresponsive to new questions about variation and geographical distribution of species. And Dana's disposition to stand up to Agassiz reappeared in the following year, after Dana had unfavorably reviewed a slipshod book on American geology by Jules Marcou, a French friend of Agassiz's. Agassiz presumed, admittedly without having read Marcou's book, to offer a defense of it on the basis of Marcou's past performance, a defense, moreover, that belittled American geologists; and he pressured Dana into printing his communication by threatening otherwise to quit the *Journal*'s board of editors. Dana followed up Agassiz's statement in the *Journal* with a sharp rebuttal that left him and Agassiz still on outwardly friendly terms but without the old warmth and, on Dana's part, with less of the old respect.

It was not Dana, however, but the unobtrusive Asa Gray, scarcely a crowd pleaser though by now the nation's foremost botanist, who presently took on Agassiz and, in a different sense from Gray's own promise of 1846, "took care of him." Agassiz rubbed his Harvard colleague Gray the wrong way in several respects. He conspicuously outdid Gray in fund raising, teaching, and public speaking, all of which not only evidenced a disparity in temperament but probably injured Gray's pride also. Furthermore, Gray (and Dana too) considered the human race one species, whereas Agassiz insisted that blacks were a distinct species and inferior to whites. This discord had overtones apart from biology. But the principal difference between Gray and Agassiz involved science—not merely the concrete question of species, but the whole nature of scientific inquiry. Were facts to be forced into an a priori mold? Or was the search for scientific truth to be open-ended, flexible, and governed by evidence?

The flood of plant specimens from army and navy explorations had kept Gray too busy to commit himself publicly to any such grandiose hypothesis as Agassiz's. Nevertheless, he had a questing mind. In the late forties, for example, plant species divergence in the Galapagos and Hawaiian Islands had caught his interest. Experience in classifying enormous numbers of individual specimens meanwhile gave him a sense of how one species might shade into another through a gradation of small variations. By the fifties, in defining the concept of species Gray relied on the principle of genetic connection, of inherited characteristics. All this pointed toward the idea of natural development of species. As a faithful Congregationalist, Gray accepted the omnipotence of God, but he thought of species creation as remote. The notion that at any moment new species might abruptly appear out of nowhere made him uneasy. And so, as Agassiz trumpeted that unsettling doctrine, Agassiz likewise made Gray uneasy.

In the mid-fifties Gray became an acknowledged ally, indeed a confidant, of Charles Darwin. This eventually won him a secure place in the history of one of the great events of science. It also gave him a chance to help kick out the main prop of Agassiz's scientific edifice.

The connection with Darwin came through Gray's old friend and correspondent, the noted British botanist Joseph Hooker. Much travelled, viewing plant distribution globally, Hooker believed each species to have been descended from a single pair and to have achieved its geographical distribution, however wide, by natural causes—the antithesis of Agassiz's views. With some reservations and questions, Gray agreed with Hooker. Hooker's friend Darwin saw their correspondence on the subject, was impressed by Gray's ability and knowledge, and opened his own correspondence with Gray in 1855. Though Gray himself had not been to East Asia, his work on the botanical specimens of the Pacific expeditions gave him, like Hooker, intercontinental expertise. This, along with Gray's intelligence, professional ability, and compatible views, made his correspondence with Darwin a long and fruitful one. Darwin's suggestive questions stimulated Gray to apply statistics to the study of plant distribution and to publish an article in 1856 which his biographer has called "one of the foundation papers in the science of plant geography." Delighted with it, Darwin in September 1857 confided his theory of evolution to Gray in a letter that was later used in establishing Darwin's priority over Alfred Wallace. Thus Gray joined Lyell and Hooker in the select group of Darwin's initiates.

Gray fired his first shot across Agassiz's bow without waiting for Darwin's great book to appear. For ammunition Gray drew upon his expert awareness that the plant species of eastern North America more closely resembled those of Japan than those of western North America. Encouraged by Darwin, Gray demonstrated this rigorously by a statistical comparison of the respective floras. Agassiz would have ascribed it all to the "thoughts of the Creator" and let it go at that. Gray sought a natural cause; and he found one, ironically, in Agassiz's own earlier discovery, the Ice Age. Once, Gray suggested, the plant life of the far North had formed a continuum between the similar eastern shores of the two continents. Then the glaciers had pushed south to the dissimilar environment of western America, separating the other two regions. The separation had lasted long enough to account for a number of species in each that were much like but not identical with species in the other—to account for them, that is, if one admitted the possibility of variations developing into distinct species. In short, Gray struck at both elements of Agassiz's doctrine: that species were created simultaneously in several places, and that they remained unchanged until wiped out by divine intervention.

Gray presented his hypothesis formally in January 1859 before the American Academy of Arts and Sciences in Boston. He did not spare Agassiz; on the contrary, he thrust to his rival's heart by declaring Agassiz's view of natural history to be "out of the field of inductive science." Agassiz seemed to be caught off balance. His rejoinder was polite but unconvincing. Nevertheless, he got the Academy to call extra meetings for an extension of the debate. Gray again was ready, while Agassiz suffered not only from the handicap of being wrong but also from the distractions of frenzied fund raising for his projected Museum of Comparative Zoology. In replying to Gray's carefully prepared paper, Agassiz had to fall back on his vast knowledge of facts and his gift for extemporaneous eloquence. He scored some points, but Gray clearly carried the day. Even Agassiz's crony Benjamin Peirce wrote after the February bout that "Gray has grown wonderfully in the last few years." In May, following up a talk by Agassiz on his view of creation, Gray played host to the Cambridge Scientific Club and gave its members a preview of Darwin's theory. To Agassiz it was a mere recrudescence of Lamarckism, and its full implications seemed lost on the other guests also. Nevertheless, Agassiz must have felt a premonitory chill. Afterward he said to his host, "Gray, we must stop this."

But it was past stopping. That fall *The Origin of Species* at last appeared. Darwin sent copies at once to Gray, Agassiz, and Dana. While Dana believed, like Agassiz, in special creations, the immutability of species, and extinction by catastrophes, he had long respected and corresponded with Darwin, had called for continuing research on species variability, and had even given Gray moral support and geological advice for the debates with Agassiz. Most important, Dana, a man of strong religious faith, had cautioned Gray not to admit any incompatibility between a developmental view of natural history and the tenets of religion. Unfortunately, Darwin's book reached Dana just as Dana suffered a nervous collapse which would affect him for years; and he could not or would not read the book before 1863. It is not clear, in fact, when he did get around to reading it. At any rate he showed no sign of accepting Darwinism before 1870.

If Dana had kept his health and long-held convictions, he might have joined Agassiz, less dogmatically and therefore more effectively, in contesting Darwin's theory. As it was, Agassiz stood virtually alone among major American scientists in his public opposition to Darwinism, though several leading scientists opposed it vigorously in England itself, and the French, still in reaction against Lamarckism, rejected it almost unanimously. Meanwhile America produced more than one potent champion of Darwinism. In four meetings of the Boston Society of Natural History during February, March, and April 1860, it was not Asa Gray but William Barton Rogers who upheld the new view against Agassiz. No other nation witnessed so extensive and organized a debate on the *Origin* carried on by scientists of such eminence. In the plant distribution debate Agassiz had scored points against Gray in geology, Gray's weak suit. Agassiz could not do that with Rogers, who turned the tables on him in a crucial dispute over New York geology and also exposed one of Agassiz's key arguments as so much scientific sleight of hand. Throughout, Agassiz made a poor impression with his dogmatic contradiction of respected scientists on no evidence but their disagreement with his views. No vote was taken, of course, but even Jules Marcou admitted that Rogers had won.

Before the debates with Rogers had ended, Agassiz carried the fight again to the American Academy in three more special meetings. Now the laymen were being drawn in, and among them Agassiz found allies in a Harvard professor of philosophy, Francis Bowen, and a wealthy benefactor of science, John A. Lowell. Agassiz let them carry the burden of debate. Their most significant tactic—an ominous one—was to question the compatibility of Darwinism with religious faith. Gray's masterful rebuttal still dealt largely with the scientific issues, but he took note of the new objection and insisted that Darwin's theory left as much room as ever for belief in design, in God's supervision of nature, a belief held by Gray himself. The intrusion of theological arguments doubtless confirmed the opinion Gray had already expressed privately, that Agassiz was "a sort of demagogue, and always talks to the rabble." Nevertheless, it was a straw in the shifting winds of the Darwinian controversy.

Agassiz did, in fact, largely confine his subsequent assaults on Darwinism to popular lectures. His only published commentary that year, appearing in the July issue of the *American Journal of Science,* consisted of an advance excerpt from the third volume of his *Contributions* which pronounced "the transmutation theory" to be "a scientific mistake, untrue in its facts, unscientific in its method, and mischievous in its tendency." Gray, on the other hand, wrote seriously, perceptively, and critically on the subject. His extended review of the *Origin* in the March 1860 *Journal* outlined Darwin's argument with clarity and understanding, comparing it with Agassiz's ideas, denying that it ran afoul of theology, and yet pointing out the gaps in both its evidence and its theoretical structure, especially in the mechanisms of variation and heredity (which were, in fact, not to be satisfactorily explained until the twentieth century). Darwin himself thought Gray's review "by far the most able which has appeared." It effectively counterbalanced the weight of Agassiz's prestige and probably forestalled a general snap judgment by Americans against Darwin, *á la française.*

Gray followed up the review with his most comprehensive statement, published by the *Atlantic Monthly* in three installments during the summer and fall of 1860. In a style effectively aimed at the general public, the three articles reviewed the whole argument, including the attacks on Darwinism by Agassiz and others. Gray again pointed out unresolved difficulties in the hypothesis. But he came closer than ever to a commitment in Darwin's favor. Darwin had the articles reprinted and circulated in England, where they impressed some of his opponents. "No one person understands my views and has defended them so well as A. Gray," Darwin wrote Dana, "though he does not by any means go all the way with me." Nevertheless, the articles still asserted Gray's faith in design, and this at a time of increasing polarization on the theological issue. Darwin himself hardened his opposition to the notion of design, as did his circle in England. The agnostic Thomas Huxley now became the great champion of Darwinism, and though Gray and Darwin kept up their friendly correspondence, Gray was no longer one of the inner circle.

With the intrusion of theology, sociology, and economics, the controversy took a new turn. In the United States, moreover, the Civil War drowned out scientific debate. For a time Darwinism ceased to be a live issue. Yet the Agassiz-Gray-Rogers debates had been revealing. The United States had produced Benjamin Franklin and Joseph Henry, and Europe had conceded them high rank in science. Nevertheless, they seemed to be flukes. In contrast, the Boston debates rested on collective enterprise as well as particular individuals—on the federal exploring expeditions in the case of Gray, on the state geological surveys in the case of Rogers, on public interest and philanthropic support in the case of Agassiz, and on the Boston scientific societies in the case of all three. In short, those debates and writings evidenced the increasing maturity of American science as a whole, both absolutely and as compared with that of Europe.

Part Two

SCIENCE, TECHNOLOGY, AND A RISING PEOPLE, 1846–1861

Chapter 9

Science and Technology in the Public Mind

Darwin and Liebig, like Newton and Franklin before them, confronted the public with the philosophical and material implications of science. And the public's reaction held implications in turn for the future of science, especially in America.

Popular approval mattered less to European scientists. "In England . . . we do not need to make science popular," the royal astronomer told Maria Mitchell in 1857. In France, Leverrier told her, one had to "make science come into social life, for the government must be reached in order to get money." "The government," however, was that of the Second Empire, and "social life" was that of the ruling circles. In England, similarly, James Joule urged in 1873 that a nobleman be made president of the British Association so as "to get at the sympathy and influence of the upper ten." But in egalitarian and republican America, scientists knew that the common man also had to be "reached in order to get money."

Some scientists deplored the necessity. "I dislike writing for the public," grumbled James Hall in 1846. To satisfy the "vulgar appetites of the people," complained James Dana, science had to be "diluted and mixed with a sufficient amount of the *spirit of the age.*" Recalcitrants were sometimes taken to task. In 1851 Commander Charles Wilkes objected to the uncompromisingly professional way in which Torrey and Gray proposed to write up his expedition's botany. "The Library Committee [of Congress] intend the work for the people," he wrote angrily, "and not for any class of scientific men." Army officers during the Great Reconnaissance of the West now and then cautioned their scientific comrades to remember Congress and the public in writing their reports.

But most scientists needed no such advice. Even James Hall angled skillfully for the favor of the people and their representatives when he had to. Joseph Henry made his grudging "oblations to Buncombe." Elias Loomis thought it "advisable to cultivate the scientific taste of the community," which he did to his enrichment through bestselling textbooks.

A few scientists did their bit through public lectures. "The course of lectures given by a celebrated astronomer during the last winter," wrote a Providence librarian in 1857, "kept our shelves bare, for a time, of all astronomical works." During the 1830s the "lyceum" movement of popular culture through lectures gave about a fifth of its platform time to science. Strongest in New England, its fountainhead, the lyceum movement slackened after the Panic of 1837, then revived in the prosperous decade between 1846 and the Panic of 1857, growing fastest in the Midwest as that region developed the necessary cities, periodicals, and transportation facilities.

As a scientific spellbinder Louis Agassiz outdid all others. With his enthusiasm, vigor, learning, and Gallic charm, his lectures won both the respect of scientists and the applause of the crowd. As heir to Benjamin Silliman's crown in the realm of scientific lectures, he drew Silliman himself up from Yale to hear his Lowell Institute lectures of 1846–47 on the "Plan of Creation in the Animal Kingdom." Five thousand jammed Tremont Temple in Boston, and Agassiz had to repeat his lectures to the overflow. He accepted many of the lucrative offers that followed from all over the East in 1847, and in succeeding years triumphantly toured the South and West. Agassiz wanted the money for his scientific projects, he loved to be lionized, but he also recognized the special importance in America of popular backing for science.

What Agassiz did for natural history, Ormsby MacKnight Mitchel did for astronomy. A small, restless, energetic Ohioan, he graduated from West Point in 1829 with a reputation for mathematical ability, left the service and settled in Cincinnati, was admitted to the bar, served as chief engineer of a small railroad, and taught astronomy at Cincinnati College. In the last role he found his calling. Even without Agassiz's stalwart physique and flavorsome French accent, Mitchel approached Agassiz in power to charm and fascinate an audience. A successful public lecture on astronomy in the winter of 1841–42 led him to give a popular lecture course. Through this he rallied the public to support a proposed Cincinnati Observatory.

The comet of 1843 that gave impetus to the Harvard Observatory may also have helped in Cincinnati. In retrospect, however, Mitchel stressed his appeal to national pride. "I am determined," he had told his fellow citizens, "to show the autocrat of all the Russias that an obscure individual in this wilderness city in a republican country can raise here more money by voluntary gift in behalf of science than his majesty can raise in the same way throughout his whole dominions." And indeed Cincinnati in 1845 acquired the finest refracting telescope in the world outside Pulkovo (until Harvard's in 1847). Mitchel took great pride in his observatory's democratic basis. There were limits, however. The people so insistently exercised their proprietorship during the first year that Mitchel himself had the telescope only one night a week. After that he reserved three nights a week for serious work, and in 1854 got the board's permission to close out the public peep show entirely.

Unfortunately the public made more visits than donations. In 1844, after scraping by from day to day on stock payments in cash and kind, improvising expedients, supervising construction, and meanwhile teaching five hours a day, Mitchel put up a building. But it had no endowment aside from his own dedication and earning power, and the latter was interrupted when Cincinnati College burned down. So Mitchel supported himself and the observatory by taking to the lecture platform. Eastern cities came to know him well over the next decade and a half. President Frederick Barnard of the University of Mississippi, a competent scientist, confessed to being "full of Mitchel and the wonders of his magical observatory." Scientists more often shared the view of John Frazer in 1859 that while Mitchel "has gained great reputation as an orator, he has not increased his fame as an astronomer." Mitchel's tragedy, like that of many other Americans, was that in earning money to support his science, he left himself little time to practice it.

Mitchel was chagrined in 1846 at not getting Harvard's Rumford Professorship. Yet science may have profited more than if he had. In time his fellow scientists had kind words for him because of the public support he won for astronomy. Even Simon Newcomb, who had called Mitchel's *Popular Astronomy* slapdash in an 1860 review, wrote long after Mitchel's death: "It is not unlikely that he sowed much of the seed from which the American astronomy of the present has developed."

Fortunately for the growing body of specialized professionals who disdained such public touting of science or who lacked a gift for it, others came forward to preach the word. By mid-century such writers of popular science books as Mrs. Almira Phelps had discovered a seller's market. In Milwaukee, still a mere village of some three thousand, booksellers of the early 1840s offered not only Mrs. Phelps's *Botany* and *Philosophy,* but also popular treatises on astronomy by Arago and Herschel, popular books by lesser writers on entomology, geology, mineralogy, natural history, botany, physics, and meteorology, biographies of Newton and Rittenhouse, and many others, even including Newton's works and the *Geological Reports of Massachusetts.* Milwaukee newspapers urged the study of science not only for practical applications but also for the elevation of mind, morality, and status. They strewed their pages with snippets of science, mostly copied from other journals. Astronomy seemed most popular, with its visual appeal, its age-old tradition of sky watching, and its religious connotations; but geology, paleontology, botany, zoology, microscopy, chemistry, and physics all were well represented. Science evidently helped sell newspapers. Magazines and newspapers everywhere catered to the public's appetite for wonder tales of science. In 1845 a weekly newspaper specializing in popular technology began publication in New York City. Significantly, it chose to call itself the *Scientific American.*

In leaving newspaper and magazine popularization largely to non-professionals, however, the professional scientists bore some responsibility for the "quackery" and "charlatanism" they so often denounced. Science items in the popular press, even when factually sound, emphasized the exotic and "curious." Genuine "wonders" of science lent credibility to fabrications. So the 1840s were marked by pseudoscientific hoaxes and fads, echoes of which linger in literature, notably in the novels of Hawthorne and the tales of Poe. Some pseudosciences or quasi sciences of the time had mixed effects. Phrenology, which claimed to measure personality traits by the contours of the skull, reached its height in the thirties and forties. It could win no better scientific endorsement than some faint praise by the elder Silliman in one issue of his *Journal.* In the end it advanced science chiefly by driving opponents to serious research. Yet it also introduced the public to the notion that even so subtle a phenomenon as the human mind might be explained by natural causes. And so it broke ground for Darwinian materialism.

Popular interest in science was shared by more notable writers than those of the press and the potboilers. The scientific and literary cultures had not yet diverged quite beyond speaking distance. The poet William Cullen Bryant as editor of the *New York Post* publicized scientific advances and wrote competently on mineralogy, botany, geology, and other branches of natural history. He had long since got over his youthful scorn for Jefferson's paleontological excursions. By 1825 Bryant saw science not as a barrier to poetry but as a new field for it. On the whole, Walt Whitman also liked science and scientists. "The paths of science present pleasure of the most alluring kind," he wrote in 1846. In laying plans for a select social club in 1856, Ralph Waldo Emerson decided that "the nucleus of the company should be *savants* [i.e., scientists] . . . a very clear, disinfecting basis." Emerson found Agassiz altogether congenial, personally and philosophically. So did Longfellow and other specimens of New England's literary flowering.

Yet Asa Gray remained outside that intellectual circle. The divergence between science and the arts was steadily widening. Emerson himself, though brother-in-law to the noted scientist Charles T. Jackson, yawned over mere catalogues of scientific fact—"we must not lose ourselves in nomenclature," he wrote—and he blithely allegorized on birds without benefit of ornithology. And Whitman, when he "heard the learn'd astronomer," became "tired and sick" and glided out to look up "in perfect silence at the stars."

Whitman spoke for the multitude. They waded happily into the shallows of scientific knowledge. They looked out over its surface with eagerness for the goods it might bear to them, with apprehension of the terrors that might rise from it, with sheer wonder as its vastness and power came more and more clearly into view. But as the professional pushed out into the deep, the layman found him increasingly hard to follow and so at last could only stand and watch. In chemistry and mathematics the professional had got beyond the American layman's depth by 1820, in botany by 1830, in still other fields by the mid-1840s—as implied by the growing demand for popularization. Lyceum lectures and popular books were not enough. "We need a more popular cultivation of science," complained the *Springfield* [Mass.] *Republican* in 1859. "A stately quarterly, that reaches only the savans of science and is intelligible only to them, is the solitary index of scientific culture in America." Still, however recondite the ways of science, its implications increasingly colored the layman's hopes for both this world and the next, the material and the spiritual.

Fewer than a third of free, adult Americans were church members in 1850. So it is not surprising that some scientists of the time cared little for religion. John F. Frazer tried in vain to achieve faith. Others never tried. Science filled their minds and satisfied their souls. "How much talk there is about religion!" wrote Maria Mitchell impatiently. Though she attended Unitarian services for years, she never became a church member, had "no settled views," and was "religious only in feeling." In 1854, chiding Joseph Leidy for his indifference, a friend revealingly alluded to "the probabilities disregarded by those who employ themselves exclusively with Natural Science."

Yet in the mid-fifties James Dana thought science had a higher proportion of believers than any other profession except the clergy. His impression cannot now be tested quantitatively. Clark Elliott, in studying about five hundred American authors of scientific articles, 1800–63, has ferreted out information on the religious backgrounds of no more than half, relying primarily on evidence of parental or ancestral faiths. Quaker backgrounds took the numerical lead, followed closely by Presbyterian, Congregational, and Episcopalian. Elliott himself doubts the validity of this ranking, since biographical sketches would more likely mention Quakerism than more common sects. At any rate, Catholic, evangelical, and fundamentalist backgrounds seem to have been rare. But Elliott ventures no statistical analysis of what his scientists believed as adults, nor does he compare them with other professionals.

So Dana's guess is as good as any and probably better than most. Many scraps of evidence tend to confirm it. Astronomers especially seem to have been religious. In both private diaries and public lectures they professed strong faith. In his classes Benjamin Peirce soared from mathematical astronomy to the solemn pronouncement, "Gentlemen, there must be a God." Ormsby Mitchel's popular lectures emphasized the heavens' declaration of God's glory, leading one hearer to remark, "What a loss to the Christian pulpit." Denison Olmsted, whose upbringing in family, community, and college had been deeply religious, preached religion along with science in popular lectures and the classroom, though not in his scientific papers.

Piety was not confined to astronomers. One of every six leading scientists was the son of a clergyman, and a significant proportion of college science teachers had studied theology formally. The religious press celebrated the devoutness of the botanist Asa Gray, the zoologist Louis Agassiz, the oceanographer Matthew Maury, and the chemist Benjamin Silliman, Sr. A lifetime of scanning the skies without seeing a cross or a cherub did not shake the meteorologist James Espy's faith. Nor did decades of chipping away at the rocks of several ages interfere with the Bible classes of George H. Cook, geologist and church elder. In 1930 an entomologist who remembered that generation in its last years asked: "Why is it that those old, simple, true naturalists, almost without exception, were men of deep religious feeling? The modern scientific man does not possess this mental characteristic." (And in fact, three-fourths of all *Who's Who* scientists in that year listed no church membership.)

A cynic might credit the piety of antebellum scientists to their instinct for self-preservation. In the South especially but also elsewhere, it occasionally behooved a scientist to pray in public. Angered by geology's tampering with the time scale of Genesis, the trustees of South Carolina College struck that science from the catalogue in 1835, and only after fourteen years of conspicuous piety and quiet string pulling could its last professor there get it reinstated. Eight years later, when Joseph LeConte took over the chair, he proclaimed geology "the chief handmaid of religion among the sciences." Looking back after twenty years, Benjamin Gould remembered such pressures about 1850 as having been "scarcely less tyrannous or unrelenting" in their way than the seventeenth-century Inquisition.

But Gould was prone to magnifying bugaboos. The religious asides of most antebellum scientists were manifestly sincere. Besides, the leaders of both science and theology were bent on preserving a shaky entente, despite certain new scientific hypotheses.

A deeply felt need underlay the alliance. The eighteenth-century Age of Reason had regarded the universe as a finished machine, a sort of cosmic, self-winding watch left ticking by its retired maker. That tidy image had evidently satisfied the gentry who contemplated such questions. But the nineteenth-century Age of Romanticism faced social and economic change of a kind and at a pace not known before. Acknowledging change, yet yearning for guidance, it came to think of nature not as a machine but as an organism, long developing under the eye and direction of a supreme being. The so-called laws of nature were seen as merely the evidence of constancy in God's watchful work. His plan had culminated in the human race, every specimen of which enjoyed his unremitting attention. Such personal reassurance was the essence of that generation's faith. "Law cannot love me," protested one religious writer in 1850.

Faith in a close, caring God encountered the "nebular hypothesis" of Pierre Simon Laplace, which asserted that the earth and other planets had not materialized full-formed at God's command but had coalesced from a cloud of diffuse matter surrounding the sun. First broached in 1796, the hypothesis did not sink into the American popular mind until the 1840s, by which time it had been accepted by such eminent (and pious) scientists as Joseph Henry and Benjamin Peirce. Meanwhile Charles Lyell in the 1830s had compellingly presented the uniformitarian thesis in geology, holding that the operation of currently observable processes explained all geological change since Day Three. Though Lyell himself insisted on the permanence of species, some theologians perceived or sensed that geological uniformitarianism might lead to the idea of biological development by natural processes, thus banishing God from the life of man to the pinpoint of an unimaginably remote past.

The nebular hypothesis was not hard to stomach, Genesis being conveniently vague about the process of creating the earth. In any case, the matter seemed to have little bearing on God's current management of human affairs. Uniformitarian geology posed a thornier problem because it extended God's aloofness to the present time. Opponents could find scientific allies in the catastrophists, represented in America most notably by Agassiz. In Agassiz's view, God had not turned his back on creation. Divinely willed cataclysms periodically reshaped the earth's surface, wiping out existing species, whereupon God restocked the earth with fresh creations. But Agassiz's supernatural catastrophism ran against the grain of contemporary science. Though the fossil record, especially as fragmentarily known in that day, was indeed riddled with discontinuities, it was also evident that certain species could be found on both sides of what the catastrophists supposed were universal extinctions. Anyway, plausible natural explanations now took precedence over supernatural.

Theologians felt the odds lengthening against them. In the popular mind, scientific knowledge, based on dispassionate reasoning from verifiable data, was steadily gaining in authority and material promise, while theological dogma was losing its credibility. Theologians hesitated to declare war on so formidable an adversary. Moreover, they themselves felt the power of reason and the attraction of predictability. So they took the fateful step of meeting science on its own terms and construing Scripture to fit it. Thus the six "days" of creation were epochs of whatever length science required. Or the creation of heaven and earth had been followed by eons of change, unmentioned by Genesis because irrelevant to man, before the remaining twenty-four-hour days of creation. Or the six-day creation applied only to the environs of Eden, leaving the rest of the globe to Lyell.

With the help of such glosses, many theologians persuaded themselves that science validated religion. In assuming the universality of natural law, science implied the existence of a single, all-powerful lawgiver. In studying nature, science illuminated his power and plan. Hence writers on religion encouraged clergymen to study science, and many did so. Even in the South, a hundred or so well-educated clergy, especially Presbyterians and those in urban churches, wrote on scientific subjects. They welcomed scientific advances, at least those not turned against faith. And scientists like Denison Olmsted gave them comfort by agreeing that true science necessarily supported true faith. "Science and religion . . . must be in harmony," wrote the respected geologist and Congregational minister Edward Hitchcock of Massachusetts in 1852.

On that premise Hitchcock devoted his life to the furtherance of natural (i.e., scientific) theology, an attempt in effect to make religion a branch of science. By the 1840s he had become the acknowledged leader in that field, and his 1851 book *The Religion of Geology and Its Connected Sciences* constituted the most notable American embodiment of its principles. Though conceding the great age of the earth, he held fast to Cuvierian catastrophism, special creations, and "special providences," by which he meant divinely willed departures from natural law on rare occasions.

Natural theology gained some ground in the 1840s. It was, for example, one of the required themes of the Lowell Lectures. But it straddled a widening gap. As it became more scientific in subject and method, its applicability to theology became more attenuated; and meanwhile its dwindling theological content became less relevant to science or even consistent with it. Each side tended more and more to ignore or dismiss the other. Edward Hitchcock himself slowly gave ground. Having first yielded on scriptural chronology, he then gradually moderated his catastrophism, though never wholly abandoning it. By 1853 he conceded that science, dealing with nature, and theology, dealing with moral philosophy, were largely incommensurable. A year later he admitted that some Biblical language could not be reconciled with science under any construction. By 1859 he had given up on detailed collation of science and Scripture. He held doggedly to the doctrine of special creations. Though he apparently read *The Origin of Species,* he died in 1863, before its ultimate triumph. And in the last year of his life, he still declared the "religious bearing" of the natural sciences to be "their most important use, as it is of all knowledge." But by then geologists saw the chief use of their science to be in pointing the way not to Heaven but to coal, oil, and mineral deposits.

The strident reiteration of theology's claims to scientific support betrayed a growing uneasiness about the reliability of that support. While not daring to repudiate scientific progress in general, theologians nervously deprecated scientific theorizing as metaphysical and urged strict Baconianism as an antidote. Most of all they feared biological developmentalism, which threatened the special standing of the human race in the sight of God.

Their scientific allies replied to mounting fossil evidence of development with the doctrine of progressionism, which clung to catastrophism and the fixity of species, explaining progression in life forms as entirely the ongoing work of a progressive-minded God. Through the 1850s, led by Agassiz, scientists generally endorsed that view. But the laity instead responded to a conflux of currents in favor of development. Romanticism's emphasis on organic growth, the era's faith in progress, the historical-mindedness of Christianity itself, and the analogy of uniformitarian geology, all prepared the public mind for biological evolution, despite the scientists' opposition. Only in the South, especially in South Carolina, where uncontrolled change was suspect, Biblical literalism strong, and the influence of Agassiz dominant, did opinion turn in the other direction, toward special creation, progressionism, and the pervasiveness of God's will in nature.

In 1844 a warning thunderclap of developmentalism sounded with the publication in London of an anonymous book called *Vestiges of the Natural History of Creation.* The public was evidently receptive, for within a year four British and three American editions appeared. The author was later revealed to have been Robert Chambers, an Edinburgh publisher, bookseller, and encyclopedia editor. His views were more eclectic than original, including his belief in biological development. Early in the century the French zoologist Jean de Lamarck had suggested that the struggle to cope with a changing environment induced physical changes in individual creatures whose offspring then inherited them, thus developing better-adapted species. Though most scientists rejected Lamarckism, Chambers picked up this and other ideas in the lively scientific and literary circles of Edinburgh and skillfully synthesized them. The author denied any conflict with Scripture (rightly construed), but his quasi-Lamarckian vision of species as having gradually developed through natural laws struck to the heart of religious faith. Nevertheless, the English public, bewildered by the ramifying of scientific specialization and impatient with arid Baconianism, welcomed the unifying concept of *Vestiges.* So did many Americans. In Springfield, Illinois, that quintessential American layman Abraham Lincoln "read it carefully" (according to his law partner), found in it a professionally congenial proof of "miracles under law," and "adopted the progressive and development theory" it embodied. Some American scientists, even in the South, were initially rather taken with it. Young Joseph LeConte of Georgia read *Vestiges* with "great interest" in the spring of 1845 and "fervently discussed it." The first few American reviews gave it casual praise.

Then its philosophical implications dawned on the scientific and theological worlds. They saw the danger to their entente and counterattacked in unison. Pious scientists, denouncing the book as ungodly, had no choice but to show also that it was unscientific. Fortunately for their purpose, factual errors, some of them ludicrous, abounded in its pages. It gave credence, for example, to the idea that complex living creatures were now and then spontaneously generated. More fundamentally the author offered no clear explanation of how evolution operated if not by divine fiat. By 1850 dozens of reviews and fifteen books had done battle for both God and science against "Mr. Vestiges." In an 1845 volume of rebuttal, amendment, and elaboration, Chambers (still anonymous) further enraged his scientific critics with the charge that they had become too narrowly specialized to judge so sweeping a thesis. This volume stung Asa Gray to a long and damaging review (though also to serious private thinking about the questions raised). James Dana had initially given *Vestiges* a brief but respectful notice in Silliman's *Journal* on the basis of a report by Silliman. Once he read the book himself, however, Dana attacked it and kept kicking at it intermittently in the *Journal* for more than a decade.

The scientific community's nearly unanimous rejection of *Vestiges* and developmentalism in favor of special creation seemed to cement the theology-science entente once and for all. Edward Hitchcock proclaimed complete harmony in 1851. But the theologians, in their eager acceptance of scientific backing, had implicitly conceded jurisdiction. And in their warnings they had incautiously conceded further that a verdict for development would relegate God to the status of remote First Cause, or even displace him altogether. That day of scientific judgment was much nearer than the theologians could have imagined in the flush of their victory over *Vestiges.*

The unreliability of scientific allies was manifested when the theologians' foremost scientific champion against developmentalism, Louis Agassiz, defected in the matter of the unity of mankind. Ironically, Agassiz's motivation, unrecognized by himself, was not scientific truth-seeking but race prejudice. He could not bear to admit that blacks were his cousins in even the remotest degree.

Racism polluted the mind of antebellum America from top to bottom, North and South, including the most reputable of scientists. "I do not like dusky company," wrote Alexander Bache. In Philadelphia, City of Brotherly Love, the racist skull-collector Samuel G. Morton stood high in the scientific community, and the noted zoologist Samuel Haldeman judged free blacks "worthless" as a whole on the basis of Morton's measurements of skull capacity (now known to have been fatally though unwittingly flawed by racial bias). All his life Joseph Henry remembered being flogged as a boy because he would not apologize to a black woman servant for his improper language. Five per cent of Albany County's people were still enslaved in 1800 when Henry was a boy there, and in his prime he moved to Southern-oriented, proslavery Princeton. For these or other reasons he often expressed his conviction that blacks were inherently inferior.

Agassiz became a crony of Bache and Henry, and he got on famously with Morton from their first meeting. But his violent antipathy toward blacks seems to have been triggered by his first close encounter with them, in his maturity, as servants in a Philadelphia hotel. In a near-hysterical letter to his mother he poured forth his revulsion at their strange, un-Swisslike features, his horror at the thought of close contact, his forecast of ruin for the United States from the mingling of whites and blacks, his pity for so degraded and degenerate a race, his instant conviction despite previous belief in human unity and brotherhood that he could not be of the same blood as they—a painful conviction, *"mais la vérité avant tout."* And so on and on.

Placing *"la vérité"* before the Bible, Agassiz began suggesting that blacks were a separately created species. By the end of 1846 Asa Gray was uneasily assuring John Torrey that Agassiz's remarks on the races had been distorted and that he was not subverting religious doctrine. But Agassiz's visit to Charleston a year later led him further down the racist road. Ignorant of anthropology and animal zoology, as John Bachman quickly discovered, Agassiz judged his cultivated hosts to be a different species from their fieldhands. By then he could write of blacks as mere specimens: "I have particularly attended to a careful examination of the negroes from Africa, as far as there are any left, & to collecting insects." At Charleston he referred publicly to blacks as a separate "species." Elsewhere in public until 1857 he discreetly used the more equivocal "variation" or "type." Nevertheless, by 1850 he had, in Torrey's words, "got himself into an unpleasant controversy with the religious part of the community" on the subject.

Agassiz was not alone in his delusion. The first American scientist to assert the doctrine of separately created human species was probably an Alabama physician, Josiah Nott, in 1844. By 1847 Samuel Morton had come openly to that point. As an ardent foe of abolitionism, Nott became a leading propagandist for the thesis. The Reverend John Bachman spoke for the scientific opposition, but the difference between his views and Nott's was narrow. Both men were proslavery. Both believed blacks to be inherently inferior. Bachman merely considered blacks a permanent variation rather than a separate species.

As it happened, Nott's thesis did not figure largely in the slavery controversy. The South held to the Bible, quotable for proslavery purposes in other respects anyway. The North considered scientific opinion irrelevant to the moral issue. So as the political issue waxed hotter, the scientific dispute cooled. Gray, Dana, and some other scientists rejected Agassiz's contention on scientific as well as religious grounds (another count against Agassiz in Gray's mind). Morton died in 1851. Bachman and Agassiz let the matter drop in the late fifties. Though most American scientists still leaned toward Nott's position, in the long run the "American school of ethnology" was significant mostly for investigating hybridization as a mechanism of variation and for weakening religious fundamentalism.

. . .

Obviously what Andrew D. White later called "the warfare of science with theology" was far from that in antebellum America. The two parties kept company as peaceably as most oddly matched couples. In their occasional spats, American theology usually played the temporizer, not the tyrant. It was theology that finally invited science to mind its own business, and science that sulked at being spurned. Such, at least, was the central issue in the most revealing squabble of that period.

The set-to had been foreshadowed in the mid-thirties by Moses Stuart, Professor of Sacred Literature at Andover (Mass.) Theological Seminary and the most influential Congregationalist of the time. He argued that divine revelation governed only the Bible's doctrinal and ethical substance, on which science could have nothing to say, and that the Bible's references to natural phenomena revealed nothing but the state of mortal knowledge in that remote past and so had no bearing on science. Though aimed at Edward Hitchcock, Stuart's remarks had little effect on the natural theology movement. But in 1855 they were echoed and amplified in *The Six Days of Creation* by Tayler Lewis, professor of Oriental languages and Biblical literature at Union College.

Though not a professional in either science or theology, Lewis had long taken a deep interest in their relationship. He offered his book primarily as a philological interpretation of Genesis, but he could not refrain from acidulous comments on the wrongheaded meddling of scientists in Biblical exegesis. He even gratuitously denied that scientific knowledge was cumulative and doubted that it contributed to social and moral progress. James Dana, devoted to science and deeply religious, took all this as a professional and personal affront. "Tayler Lewis's work is one of the worst Infidel Publications that has appeared," he wrote Asa Gray in 1856. "His arrogant system pronounces Bacon's Novum Organum nonsense, inductive philosophy a road to error, and men of science chasers after bubbles." So Dana fired a counterblast of 166 pages, "Science and the Bible," in four issues of the theological journal *Bibliotheca Sacra,* 1856–57.

On specific points Lewis and Dana took much the same position, though each called the other infidel. Both construed the "six days" of Creation as long epochs. Both accepted the unity of the human species. They even agreed that science could not illuminate first causes nor the origins of life. Lewis, however, implied that a theory of natural biological development might be true and would not conflict with faith, though he did not endorse it. He did flirt with developmentalism in suggesting that God intervened only on rare occasions, letting nature develop on its own otherwise. To Dana, who insisted that God's control was direct, ceaseless, and necessary, this smacked of *Vestiges* (though Lewis expressly rejected that book). So did Lewis's assumption of a continuing spontaneous generation of life. Dana scoffed at Lewis's "peculiar notion of a huge self-acting something, now and then aroused to progress by God." But the heart of the debate was the question of whether or not science had anything to offer religion. Not so much explicitly as implicitly in what he wrote, Dana maintained that it offered much and that religion needed it.

Dana's assault on Lewis was cheered by his fellow scientists, even Lewis's Union College colleague, the chemist Charles Joy. In 1859 the retiring president of the American Association for the Advancement of Science took pains "to exonerate this Association from all suspicion of undermining . . . faith." But the local newspaper remarked during the meeting:

> The masses have . . . a wide and deep belief in the power, the beneficence, and what we may venture to call, the divinity of science. . . . This is the beginning of a new reign of knowledge. . . . The light that will shine forth from [the Association], as the painters of the Nativity illumined [that] scene . . . , may be the advent of that era.

The star of Darwinism stood in the east. More than contradicting religion, science already showed signs of supplanting it.

Dana's furious defense of natural theology had been in a lost cause. The nebular hypothesis, the predicted discovery of Neptune, Lyell's uniformitarianism (which even Dana accepted in geology), *Vestiges of Creation,* and parallel tendencies in society all had paved the way for quick acceptance of Darwinism by both the public and the scientists, with the clergy close behind. In 1873 Agassiz conceded that it had won "universal acceptance." Dana reluctantly came round to it, though like other American scientists he placed neo-Lamarckian purpose and will above mindless natural selection as a mechanism. In 1871 he wrote Lewis an almost apologetic letter accepting Lewis's view of God as First Cause and lawgiver, and Lewis graciously forgave him his earlier onslaught. As late as 1885 Dana still insisted that science could throw light on the Bible; but long before that, in reviewing Dana's 1863 *Manual of Geology,* Asa Gray had stated the position that would be taken by modern science, one much like that of Moses Stuart and Tayler Lewis. "We have faith in revelation, and faith in science, in each after its own kind," wrote Gray, but scientists had no call to reconcile the two. The public seemed to agree.

Science appealed to the public's yearnings for national prestige, cultural betterment, tales of wonder, and philosophical light. It also held out a more tangible lure, that of limitless material benefits. Earlier in the century the general public in both England and America had been skeptical on that point. As late as 1843 in the scientific center of Philadelphia one devotee complained that "sordid and ignorant men" could still "superciliously demand . . . what is the use of all this," and in the England of 1850 some "practical men" still scorned science as "speculative rubbish." But by 1855, even in the scientifically backward South, at least one scientist was encouraged to rejoice that "the progress of practical science" had finally disabused all but "the unreflecting & ignorant" of such doubts.

The new faith owed much to popular identification of all technology with "practical science." The notion that science might *sometimes* yield material benefit was, of course, far from new. Francis Bacon took it for granted in the seventeenth century. It became a tenet of the eighteenth-century Enlightenment. What was new was the dubious but spreading popular assumption that *everything* in technology was rooted in science. Daniel Webster sweepingly ascribed the "progress of the age" to the application of science, and an Indiana congressman gave credit to "scientific investigations" for "the most valuable and productive of the arts of life, the most important and wonder-working inventions of modern times." In 1841 the eminent Joel Poinsett solemnly averred that a half-century's improvements in "all that contributes to civilization and to the comforts and conveniences of life, are due altogether to the application of science."

Professional scientists and technologists, who surely knew better, heartily endorsed the delusion, presumably because they wished it were true, hoped it might be some day, and in any case knew it to be in their interest. Thus the head of the Rensselaer Polytechnic Institute claimed in 1855 that "science has cast its illuminating rays on every process of Industrial Art." Since the Franklin Institute of Philadelphia had been founded in part to demonstrate "the intimate connection which exists between science and skill in the mechanic arts," it is not surprising that its spokesmen sometimes stretched the truth in that direction. A reputable scientist, Frederick A. P. Barnard, may have had in mind the interest of the university he headed when he asserted in 1868, "The applications of truths of science to the industrial arts are so numerous, that there is scarcely any article we handle in our daily life, which does not furnish an illustration."

Such claims did not pass unchallenged. In 1852 a commentator on the Crystal Palace Exhibition in London put them to the test. After considering coal, ore analysis and refining, iron and steel, alloys, chemical exhibits, canning, vulcanizing, steam engines, waterwheels, pumps, printing presses, machine tools, firearms, textile machinery, gas lighting, naval architecture, civil engineering, and the Crystal Palace itself, he reported having found almost nothing useful derived from or even suggested by professional science. Unabashed, the industrial chemist Benjamin Silliman, Jr., in 1853 decried "the custom, in quarters where a better spirit might be looked for, to ridicule the claims of science, and deny the obligations it has conferred upon industry. . . . It needs only a slight investigation to discover how groundless such denials are in reference to the past." Silliman's investigation must have been slight indeed, since as his leading example he offered the cotton gin.

Silliman need not have worried, however. By then the public had taken the lure and was solidly hooked. To understand its attitude toward science, therefore, we must consider its reactions to technology.

There were dissenters from the worship of technology, to be sure. "*Useful Application* is the name of the great engine working on the railway of civilization," commented an English periodical in 1848. "Men . . . bestride the colossal machine . . . and in ignorance sit upon the safety valve. . . . Carried onward with precipitous speed, and in delirious joy . . . they fancy they are moving a world, [only to end in] some miserable wreck." In the same year an American magazine put it more jauntily:

> We're rushing on at rail-road speed,
> (Look out, friends, when you hear the bell!)
> But where the terminus will be
> 'Twould take a conjurer to tell!

And in 1863, with further evidence in hand, young Henry Adams tried some conjuring:

> Man has mounted science, and is now run away with. I firmly believe that before many centuries more, science will be the master of man. The engines he will have invented will be beyond his strength to control. Some day science may have the existence of mankind in its power, and the human race commit suicide by blowing up the world.

Moreover, some historians, notably Leo Marx, have surmised widespread subconscious popular misgivings, citing as evidence the frequency of terror and menace in antebellum metaphors of technology. During the 1840s, for example, Americans and visiting Europeans sometimes referred to locomotives as "iron monsters" or "dragons." In his *American Notes* of 1842 Charles Dickens saw one such engine as a "mad dragon . . . screeching, hissing, yelling, panting." In rebuttal John Kasson has suggested that such metaphors may have been merely in the Romantic convention of evoking the "sublime" through terror. The words "awesome" and "awful" are close kin, after all, and "God-fearing" was a stock term for "pious." Even some scientists resorted to such conventional rhetoric. Agassiz commented on "something infernal in the irresistible power of steam." Dana remarked "there is terror in its strength even now," though knowledge and care could tame it. John Frazer wrote in 1845 of a locomotive train "rushing over river and valley and through the very bowels of the mountain, making its scared echoes reverberate with its warning scream!" None of these seems to have otherwise recoiled from technology, let alone science.

But, as Kasson admits, real fears may also have prompted the metaphors, even in the scientists' case. And quite explicit apprehensions were voiced by certain literati whose works, like all great literature, belonged to the telltale dream life of their times. American editions of *Frankenstein* and *Faust* came out in the 1840s, and those motifs haunted Nathaniel Hawthorne and Herman Melville. The forces released by technology evoked in both writers a sense of the mystical and supernatural, an instinctive resort to Satanic imagery, an echoing of folklore and classical mythology in their dread of the knowledge that outruns wisdom and challenges divinity. The tale of Pandora and the forbidden box lodged in Hawthorne's mind for years before he exorcised it as a children's story. "Always," wrote Melville in the fifties, "machinery strikes strange dread into the human heart, as some living, panting Behemoth might." In his short story "The Bell-Tower," a mechanical clock-figure struck its maker. "So the blind slave obeyed its blinder lord; but, in obedience, slew him," wrote Melville. "So the creator was killed by the creature." To Hawthorne and Melville, nineteenth-century technology seemed to be turning men into machines, and the machine into God—acts of ultimate destruction, temporal and spiritual.

Less grandly, Hawthorne and Melville saw technology as a violation of tranquillity, an intrusion of the machine into the garden, a threat to the pastoral ideal of what America should be. Eating the fruit of knowledge would bring expulsion from Eden. Both Melville and Hawthorne also had seen in England what Blake called its "dark Satanic Mills," and Melville realized better than Hawthorne how completely such factories would transform the social order as well as the landscape.

Ralph Waldo Emerson wavered. At first entranced by technology, especially railroads, he began to wonder who rode and who was ridden. He granted that technology was in some sense a part of nature. Its works stirred the imagination, unified the nation, strengthened social control, and made more land accessible and fruitful, thus extending rather than blighting the pastoral ideal. And yet it might tend toward materialism and dehumanization. "The machine unmans the user," he wrote. In the 1850s he grew surer that it debased humanity's vision and lowered its moral stature. Still, he called for moral and spiritual leadership to govern technology, not reject it.

Over and against such forebodings ran a swelling current of American faith in the unalloyed beneficence of technology. Another mythic figure thrust Eve and Pandora, Midas and Icarus, Faust and Frankenstein into the background. Emerson himself could not help exclaiming, "What a range of meanings and what perpetual pertinence has the story of Prometheus!" To Americans the stealer of fire from the gods triumphed in the final reckoning. The eternal homage of mankind more than made up for the eternal torment to which he was condemned. And they paid homage to the successors of Prometheus among themselves. The editor of the *United States Magazine and Democratic Review* opened his January 1846 issue with a paean to the physical sciences and their benefits to mankind. For "the Prometheus of the nineteenth century," he declared, "no rewards are esteemed excessive, no dignities too exalted."

Scientists, being human, snatched at the homage with little thought for the penalty. James Dana spoke for them when he assured the Yale alumni in 1856: "Science is an unfailing source of good. . . . Every new development is destined to bestow some universal blessing on mankind." His only concession was to add: "If evil appears mixed with the good, . . . it is so mixed in the heart of man, and this is its only source." (That, of course, was a fatal concession.) The scientists dimly saw but shrugged off, disavowed, or explained away the looming technological demons of our own time, environmental disaster and military annihilation.

Lyell and Agassiz lamented the industrial pollution of Europe's air as compared with that of America. Another scientist fled Cincinnati in 1851 to save his health from the industrial smog already beginning to blind Ormsby Mitchel's observatory. A Cleveland naturalist that year sadly reported the extinction of whole species of Mollusca in the industrially polluted Cuyahoga and other rivers in Ohio. None of these carpers blamed science, though their brother scientists in geology and chemistry were quick to claim credit for advancing the industries that generated the pollution. (It must be said, however, that an ardent amateur scientist, the American diplomat George Perkins Marsh, has been called by a modern historian "the first in any country to raise his voice in protest against the abuse of nature" with his classic book of 1864, *Man and Nature.*)

As early as 1835 President Francis Wayland of Brown University wrote of "science and the arts [i.e., technology] furnishing means of destruction before unknown and capable of gratifying to the full the widest love of slaughter." Yet scientists evinced no qualms when scientific methods—systematic testing, controlled experiments, laboratory research—proceeded to turn the art of war into military science. Metallurgy and ballistics joined the sciences. In 1843 Joseph Henry introduced the electrical method of determining projectile velocity, while Daniel Treadwell, Rumford Professor at Harvard, busily promoted his newly designed steel cannon. An Anglo-American quarrel in 1846 moved young Wolcott Gibbs to suggest applying John Ericsson's new screw propeller to submarine vessels. Other scientists that year were fascinated by Schönbein's discovery of guncotton. Silliman's *Journal* and the *Journal of the Franklin Institute* reported on the new explosive at length. James Rogers experimented with it. Charles Jackson claimed credit for it, along with surgical anesthesia. Even the pious and peace-loving Reverend Moses Curtis, a respected botanist, liked to "flash a little now and then for the curious." The scientists were not war lovers, but neither were they principled pacifists, though the reform movements of the 1840s included an active and eloquent antiwar crusade. If the scientists had felt the need of apology for "furnishing means of destruction before unknown," they might have echoed Benjamin Franklin's hope of 1784 that invasion by balloons would make a future war unthinkable—a line of argument regularly advanced between wars ever since. Or they might have pointed out that some wars are just and necessary, as did an English scientist who credited science with saving Christendom from the infidels by coming up with Greek fire.

It did not occur to the scientists or to the public at large, however, that any apologies were in order. Unreserved praise of technology or "practical science" went far beyond its contributions to material wealth and comfort. Scarcely any aspect of society or culture, it seemed, had failed to benefit from it. The telegraph, for example, would lessen "the cares, anxieties, and expenses of a government." Today that might suggest efficiency in suppressing individual liberties, but the point was not made then, though tyranny flourished in many nations. On the contrary, Denison Olmsted insisted that improved transportation, communication, and production would elevate the masses intellectually and politically as well as materially and thus strengthen democracy. Democracy in turn, as Tocqueville and others maintained, gave applied science freest play and greatest incentive. Scientists predicted, moreover, that technology would further aid democracy in America by overcoming sectionalism and prejudice, hence preserving the Union.

Technological unemployment held no terrors for Americans. American workers welcomed the labor-saving technology that Europeans resisted. The elastic demand of American consumers, untrammelled by class-bound inhibitions and freely expanding in a broad and richly endowed territory, absorbed increased production without worker layoffs. Furthermore, as the *Scientific American* liked to point out, new technologies created whole new callings.

Technology would even work for world harmony, democracy, and peace, it was believed. Alexander Bache viewed the pursuit of science in itself as a bond of international fellowship. Others saw science-based industry as making worldwide interdependence profitable and science-based communications as making it feasible. Improved communications would also spread religious and moral light, mutual understanding, and democratic aspirations throughout the world, they supposed.

By mid-century, popular enthusiasm for technology (and by extension, science) was being further whetted by ritual celebrations in the form of public exhibitions or fairs. At New York, Philadelphia, and Baltimore, technological or "mechanics" institutes held annual exhibitions of manufactures and inventions. In 1846 a group of manufacturers got up a "National Fair" of industrial products at Washington. President Polk himself joined the crowds, despite the pressures of war with Mexico, and left impressed with "the genius and skill of our countrymen." The year

1850 brought a spate of articles and sermons acclaiming the half-century's progress in science and technology, thus priming the American mind for what Lewis Mumford has called the machine age's "cock-crow of triumph"—the great Crystal Palace Exhibition of 1851 in London, the first of the modern world's fairs.

The Great Exhibition signalized England's shift from distaste for industrialism to confident acceptance of it. No dark Satanic mill enshrouded the machinery exhibits. The light of day shone upon them through Joseph Paxton's great glittering cage of iron and glass called the Crystal Palace. Some fourteen thousand exhibitors, half of them from other nations, displayed a hundred thousand products. Dedicated to international peace, the Exhibition purposely limited military items. Only one small cannon lurked amid the Krupp exhibits. The lately discovered element called uranium figured in an oxide used to color glass, with an effect pronounced "charming."

Scientists—Joseph Henry, Walter Johnson, and Charles Wilkes—formed a majority of the five-man unofficial commission set up to plan American participation. Henry had misgivings. Congress provided no money, and so as its agent in London the commission settled for a Boston horse-trader and auctioneer who served unpaid in order to peddle his own goods and, for a consideration, promote those of others. Delays, inefficiency, and sundry mishaps threatened to end in a fiasco. One American visitor sourly remarked that the American section "well represented our country, being a large space only partly filled." But among the thinly scattered American exhibits were a few eye-catching inventions that turned initial English scorn into admiration—Samuel Colt's revolver, Alfred Hobbs's lock, Gail Borden's dehydrated "meat biscuit," and especially Cyrus McCormick's reaper. Only 6 of the 166 Council Medals went to Americans, but they included medals to Borden, McCormick, and Charles Goodyear. William and George Bond won a Council Medal for American science with their astronomical chronograph.

So American pride was salved, and the Exhibition as a whole overwhelmed those Americans who saw it. Young Josiah Whitney could not find words for its "magnificence & extent." And American stay-at-homes were deluged with newspaper accounts, woodcuts in illustrated papers, panoramas and lithographs, and a stream of magazine articles.

Since among its other glories the London exhibition turned a profit, the American commissioner promoted another world's fair for New York City in 1853, state-chartered but privately financed. Rejecting designs by James Bogardus, the pioneer of iron-frame buildings in the United States, and by Joseph Paxton himself, the building committee successfully adapted the technique and spirit, though not the heroic scale, of Paxton's Crystal Palace. The New York Crystal Palace opened so far behind schedule that it ultimately lost money. The delay brought sharp criticism, and Joseph Henry looked for nothing but "humbug" from the affair. Yet in the end it made an impressive showing.

European visitors and commissioners found Americans leading the world in the variety and specialization of their machinery. The night spectacle of the glass and iron building ablaze with gaslight entranced visitors, such as the young geologist William Brewer. Scientific instruments and "products resulting from their use" made up a large class of exhibits, including optical instruments, pumps, measuring devices, clocks, microscopic slides, daguerreotypes, dental apparatus, and patent eye-cups. In Charleston, South Carolina, John Bachman wrote that "the city is travelling & Crystal Palacing." Young Samuel Clemens, at seventeen some years from becoming Mark Twain, made his way from the Mississippi just to see the New York fair. "A glorious sight," he wrote home in his earliest surviving written composition, "a perfect fairy palace—beautiful beyond description."

By mid-century, in short, technology was riding high and carrying science with it. The professionals themselves verged on hubris. "Never was there such an age of progress," wrote a leading civil engineer in 1849, "and it has hardly commenced yet." From Pulkovo Observatory a young American astronomer wrote home in 1866 that "if we do not hitch onto the moon and quarry our granite there, it won't be the fault of the Yankees."

So ran hopes and fears, premonitions and dreams in that exhilarating time. The realities, as near as history can get to them, make another story: of how antebellum science actually earned its keep, of how far technology paralleled, leaned on, and supported it.

Chapter 10

The Wherewithal of Science

Some scientists, like Louis Agassiz, courted public favor because it warmed and reassured them, but all of them needed it to be professionals, to live by their work. Scientists appealed to philanthropy, peddled their expertise as teachers, offered their science for pay to industry, agriculture, and government, and occasionally even exploited it as entrepreneurs.

The day of the self-financed amateur was passing. In 1846, 15 per cent of the leading scientists (those in the *Dictionary of American Biography*) were simon-pure amateurs, drawing no income at all from science-related work, but by 1861 the proportion had fallen to 9 per cent and by 1876 to 4 per cent, while those dependent wholly on science-related work rose from 60 per cent in 1846 and 1861 to 70 per cent in 1876.

A successful businessman could spend both time and money on science and might even retire early to give it his full time. In a family enterprise one member might be allowed time off for science, as with Samuel Haldeman and Leo Lesquereux. A doctor such as Franklin Bache might scant his practice to pursue science, or he might turn his duties to account as did Thaddeus Harris when as a country doctor he paused in his rounds to incarcerate insects in his pillboxes.

But conscience, need, or cupidity held most such cultivators more strictly to their other duties. Dr. George Engelmann had to let his botany go by the board whenever cholera visited St. Louis. The Reverend John Bachman got up at four in the morning to serve both God and mammalogy. "I am not sure that I can stand it this way," he wrote in 1845. The double life of business and zoology killed Amos Binney in his early forties. An Ohioan in 1849 complained that "entomology with me encroaches fearfully on business." And the growing complexity of science demanded more and more time. A New Jersey lawyer mourned in 1853 that he could no longer keep up with his beloved science; "as facts are multiplied and hypotheses increased . . . I discover that I am left hopelessly behind." After retiring, Isaac Lea found himself working longer and harder in natural history than he had in his business.

In a sense, those who spent time and money on science without compensation might be called philanthropists. The nation's economic growth seemed to promise weightier benefactions. Millionaires multiplied—eleven in Philadelphia in 1846, rising to twenty-five in 1858; fourteen in New York in 1846, rising to twenty in 1855. Boston, with eighteen millionaires in 1850, did its duty nobly, urged on, as we have seen, by civic pride, personal taste, economic interest, and a strong sense of stewardship. Elsewhere, however, philanthropy remained coy and elusive. Though the banker George Peabody eventually gave half a million dollars to science, before the Civil War he spared only a thousand for a chemistry school and lab at the Maryland Institute.

The most revealing fund-raising foray was that of the young Arctic explorer Isaac Hayes in 1859 and 1860. Seeking thirty thousand dollars for a new expedition, he found it "exceedingly hard . . . to get at men's pockets." Federal aid was not forthcoming, and though Alexander Bache chipped in a thousand dollars of his own money and Joseph Henry matched it from Smithsonian funds, Hayes had to look mainly to philanthropic laymen. Philadelphia businessmen thought the scheme too risky and none of their concern anyway; the scientists, they said, "ought to get it up" by themselves. In St. Louis, benefactors of science felt that their charity should begin at home. In New York, August Belmont and other millionaires gave only one hundred dollars apiece. Only the munificence of the New York shipping millionaire Henry Grinnell permitted the Hayes expedition to sail at last for regions that could not have seemed much colder than the American public.

Teachers made up nearly half of all *DAB* scientists in 1846 and slightly more than half in 1861 and 1876. Their salaries varied widely. In the mid-forties small colleges like Bowdoin or Wabash paid $600 a year, while larger, better-endowed, or state-supported colleges ranged from $1,200 to $2,300. At Harvard, Louis Agassiz and Eben Horsford got $1,500. In comparison, leading Boston physicians in 1845 made about $9,500, while Massachusetts grammar school masters got $1,500 in Boston and $1,000 in other large towns.

In the prosperous mid-fifties, college salaries rose, especially for science teachers, who were in growing demand. Salaries in general now ranged from $1,200 to $3,000 in New England. Asa Gray got $1,800 and a rent-free house, and Agassiz's pay was raised to $2,000 in 1859. In the Middle Atlantic states, salaries ran a wider gamut from $750 to $3,000, Columbia offering $3,000 and a rent-free house. In the South Atlantic states, salaries ranged from $500 to $2,500, the higher levels being at large, state-supported colleges, which catered to a politically dominant, landed elite. In the Southwestern states the high was about $2,000, and the Midwest stood lowest with a high of only $1,200, though the region's rapid growth brought improvement after the mid-fifties.

The differentials between regions were tempered by variations in living costs. In New York City during the mid-fifties a professor and his wife needed $2,500 a year to live comfortably. Columbia counted a rent-free house as equivalent to $1,000 a year, whereas Virginia figured it at $300. In Ann Arbor, Michigan, room and board for one man cost only $3 a week as compared with $15 in New York City. Though most colleges were in small country towns, these also varied somewhat in living costs, Princeton being more expensive than Charlottesville, for example.

Wherever they were, science professors complained of being underpaid (though in fact they often fared better than professors in other fields). At Princeton in 1846 Joseph Henry's pay fell $250 to $300 short of what he needed to live on. At Harvard in 1849 John W. Webster's financial desperation led him to brain his most bothersome creditor and render the corpse in the chemistry lab; the hangman subsequently put an end to Webster's living expenses. Agassiz warned his students to learn thrift if they wanted to be scientists. At Rutgers in the early fifties, George Cook earned $1,600 but rarely got paid on time; once he had to wait six months. To make ends meet he took a second job with the New Jersey geological survey. Among other moonlighters, John Torrey held two professorships and a job in the New York Assay Office simultaneously.

Some colleges experimented with the medical school system of letting professors collect fees from their students in lieu of salaries. The Universities of Missouri and Virginia and the Lawrence Scientific School at Harvard tried a mixed system of student fees plus a small salary, then abandoned it. In retrospect Charles W. Eliot called the Scientific School's fifteen-year trial of the system not only "theoretically unwise" but also "practically disastrous." Professors found it both uncertain and demoralizing, given that some students would pay for knowledge, others for exemption from it.

Teaching loads varied from ten to fifteen hours of lectures and recitations a week, ten being more common in the stronger schools. In small colleges especially, the teacher shouldered a greater variety of courses than now, but they were correspondingly more superficial and so took less time to prepare. Teachers with the heavier loads nevertheless complained of being "overworked bodily & mentally" or "actually used up in the immediate labor of teaching." They also did work now relegated to lab or teaching assistants or even janitors. And they often acted as campus police, like the math professor at Davidson College who in 1854 quelled a mob of rock-throwing students by charging them with his Mexican War sword upraised.

All this left most science professors with little time, energy, or stimulus to do research. In 1845 Joseph Henry chafed at having to break off "an interesting course of experiments" at vacation's end and return to the drudgery of teaching and exam grading. With few if any advanced students, professors grew jaded with drilling students in the bare elements of several sciences. Some relief came when increases in teaching staffs permitted more specialization. By 1860 most colleges had four or more science teachers. At Union College in 1854 Charles Joy exulted that his duties, "to teach *only Chemistry,"* were "just what I should like to fulfill were I worth a million." His luck held when in 1856 Columbia divided one of its chairs into physics and chemistry and named Joy to the latter. But even this boon offered only limited opportunity for serious research.

In 1844 and 1851 Alexander Bache exhorted American colleges to allow professors more facilities, time, and money for research—in short, to recognize research as one of the professor's regular duties. Then as now, however, academicians were themselves divided on the proper place of research in their profession. Josiah Cooke of Harvard insisted that good teaching took "a great deal more thought . . . than nine tenths of your 'original investigations.' " In an address later published, Denison Olmsted of Yale said that the ideal teacher should be good at both research and teaching, but that the breadth of knowledge required for good teaching made it necessary to forego some depth and single-mindedness in research. For years Olmsted's pamphlet rankled in Joseph Henry's mind. He called it "a plea for stupidity or an apology for dunces." Henry believed that no "second hand teacher" could inspire a class with the "enthusiasm" and "breadth of thought" of an "original investigator." Besides, he wrote, a professor's outside reputation for research would give him power within the college and reflect honor on his students.

Whatever the pros and cons, it remained largely true that, as the *New York Quarterly* observed in 1853, American colleges did not expect professors to do original research. With rare exceptions like Asa Gray, Eben Horsford, and Louis Agassiz, professors were not hired on their reputations as researchers. Charles Jackson noted in 1844 "how often their standing is very low in science" and thought that personal influence, cliques, and religious affiliations had most to do with their appointments. Throughout the antebellum years it helped to be an alumnus, especially one favored by the president or trustees. One man appointed to a science chair by Harvard on that basis had turned out by 1846 to be totally incompetent. Southern colleges were partial to Southerners. In a few cases like those of Gray and Horsford, teaching loads were lightened to encourage research, and Benjamin Gould in 1849 saw such cases as "favorable omens." But antebellum colleges proved slower than he had anticipated in recognizing research as a normal function of professors.

Still, as Bache pointed out in 1844, the colleges did help to develop public appreciation of science, and they did "give their daily bread . . . to its cultivators." Thus sustained, even without encouragement to do research, college teachers of science managed, as earlier noted, to account for about half of all antebellum scientists memorialized in the *DAB.*

Science professors occasionally found ways to earn money on the side through their science, depending on their location, reputation, scientific field, and aptitude for the specific sideline. The proportion who did so was probably small, however. As early as the 1830s some augmented their salaries by serving as expert witnesses in court, as advisors to government on boundaries, road-building, and boiler safety, or as part-time or temporary employees on state or federal geological and natural history surveys. Agassiz earned nearly six thousand dollars from public lectures in 1847; but lecturing did not pay off for most, John Frazer complained, unless they could "read Shakespeare or speak broken English." A few made good money by writing textbooks. The chemist John W. Draper and the botanist Asa Gray earned several hundred dollars a year in textbook royalties at ten cents a copy during the fifties. Yale professors did especially well in that line. The younger Silliman's chemistry text sold more than fifty thousand copies all told, and Denison Olmsted's texts about two hundred thousand. Elias Loomis's texts in several sciences had a worldwide sale and enabled him to bequeath Yale three hundred thousand dollars in 1889. Yet James Dana's *Mineralogy,* the best of the Yale texts, sold less than five hundred copies a year in the mid-fifties. And serious scientific books earned little or nothing. Even so practical a work as the chemist Charles Wetherill's standard treatise on the manufacture of vinegar sold under a thousand copies, short of the minimum required for royalty payments to commence.

Part-time income from private enterprise was nevertheless becoming significantly more common and lucrative, especially for professors of geology and chemistry. Indeed, in the antebellum period about one in every seven *DAB* scientists earned his livelihood chiefly from private enterprise rather than education or government. Leaving those in government (slightly more than a fourth of the total) for the chapters on science-government relations, let us now consider those who worked for industry, agriculture, or commerce. We may thereby also gauge, field by field, the extent to which science was beginning, though still only beginning, to meet the public's soaring expectations of material benefit.

Of the scientists who served private enterprise, the geologists were the most conspicuous. They seemed at first to have a map to the buried treasure that had once drawn Europeans to the New World and was now beckoning again. Mining fever did not wait for the days of '49. In the mid-forties Americans rushed to the new-found copper deposits of Michigan's Upper Peninsula. From there young Josiah Whitney wrote in 1845, "it is enough to sicken one to hear the everlasting copper copper copper from all tongues." It was also enough to give him a summer job as geologist for a mining company. In New York City three years later, a lecturer on popular science reported that California "gold fever rages here dreadfully" and had brought him an engagement to lecture on gold-bearing rocks. The mere rumor of a new tin mine in North Carolina in 1860 turned one young man's thoughts away from chemistry and toward geology.

A like turn gave Whitney a career. In 1847 Charles Jackson, as director of a federal geological survey of Michigan copper lands, brought his young friend Whitney back from Liebig's chemistry lab as first assistant at five dollars a day. During the next three years Whitney served with such future luminaries as Wolcott Gibbs, Charles Joy, John Locke, and John W. Foster. Forsaking chemistry, Whitney then became a leading consultant for mining companies, with more job offers than he could begin to fill. Making five hundred dollars a month, he remarked in 1853, he could not afford to be a Yale professor. A year later he published his widely known book *The Metallic Wealth of the United States,* the first comprehensive work on American ore deposits.

Not all geologists fared so well. A Midwestern mining company paid the brilliant but eccentric James G. Percival only two thousand dollars for a year's work searching out lead deposits that added a million dollars to the value of the region. And after 1854 the commercial demand for geologists levelled off, turning even Whitney to state survey jobs.

In California a recurrent pattern showed itself: unrealistic public hopes for applied science, followed by disappointment and then disdain. As early as July 1849 a Sacramento newspaper smugly reported that unskilled men were more effective in finding gold than "many geologists and practical scientific men." As pickings grew slimmer, some Californians turned back to the scientists, urging a state geological survey or setting up ephemeral societies to study and exchange scientific lore, but little came of this during the fifties. American miners preferred immediate profit to long-term productivity. They scrabbled for glamorous metals and ignored others of greater potential value. They trusted their own luck more than hifalutin experts. And so the Europeans far outdid them in applying science to mining.

So also with the "oil fever" of 1859 and after, though it produced a famous example of science playing midwife to industry. In 1854 a group of businessmen retained Benjamin Silliman, Jr., newly appointed professor of chemistry at Yale, to report on the commercial possibilities of petroleum, till then little used except as a medical nostrum. After his bill of $526.08 was paid, Silliman released his report, published in 1855. It synthesized existing knowledge and supplemented it with Silliman's own analyses. Silliman's one brilliant insight was in surmising that the oddly variable boiling points of some petroleum distillates signified their breakdown by pressure and temperature into several new compounds. In short, he saw the possibilities of "cracking" petroleum. Beyond that he contributed his prestige, his optimism about obtaining illuminants, paraffin, and gas from petroleum, and his systematic compendium of information. It was enough to whack the infant petroleum industry into lusty, squalling life.

Yet in the industry's early years, professional geologists could not agree on the origin and nature of petroleum deposits, their relation to coal measures, and other aspects of their location and working. In consequence, much consulting money went instead to oil smellers, spiritualists, and wielders of divining rods.

Many scientists scorned the profit motive as out of keeping with professional ideals. Joseph Henry high-mindedly refrained from patenting his telegraph before Samuel Morse took it up. Such scruples may have tended to reinforce themselves, since inveterate money grubbers presumably did not go into science or else soon left it. But the hectic growth of the 1850s, the dawn of what came to be known as the Gilded Age, tended otherwise, especially among the economic geologists. Mining was a gamble with big payoffs, and the geologists sometimes knew the results (or thought they did) before the betting closed. So they ran a greater risk than other scientists of contracting the itch to be rich.

Most resisted the temptation. At one extreme, James Percival was too unworldly even to think of making money. At the other, Josiah Whitney suspected everyone's motives including his own; and so he made it a lifelong rule never to hold an interest in any mine he might be asked to give an opinion on, or otherwise use professional information for profit. Between the extremes, geologists usually held their avarice in check. Yet there was a moral loophole through which some squeezed. At the age of seventy-four Joseph Henry himself repented of his early squeamishness, reasoning that the money he had spurned could have done much for science. Geologists sometimes rationalized moneymaking schemes on the same grounds. James Hall (according to his biographer) "allowed himself to be willingly led into an endless variety of land and mining ventures"—Ohio coal lands, North Carolina mining projects, schemes for producing salt, coal, and wine in western Virginia, and so on—in order to pay his assistants and keep building his paleontological collections. In such ventures he sometimes allied himself with Benjamin Silliman, Jr., the most conspicuous example of a scientist distracted and at last nearly destroyed by recurrent money fever.

Silliman's father had been generously paid for geological consulting and had not been above trading on his prestige by accepting the nominal presidency of an insurance company. Young Ben, moreover, marvelled at the wealth generated by the rubber-vulcanizing process of Charles Goodyear, a family friend. Perhaps all that put the taste in his mouth. In 1853, at any rate, Silliman wrote: "I am willing to take any amount of trouble to secure a good fortune, provided I can do so with strict regard to professional etiquette & the highest honor." A few months later he and James Hall joined with others in a Kentucky coal-land venture that strained their high principles. Though Silliman conceded that the field of mining was also an ethical mine field, he tried unsuccessfully to recruit the geologist and ex-clergyman J. Peter Lesley with free stock and the prospect of getting "plenty of sea room hereafter for pure science." Silliman and Hall planned to write a glowing geological report on the coal lands and then sell their interests in England, using a trustee as a screen. In that way, wrote Silliman, "our reports will appear much better, beside the general fact that a man of science loses caste by having his name connected with anything of the sort in a public way." But a few days later, having caught "the principal party" in "serious moral dereliction," Silliman and Hall seized an opportunity to sell out immediately. "Meanwhile," the unchastened Silliman wrote, "if a chance turns up for some such speculation we may still avail ourselves of it."

Silliman plunged at once into his petroleum analysis, then served briefly as president of an oil company. "B. S. Jr. has too many strings on hand . . . to do much chemical work," complained his brother-in-law James Dana in the very month of the famous report. That summer Silliman accepted the presidency of a copper mining company, hoping to make enough so that "I can devote myself to science for its own sake as in days gone by." But Dana presently noted that Silliman was still "on the constant go . . . and alas for science."

The American public expected more of applied chemistry than of any other applied science. They saw Europeans beginning to apply the fundamental principles of chemical action as those grew clearer. Perhaps also there lingered some folk memory of the magic attributed to alchemy. But American circumstances differed from those in Europe. Europeans used chemistry to conserve or supplant scarce natural resources, whereas Americans had raw materials in plenty. The American tanning industry had so much bark at hand that it ignored European use of chemical salts as substitutes. Where Europeans might try using chemistry to stretch out the productivity of old mines, Americans were more apt to look for new mines. Though Silliman's *Journal* consciously tried to give applied chemistry more play in the late 1850s, that field accounted for only about one in fourteen chemical articles in American journals. Furthermore, public perceptions exaggerated the achievements of applied chemistry in both Europe and America. Even in industries based on chemical processes, major advances usually owed more to blind groping.

One of the most spectacular of the century's new industries is a case in point. For almost a century scientists had struggled in vain to make cheap steel. In 1847 a neophyte Kentucky ironmaster named William Kelly found the way not through science but from chance observation. Eight years later the Englishman Henry Bessemer did the same thing in the same way. All that science contributed to the birth of the new industry was a logical explanation after the fact which helped persuade customers to try the process. In antebellum America the iron industry had little use for trained chemists, even for routine chemical control. Not until 1860 did the Cambria Iron Company of Johnstown, Pennsylvania, hire Robert W. Hunt to set up the first permanent analytical laboratory in any American ironworks. In that year Hunt's teacher, James C. Booth, could not get eastern Pennsylvania ironmasters to put up a total of twelve hundred dollars for control analyses of ore.

Other major industries in those years owed little or nothing to chemistry. In the first quarter century after Silliman's report, only about twenty-five American articles appeared on the chemical analysis of petroleum, and most improvements in refining were the work of practical refiners, not trained chemists. A book on American glassmaking in 1865 predicted great advances through "progressive chemical development," but reported little to that time. In 1858 a maker of blasting powder turned not to a chemist but to "a cousin—an old powder maker in this market & he gave me the proportions he used when his powder gave the best satisfaction." Chemists had worked for decades in vain to make rubber commercially useful; but the man who finally turned the trick, Charles Goodyear, admitted that his discoveries "were not the result of scientific chemical investigations" but of random trial and error. Gail Borden used a friend's laboratory equipment in developing his condensed milk, but in total innocence of scientific theory. And neither chemical nor biological science gave any help to the brewing industry until after 1880.

Still there were enough successes to keep hopes alive. James C. Booth and a former pupil, Campbell Morfit, waved the banner in teaching applied chemistry—Booth at the Franklin Institute (1836–45) and the University of Pennsylvania (1851–55), Morfit at the University of Maryland (1854–58)—and in carrying on consulting work. In collaboration they reported on gun metal for the Army Ordnance Department, brought out a chemical encyclopedia in 1850, and published a treatise, *On Recent Improvements in the Chemical Arts,* in 1852. In 1847 a Franklin Institute reviewer of Morfit's *Chemistry Applied to the Manufacture of Soap and Candles* looked for great things from applied chemistry. That year another of Booth's pupils, Richard Tilghman, presented the first systematic study of hydration and later applied it to candle making. Tilghman successfully sued Procter & Gamble for using it in soap making, but the firm simply drew on French chemical findings to the same end.

French chemical studies also gave new life to the old American rosin industry. One of Robert Hare's students established the first American platinum works in 1842. In the mid-fifties Eben Horsford, Harvard's Rumford Professor of Applied Science, applied his science in developing an improved baking powder. With a chemical dealer, he founded the Rumford Chemical Works to make it, and in 1863 he quit his chair to run the company.

In sugar refining, chemical laboratory techniques began to pay off in the mid-forties. In 1845 Richard S. McCulloh's "Sugar and Hydrometers" appeared as a Senate document, discussing not only refinery equipment but also sugar analysis and the construction and use of apparatus, especially the French polariscope. A Philadelphia refinery was already using the polariscope, and a Franklin Institute committee that year praised the firm for its application of science. On a Louisiana sugar plantation, 1845 also brought the first use of Norbert Rillieux's multiple effect evaporator, which a century later was called "probably the greatest invention in the history of American chemical engineering."

The expansion of textile and paper manufacturing in Massachusetts during the 1840s created the nation's most important market for industrial chemicals. The textile industry turned to French-developed chemical bleaches, which in a few hours at any season did the work of months of summer sun. Chemical factories sprang up beside the great print works and dye houses of New England. A Massachusetts physician named Samuel L. Dana, drawn into the work by neighboring textile mills in Waltham, began making sulfuric acid and bleaching powder in 1826 and seven years later joined a Lowell print works as the first trained chemist regularly employed by an American textile firm. There he devised a continuous-operation bleaching process that became world-known as the "American system." To the presumable relief of the workers, he also found that sodium phosphate could replace the cow dung till then used in the calico-printing process.

As more and more industries used artificial compounds, they gave work indirectly to chemists in the manufacture of chemicals, even when employing no chemists themselves. Some European-born chemical manufacturers had arrived with training or degrees in chemistry, and at least one in Philadelphia hired other chemists. In 1846 a Baltimore chemical manufacturer hired William P. Blake, then a Yale student, as a full-time chemist. When Blake actually started work in 1850 he became, it has been said, the first college-trained chemist employed full-time in American industry. He certainly was an early specimen of the native breed.

Most antebellum American chemists who worked for industry, however, did so as occasional consultants, as had Robert Hare, the elder Benjamin Silliman, Samuel Dana and his brother James, Charles Jackson, and James Booth in earlier years. In 1849 the New York consulting firm of Doremus and Harris offered to analyze ores, soils, mineral waters, and "all articles employed in the Arts and Manufactures," as well as to teach students and sell apparatus. Another New York consultant did well by testifying in chemical patent cases.

The late forties and early fifties were good times for chemical consulting. Jackson got twenty dollars a day from mining companies, John Torrey made several hundred a year by what he called "jobs analysis," J. Lawrence Smith took in two thousand dollars a year at Louisville, and John P. Norton at Yale counted two jobs available in consulting or teaching for every graduate chemist. By 1856, however, supply was outrunning demand. With a Göttingen Ph.D., Charles F. Chandler had to work as a janitor at Union College for four hundred dollars a year. Another Göttingen Ph.D., Charles M. Wetherill, thought seriously of quitting science altogether. At Boston in 1857 Francis H. Storer, back from two years of European study, found a Dartmouth graduate monopolizing the consulting business and so had to settle for a job as chemist with a gaslight company.

A few chemists could earn money by opening a private teaching laboratory. James Booth ran one in Philadelphia for half a century with various partners, and Charles Jackson operated one as a sideline in Boston. But even Philadelphia, the leading center of American chemistry, could not support two such enterprises, as young Wetherill found out in the early fifties and Frederick Genth in the late fifties.

Though Booth and Jackson wanted to believe that industrial employment did not rule out contributions to science, industry in fact did not hire chemists for full-time research until the late sixties in Germany, the early seventies in England, and later still in the United States. Until then industrial chemists worked mostly on routine analyses or at best on the technology of production—chemical engineering in embryo. A Massachusetts paint company chemist in 1854 fretted at being "cut off from the scientific world, and from the companionship . . . of enthusiastic fellow-workers." When J. Lawrence Smith laid plans to become a chemical manufacturer in 1856, he saw himself as giving up scientific work temporarily so that he could make enough money to pursue it later.

In agriculture as in industry, the public expected too much from chemistry. The Midwest, gentle in terrain, rich in good land, short of labor, paid more heed to machines. But the East, populous, hilly, much of its soil worn out, eagerly welcomed the renewed fertility that chemistry seemed to promise. The Englishman Humphrey Davy's *Agricultural Chemistry* had raised the hopes of Eastern farmers in the 1830s, and the occasional successes of industrial chemistry encouraged them further. Then in 1841 John W. Webster of Harvard brought out two American editions of Justus Liebig's *Organic Chemistry in Its Application to Agriculture and Physiology,* a great advance over Davy. Its appearance coincided with a rapid efflorescence of agricultural journals and local agricultural societies, all of which helped spread Liebig's doctrines, as did the published letters of Eben Horsford from Liebig's Giessen laboratory. The texts of the Scottish agricultural chemist James Johnston also sold well in the 1840s, and the press fully reported Johnston's American lectures in 1849–50.

Johnston's American student John P. Norton, after becoming professor of agricultural chemistry at Yale in 1847, followed his teacher's practice of addressing farmers directly through lectures and writings. Norton's 1850 textbook, *Elements of Scientific Agriculture,* went through at least four printings; and his lectures to farmers, lucid, persuasive, and couched as simply as possible, brought many doubters around. "I have talked to hundreds of farmers," he remarked offhandedly in one such lecture. He pressed the claims of science, urging farmers to seek help from animal physiology, meteorology, entomology, geology, mineralogy, and above all, chemistry. The state, he maintained, should set up agricultural colleges with research facilities and experimental farms.

In the early forties Liebig had shifted his emphasis from ammonia to phosphates and minerals. So at less than cost, Norton offered chemical soil analyses as a simple and reliable guide to fertility. Not everyone had Norton's public spirit and private means, however, and anyway his offer was to Connecticut farmers only. Self-styled "chemists" and "professors" accordingly began speckling farm papers with ads offering mail-order soil analysis and advice for five or ten dollars. Soil analysis seemed a panacea. State geological surveys gave it further credibility by publishing soil analyses, often copied by farm journals. Then the bubble burst.

Norton's—and Liebig's—claims for soil analysis were simplistic and overly sanguine. Agricultural science faced the immense complexities of living things, whereas industrial science did not. Neither farmers nor scientists, including Liebig, fully comprehended how many factors besides mere chemical composition determined the yield of a given soil and the quality of its produce. Even the soil analyzed was difficult to sample properly, varying unpredictably on the same farm. Samuel L. Dana, who had shifted his attention from industrial phosphates to fertilizers, contested Liebig's narrowly chemical theories as early as 1842 in a bestselling manual for farmers. For years scientists accepted Liebig, who was precise and chemically sophisticated, and rejected Dana, whose ideas were vague and muddled, though on the right track in some respects. But the inadequacy of soil analyses, by quacks or otherwise, became inescapably evident in practice. An article in Silliman's *Journal* acknowledged as much in 1852. Weakened by overwork, Norton was already too sick to reply, and he died a couple of months later at the age of thirty. The farmers' reaction against Liebig and the quacks extended to all chemists, who in turn drew back from agriculture. Not until the 1870s was confidence fully restored.

Nevertheless Liebig had synthesized thirty years of agricultural chemistry, contributed brilliant insights, and laid the groundwork for another generation of progress. The unfortunate Norton had himself helped create an enduring institution in Yale's scientific school and a tradition of public service in agricultural science. And Samuel W. Johnson, Norton's pupil and eventual successor at Yale, soon set to work rebuilding the reputation of agricultural science on a more solid footing.

Norton's immediate successor, John A. Porter, a former Liebig student, helped keep Yale ahead of other schools in agricultural science until 1857, when Johnson took over the agricultural chemistry course and lengthened Yale's lead. Johnson had studied at Leipzig, where he was deeply impressed by the world's first agricultural experiment station, and with Liebig, who was by then jaded and dispirited. Determined not to raise false hopes again, Johnson warned farmers that soil analysis was "a chance game." By 1861 he was calling it "insignificant" as a guide. His prescription was to study the whole science, not just the chemistry, of each plant thoroughly in laboratory and field through rigorous observation and systematic experiment. He held up the German agricultural experiment stations as a model. Johnson established high credentials among scientists in 1860 by correcting Robert Bunsen's figures on certain atomic weights, thus beating a leading German scientist at his own game, and among farmers in 1856 through exposing misrepresentation and overcharges by makers of commercial fertilizers. But his main objective was to update and promote agricultural science in America, not only by teaching but also by articles and reviews in Silliman's *Journal* and *Country Gentleman,* a leading farm journal.

Johnson lacked time and facilities for his own research during the late fifties, while a German-born geologist and chemist, Eugene W. Hilgard, a Heidelberg Ph.D., was making a major contribution to the multifaceted approach Johnson called for. Hilgard's state survey report on the geology and soils of Mississippi made Hilgard a founder of pedology, the branch of soil science that deals with the natural position, forms, and origins of soil. But the Civil War delayed the distribution of Hilgard's report until 1866 and thus obscured its significance.

For agricultural as well as industrial chemists, antebellum America had few full-time jobs. The enthusiasm of the 1840s had not moved the public to fund research positions or experiment stations, nor did philanthropy come through. Beginning in 1847 several states employed agricultural chemists, sometimes with no compensation other than the use of a laboratory in which to earn analysis fees. A few colleges hired them. Before his Yale salary started, Samuel Johnson pieced together a meager living from $615 in student fees at his Yale lab, a $400 salary from the State Agricultural Society, and $155 from articles. But none of these employments, including teaching, left chemists with much time or energy for research.

Of the remaining sciences on Norton's shopping list for farmers, only entomology offered some limited job opportunities before the Civil War. Hoping for state or college jobs, entomologists touted their knowledge for its possible benefits to crop yields. "Much as I despise this sordid test," wrote one, "I am forced to own its necessity when the public is to be enlisted." It was Thaddeus Harris, the Harvard librarian, who made entomology more useful in America than anywhere else. He had little down-to-earth knowledge of farming and little time to experiment. His remedies for insect depredations came from his reading, though his insect descriptions were firsthand. Nevertheless his report on Massachusetts insect pests, published by the state in 1841, became the leading manual in the farmer's war on insects for many years afterward.

The $175 Harris got for his report made him the first American economic entomologist to be paid as such. The first permanent, full-time appointment in the field went to Asa Fitch, New York state entomologist from 1853 until his health failed in the early seventies, at a salary of $1,000 a year. His annual reports on insect life histories were lasting contributions. In his 1861 report he urged control of the wheat midge by introducing its European parasite, but the tactic was not tried. The first federal job went to Townend Glover as entomologist of the Bureau of Agriculture from 1854 to 1878. An eccentric though dedicated man, he scorned systems of classification, merely describing and drawing specimens, then throwing them away. His immense labors consequently left no lasting mark. Beyond the work of Harris, Fitch, and Glover, little more was done in applied entomology during the antebellum period.

If geology, chemistry, and entomology had spotty records as adjuncts to private enterprise, the records of other sciences were nearly blank. Those sciences depended for support almost entirely on teaching, government work, philanthropy, and amateurs. Except for actuaries, mathematics seemed irrelevant even to business, and most life sciences were ignored even by farming and fishing. Many laymen seemed to feel that astronomy had long since done all it could for navigation and now occupied itself with mere "laborious trifling." The new age of railroads did suggest astronomical time checks as a marketable product, and Elias Loomis preached their value. The Harvard Observatory provided such a service for Boston clocks without fee. But not until after the Civil War did astronomers refuse to give the time of day and begin charging for it.

Physics had brought the world electricity, which by 1865 had generated an American telegraph industry capitalized at more than forty million dollars. But the industry hired no scientists and indeed used no science, solving its technical problems through trial and error. Only one electrical motor showed up at the 1851 Crystal Palace exhibition in London, though Joseph Henry had devised a prototype in 1831 and Thomas Davenport of Vermont (an untutored blacksmith) had built a practical model in 1834. When a bright young physicist lectured in 1859 on "the practical value of physical science," all he credited to electricity besides the telegraph was electrotyping and electroplating.

Federal science benefitted the American merchant marine through hydrography, astronomy, and the wind and current charts of the navy's Lieutenant Matthew Maury. Maury's work also paid off spectacularly by locating the most promising whaling grounds and a feasible route for an Atlantic cable. But kerosene lamps put the whalers out of business, while steam, iron, and the Civil War reduced the American merchant marine to insignificance.

Whatever their private grumblings about "sordid necessity" and "oblations to Buncombe," the scientists kept on proclaiming science to be a horn of plenty; and however scant the evidence, the public remained eager to believe. The scientists' motives in this are not easily untangled. Some scientists were true believers; some were selling a bill of goods. Some wanted to make a living; some, like Silliman, wanted to make a killing. Some were seized by the problem-solving passion that had drawn them to science in the first place. Some, like Norton and Johnson, genuinely felt a call to public service. In some, quite possibly, all these motives were mingled, consistent or not. In most, certainly, more than one motive existed.

Most scientists, perhaps all, had the good of science at heart as well as their own. On his arrival in America, Agassiz was struck by the scientists' lack of leisure to *do* science, not just *use* it, whether they lived by teaching, government service, or private employment. This handicap, he shrewdly suspected, accounted in part for their characteristic data-gathering and short-range researches instead of those requiring deep and sustained thought. By striving to show that science could be useful, the scientists consciously prepared the public mind for the corollaries that more science could be more useful, and that science should therefore be enlarged as well as applied. Thus they advanced the day they all dreamed of, the day of untrammelled research, of support without strings.

Furthermore, some of them glimpsed the prospect of mutual support and enrichment, of synergy and symbiosis, as science and technology drew closer together. In his late teens James Booth envisioned the laboratory as a miniature factory and the factory as a mammoth laboratory. Though scientists in industry might have to do "much drudgery which is of no interest to them," wrote Charles Jackson in 1844, they were where they could "pick up some new & useful & scientific information." That insight may have come more readily to Americans than Europeans. The line between science and technology seems to have been less sharp and invidious in America than in Europe, where rigidly defined hierarchies were traditional. Practical work carried no public stigma in the New World, and science (as some scientists complained) had no aristocratic standing. At any rate, American scientists saw science and technology as sharing a continuum, and from Benjamin Franklin to Willard Gibbs, individual American scientists were capable of running the full circle in their own careers. Franklin's great-grandson Alexander Bache argued in 1856 that "physical science" needed "the facts which the arts furnish to build upon," as technology needed scientific principles to apply. "This," he concluded, "is a debtor and creditor account which it would be difficult to adjust." He would not have been surprised by Menlo Park, the Bell Laboratories, or NASA.

The story of science in nineteenth-century America, therefore, requires some account of American technology: its motivations and mores, its social composition and status, its self-image, its professionalizing, and —to coin a word by analogy—its scientizing.

Chapter 11
The Technological Connection

If one tries to picture the relationship between science and technology, the most natural but least satisfactory image is of a straight-line spectrum, with science at the theoretical end gradually blending into technology at the practical end. A somewhat better metaphor may be borrowed from technology itself. In this, applied science and technology are connected not in series but in parallel. They run separate courses from theory to practice, coming closest to each other at the start and the finish. In applied science, however, the basic theory is the child of curiosity, whereas that of technology stems from hope of material gain.

From opposite sides of the configuration, the scientist sees theory and the technologist sees practice as the higher order. Since this book adopts the scientific orientation, it will follow the technologists "upward" from practice to theory, from professionalizing through scientizing to conjunction with theoretical science in the higher education of engineers. Scientizing and higher education have obvious relevance to science in general. As a prime factor in both, so does professionalizing. But it has a more direct significance as well. Since the public regarded science and technology as Siamese twins, in raising the prestige of technologists professionalizing helped raise that of scientists too.

Professionalizing begins with the deepening of shared expertise. From that its other attributes flow. As the expertise grows more extensive and abstruse—in the case of technology, as it becomes more scientized—training in it shifts from the job to higher education and specialized literature. This raises group consciousness and pride, which in turn give rise to group-enforced certification, ethical standards, and responsibility to the helplessly inexpert public. To further these lofty ends, as well as to promote the legislative and economic interests of the group, an association is organized. We have followed the scientists part way along that road. Now let us detour to track the technologists.

Like the scientists, the technologists responded to their social and economic environment. They showed a similar pattern of geographical distribution, with the Northeast as the region of heaviest concentration. New England, with 15 per cent of the nation's population, produced nearly 40 per cent of leading technologists active between 1846 and 1876, as defined by inclusion in the *Dictionary of American Biography.* Massachusetts and Rhode Island each produced five times as many per capita as the rest of the nation. The Middle Atlantic states, with a third of the nation's population, produced about 45 per cent. Thus the great majority of technological leaders grew up in science-conscious areas, and many carried that probable mind-set to the booming North Central states.

Like their scientific counterparts, the *DAB* technologists tended to be middle- or upper-class in origin. More than a third of the whole, and fully half of those beginning their careers between 1865 and 1876, were sons of professional men. The category of entrepreneurs' sons and that of small farmers' sons each accounted for a quarter of the total, though only a fifth apiece of the later entrants. Technologists differed from scientists, however, in being led to their calling less by family or schooling and more by job experience.

The foregoing statistics lump engineers and inventors together, since many qualified for both categories. Leading inventors, as distinct from engineers, were even more likely to be New Englanders, especially Bay Staters. They were mostly of middle-class origin, though only 7 per cent were sons of professionals. Three-fourths were sons of farmers or businessmen, about equally divided. Inventors shared some traits with scientists. They were born puzzle-solvers, sought glory in priority, and prided themselves on helping mankind. Some, like Charles Goodyear, saw themselves as doing God's work. Yet they had trouble professionalizing. For most, profit came first and altruism a distant second. Ethical standards were shaky. Corrupt influence pervaded the Patent Office, at least in the view of inventors. A private diary reveals, for example, that in 1859 the patent examiner and sometime chemistry professor Leonard Gale took a bribe of several hundred dollars for issuing a cotton-baling patent. So antebellum inventors were loners, more suspicious than fraternal.

Furthermore, the inventors' expertise was not as esoteric, certifiable, nor even communicable as that of engineers. Leonard Gale estimated in the mid-fifties that seven-eighths of American patents were for mechanical inventions and therefore largely unbeholden to book learning, journals, or professional associations. The *Journal of the Franklin Institute* began in 1826 with the "intelligent mechanic" in mind, but in the 1830s Alexander Dallas Bache and his coterie in the Institute turned the *Journal* toward scientists and professional engineers instead. In 1846 Orson

D. Munn and Alfred E. Beach took over the newly founded *Scientific American* and made it hugely successful among inventors, mechanics, manufacturers, and the general public. It served as a spokesman and forum for inventors. But despite its name it had no standing as a research journal. When some prominent technologists organized the National Association of Inventors in 1845, their creation lasted only a couple of years, though two conventions were held, an ambitious program was laid out on paper, and a feeble journal appeared briefly.

In contrast, civil engineers had a head start toward professionalism. After the military, their field of engineering claimed seniority. In 1846, 40 per cent of the *DAB* civil engineers were sons of professional men and 70 per cent were college graduates, in each case twice the proportion of other *DAB* technologists in that year. And nearly 40 per cent were primarily salaried, as against only 6 per cent of other technologists. In short, a large proportion of civil engineers learned about professionalism from family and schooling and about organization and teamwork from their jobs.

The canal system had been the nursery of American civil engineering, but by mid-century civil engineers preferred railroads, which employed more than a third of the *DAB* leaders. The engineers reconnoitered terrain, proposed routes, made maps and profiles, recommended modes of construction, and upon management approval made final plans. Contractors built the road, but the engineers advised on letting contracts, checked progress, certified work for payments, and sometimes changed specifications during construction.

The profession had its drawbacks. In the field, civil engineers worked from dawn to dusk, in heat and cold, tormented by insects, sometimes waist-deep in briery swamps, or struggling through tangled brush with equipment and supplies on their backs, or even dying from malaria. When work stopped in bad weather they fretted in their isolation, longing for their families and despising frontier coarseness and brutality. The young rodmen often sank into drunkenness and promiscuity, to the disgust of the engineers.

Away from the field another kind of corruption touched professional pride, that of the contractors who bribed engineers to make false estimates and fired those who refused. The "Canal Ring" bled New York State for a quarter century before it was exposed in the 1870s. Nepotism and, on state or federal projects, party politics sometimes governed professional advancement. Pay varied with the job, the region, the company, and the state of business, but it was seldom lavish. Regular engineers earned from less than $1,000 a year up to $1,700, assistant engineers from $800 to $1,200. Moreover, "not knowing whether you are to be in or be out of business more than one month at a time" sapped morale. Hard times threw many out of work in the mid-forties, and again for a year or two at a time after 1850, 1854, and 1857. "For God sake," wrote one engineer to another in 1854, "let us turn out and make a fortune for our interesting offspring in some respectable way and leave Engineering and Contracting to the dirty crew it belongs to."

Yet most found enough pride of calling to keep them faithful. Even the hardships let them brag of "that important attribute of the Bull Terrier, *hang on pluck,* which is so necessary in the composition of an Engineer." Finding a crooked river hemmed in by mountains, with rocky points running down to the water, the engineer John Childe swung his hat and cried out in exhilaration, "This is the place for engineering!" And he ran his railroad through. His "enthusiasm for the [railroad], and his devotion to it, were like that of the artist," wrote his biographer. They gloried in what their labors gave the world. And some paid their profession one of the sincerest of tributes when they urged their sons to follow it.

They also found in it honor among their fellowmen, notwithstanding dissolute rodmen. Even the assistant engineers were called "young gentlemen" in reports. The cream of local society welcomed their company, and in a barroom a noted builder of suspension bridges overheard himself referred to as "Mr. Ellet, the great engineer." On railroads, chief engineers now and then moved into managerial positions, among them no less a tycoon than J. Edgar Thomson of the Pennsylvania.

As early as 1843 the prominent civil engineer John Trautwine scolded a turnpike company that asked him to submit terms in competition with others. "I will have nothing to do with making a huckstering business of the profession," he informed it. "Suppose you advertised in the papers for the cheapest lawyer to undertake an important suit, do you think a single decent member of the bar would propose?" To some the time seemed at hand to embody this spirit in a professional society. A couple of tries in the 1830s had come to nothing. Civil engineers were still too scattered and provincial, not yet having made travel fast and cheap enough for national organization. But letters and editorials in the *American Railroad Journal* kept the idea alive. In 1848 a local society sprang up in New York and claimed a third of the state's three hundred or so civil engineers before its members scattered to jobs elsewhere a year or two later. In 1848 also the longer-lived Boston Society of Civil Engineers came on the scene with its own quarters and library, though its members likewise moved away during the 1850s, leaving it moribund for some time.

At last, in 1852, a dozen civil engineers in the New York City area formed the American Society of Civil Engineers and Architects, to include "Civil, Geological, Mining and Mechanical Engineers, Architects and other persons who, by profession, are interested in the advancement of science." During the fifties it had no permanent habitation, published nothing, met a few times and heard a couple of papers, but with an attendance averaging six the first year and fewer the second. After 1855 it did not meet for a dozen years. But the dormant seed turned out eventually to be viable.

. . .

"Mining, as a profession, is unknown to us," said Joel Poinsett as late as 1841. But the Lake Superior copper rush of 1845 and the California gold rush of 1849 opened a field for it. Before a convention of geologists in 1846, Charles T. Jackson outlined "the qualifications of a Mining Engineer." They were formidable, though Jackson conceded that any one man would seldom need them all. They included knowledge of geology, mineralogy, chemical analysis, surveying techniques and mathematics, the history of mining in America and elsewhere, and the practical operations of mining, including machinery, management of workers, and terminology. A dozen years later, an American student at the Royal School of Mines in Freiberg, Saxony, drew much the same picture. Jackson's remarks further touched on another attribute of professionalism, ethical standards. The mining engineer, he said, should stand "between the company employing him & the public and ought to hold no interest directly or indirectly in the mine he reports upon." (The warning was needed but not always heeded. From 1863 to 1865, for example, James D. Hague, an eminent mining engineer and graduate of Harvard and Freiberg, fed inside information about the copper mine he superintended to a Boston friend in return for surreptitious stock purchases in Hague's name on credit.)

The range of expertise called for in mining engineers ought to have given impetus to professional organization. But numbers were lacking. As late as 1863 a Freiberg alumnus in California reported that "the field for mining engineers, educated ones, I mean, is much smaller than one would suppose." Encounters with charlatans had turned Californians against engineers as well as geologists. "The less of this scientific tomfoolery," wrote one self-taught man that year, "the more money will be made." Of nearly six hundred *DAB* technologists, 1846–76, only twenty were primarily mining engineers. Such specialists were too few and scattered for a viable organization. Still, those few were proving their practical worth and preparing the ground for solid professionalism.

Mechanical engineering had numbers in its favor. It arose from the fields of power and machinery, each of which accounted for about a seventh of the *DAB* technologists, 1846–76. But only 31 per cent of those in power and 11 per cent of those in machinery were sons of professional men, as against more than half of the *DAB* civil engineers; and only a fifth of those in machinery and power attended college, as compared with two-thirds of the civil engineers. Only a fifth were primarily on wages or salary, most of the rest depending on entrepreneurial income or royalties. In social origins, education, and terms of employment, therefore, those in the fields that spawned mechanical engineering had less of a professional orientation than did the civil engineers. And their fields required less formal scientific knowledge than did mining engineering.

Mechanical engineering came out of the machine shops of New England textile mills. In the 1840s those shops set up for themselves, growing larger and more specialized as demand rose for marine and factory steam engines as well as machine tools. Their focus on custom work tended to minimize competition and thus to encourage the free exchange of techniques and improvements, an attribute of professionalism. But a tradition of family- or class-oriented elitism in their management perpetuated what has been called "shop culture," the training of future leaders in the shops of friends or relatives, as distinct from formal education or "school culture."

Other nurseries of mechanical engineering developed by the 1840s. As the navy turned to steam power, it appointed civilian steam engineers with the social standing of regular navy men. More important, railroad shop superintendents emerged as mechanical engineers in effect. The railroad shop element in the fifties helped establish the very term "mechanical engineer," which had first been heard in the forties. In 1853 the brilliant Zerah Colburn began a mechanical engineering department in the *American Railroad Journal,* emphasizing the use of science and its methods. In 1854, for the first time in his career, he signed himself "Zerah Colburn, Mechanical Engineer."

But the time had not yet come for the capstone of professionalism, an association. In 1860 the editor of the newborn journal *American Engineer* promoted an "American Engineers' Association" to be centered on steam engineering. No engineers of standing deigned to join, however, and so it collapsed almost at once.

Antebellum inventors were no more inclined to scientize than they were to professionalize. Most being devisers of mechanisms, they carried on no experimental research to derive new principles or generalizations. Instead they used well-known mechanical principles and counted it success when their models worked as envisioned. Unlike scientists, inventors proceeded from personal experience and thought, scorning the finespun theories of others. Europeans were already developing the theory of kinematics (though not yet the significant application of it) and had begun analyzing gear teeth mathematically to improve efficiency, but Americans paid no heed to either development. Besides, the mechanical inventors' mode of thought tended to be nonverbal, a rounded imaging of spatial relationships and of the interplay of diverse elements, whereas scientific thought tended to be verbal, symbolic, linear, and sequential. (Of course, from Kekulé's benzene ring to the double helix of Crick and Watson, scientists have also thought spatially. But for the sake of communication their concepts were promptly verbalized and quantified.) In any case the lone-hand inventors could not afford experimentation solely to enlarge the world's understanding. Knowledge as such was, after all, not patentable. So they left such things to those supported by government, colleges, societies, or business, to wit, the scientists and the engineers.

Quickened by the rising tempo and scale of industry and transportation, the scientizing of technology was most evident in the fields of power, materials, and construction. Philadelphia's Franklin Institute, which brought scientists and technologists into close association, set an example in 1829 by sponsoring a program of elaborate and ingenious experiments on the most efficient means of using waterpower. Manufacturers and others all over the Northeast responded to the Institute's campaign for funds and were rewarded by empirical tables that told millwrights exactly what efficiency could be hoped for in any conceivable combination of elements—wheel velocity, the shape, size, and number of buckets, head of water, wheel size and type, inlet aperture, and gate form. The young Alexander Dallas Bache seized on that success to lead the Institute during the early thirties in a federally backed program to determine the causes of steam-boiler explosions, a step toward making a science of steam engineering. After Bache moved to Washington, his protégé John Frazer carried on the movement to scientize technology as editor of the Institute's *Journal*.

As the Franklin Institute had demonstrated, waterpower technology in particular lent itself to the scientific approach. Engineers began taking over from millwrights in the early 1840s when Franklin Institute *Journal* articles on a French hydraulic turbine caught the attention of Uriah Boyden, a Massachusetts civil engineer, largely self-trained but with a strong taste for pure science. Backed by the Locks and Canal Company, which controlled Lowell's waterpower, Boyden achieved remarkable improvements in turbine efficiency not by applying scientific theory but by using scientific methods in systematic experiments and elaborate calculations (which would have been far easier if he had known calculus). In the early fifties he was joined by the English-born James B. Francis, the company's chief engineer and superintendent, who shared Boyden's professed (though not entirely genuine) disdain for mathematical theory and his reliance on experimental research and creative design—attitudes that became characteristic of American hydraulic engineering generally. Using dynamometers, testing flumes, and full-size apparatus, all financed by the company, they advanced the development of the "American mixed-flow turbine," still one of the three main types of hydraulic turbines. With the same resources and techniques, Francis also worked out weir formulas for water flow that remained standard well into the twentieth century.

Contrary to popular belief, nineteenth-century steam engines owed little to basic science, even the science they had given birth to, thermodynamics. Practical experience had already suggested what thermodynamic theory would prescribe, high pressures and maximum use of steam expansion. Experienced steam engineers in 1846 scoffed at a neophyte who bought books on the steam engine. One of the principal advances in that era, John F. Allen's "variable cut-off," came (as a co-worker put it) from "the mind of a man who had no knowledge of mechanics except what he had absorbed in engine-rooms." The indicator card, graphically recording the relation of steam pressure to volume during a stroke, came to be regarded in the 1870s as the engineer's stethoscope. Yet as late as 1861 only the Navy Department and a few civilians like George Corliss, John Ericsson, and Frederick Sickels used it in the United States. Machine shops determined the sizes of piston rods, cylinder heads, and crank pins by rule of thumb rather than calculation from the known strength of materials. Engineers differed widely on such questions as the advantages of high pressure and cylinder expansion of steam, the influence of piston speed, the laws governing salt deposits in boilers, and the effects of condensation on efficiency.

It was neither a private society nor an industrial complex but the federal government that at last took the lead in scientizing steampower technology through the work of an industrious, methodical, combative navy engineer fittingly christened Benjamin Franklin Isherwood. His two-volume *Engineering Precedents for Steam Machinery* (1859) collated a formidable mass of systematic data on the actual working of various steam vessels, discrediting simplistic theoretical assumptions by engineers. He described his two massive volumes of *Experimental Researches in Steam Engineering* (1863, 1865) as simply "collections of original engineering statistics with the general laws deduced from them" and added that "science is nothing but a similar collection of statistics." But this Baconian manifesto merely echoed what such scientific counterparts as James Hall and Joseph Leidy were saying about their own researches.

The army made its contribution in another field of power. Small-arms development continued to lean on mechanical ingenuity and production techniques. But heavier weapons by the 1840s increasingly demanded a quantitative, experimental approach in such matters as metallurgy, ballistics, design, and propellants. Fortunately West Pointers had the best engineering education then available, making them scientific technologists first, military men second. Major Alfred Mordecai of the Ordnance Corps had close ties with the scientific community. In the forties he made the first accurate velocity measurements of American projectiles and the first large-scale, controlled American gunpowder experiments, distributing the results to scientists as well as soldiers and gun makers. In 1839 Mordecai and his West Point comrade Benjamin Huger became mainstays of the new permanent ordnance board, which periodically tested and evaluated small arms, set standards, and systematized American military armament.

In the fifties Captain Thomas J. Rodman of the Ordnance Corps made some of the earliest industrial applications of calculus in his researches on gun endurance. His pressure gauge became indispensable for interior ballistics. His concept of "initial tension," cooling gun castings from the inside out and thus compressing the inner layers, and his perforated-cake powder, increasing in surface area as it burned, were major advances. His naval counterpart, Commander John A. Dahlgren, who had mathematical ability and experience in Coast Survey duty, took over the Washington Navy Yard in 1847 and changed its emphasis from shipbuilding to ordnance research and development, with elaborate experiments (though less use of calculus than Rodman's).

As late as the mid-1840s civil engineers still designed timber truss bridges, the most common type, out of experience and intuition. In 1847, for example, an engineer about to bridge the Androscoggin wrote one who had just bridged the Merrimack, offering to buy "all your plans . . . with the *dimensions of the parts* . . . full in every particular." In that year Squire Whipple (so christened), an upstate New York engineer, published *A Work on Bridge Building,* for the first time in America offering a system for precise analysis of stresses in the members of a truss bridge. Mindful of its readers' limitations, the book required no calculus or even trigonometry, but provided useful mathematical and graphical methods. Thus an engineer could save any material not needed for a specified safety factor. Whipple took lifelong pride in his contribution to "the science of Bridge Construction." Another civil engineer, the West Point graduate Herman Haupt, missed seeing Whipple's book but developed some of its ideas independently in his *General Theory of Bridge Construction* (1851). Though inferior to Whipple in its presentation, it won prompt and lasting success.

In building the suspension bridge at Wheeling, 1846–49, Charles Ellet, Jr., tested parts systematically during construction but used little scientific theory. The German immigrant John A. Roebling, however, had been trained in Berlin at one of the world's foremost engineering schools. His Monongahela bridge at Pittsburgh (1846) has been praised by a modern authority as "the first suspension bridge scientifically designed and constructed to provide against all the forces with which a bridge must battle." A contemporary described his reasoning as "always clear, simple, and explicit, and sustained by philosophical and scientific facts," and his arguments as being "drawn from his store of scientific knowledge." Roebling went on to become one of the most notable of American bridge builders.

The iron framing of buildings came belatedly to America, where timber was cheap. But James Bogardus's patented system of precast members produced hundreds of iron-framed buildings in Eastern cities by the end of the 1850s. This broadened the scope of American materials-testing from cables and timbers to iron beams and columns. In the 1830s the *Journal of the Franklin Institute* reprinted English studies of the subject, and an Institute committee designed and used a sophisticated testing machine for iron and copper at high temperatures. The committee's leader, Walter R. Johnson, won an international reputation for his later research along those lines. At West Point, Dennis Mahan's teachings and writings carried the word of French advances in strength-of-materials science. And Major William Wade's testing machine, the second of its kind built in America, was put to use by Wade and Captain Rodman in experiments with cannon.

Industry took up the pursuit. In the 1850s the Trenton ironworks of Cooper, Hewitt & Company began making wrought-iron I-beams. Abram S. Hewitt bragged about his experiments to ascertain the best distribution of iron in the beam. "The highest mathematical knowledge and skill were required to determine the laws which governed the strains upon wrought iron," he wrote, and this achievement was "one of the proudest triumphs of modern science." However that may have been, R. Hoe & Company did make Cooper & Hewitt a hydraulic press in 1856 for testing the deflection of iron beams, with a ram capable of sustaining 350 tons.

Unfortunately most antebellum builders persisted in the cut-and-try tradition. In consequence their experimental data all too often came from the actual collapse of bridges and buildings.

In antebellum America, the scientizing of technology was only beginning. Despite the mystique of Yankee ingenuity, Europe led in both using science and scientizing technology throughout the century. But Americans had begun to see the light. In 1855 Daniel Treadwell, formerly Harvard's Rumford Professor of the Application of Science to the Useful Arts, declared that technology did not advance solely through applying scientific knowledge but also used "precepts of art . . . derived from scientific research." Treadwell himself was prominent as both an inventor and a scientist. Being a professor also, he presumably thought it unnecessary for him to point out the further coupling of science and technology in higher education. As engineering science grew more sophisticated, it inevitably joined hands with physical science in academe. The "scientific schools" and "polytechnic institutions" sprouting in the 1840s and 1850s were precursors of such fruitful science-cum-technology hybrids of our own time as MIT and Caltech.

There were doubters. One "distinguished engineer," himself a Harvard graduate, was said to have grumbled in 1853 "that all Engineer Schools only made students conceited, and that they were worse than useless." Shop culture dominated mechanical engineering until after the Civil War. Civil engineers could, by a sort of apprenticeship, still rise professionally without college training. James B. Eads, who left school at thirteen, became a civil engineer of world repute. But most professionals and laymen alike were coming to a different view.

As in science, Europe pointed the way. Americans knew and respected its technological achievements. In 1846, 10 per cent of the *DAB* technologists were European-born, and the proportion increased thereafter. Like the scientists, American technologists made European pilgrimages, not only to visit foundries, mills, and engineering works, but also to study in technological schools and colleges, especially in France and Germany. The *Journal of the Franklin Institute* and other periodicals carried accounts of such schools, most of which gave two years to basic sciences and a third year to specialized applications. In 1856 young Daniel C. Gilman of Yale published an extensive and enthusiastic firsthand report on them in the *American Journal of Education* which Yale reprinted and circulated in a fund drive for its own scientific school.

Gilman's report included the Royal School of Mines at Freiberg, Saxony. American engineering students in Europe almost always went there. Freiberg was a drab, crude mining town, and the school's academic advantages turned out to be somewhat exaggerated. The chemistry lab was small and crowded, the big mineral cabinet was not freely open to students, and there was practically no lab work in physics. Some of the professors lectured by reading stupefying masses of useless details. Still, travel costs, living expenses, and tuition for a Freiberg education came to far less than for college study in America, which had no equivalent institution anyway. The school's curriculum ran not three years but four, its library resources were rich, most of the professors were able and helpful, and even the town's dullness meant few distractions. Besides, as one American wrote later from experience, Freiberg was better than Paris as a rehearsal for life on the mining frontier. The small American enrollment grew slowly in the fifties, but Americans comprised more than a quarter of total enrollment in the early sixties and half in the late sixties.

In civil engineering the American movement for higher education had long been afoot. West Point had stressed science and civil engineering so early as to be in effect the nation's first engineering college. French influence predominated there, especially after Dennis Mahan became professor of engineering in 1832. Prepared by four years' study in France, Mahan did much to shape the curriculum, and his civil engineering text was standard in American colleges for a generation after it came out in 1837. As late as 1864 the University of Michigan modelled its scientific school on West Point and took pains to get West Pointers as professors.

But West Point itself passed its zenith in engineering during the 1830s. More than a hundred of its 1830s graduates became civil engineers at some time but only eighty of its 1840s graduates and sixty of its 1850s graduates. By the mid-forties its board of visitors was showing impatience with its engineering emphasis. In 1852 Mahan complained that the cadets had only four months of civil engineering. His colleague John M. Schofield took himself seriously as a physics professor but recognized that his teaching did little for cavalry commanders. "This is, in fact," he wrote, "generally and perfectly well understood at West Point." The superintendent told Congress in 1860 that the Military Academy was "no longer a school of engineering" but "an institution for the purposes of national defense."

The Naval Academy, founded at Annapolis in 1845 largely through the efforts of the mathematician William Chauvenet, came too late to repeat West Point's role. The navy's own Corps of Engineers was recruited chiefly from among civilians like Benjamin Isherwood. Disheartened by the academy's indifference to science, Chauvenet, its most distinguished scientist, moved to Washington University in 1860, commenting ruefully that no one at Annapolis seemed to regret his leaving.

Among civilian schools, engineering education developed in two ways: from an independent seedling of science instruction for farmers and workingmen, or by grafting an engineering branch on an existing college. The merits of the two approaches were debated for many years, but American pluralism still accommodates both.

Founded at Troy, New York, in 1824, the Rensselaer Institute pioneered in the first mode with a one-year course for farmers and mechanics. The canal milieu of upstate New York led it to award the nation's first formal degrees in civil engineering in 1835. Nearly killed off by the depression of the forties, it was rescued by a new director in 1847, the New Hampshire-born Benjamin Franklin Greene, aloof, arrogant, largely self-educated, but full of ideas and ambition. He reorganized the Institute into a three-year engineering school of college rank, drawing inspiration from the Ecole Polytechnique in Paris. Thereafter the school was known as the Rensselaer Polytechnic Institute. Three of its ten teachers were in basic sciences. One parent complained that too much was crammed into three years' study, but he considered the professors able and the standards high. By 1856 enrollment reached 123, drawn from throughout the nation and the world. Greene's grandiose plans set him at loggerheads with the trustees and led to his resignation in 1859. Enrollment fell. But RPI survived and in 1862 even began offering the nation's first four-year course in engineering.

Other independent engineering schools were established at Philadelphia in 1853 and Brooklyn in 1855, though the former petered out by the end of the century. Inspired by RPI and the French schools, Peter Cooper founded his Cooper Institute in New York City during the late fifties, but it did not become a full-fledged engineering college until the twentieth century.

The most successful independent school was the brainchild of William Barton Rogers. He had pondered it for years before his brother Henry wrote him from Boston in 1846 that the industrialist John Lowell might back it. At Henry's prompting William conjured up a shining vision for Lowell. Boston, he wrote, would be the best place in the world for such a school because of the city's "knowledge-seeking spirit," its "intellectual capabilities," and its leadership in industrial technology. He enclosed an eloquent plan. The school would have "an entirely practical department" of applied chemistry, practical mathematics, drawing, architecture, and engineering. But first and foremost would be a department devoted to basic science, upon which (Rogers claimed) rested every technological advance in the previous fifty years—including, he shrewdly remarked, the dyeing and printing works at Lowell. Such a school would, he predicted, "finally expand into a great institution . . . of physical science and the arts . . . and would soon overtop the universities of the land."

John Lowell decided against the project. But William Rogers persisted in his dream. In 1853 he forsook the University of Virginia and settled in Boston to live by lectures, consulting fees, and income from savings. In Massachusetts the industrial growth of the fifties enhanced the appeal of technological education. Rogers successfully lobbied the legislature for a share of proceeds from the state's new-made land in the Back Bay of Boston, earmarked for "public educational improvements." On April 10, 1861, Governor John Andrew signed an act incorporating the Massachusetts Institute of Technology and granting it a block of Back Bay land, conditional on its supporters' raising a hundred thousand dollars. Three days later the Civil War began. William wrote Henry that "there is little of science or letters thought of now." But in the end, the war meant only postponement, not defeat.

Meanwhile, existing colleges had begun to sprout engineering schools. By the late forties, a score or more had added courses called "civil engineering" to their catalogues, but some of those courses appeared only briefly, some were never taught, and most were superficial. The first true college-affiliated schools of technology grew out of courses in chemistry, which had long drawn on formal science. It was in these college-affiliated schools that science and technology mingled most intimately. Such schools emphasized engineering and applied science, while their cognate colleges went on giving basic science courses as before. Yet they called themselves "scientific schools," and the same students and faculty were apt to show up in the classrooms of both school and college.

Yale set up the first of the scientific schools in 1846 at the instance of the two Sillimans and with the help of five thousand dollars from John P. Norton's father for a chair in agricultural chemistry. Classes began in the fall of 1847. No other gifts came in. For years the school barely scraped by. Its unsalaried professors, John P. Norton and Benjamin Silliman, Jr., teaching agricultural and industrial chemistry respectively, depended on student fees and what they could make by commercial analyses on the side. The cost of supplies and the rent and upkeep of the building (a two-story, white frame house formerly occupied by Yale's president) ate up most of that income, leaving little for Norton and Silliman. Yale College proper gave them neither help, hindrance, nor heed.

Still, the new school could offer access to Yale College science courses, books, and collections. James Dana gave his sympathy, advice, and influence. Previous college study was not required, and instruction had to be watered down for some students. But future leaders like Samuel W. Johnson, William Brewer, and George Brush enrolled, and all the students were earnest and enthusiastic—fortunately, since they were exempt from college rules, including compulsory chapel. The regular college's faculty and students looked down on them. Nevertheless, "we were a happy little community," recalled one of them years later. He remembered with nostalgia "the kitchen & central room of Pres. Day's House, with its array of tables & re-agents & happy, hard-working scientifics—with Prof. Silliman's fat, jovial face popping in occasionally & the beloved Norton ever working faithfully & laboriously at his desk."

Beginning in 1849 Silliman popped out each winter to teach chemistry for pay in Louisville, leaving Norton to carry on alone. Norton's death at thirty in 1852 threatened to knock the new school into the grave also. But at Brown University, President Wayland's moving and shaking had just then dislodged his Liebig-trained professor of applied chemistry, John A. Porter, and his West Point-trained professor of civil engineering, William A. Norton. William Norton had already agreed to open a civil engineering "school" at Yale, Porter took over agricultural chemistry, and their Brown students followed them in a body to give the Yale Scientific School a new lease on life.

Soon afterward the younger Silliman returned and two brilliant young teachers were hired, George J. Brush in metallurgy and Samuel W. Johnson in analytical chemistry. Financial salvation came from Porter's father-in-law, Joseph E. Sheffield, who had made a fortune in cotton and railroads. His gifts and bequests ultimately came to more than a million dollars. The school gave its first civil engineering degree in 1859; and by 1861 Sheffield had provided a large building, a fifty-thousand-dollar endowment, and the name the school bears to this day: Sheffield Scientific School.

Harvard's scientific school followed a more troubled and tortuous path. In 1846 Harvard's new president, Edward Everett, prompted by his own German university experience, called for a school of science and a graduate program. Benjamin Peirce drew up a plan at Everett's request, but played down the graduate school idea and stressed the newly vacant Rumford Chair as a peg on which to hang the scientific school. In 1847 the corporation approved the plan and named Eben Horsford to the chair. Flourishing a strong endorsement by the great Liebig, Horsford then got the technology-conscious textile millionaire Abbott Lawrence to put up thirty thousand dollars for a Liebig-style laboratory and a twenty-thousand-dollar endowment to help support professorships of engineering and geology. Harvard christened the new enterprise the Lawrence Scientific School.

So far, so good. But matters now began to turn awry. Avid for a prestigious faculty, Abbott Lawrence brought in Louis Agassiz as professor of geology and zoology with carte blanche to teach what he chose —or nothing at all, if he preferred full-time research. Agassiz elected to concentrate on museum, lab, and field work, teaching geology and zoology in alternate semesters to a small class, which did little for engineering students. No civil engineering professor could be found until Henry L. Eustis consented to leave West Point in 1849. He delegated math instruction mostly to student assistants and gave his attention to drawing (including landscapes), drafting, and field work. With the demand for engineers strong, Eustis had trouble holding his students through the full five-semester course. They preferred drawing pay to drawing landscapes. As for Horsford, he opened what was touted as the best-equipped chemistry teaching lab in the world, but being unendowed it depended for support on enrolling far more students than could be expected in the America of that day. Both supplies and assistants cost more than in Germany, and so Horsford had to cut corners and suffer the drudgery of teaching the bare elements to students who wanted little more anyway.

The Lawrence School struggled too desperately for mere existence to set and hold a steady course. Donations were few, enrollment was humiliatingly low, and tuition barely covered day-to-day expenses, leaving little to augment the faculty's meager salaries. At Harvard as at Yale, the "scientifics" were of low caste. The library ignored their needs, leaving them dependent on their professors' books. In the early 1860s firmer hands took the tiller, and enrollment reached a high of seventy-nine in 1865. But across the Charles in Boston, MIT opened its doors that year, and Lawrence enrollment thereafter fell off sharply.

At Dartmouth, though its president shied from such novelties, Abiel Chandler's irresistible bequest of fifty thousand dollars led to the opening of the Chandler School of Science and the Arts in 1852. But other such schools encountered the same problems as Lawrence, or worse. At the Universities of Virginia and North Carolina, they lived out their short and sickly spans before the Civil War came. Half a dozen other institutions also tried their luck in those years. But there were still too few paying students to support so many schools, and too few such benefactors as Lawrence, Sheffield, and Chandler. As for public funds, the University of Michigan opened the only state-supported engineering school before 1860, and it granted no degrees until 1861. Voices were raised in behalf of public grants to agricultural and engineering colleges, but the slave-state oligarchy, which saw no profit and some peril in public enlightenment, had its puppet President Buchanan veto a bill to grant federal land to the states for that purpose.

Nevertheless, though many educational projects of the fifties turned out to be mere wind, it was evident which way that wind was blowing. In 1859, the year of Buchanan's veto, the Comstock Lode, and Drake's oil well, the *New York Times* gave a clear reading of direction:

> The Professor is no longer the near-sighted, thin-faced and clean-shaven man, in rusty black, that he was. Now he wears a beard, drives possibly a fast horse, is discovering gold-diggings, indicating new mines, and pointing out to capitalists the wealth of unannounced coal measures. The change has been wrought by the spirit of the age.

It called its editorial "College Progress."

Chapter 12

The Public Purse

Vital as were academe, private enterprise, and philanthropy in shaping modern American science, they did not do it all. Government also played a major role.

Briefly it seemed that the federal government might even subsidize technological research simply for the general good. Backed by Senator Thomas Hart Benton of Missouri (who wanted to expedite Western travel and promote Missouri zinc for batteries), the brilliant electrical experimenter Charles G. Page in 1849 won a twenty-thousand-dollar grant from Congress to develop a locomotive powered by wet-cell batteries. But the precedent worried some, and Page's exhaustive tests of motor designs impressed scientists more than congressmen. Denied a further grant, Page managed to complete a prototype, which proved wildly more expensive and unreliable than an ordinary steam locomotive. Congress made no more such grants. Except in weapons, government thereafter contributed little directly to technological research.

Science was quite another matter. Government employed nearly a third of the leading antebellum scientists. And its centralized organization amplified its economic impact. While college professors like the Sillimans, Norton, and Rogers might build academic bases for science, none (perhaps excepting Agassiz) wielded such concentrated fiscal power and held such commanding positions in the scientific community as did certain government scientists.

State governments in the antebellum years made a substantial contribution to science indirectly by supporting higher education. More directly they helped by distributing meteorological instruments, financing entomological studies, subsidizing museums, and hiring agricultural chemists. But their most conspicuous direct aid was through geological surveys.

The significance of the state surveys may be gauged from George P. Merrill's massive study, *The First One Hundred Years of American Geology* (1785–1880). More than half of it is given over to the "Era of State Surveys," 1830–80. The states supported not only the geologists and their field work, but also the publication of their findings. James Hall, the most prolific and tenacious of the lot, published fifteen quarto volumes of more than forty-five hundred pages and a thousand full-page plates at the expense of the State of New York.

After a burst of state activity in the 1830s, when fifteen surveys were authorized, came a slackening during the 1840s, which brought forth only three new ones. Merrill speculates that economic depression and a shortage of competent geologists accounted for the lull. But the fifties brought fourteen more surveys, six of them in states reviving earlier ones and eight in states establishing their first. Boom times had come again, the gold rush had dramatized mineral possibilities, and more geologists had been trained. Moreover, new states had come into the expanded Union, and some of them were quick in catching up. By 1860, twenty-nine of the thirty-three states had sponsored surveys at one time or another.

In almost every case, beginning with the first state survey in North Carolina in 1823, the geologists themselves had to lobby for their appropriations, directly or through influential sympathizers. They did so with skill and effect, forecasting rich returns not only to mining but also to industry, agriculture, transportation, and education. Such tactics, one scientist wrote defensively in 1861, were "not only politic but right" in a democracy. The geologists wrote letters in support of one another to governors and legislative committees. They ran survey-puffing editorials in scientific journals. In the early thirties Massachusetts became the first state to publish its survey's report. In it Edward Hitchcock emphatically proclaimed the economic promise of his work. Those who followed did likewise. Some, like Hitchcock, kept the economic commitment conscientiously in mind during their work; others, like James Hall, gave it only lip service while attending mostly to "pure" science—in Hall's case, paleontology.

In Southern states, after California gold eclipsed their mineral deposits, the utilitarian argument shifted to agriculture, to which the region was increasingly committing itself. In order to market crops, the South wanted roads and railroad lines, the routing and building of which required geological data. Cotton was exhausting the soil of older plantations. And so planters, especially when pinched by low prices, worked through agricultural societies to lobby for surveys. They called for agricultural censuses, soil analyses, and studies of erosion, and the state geologists usually complied. In Mississippi, state geologist Eugene Hilgard made notable contributions to the science of soils.

North or South, legislators could not always be easily won over with fine illustrations of fossils. State money, especially in hard times, might suddenly dry up. Politics and even religious sectarianism could intrude into state survey appointments. Nevertheless, the public and their representatives gave surprisingly ready approval even to pure science. So in the antebellum years, state surveys did for much of American geology what national governments had done for European.

Of all the state geologists, James Hall of New York went farthest in putting government to work for science. His techniques of bullying and overawing appropriation committees, of citing past expenditures as grounds for future ones, of promising economic benefit without much interest in achieving it, do not in themselves explain his unparalleled career as a state geologist from 1836 to his death in 1898. He could not have done what he did without extraordinary energy and endurance, unshakable faith in the importance of his scientific output, and the ability that won him an early and lasting international reputation. In justice to his legislative supporters it must be said that their genuine respect for science, regardless of its economic utility, explains something too.

Hall's influence went far beyond New York's borders. He was looked to for guidance in organizing many state surveys, and in some he had a more direct hand. For a time in the late 1850s he was officially the state geologist of Iowa and Wisconsin as well as New York, though his young assistants did the field work and Hall showed up only in winter when the legislature met. This sideline gave him some patronage power outside New York. The distinguished geologists who had been his assistants might have given him still more influence if he had not been so quick to quarrel with them.

The fact remained that Hall's power base was a single state, and the needs of science were outgrowing state support. Perhaps Hall himself sensed this. "I feel too isolated here," he wrote in 1846, adding that he would like a position with the Smithsonian Institution. In 1857 he welcomed Joseph Henry's suggestion that he take charge of a proposed federal geological survey of New Mexico Territory. "I could," he wrote Henry, "give several younger men than myself an opportunity of earning a reputation in geology." (The appointment did not materialize.) By then he must have recognized that even a James Hall in the Empire State could not match the resources and power of an Alexander Dallas Bache in Washington. After all, twice as many leading scientists worked for the federal government, 1846–76, as worked for the states, and the reins were gathered into fewer hands.

Since a cardinal goal of scientists was unfettered research, they had occasional qualms about dependence on government. In 1844 Bache warned his colleagues that federal grants might have strings attached, that government-supported science might be constrained and misdirected by scientifically ignorant politicians and bureaucrats. He thought the safest channel for government funds would be some independent agency, such as might come out of the still unsettled Smithsonian bequest. In 1849, however, Joseph Henry, as director of the new Smithsonian Institution, looked back on his first three years on the job and concluded that his institution should "ask nothing from Congress except the safekeeping of its funds."

But Bache, Henry, and their confreres were shying from the bit, not the feedbag. The needs of science were outgrowing what philanthropy, private societies, and even the colleges could offer. The societies themselves occasionally petitioned Congress for aid in specific projects, and in 1851 the American Association for the Advancement of Science set up a standing committee for that purpose. Experience had made Congress wary lest short-term aid turn into a long-term commitment or even a permanent bureau, and so not much came of these efforts. Nevertheless the scientists persisted in seeking government support.

In 1838 Walter Johnson summed up most of the arguments. He cited the European example—the Royal Institution in London, the Ecole Polytechnique in Paris, the Gewerbverein in Berlin—and urged America to follow it and thereby shake off humiliating dependence on foreign science. The federal government, he believed, needed more and more scientific advice and protection against charlatans. Science could help the government improve agriculture and develop natural resources. He suggested an institution for physical research, pointing out that Congressional committees on military and naval affairs, public lands, agriculture, patents, roads and canals, public buildings, commerce, and manufactures all had need of expert scientific or technological counsel.

Despite his misgivings of 1844, Bache also took up the cry at intervals for more than twenty years, emphasizing the needs of the executive branch more than the legislative. To Johnson's list of governmental functions involving science and technology, Bache added coinage, weights and measures, military and naval colleges, coastal and interior surveys, astronomical observatories, and lighthouses. In 1851 he declared that neither executive nor legislative branches could avoid science-related decisions any longer. And in 1856 he called the roll of agencies already involved in such matters: his own Coast Survey, the Patent Office, the Nautical Almanac, the Naval Observatory, the Army Ordnance and Engineering Corps, the Surgeon General, the Topographical Bureau, and several recent scientific expeditions by the navy.

Bache's 1851 speech implied that the march of events would by itself force government to seek help from science and pay for it with support. But Bache knew better than anyone else how to speed and guide the process. Not for nothing was he the great-grandson of Benjamin Franklin. Like his famous ancestor, Bache was not only a scientist and organizer but also a consummate politician and diplomat. He had, wrote one admirer, a "marvelous knowledge of human nature and . . . unrivalled skill in using it." Bache once remarked privately that he studied men as he would "study physical phenomena." With his crony Joseph Henry, whose high cards included integrity, dignity, good sense, a towering scientific reputation, and administrative independence, Bache created the most potent duumvirate in government science.

Some politicians could be as balky as Simon Cameron, who told the Senate in 1861, "I am tired of all this thing called science here." But the vote on a Smithsonian appropriation went against Cameron twenty-nine to six. In the debate William Fessenden of Maine spoke humbly for the majority of politicians: "As to . . . [what the Smithsonian] has done for science . . . I have no doubt that it is doing much. . . . I ought to know more about it. . . . It is my own fault." As Joseph Henry had written privately a dozen years earlier, politicians tended to be "timid" in voting aid to science unless they were reassured by acknowledged experts. But few of them were against science on principle. Unlike James Hall, Bache and Henry had their way as much by educating politicians as by fighting them.

How Bache, Henry, Matthew Maury, Spencer Baird, and other scientist-politicians maneuvered in concert and in competition to harness the power of the federal government, and—more important—what they achieved with that power, are matters too large and complex for this or any other single chapter. Those matters will be the chief concern of the three chapters that follow.

Chapter 13
Bache and Maury, Barons of Bureaucracy

When Alexander Dallas Bache became superintendent of the Treasury Department's United States Coast Survey in 1843, he did not start from scratch. His predecessor, the Swiss geodesist Ferdinand Hassler, had left him an establishment with European scientific standards, a tradition of civilian control, an auxiliary of scientifically trained army and navy officers, and Congressional support by then reaching a hundred thousand dollars a year. Nevertheless, Bache—and those who knew him—saw all this as a mere beginning. And so it was. Ten years after Bache took over the superintendency, Benjamin Silliman, Jr., rated that position "the most honorable which a scientific man can reach in the United States, and also the most laborious and responsible." And in 1858 a letter to the *New York Times* complained that Bache had become "the most powerful functionary under this Government."

Bache was above all a great organizer, as he had already shown in organizing Philadelphia's school system. Within six months of Bache's appointment the initially hostile navy secretary was praising "the system, order and regularity" Bache had brought to the Survey's "complicated and difficult operations." Bache gloried in the struggle. Looking back over his first five years as superintendent, he called them the best of his life, full of "new responsibilities . . . new ties . . . new designs." Opposition, he felt, only made him stronger.

Bache met with opposition of several invigorating varieties. By 1854 he was complaining of "perpetual worry, & over-occupation." At the outset he charmed most of the existing staff into loyalty and brought the rest to heel, like it or not. He wanted the affection of his subordinates, he wrote, "but I *can* do without it." His subsequent appointments raised charges of nepotism and of favoritism toward Philadelphians, especially graduates of the Central High School, which he had formerly headed. Some grumbled about his highhandedness. Yet at one time or another, Bache enlisted a remarkable number of the nation's top scientists in the work of the Survey.

As a civilian agency, the Coast Survey generated some friction by using army and navy officers in its operations. One naval officer, after eight years' Survey duty, complained that navy men got little of the credit and most of the blame. But such complaints were not common. Like Hassler, Bache won over young navy officers by keeping them long enough for thorough training in geodesy and hydrography, better than the fledgling Naval Academy could yet offer, and useful to them in peace as well as war. Navy secretaries learned to waive rules about rotation of duty and consult Bache before detaching officers from the Survey, lest important and delicate work be interrupted. A civilian employee who later became a civil engineer looked back on the Survey in the fifties as "the best school in the world" for his profession, teaching him mathematics, the use of the finest instruments, and the habit of precise observation. Many of the best Civil War naval officers likewise credited their success in part to their Survey experience. As for army officers, Bache's West Point education helped win their good will, and so did his considerate treatment of them. He did his best to give their Survey duty its due weight in their service records, not stinting on official praise for good performance.

The civilian-military disjunction thus made less trouble within the Coast Survey than outside it. One naval officer who had done duty with the Survey blamed a subsequent threat to his career on anti-Survey feeling among line officers. Like Hassler before him, Bache contended with sporadic efforts to transfer the Survey from the Treasury to the Navy or War Department. Senator Thomas Hart Benton of Missouri, unconcerned with shipping interests or coastal surveys, led the transfer movement in the late 1840s on grounds of economy, suggesting that the navy might do all the work as part of its regular duties. The success of army and navy scientific exploring expeditions in the fifties provided another argument.

Bache and his supporters, including the scientific establishment and Senator Jefferson Davis, beat off the besiegers. The Survey, Bache insisted, had a mainly civilian purpose and an amphibious domain. The navy had not done well when it controlled the Survey for a time in the 1820s. Military and naval assignments tended to be transient, whereas a civilian survey could count on a "permanent nucleus," as Bache put it, and therefore experienced, long-range planning.

Bache still had to battle annually for the Survey's appropriation. His adversaries played upon personal and bureaucratic jealousies, the theme of economy, and the Survey's anomalous status as an officially temporary project that somehow seemed to be further from completion every year. The New York City press, especially Horace Greeley's *Tribune,* sniped at Bache as a champion of government largesse to toplofty scientists rather than down-to-earth inventors.

But Bache had good cards and played them well. He gave an impression of "perfect directness, truthfulness, disinterestedness, and good temper," wrote one supporter. That helped, and so did the fact that he was not really so nobly ingenuous as all that. In broadening Survey operations to cover fourteen coastal states, for example, he purposely broadened his base of support in Congress; and he cultivated landlocked politicians as well. Meanwhile he mustered a host of interested allies: shippers, insurance companies, chambers of commerce, boards of trade, businessmen generally, humanitarians concerned with saving lives and ships, and not least, the scientific establishment he sustained and led.

Unlike his bureaucratic counterpart and rival Matthew Maury, who sounded his own horn too blatantly, Bache struck a public pose of mute modesty while privately orchestrating a chorus of praise by others. The editor of Silliman's *Journal,* James Dana, who prized the Survey as a dike against "a flood of quackery and ignorance," invited Bache to plant unsigned encomiums on it in the *Journal,* which Bache did. Bache's coterie, the Lazzaroni, dominated the American Association for the Advancement of Science and made it a cheering section for the Survey. Bache also called the attention of Congress to praise of the Survey by eminent European scientists.

With all this massed support, and with the redoubtable Joseph Henry at his side, Bache still went into every autumnal funding campaign with nerves stretched taut. Yet the appropriations grew year after year. Bache had expected that the Mexican War would mean a cutback. Instead, by taking army officers away from the Survey, the war gave Bache a further argument for keeping the civilian component. The peace that followed gave the Survey the entire Pacific Coast to work on, along with an argument for still more money. Appropriations now regularly exceeded four hundred thousand dollars.

Congressmen grumbled, and in the fall of 1858 a final concerted assault began with press attacks, pamphlets, and efforts to rally all of Bache's opponents and enemies. Bache was accused of lusting for power in both science and government, power to be won with taxpayers' money, patronage, intrigue, and a clique of elitists. The Bache side counterattacked in force, armed with a glowing AAAS report of eighty-eight pages on the Survey's efficiency, economy, commercial and humanitarian benefits, and internationally acclaimed scientific achievements. In March 1859 the *New York Times* reported that "an appropriation of $452,000 for the Coast Survey . . . was agreed to with singular unanimity." Never again was there another serious attack.

Appropriations for the Survey had thus reached a level not to be equalled by any other federal scientific agency until 1884. But the scientific Alexander in command looked for new worlds to measure. "I only wish we were surveying the U.S.!" Bache wrote privately in 1854. "See how the appetite grows with what it feeds on." Having Congress well in hand, by the spring of 1860 he was eager to absorb all the interior surveys and asking, "Is it not nearly time to begin the discussion?" Events that year on a larger stage made clear that it was not.

Our survey of Bache's domain does not end there. From Hassler he also inherited the post of superintendent of weights and measures (which with his Coast Survey pay made him the highest-salaried scientist in the nation). While standardizing measures and instruments for customhouses and states, he dreamed of an international congress to establish a worldwide system. Apart from his official Coast Survey work, he made numerous studies of harbor improvements, often with General Joseph G. Totten of the Army Engineers and Captain Charles H. Davis of the navy; and with Joseph Henry he served on the Light House Board, again alongside army and navy officers. One of his enemies within the survey paid this scientific factotum the most convincing of compliments, a grudging one, when he wrote in 1850: "Why does Bache labor with such devotion in the cause of the C.S.? Why does he so economize to the last cent? Why does he work so intensely himself and screw and drive his assistants to the utmost extent of their powers? . . . He is laboring with all the powers of a Jesuitical mind of no mean order to attain scientific reputation." This raises the question of Bache's actual contribution to science in America.

Bache's engineering background and his work in applied science did not signify indifference toward basic or "pure" science. While in charge of the Survey he wrote numerous purely scientific papers. He served as vice president and president of the American Philosophical Society during the forties and fifties. In his 1849 report he was even bold enough to call basic science the real justification for the Coast Survey.

The Survey's official assignment imposed limits on its scientific range, but they were broad limits. "The range of the Survey," wrote Benjamin Gould in retrospect, "was made to cover almost the whole range of physical science, from the structure of the microscopic dwellers in the bed of the ocean, up to the improvement of lunar tables and the determination of positions of fundamental stars." The significant words are "was made to." Bache saw no impropriety in furnishing vessels to Agassiz for work in marine biology off Cape Cod in 1847 and Florida in 1851, the latter junket yielding important information on reef building for the government. Bache got Jacob Bailey of West Point, the nation's foremost microscopist, to study organic remains in samples of sea bottom collected by the Coast Survey. Work in natural history could have been taken on by the Coast Survey's land parties, as in the Western surveys, if Bache had chosen to push that aspect.

But Bache from the start was a scientist-engineer, a geophysicist, with a predilection for measurement, instrumentation, and mathematical theory rather than natural history, and the Survey "was made to" take its tone from him. He held even the humdrum work of surveying to more precision and sophistication than mere utility demanded, in quest of what has been called "the geodesist's holy grail, the exact figure of the earth." (When the triangulation from the Canadian border to the tip of Florida was found to be off by eighteen inches, a review went on until the discrepancy was accounted for by an error made near Cape Hatteras.) For observations of terrestrial magnetism, his own longtime specialty, Bache set up more than a hundred stations along all three coasts by 1858, and he considered the Coast Survey table for magnetic variation "one of the contributions to general physics."

In 1876 the noted astronomer Charles A. Young called the Coast Survey the single organization to which American astronomy owed most, not so much because of its direct scientific contributions—although those were significant—as because of its indirect encouragement of basic astronomical research. For making the observations that were essential to an accurate survey, Bache subsidized both individuals and observatories, and he knew enough to pick the best. Many stayed in astronomy who would otherwise have left it for better pay in other fields. Instead of relying on the Naval Observatory under his adversary Matthew Maury, however, Bache preferred to employ his crony Benjamin Peirce at Harvard. When Benjamin Gould took over the Survey's longitude department, he too made his headquarters at the Harvard Observatory. This patronage helped make Cambridge a center of astronomical research in the 1850s.

Though the Survey never followed up the work done for it by Bailey and Agassiz in marine biology, it plunged deep into oceanography. Bache established stations for tidal observations on all three coasts, and the findings helped in elaborating on existing tidal theories. Ocean winds, currents, and temperatures were studied, with special attention to the Gulf Stream; and new techniques for deep-sea soundings were developed.

"Various other institutions took the name 'national'," wrote Gould in retrospect, "but there was only one really national scientific institution for the first twenty years of his administration, and that was the United States Coast-Survey." Gould referred, no doubt, to the Survey's scale, scope, and able personnel, but his remark raises the question of character. Was the Coast Survey "national" in the thrust and manner of its work?

Certainly it epitomized two related American scientific tendencies: precision of measurement and skill in instrumentation. Bache, said Gould, "never shunned a tenfold labor, if it was to be repaid by double precision." Not only by strictness in routine work, but also in adopting and often improving the best available instruments and techniques, Bache raised the precision of the Survey's data to a high degree. His superintendency of weights and measures thoroughly suited him, and it encouraged his growing interest in the purely scientific problem of determining physical constants with maximum precision. He introduced electrotyping and photography to the reproduction of charts. Under his aegis, Sears C. Walker worked out techniques and organization for telegraphic determination of longitude by synchronized observations in different places, a notable advance in geodesy. Walker even made systematic studies of the human factor in observing and recording signals. The Survey sent out chronometric expeditions in 1849 and 1855. And in the New York Crystal Palace Exhibition of 1853, the Survey's exhibit of "conditions, methods, and instruments" stood out (according to the younger Silliman) as the one "most honourable to our country."

The Survey was "American" also in its assumption that theory and practice were mutually sustaining. It was "American" in its passion for fact gathering. "The ordinary labors in magnetism are like those in astronomy," Bache wrote approvingly in 1843; "they yield no point of discovery, but go to the general accumulation of facts." His tone was much like that of Agassiz or Baird in natural history.

Bache and his ally Henry diverged from the natural history mainstream of American science chiefly in their quantitative or mathematical approach, even to biological phenomena. The Coast Survey differed from the Western surveys not only in its greater precision and its longevity, but also in the more theoretical structure of its operations.

Nevertheless, to say that even in those respects Bache's Coast Survey was not characteristically "American" raises a question about the definition of the term. So large did the Coast Survey loom and so powerful was its influence that what the Coast Survey did could be said to be American simply because the Coast Survey did it.

Though Bache's professional brethren accepted him as the high priest of government science, the lay public accorded that eminence to Lieutenant Matthew Fontaine Maury of the United States Navy. To the historian, Bache and Maury typify two varieties of the nineteenth-century American scientist, each with his legitimate place (not necessarily equal) in the scheme of things. To themselves, however, Bache and Maury seemed irreconcilable antagonists.

Maury was a short, stout man, whose balding pate accentuated the fullness of his clean-shaven face and domed forehead. His full lower lip and the spots of color in his high cheekbones redeemed his face from stolidity, however, and suggested the fire within. Though Maury had a dreamy look when deep in thought, spoke in a soft Southern voice, and worked unruffled among his romping children, he quarrelled readily and fought hard.

Born in Virginia, but reared on a small Tennessee farm where his father grew cotton with the help of five sons and three slaves, Maury became fascinated with the sea before he ever saw it, on the strength of his older brother's adventures as a navy midshipman. In 1825, at nineteen, he too wangled an appointment as midshipman. Maury's only formal education had been some spotty elementary schooling and seven years at a local academy, and he got little instruction from others during his ensuing nine years of sea duty around the world. But he studied on his own. In keeping the log of a ship on patrol off the west coast of South America, Maury delighted in recording data on wind and weather, compass variations, currents, coastlines, and tides. This led him in 1834 to publish an article in Silliman's *Journal* on "the Navigation of Cape Horn." Assigned to shore duty, he published a navigation text for midshipmen, watered down from Bowditch's *Practical Navigator.* Promoted to lieutenant in 1836 on the strength of his book, Maury buttressed his reputation as a scientist by studying geology and astronomy.

Naval officers in that day were well advised to develop some such sideline. With few ships in commission, most officers had to wait idly on shore at half-pay for years unless they found something to do at navy yards or with the Coast Survey. Partly to alleviate the glut of idle officers, Congress in 1836 authorized the navy to explore the South Pacific. A year later Maury was appointed astronomer of the projected expedition, but when the original commander was replaced by Charles Wilkes, with whom Maury had just quarrelled, Maury begged off. In 1839 a stagecoach accident left him with a permanently lame leg and therefore permanently on shore duty. Two years later, his scientific bent finally paid off. He was appointed superintendent of the navy's Depot of Charts and Instruments in Washington.

Maury's predecessor, Lieutenant James M. Gilliss, who had studied astronomy in Paris, had made the most of orders to undertake some modest astronomical observations in conjunction with the Wilkes Expedition. Not only had Gilliss triumphed over inadequate instruments to win the admiration of professionals for the quality of his work, but he had also talked Congress into authorizing a respectable observatory in place of the existing thirteen-by-fourteen-foot wooden building—thus "smuggling" a national observatory into the domain of government science. After Maury arrived in the summer of 1842, Gilliss went off to Europe to get instruments for the establishment.

Maury meanwhile began extracting and collating data from the mass of ships' logs stored at the depot, and systematizing new data by means of a standard blank form to be filled in by all government vessels cruising the high seas. He planned to develop a new kind of chart indicating probable wind, weather, and currents at any given point and season. As the new observatory building neared completion in the spring of 1844, Maury delivered well-publicized papers on the Gulf Stream and ocean currents to two scientific meetings; and that summer the secretary of the navy, a fellow Virginian, made his navigation text required reading for all midshipmen. Maury's name was becoming known to the public, and his friends were busy in his behalf. On October 1 Lieutenant Gilliss was stunned by the announcement that not he but Maury was to be superintendent of the new Naval Observatory.

Though Gilliss accepted his superiors' decision without complaint, the scientific community tended to sympathize with him. And Maury's course at the observatory did nothing to change their minds. Maury started out bravely enough. He studied the works of leading European astronomers, rhapsodized on the joys of stargazing—"with emotions too deep for the organs of speech, the heart swells out with unutterable anthems"—and began an ambitious star-cataloguing program. In January 1846 he discovered and reported that Biela's Comet had split in two (a discovery that the historian Bernard de Voto almost a century later seized on as having been symbolic of impending disunion).

But Maury's enthusiasm soon cooled. By the end of 1846 he was complaining of insufficient facilities and manpower. The drudgery of reducing observations to proper form for the star catalogue fell years behind; and in the end the overambitious project petered out with the fifth volume in 1859, covering observations for 1849 and 1850. As Maury's passion for the study of the sea reasserted itself, he let the astronomical instruments deteriorate and the observatory staff slack off. By the time the splitting of the Union dislodged the pro-Confederate Maury from command, those who remained, even the naval officers, had come to feel that thorough scientific training and single-minded commitment to astronomy should have been given more weight than a navy commission as qualifications for running the establishment.

Maury's downgrading of astronomy at the Naval Observatory may explain why the Nautical Almanac, though established in 1849 under the Naval Observatory budget, was domiciled in Cambridge, Massachusetts, and reported directly to the secretary of the navy. The new bureau proposed to raise the nation's scientific standing, promote astronomy, and improve navigation, all this by producing an annual ephemeris or astronomical almanac of five or six hundred pages, more accurate in data and advanced in theory than the British Nautical Almanac, on which Americans had hitherto depended. The secretary of the navy, jealous of Maury's celebrity, went along willingly with the unusual bureaucratic arrangement, but the impetus for it had come from Bache and his Coast Survey protégé, navy lieutenant Charles Henry Davis, who was appointed to head the new office.

Born in 1807 in Massachusetts, Davis interrupted his Harvard studies at sixteen to accept appointment as a midshipman, then returned after seventeen years of sea duty and got his degree. Meanwhile he had struck up a lifelong friendship with Benjamin Peirce, who gave him solid grounding in mathematics. They became brothers in spirit and, when Davis married Mrs. Peirce's sister, brothers-in-law also. Davis's appointment to head the Nautical Almanac in 1849 followed seven years of Coast Survey duty, during which he had made contributions to the knowledge and theory of tidal action and won the hearts of Bache and Henry.

Davis was a tall, spare man of military bearing, with a high forehead, an aquiline nose, and kindly eyes. The naval officer Samuel Du Pont in 1852 found him "a remarkably clever talker, giving him great personal influence over those he is associated with." One of the nine young men in the Almanac office in 1851, the North Carolinian Benjamin Hedrick, thought Davis "one of the finest and most gentleman-like men I ever met with." Even a passing newspaper reporter remarked on Davis's "chivalric courtesy." With all this, Davis had intellectual ability, speaking several languages well, reading widely and deeply in literature, making close friends in the literary galaxy of Boston, and demonstrating both capacity and love for science.

Lieutenant Davis set up shop in the little brick building in Cambridge because, he said, printing could be done thereabouts most cheaply, quickly, and accurately, and because the Boston-Cambridge area had "the best scientific libraries of the country." Surely also he was influenced by his family and social ties there, the availability of Peirce (already doing Coast Survey work), the proximity of the Harvard Observatory, and the distance from Maury. Davis was a happy man in Cambridge, and he ran a happy office, with a cheerful freedom from discipline and restraint that might have demoralized a ship but was the ideal regime for a crew of mathematicians and astronomers.

It was no wonder that, with this man and this environment, the Nautical Almanac became a notable training ground for bright young scientists. Hedrick, for example, with Davis's encouragement, took afternoons off to study with Peirce, Horsford, and Agassiz, and went on to a professorship at the University of North Carolina (and eventually to a place in the history of the struggle for academic freedom). With Chauncey Wright in the Almanac office, Simon Newcomb later recalled, "philosophic questions were our daily subjects of discussion." Among a dozen others of his eminent fellow alumni, the noted astronomer Joseph Winlock mentioned John D. Runkle, William Ferrel, Truman Safford, and George W. Hill. "Few indeed," wrote Winlock, "have ever been so employed who have not retained a taste for scientific pursuits," whereas he knew of at least four promising men who had not only left Maury's observatory but the field of astronomy as well.

Maury himself complained now and then about Davis's delay in returning men borrowed from the Naval Observatory. Davis in turn flared up when in 1852 Maury began declaring that Davis merely put together mechanically the data collected by the observatory and that any passed midshipman could do the job. "To these remarks," Davis wrote bitterly to Henry, "I can trace my loss of salary." Davis was sure Maury knew that the work required German, Latin, French, higher mathematics, and much reading in the field.

As consolation, Davis had the success that year of the first volume of *The American Ephemeris and Nautical Almanac,* giving tables for 1855. Davis reported proudly that the American figures were significantly more accurate than the British. He also pointed out the book's value for geographers and surveyors and as an astronomical ephemeris for observatories.

One of Maury's astronomical lapses had meanwhile dealt his predecessor Lieutenant James Gilliss another blow. For years Gilliss's great ambition had been "to win the confidence of scientific men." Denied command of the observatory he himself had brought into being, he caught up the suggestion of a German astronomer that the distance of the sun from the earth might be determined by simultaneous observations of Venus and Mars from similar degrees of north and south latitude. European astronomers proved reluctant to cooperate, so Gilliss pinned his hopes on the Naval Observatory as the principal northern base and made much of the all-American character of the project as a contribution to national prestige in science. With the help of Bache, Henry, the American Academy, and the American Philosophical Society, he got a five-thousand-dollar appropriation from Congress and a Navy Department assignment to Chile with two midshipmen and a civilian clerk. The Smithsonian came through with five thousand more for a telescope and astronomical clock. And so the little party made its way to Chile by private passage in the fall of 1849.

At Santiago, where he built an observatory, Gilliss persevered against handicaps. One of the midshipmen was injured in a riding accident; the other had a poor opinion of Gilliss's scientific attainments. An unsympathetic navy secretary came near cutting off funds. Bache, suspicious of Maury's good faith, had persuaded Gilliss to make magnetic, meteorological, and seismological observations as a hedge against astronomical miscarriages; and Congress insisted on a commercial, agricultural, and mineral report on Chile. From 1849 to 1852, the Gilliss party not only carried out these charges along with its Venus and Mars observations, but also made thirty-three thousand star observations. It even bought, begged, and collected a number of natural history specimens, organized a plant and seed exchange program, and trained three young Chileans in astronomy. (Before returning, Gilliss sold the building and instruments to the Chilean government, which then set up its own program—the first such in South America.)

When he got back, Gilliss was shocked to find that Maury had neglected to make the supporting observations from Washington. The Harvard College Observatory had made some of Mars for its own scheme of determining solar distance, and George Bond, the son and assistant of its director William Bond, later denied that negligence at home was to blame for the failure of the Gilliss expedition in its main purpose. But

Bache, Henry, and their circle were annoyed with the Bonds and furious with Maury, who had publicly called the Gilliss expedition useless even before it began. Bache's shrewd advice to Gilliss to hedge his bets with other data made it possible for Gilliss, put on the reserve list at full pay, to write six volumes of reports on the expedition and get four of them published by the government. They included much of value to both science and history. But Gilliss brooded for years over what he considered Maury's treachery in robbing him of still higher scientific honor.

Maury's dereliction need not be laid to malice, however. His course in the late 1840s was another demonstration of how scientists could bend government support to their own predilections. Maury was simply letting astronomy slide in favor of oceanography. His *Wind and Current Charts* began coming from the press late in 1847 and made their first big splash in the following spring, when a Baltimore bark used them to cut the time to Rio de Janeiro by a third. The day of the clipper ship, dedicated to speed, had just dawned, and Maury's fame brightened with it. His charts helped make possible the *Flying Cloud*'s famous record run from New York to San Francisco in eighty-nine days—less than half the usual time—as well as other celebrated sea sprints. For sailing ships generally, improved design probably did more to cut sailing times. Nevertheless, the demand for charts ran into the thousands, and in return for them Maury exacted cooperation in a still vaster program of data gathering. A thousand ships were reporting by 1851. He also collected data on the breeding habits, migrations, and gathering places of whales, and his resultant whale charts thus had some zoological as well as decided commercial value. (Fortunately for the whales, Benjamin Silliman, Jr., another scientist with a nose for the commercial, came along with his petroleum report of 1855 before Maury's charts had time to wipe out the species.) With the *Wind and Current Charts* came *Explanations and Sailing Directions,* 18 pages in the first edition (1850), 772 pages in the sixth (1854), and two volumes totalling 1,300 pages in the eighth (1858–59).

Seizing on a British proposal for meteorological cooperation, Maury organized an international conference at Brussels in 1853 at which he was the most conspicuous of twelve delegates from ten nations. It worked out a uniform system of marine meteorological observations based largely on the Maury model. This coup raised Maury to the pinnacle of his international celebrity and brought him a shower of medals, orders, and honors from monarchs and learned societies.

Such trinkets did not turn aside the barbs of professional scientists in his own country. Major scientific journals ignored the Brussels conference, at which ten of the twelve delegates had been naval officers. Almost from the beginning of his years in Washington, Maury had seemed fated to run afoul of the professionals, fated by ambition, touchy pride, a craving for popular acclaim, deficiency in professional credentials, and a rashly speculative turn of mind.

Enmity between Maury and the Bache circle grew deeper and more open as the fifties advanced. Maury's oceanography encroached on the sphere claimed by Bache's Coast Survey, and both competed for the shipping constituency. Maury collided with Henry by proposing to enlist farmers along with sea captains in his network of meteorological observers. Early in 1856 Henry pointed out publicly and tartly that the Smithsonian already had a network of weather observers throughout the land, and that Maury's wind charts were neither scientifically sophisticated nor likely to be useful to farmers. Maury's lobbying for Congressional authorization stirred up a campaign of letters to the press ridiculing Maury's scientific pretensions. Defeated in the Senate, Maury fought on for his farmer-network scheme with articles and extensive lecture tours, but the coming of civil war ended his efforts in that direction.

To Bache's annoyance, Maury had carried on a navy program of transatlantic soundings that revealed a relatively shallow submarine plateau between Newfoundland and Ireland, one which, Maury wrote early in 1854, "seems to have been placed there especially for . . . a submarine telegraph." That report gave Cyrus W. Field the signal to launch his famous Atlantic Cable venture, and in gratitude Field promised Maury first use of the cable to determine longitude. Then in 1856 Bache got the navy to order new soundings by the officer who had done the previous ones for Maury but had since been working for the Coast Survey. A quarrel ensued as to Bache's jurisdiction over deep water, the superiority of the Coast Survey's sounding techniques, and whether Bache or Maury should have first crack at the cable for longitude determination. The secretary of the navy was rather relieved when the cable of 1858 went permanently dead and rendered the longitude issue moot.

Thus, by the end of the fifties, the bureaucratic barons were having at each other amain. When Maury heard in 1856 that Henry was arranging for telegraphic weather reports, he raged, "Do you not see what villains men are? They'll steal your brains, and would your money too but for the hangman's whip." And in 1858, "Bache's jealousy & envy surpass my comprehension. He is a bad man." Henry's private comments were more sarcastic than savage. Those of some others were less restrained. The navy's leading mathematician, William Chauvenet of the Naval Academy, took Bache's part. Alarmed by hints of bureaucratic expansionism, Chauvenet asked rhetorically in 1858, "Shall Chauvenet & Coffin be put under Maury? The first named individual answers !!!*NO*!!! . . . He will resign first—but not without showing fight." English scientists had accepted Maury as a provider of data, a role they thought proper for Americans. But in the summer of 1860 Benjamin Peirce wrote Bache with fierce relish, "I killed the great Mauritanian with several of the best English savans."

. . .

To Bache and his allies, Maury had come to stand for the demon of charlatanism they were struggling to cast out of American science. "Our real danger," Bache pointedly warned the AAAS in 1852, "lies now in a modified charlatanism, which makes merit in one subject an excuse for asking authority in others, or in all."

The scientific philosophies of Bache and Maury were not totally opposed. Like Bache, Maury at first paid lip service to Baconianism. "To the philosopher," he wrote in 1844, "every newly discovered fact . . . is a gem. Our knowledge of nature and her laws is but a number of such facts, brought nigh and placed side by side." Like Bache, Maury saw pure and applied science as gradations in a continuum; and like Bache, he turned that idea to good account in seeking public support. Like Bache, of course, he preached the legitimacy of scientific research as a function of government.

But the differences between them were wide and significant. Their choices of European models reveal this. In his office, Bache kept a bust of Arago, the French physicist and astronomer; and he drew upon the German mathematician and physicist Gauss for the theoretical basis of his work in geodesy and terrestrial magnetism. Like Bache, Arago and Gauss thought in quantitative terms, emphasized precision of measurement, and focussed their work on specific, well-defined fields. Even in an oration on the death of Humboldt, Bache made a point of praising Arago. For Humboldt was different in spirit, engaged in a romantic, almost mystical search for the unity of all creation. The great German explorer and naturalist took the whole world as his field and tried to embrace it in his massive work *Kosmos.* Humboldt was reaching for what was unattainable by any one man in science. Nevertheless, Humboldt was Maury's model. Indeed, after the Brussels Conference of 1853 Maury paid a worshipful visit to Humboldt, who subsequently induced the King of Prussia to award Maury the Kosmos Medal.

The half-dozen years before that visit had been Maury's best as a scientist. In oceanography, a field largely inaccessible to the unaided individual scientist, Maury's elaborate and far-flung data-collecting system constituted in itself a major contribution. Scientists like Dana and Bailey seized on Maury's data—isothermal lines, sediments, species distribution—and asked for more. Maury happily obliged them. He used his own data to valid scientific purpose, as in his maps of ocean surface temperatures and bottom contours. But as his data gathering became more extensive and systematic, he became increasingly conscious of anomalies and of outside influences on any given area under study. So he widened his scope to the Arctic, to river drainage basins, and so forth. His purview became global, even threatening (as Henry saw it) to overrun the land. Meanwhile, just as he had once moved brashly into astronomy, so Maury now extended himself into other disciplines, such as physics and biology. In this, Maury foreshadowed the coming of oceanography as a new interdisciplinary science. And in the large hypotheses he began venturing, Maury raised questions and sketched lines of approach for that field.

But even in his practical *Sailing Directions* Maury began to show delusions of Humboldtean grandeur, to hazard pure guesses about the grand global dynamics of the ocean and its weather. For years, Bache, Henry, and their younger adherents had dedicated themselves to "putting down quackery" before "quackery put down science." Dana had seen the Coast Survey as "our protection against a flood of quackery and ignorance." To these, Maury's oceanic effusions began to pose the most dangerous threat of such a flood, and presently the threat seemed to be confirmed. Maury knew his Bible well. Surely he mused over the thirty-eighth chapter of Job, full of oceanographic allusions—"Hast thou entered into the springs of the sea? or hast thou walked in the search of the depth? . . . Hast thou perceived the breadth of the earth? declare if thou knowest it all." He might have done well also to ponder a hint from the thirty-first chapter: "Oh . . . that mine adversary had written a book."

Evidently he did not take the hint. For early in 1855 Harper and Brothers published the first edition of Maury's *The Physical Geography of the Sea,* a book that made him famous in his own time but also harpooned his scientific reputation. It had germinated when the publishers of Maury's *Sailing Directions* urged him to do a popular version of that work before "some Yankee bookmaker" beat him to it. Scribbling at top speed every evening, Maury wrote the entire manuscript of what he hoped would be his magnum opus during the spring of 1854, relying on his wife and children as critics.

The result bore the marks of haste. *The Physical Geography of the Sea* was a disorganized, repetitive grab bag of material from the *Sailing Directions,* which itself had been a receptacle for earlier articles and papers. In dealing with the circulation of the atmosphere, Maury did not even trouble to rework earlier material into harmony with later. Besides its errors of fact, the book's omissions betrayed Maury's unfamiliarity with recent work by others in the field. Worst of all, from the standpoint of professional science, Maury now advanced as sober hypotheses a number of sweeping, sometimes preposterous statements that had appeared in the *Sailing Directions* as mere tentative speculations, clearly distinguished from fact. Yet the book was an immediate popular success, remaining in print for twenty years and being translated into half a dozen European languages.

There were several reasons for this. The savings in sailing times made possible by Maury's hard data predisposed the lay reader to take his wild theories seriously. The book embodied popular misconceptions about weather, climate, winds, and currents. It quoted copiously from the Bible, even using it to support scientific hypotheses, and harped on natural theology's "argument of design," the thesis that an intelligent being must have designed the intricate machinery of the cosmos. Perhaps most stimulating to sales was Maury's striving for popular effect by dint of a lively imagination and a romantically oratorical style. Several chapters opened strikingly, especially the first, which began:

> There is a river in the ocean. In the severest droughts it never fails, and in the mightiest floods it never overflows. . . . The Gulf of Mexico is its fountain, and its mouth is in the Arctic Seas. It is the Gulf Stream. There is in the world no other such majestic flow of waters. . . .

In short, the book thoroughly suited the popular taste of the times.

Popular approval did not impress Bache and his friends. More likely it had the opposite effect. "Maury's work on the sea," wrote Henry privately in 1857, "contains more absurd propositions than are to be found in any book ever published by a person in such high position." A review of the book, wrote Gould in 1855 to a Coast Survey stalwart, "*must* be done & I had a long talk with Agassiz about it. . . . I will . . . collect a big mass of materials to help . . . at the Augean cleansing." All that came of this was a brief notice in Silliman's *Journal* that Maury had brought together material from his *Sailing Directions* "under a popular form . . . and the work cannot fail to find many interested readers." Then followed merely a list of chapter titles and the terse comment that "while the work contains much instruction, we cannot adopt some of its theories, believing them unsustained by facts." Three years later, after two enlarged (but little improved) editions had appeared, Dana complained to Bache that the *Journal* was still waiting for "the review of Maury that Gould was to furnish us." John LeConte published a scathing, eighteen-page review of the 1858 edition in the *Southern Quarterly Review,* remarking that "something should be done to check the torrent of pseudo-science which is sweeping over our country." But the perfunctory jab of 1855 was all that Silliman's *Journal* ever published by way of a review.

The most notable response to Maury's "pseudo-science" came in the end not from Bache's elite circle of professional scientists but from an amateur, a Nashville, Tennessee, schoolmaster named William Ferrel. A shy bachelor, born in a Pennsylvania log cabin and largely self-educated in science, Ferrel had published his first scientific paper in Benjamin Gould's *Astronomical Journal* in 1853 at the age of thirty-six. That and another in 1856 were significant contributions to the theory of tidal effects on the earth's rotation. The editor of a Nashville medical journal, aware of Ferrel's disagreement with Maury's wild notions, invited the schoolmaster to write a critical review of *The Physical Geography of the Sea.* Scorning Maury's absurd theory that global winds crossed each other at the equator and converged on the poles, Ferrel presented his own ideas in "An Essay on the Winds and Currents of the Ocean," the first modern analysis of the general circulation of the atmosphere and ocean. It offered the hypothesis that the earth's rotation deflected winds and currents away from a great circle path. Joseph Henry, for one, was rightly impressed by Ferrel's essay, though Maury paid it no heed.

Ferrel's path-breaking essay was precise but not mathematical. He elaborated on it mathematically in a series of articles, 1858–59, using observed data. In these he developed a general quantitative theory of the relative motion of bodies on the earth's surface and applied it to winds and currents, with a result now known as "Ferrel's Law," to wit, that the earth's rotation deflects a moving body to the right in the Northern Hemisphere and to the left in the Southern Hemisphere. Meanwhile Gould had got him his first scientific job in the Nautical Almanac at Cambridge, 1858–67, after which Ferrel joined the Coast Survey in Washington. Now regarded as the chief founder of geophysical fluid dynamics after Laplace, Ferrel pioneered in advancing meteorology beyond mere description to the use of mathematical physics.

Matthew Maury undeniably also made contributions to the advance of American science, but they were primarily organizational and empirical. His success in capturing the public's imagination doubtless helped increase public readiness to support science through government. The data he accumulated helped others to make genuine scientific progress, as in Jacob Bailey's marine micropaleontology. Even *The Physical Geography of the Sea,* with all its absurdities, inconsistencies, and irrelevances, had a lot of useful information, especially about the deep sea bottom.

Nevertheless, Maury's inability to accept and accommodate to valid criticism, his flowery prose, his airy theorizing, his lack of mathematical and logical rigor, his inadequate training, and his overblown scientific pretensions all marked him as part of a passing era. Buoyed up by popular acclaim, Maury never understood the scorn of the professionals, nor would he have understood the verdict of a recent historian that his most significant contribution to science was the stimulus he gave Ferrel.

Chapter 14

The Smithsonian, Seedbed of Science

The Smithsonian Institution, that fulcrum of scientific power in antebellum America, owed its existence to the death in 1835 of a childless young Englishman. Under the will of his late uncle, a wealthy English bachelor named James Smithson, a sizable estate that had been held in trust for the young man and his future offspring thereupon passed instead "to the United States of America, to found at Washington, under the name of the Smithsonian institution, an establishment for the increase and diffusion of knowledge among men." Beyond this bare statement, nothing was or is known for sure about Smithson's motives and intentions in making the bequest. Except for the fact that Smithson himself had been an amateur chemist and naturalist, the precise definitions of "increase," "diffusion," and "knowledge" remained an open question. For years, therefore, Congress debated the disposition of the half-million dollars involved. Some congressmen favored an agricultural school, others variously proposed a college, a German-style university, an observatory, a library, a museum, a lecture foundation, a research foundation, or some combination of these.

At last, in August 1846, Congress passed a bill providing for a secretary, a board of regents, and a building. The building was to have space for a library, natural history cabinets, a chemical laboratory, an art gallery, and lecture rooms. A copy of every copyrighted publication was to be deposited in it. And whenever space permitted, the building was to house federally owned objects of art and natural history. From all this it was apparent that Congress had still not been able to make up its mind about what the Smithsonian was to be and do. But in authorizing the secretary and regents to spend the income otherwise "as they deem best suited for the promotion of the purpose of James Smithson," Congress at least devolved the decision upon a less cumbersome and many-minded body.

A course was promptly and adroitly set by the youngest regent, Alexander Dallas Bache, the only professional scientist on the board. He knew not only where he wanted the Smithsonian to head but also the man who could best guide it in that direction. So he carried through a board resolution specifying the qualities wanted in a secretary: "weight and character and a high degree of talent . . . eminent scientific and general acquirements . . . capable of . . . original research . . . well qualified to act as a respected channel of communication . . . [with] the world of science and letters." This, it will be noted, assumed that "knowledge" primarily meant science. And its leadership specifications fitted only one American scientist of that day—Bache's close friend, Joseph Henry of Princeton.

Henry dismayed Bache by at first hesitating to accept the position. Some years earlier Henry had begun to fear the loss of his talent for research, but he had regained confidence and was still publishing at forty-nine. With his lectures thoroughly worked out, he looked forward to more money and research time at Princeton, whereas the Smithsonian job would burden him with administration and politicking. Henry's Princeton friends eagerly seconded these misgivings.

Bache rose to the occasion, however, arguing that the success of the Smithsonian and therefore the hope of timely progress in American science depended on Henry's acceptance. When Henry tried to price himself out of the running by demanding three thousand dollars a year, a free house, and permission to teach part-time at Princeton, Bache got him thirty-five hundred, the house, and the Princeton dispensation. On December 3, 1846, Bache informed him of his election—"Science triumphs. Io Paean!!" Henry now felt bound to accept. And though he soon discovered "the annoyances, the difficulties and the uncertainties" of his new job, he sealed the bargain in June 1847 by refusing the biggest plum in American academic life, the only college position that could have won him from Princeton, Robert Hare's professorship at the University of Pennsylvania. By then Henry felt irrevocably committed to the Smithsonian by his obligation to Bache, his duty to American science, and his own pride. "I do not relish the idea of giving up," he wrote. And in the end, more than thirty years later, it was to be death that took him from the Smithsonian.

Henry's personal sacrifice turned out to be as heavy as he had feared. For a couple of years he taught part-time at Princeton, giving a few lectures "in a perpetual flurry" and then dashing back to Washington, to the disgruntlement of his students. Meanwhile he dreamed of getting the Smithsonian under sufficient headway to let him resume his researches. He experimented briefly in his old quarters at Princeton, hoping to be inspired by their associations. In 1858 his daughter found him wistfully looking over his old notebooks. But Henry's days both of teaching and of basic research had effectively ended with his arrival in Washington.

One of Henry's old Princeton friends wrote of him, "I know few men so ill qualified for getting through the ordinary details of business." Henry complained that visitors and trivial correspondence took up much of his time, yet he insisted on dealing with both in patient detail. Practice may have improved his efficiency, but in May 1850 he wrote that he was working from nine in the morning to eight at night. In January 1853—his busiest season, what with Congress in session and an annual report to get out—his workday ran from eight in the morning until ten at night.

Apart from sheer drudgery, Henry bore another cross, that of contending with sundry individuals who had their own ideas for spending Smithson's legacy. Eight years after his appointment Henry wrote privately, "Had I before entertained any doubt of the Presbyterian doctrine of the depravity of the human heart, I should have been entirely convinced of it before this time. . . . The property of the government is everybody's property and everyone considers himself entitled to a share, or, in [other words], the public is a goose and he is a fool who does not pluck it."

Long before the Smithsonian was born, Henry and Bache had agreed on where American science should tend and what it needed. It should rise above Baconian fact gathering to search out principles. It should strive for better balance by more work in physics, chemistry, and mathematics. It should not be too narrowly utilitarian. And it should have financial support for sustained, basic research. In a letter of September 1846 to Bache, Henry had spelled out privately why and how Smithson's money should be applied to these ends. Thousands of institutions were *diffusing* knowledge in America, while not one directly supported its *increase.* Such basic research as did get done was "chiefly not in the line of science properly so called which is a knowledge of the laws of phenomena but in that of descriptive natural history." So Henry envisioned the Smithsonian as a foundation for subsidizing basic research by a peer-selected group of leading scientists and for diffusing their findings by publication.

This was Henry's ideal, but he knew it faced obstacles. True, the act establishing the Smithsonian was intentionally vague and therefore flexible to some extent. Under its terms, not more than an average of twenty-five thousand dollars a year was to be spent on the library, which implied that less might be spent. No deadline was set for receiving natural history collections. The balance of the income might be used as seen fit to promote the testator's purpose. Nevertheless, certain provisions and implications got in the way of Henry's grand plan.

To begin with, Henry resented the requirement that money be sunk in a building. As Asa Gray reminded him, the Lowell Institute had been better off for having no building at all. But, Henry wrote sarcastically to his wife, "the very salvation of the integrity of the union of the states is thought to be connected with a large building at Washington." Representative Robert Dale Owen of Indiana, one of the regents, had conceived the notion that what America needed was medieval Norman architecture, and so he encouraged young James Renwick, Jr., architect of the just-completed Grace Church in New York, to design a "Norman" pile for the new Smithsonian. Three days before Henry was elected secretary, the Institution's building committee recommended Renwick's elaborate and expensive plan.

Henry at once "exerted all my talents and influence to prevent the expenditure," enlisting the vice president and the chief justice (both regents ex officio) among others. But he had to compromise with Owen's "architectural mania," the local pride of Washington's citizenry, and the danger that with the Mexican War in progress an appeal to Congress might undo the whole bill. On Bache's canny motion, the building was permitted to go up in stages over a five-year period, thus (as Henry put it) "gaining interest and postponing the drain of maintaining a museum." So, in dramatic isolation on the Mall, the nine red-brown freestone towers and the chapel-like wings of Renwick's and Owen's eccentric vision gradually began to rise. Henry mourned the apparition as a "Norman cenotaph over one half the buried funds of the Smithsonian legacy," and he probably was the author of an item in Silliman's *Journal* protesting that the true monument to Smithson's memory was "to be built in the hearts of the people . . . and not in this land only, but everywhere." Nevertheless the masonry mounted. By the end of 1855, having devoured $303,000 of the $538,000 spent by the Smithsonian to that date, the structure was essentially complete. And at whatever cost to his sensibilities, Henry and his family lived in its east wing from 1855 to his death in 1878.

Though Henry never ceased hoping to get rid of the "cenotaph," he could have consoled himself by reflecting that Owen and his Congressional colleagues were not called "representatives" for nothing, and that one effective way to "the hearts of the people" may indeed have been through Renwick's eye-catching tribute to the power and glory of science.

Henry waged his noisiest and most successful battle over the library. The humanist faction in Congress had raised the issue by mandating a library that as a copyright repository was ipso facto not confined to science. Bache unwittingly increased the threat by persuading Henry in December 1846 that it would be politic to appoint as librarian a man of letters, Charles C. Jewett, a young professor of modern languages at Brown University.

Though a long, frank talk with Jewett impressed Henry favorably, he hoped the young man would not accept, since their views of the Institution's proper course differed so widely. The regents, however, were equally impressed, and Jewett did accept. It was precisely Jewett's drive and ability that made him dangerous to Henry's hopes. Jewett also had a formidable ally in Rufus Choate, a regent zealous for a "grand and noble public library," who got the board to earmark half of the endowment income for library purchases in "all departments of human knowledge."

This arrangement was not to take effect until the building was finished. Jewett, who did not begin full-time work until 1848, therefore had to make do with four or five thousand dollars a year for the next five years. Nevertheless, he built up a collection of thirty-two thousand volumes with the help of exchanges, developed a plan and process for printing joint catalogues of several libraries, and projected a comprehensive bibliographical collection, including a union catalogue of all American libraries. In 1854, when Henry proposed to cancel the 1847 agreement, Jewett countered by demanding the original maximum of twenty-five thousand dollars a year. That spring, Henry wrote Joseph Leidy:

> The whole affair has reached a crisis.... Prof. Jewett or myself must leave.... Prof. Jewett is making a desperate effort to get up a hostile feeling against me in Congress and will attempt to array the city of Washington in opposition to my views. . . . I have the most cordial support of Bache and a majority of the Regents.

What Henry presently called "the Smithsonian war" was on.

Jewett tried to get independent status for himself and the library, but Henry promptly dismissed him and was upheld in this by the regents. Choate resigned in a rage, and both houses of Congress began investigations. Jewett had strong support among the nation's librarians, and the literary journals mostly took his part. But, as Henry remarked of a diatribe in the *North American Review,* "the noise of the greatest gun from New England makes but a small impression on the ear at Washington." Henry rallied the scientific community solidly behind him, including Louis Agassiz, Asa Gray, and Silliman's *Journal.* With Bache's aid he made headway among the politicians, though Senator Stephen Douglas, a regent, took the library side. In the end, the Senate vindicated Henry, and the House investigation petered out inconclusively. In 1859 Henry got the copyright-depository provision repealed, and in 1866 he sent forty thousand volumes to the Library of Congress on permanent deposit. Over the following century, further Smithsonian deposits, mostly of scientific works acquired through exchanges, built that collection to more than half a million items.

Jewett went on to a distinguished career at the Boston Public Library, while the Smithsonian's own library lapsed into disarray. A visitor in 1857 found that the clerk had left, the cataloguer was about to leave, and the work seemed likely to fall to the janitor along with kindling fires, sweeping, and dusting. Meanwhile books coming in from all over the world through the exchange system were "piled up pell mell behind the shelves." A conchologist complained in 1860 that "unfortunately there are very few books here to name by. I shall have to take a great many [shells] over to England with me to identify." Henry's triumph was not without cost, even to science.

Henry, like Agassiz and other scientists, of course appreciated the importance of scientific libraries. His objections were to Jewett's bibliographical catholicity, the purely local availability of the library, and especially its drain on the Institution's limited funds. Still more fundamental were the issues of who was to run the Institution and whether it was to concentrate on science or be intellectually and financially diffused into the wide open spaces, American style. The leaders of the scientific community recognized that deeper crisis. Some, like Dana and Agassiz, took exception to Henry's antipathy toward a Smithsonian natural history museum. But they put the issue aside to help save Henry for the Smithsonian, and the Smithsonian for American science.

In 1870 Henry asserted that "the Institution is not a popular establishment and . . . does not depend for its support upon public patronage." But the building and library fights showed otherwise. Congress had created the Institution, laid down its guidelines, and provided for a government majority on the board of regents. Even in the narrow fiscal sense, Congress had by 1870 made supplementary contributions to the program, and in the next century its annual appropriations would eventually far overshadow the income from Smithson's endowment. So Henry in fact had to deal with the politicians and therefore court their electors.

Despite his 1870 boast, Henry acknowledged those political realities at an early date. "In order to render the Smithsonian Institution popular," he wrote privately in 1848, "we shall be obliged to do something for the application of science to the useful arts." At the moment he was thinking of a prize for the best treatise on agricultural chemistry—good political strategy, given the overwhelming rural voting majority. Likewise mindful of that majority, Stephen Douglas in 1852 tried to raid the Smithsonian endowment to form an "Agricultural Bureau." After Douglas addressed an enthusiastic meeting, Henry rose unexpectedly, denounced the threatened breach of Smithson's trust, and reminded the audience of what science had done and would do for farming. Some years later Henry was described as looking like a farmer himself, with his rumpled clothes and stout, vigorous form. Perhaps that semblance helped in the Douglas encounter. At any rate, the press made Henry the hero. "You have achieved a noble triumph," wrote Lieutenant Davis from the Nautical Almanac in Cambridge. And indeed, Douglas dropped his proposal, was elected a regent himself in 1854, and became a good friend of Henry.

In his annual reports, widely circulated to libraries, societies, colleges, and individuals, Henry's drum beatings for basic research were meant for the puissant laity. "The study of abstract science," he explained to them in 1860, "offers unbounded fields of pleasurable, healthful, and ennobling exercise to the restless intellect of man, expanding his powers and enlarging his conceptions of the wisdom, the energy, and the beneficence of the Great Ruler of the universe." And Henry meant it. But he further claimed that "nearly all the great inventions which distinguish the present century are the results, immediately or remotely, of the application of scientific principles to practical purposes." In this, one is reminded of his private comment in 1847: "in a country like ours . . . since it will probably be found necessary to make a few oblations to Buncombe, practical science must have a share."

Representative James Meacham of Vermont, a Smithsonian regent and advocate of the library, was not impressed. Speaking to the House in 1855 he asked belligerently, "Where are the Smithsonian researches? Where are the 'new truths' which have been developed at the Smithsonian?" The question may have given Henry a pang. Through no fault of his, the Smithsonian never did become the sort of institute for advanced study that he had dreamed of. But a fairer question would have been: what did Henry, through the Smithsonian, do for science and the nation within the limits imposed on him by Congress and the endowment?

Because of those limits, the plan Henry presented at the outset was, he confessed privately, "by no means the best which might have been adopted." But it won enthusiastic approval when circulated among the nation's scientists. This was not surprising. Henry's "programme of organization" conceded that all branches of knowledge deserved attention, including literature, philosophy, and fine arts, but in disdaining "unverified speculations" and in suggesting proper subjects it leaned heavily toward the sciences.

Officially adopted by the regents in December 1847, the "programme" was reprinted in every annual report until the 1870s. It was aimed at benefitting "mankind in general" rather than the local populace. Funds being limited, it proposed doing only what other American institutions could not do and shunning expensive, long-term projects—becoming, in other words, a seedbed of science. To *increase* knowledge, Henry proposed offering "suitable rewards" for research papers and making annual research grants to "suitable persons." To *diffuse* knowledge, the Institution would publish periodic reports on recent scientific progress, as well as occasional treatises on "subjects of general interest." Henry's grudging nod to Congressional requirements went no further than museum collections to "verify [the Smithsonian's] own publications," a collection of experimental apparatus, an art gallery of casts and copies, and a bibliographical library (this last, of course, went out with Jewett).

Even before Henry formally presented it, his idea of systematic research grants had not been warmly greeted. The Smithsonian Committee on Organization in January 1847 ventured to suggest that "special appropriations may occasionally be made" for that purpose, but "with great care, for important objects only, and where there is fair promise of speedy result." A year later Henry ruefully explained to a fellow scientist that Smithsonian income would be too small for a continuing grant or fellowship program without government aid. And in 1851 Bache wrote Henry wearily, "the naked thing cannot go, of that I am persuaded." So direct Smithsonian aid to scientific research remained sporadic and catch-as-catch-can.

Under the Smithsonian program, the experimental and mathematical sciences did not fare as well as Henry and Bache would have liked. As Bache's crony, Henry did what he could for terrestrial magnetism, enlisting the aid of government surveys. In astronomy, Henry could cite aid to Sears Walker's work on Neptune, to the Gilliss expedition, and to the Coast Survey through star tables. Direct aid to chemistry and physics rested largely on the Smithsonian laboratory, for which Henry initially ordered the latest European apparatus in hopes of inducing "a pilgrimage to Washington of all the *quid nunc* professors in our country to enlighten themselves." The laboratory survived until the fire of 1865, and from time to time able men made use of it. But it was so small that little more was done than sundry investigations of a practical nature for the government.

The Smithsonian made more of a mark in meteorology, which straddled theory and empiricism, pure research and practical application. So large did that program loom that Henry labelled one major category of expenditures, totalling more than sixty thousand dollars during his long tenure, "Meteorology and Researches." To start with, the regents let him have a thousand dollars for instruments; these he doled out carefully to voluntary observers throughout the nation, more than two hundred of them by 1852. Once engaged in the work, most of the volunteers were left to supply whatever they needed thereafter. One dedicated Arkansan, whose twenty years of observations were published by the Smithsonian, made do with a thermometer and a tin cup for rain. Along with the volunteers, Henry enlisted the cooperation of two state programs, a hundred or so military posts, various exploring and surveying parties, and an existing network in the British dominions of North America. In creating this continental—indeed, with some South American observers, hemispheric—operation out of such slender means, Henry showed talents for promotion and organization worthy of the Age of Enterprise.

The meteorology program had obvious practical applications in farming and maritime commerce, especially after Henry organized the first telegraphic network for gathering information and issuing forecasts and warnings. Beginning in 1858, a daily weather map was posted at the Smithsonian. In return for facilities and funds, Henry wrote a series of articles on the relation of meteorology to agriculture for the Patent Office's annual agricultural reports. Having proved the practical value of the program, Henry was able in 1874 to get it transferred to the army's Signal Service; by then he had surely earned the title of "Father of the Weather Bureau."

But beyond all that, Henry wanted to lay the basis for a scientific theory of weather. He enlisted the support of leading American meteorological scientists—Elias Loomis, James Coffin, William Ferrel, and the "Storm King," James Espy (though Henry questioned some of Espy's theories). Most of all, the Swiss immigrant scientist and friend of Agassiz, Arnold Guyot, contributed his expertise, selecting instruments, preparing a detailed guide for observers, and publishing tables. Henry himself, though chary of so contentious a field, used his own Patent Office articles to pioneer in wedding physics to meteorology. Not even today's sophisticated computers have yet conquered the enormous complexity of that task. Nevertheless, Henry's work faced more truly toward the ultimate goal than did Maury's speculative sallies.

The line between increasing and diffusing knowledge is hard to draw for some *indirect* effects of Smithsonian activities. The stipulated program of free lectures, which Henry dutifully maintained until the fire of 1865 burned out the hall, was a clear case of diffusion (and local at that, to Henry's distaste). Yet in publicizing the talents of able scientists like Agassiz, Guyot, Benjamin Gould, and the two Sillimans at the seat of political power, the lectures must have advanced basic science, not only by adding to the speakers' income and celebrity, but also by impressing certain men who had a say in allocating public money.

Henry knew that knowledge had to be diffused in order to be increased. He needed no prodding, therefore, to develop an international system of exchanging scientific publications. Beginning with the exchange of the Smithsonian's own publications for those of foreign societies, and eventually with the help of reciprocal waiving of import duties, free shipping by public-spirited companies, and Congressional funds, the Smithsonian became a clearinghouse for exchanges between many American societies and their foreign counterparts. This stimulated at least one society to liberalize its own exchange policy. "The Institution," Henry proclaimed in 1853, "is now the principal agent of scientific and literary communication between the old world and the new."

The Smithsonian's publication program benefitted basic research still more directly. Henry's first concern was to do only what existing agencies could not, which in that day included the publication of long or expensively illustrated scientific works. His second concern was to exclude "unverified speculations on subjects of physical science." In 1847 the versatile politician, diplomat, and scholar George P. Marsh brought Henry the ideal manuscript, a study by two Ohio archaeologists, Ephraim G. Squier and Edwin H. Davis, of more than two hundred Indian mounds in the Mississippi Valley. Henry tried it on a committee of the New York Ethnological Society, which pronounced it outstanding but beyond the society's means to publish—precisely what Henry wanted. He also saw it as countering complaints about his leaning toward physics. And so in 1848, *Ancient Monuments of the Mississippi Valley* inaugurated the Smithsonian Contributions to Knowledge with the éclat of a major work in anthropology.

By the time of Henry's death, the Contributions series had brought forth twenty-one fat quarto volumes. Even in these, the experimental sciences were not heavily represented. In physics, for example, aside from some Contributions on terrestrial magnetism, the Smithsonian's most useful publications were reprints of key European articles in the annual *Reports.* Henry himself, on rather lofty grounds of propriety, declined to allow any of his own work to appear in the Contributions, though after becoming secretary he did some good things in applied physics. Nor did any indisputably classic Contributions in other fields appear during Henry's time, though a number were of high quality.

Still, Henry's own intelligence, knowledge, and industry as editor helped win respect for the Smithsonian Contributions series. Its volumes played a prominent part in the exchange program. So did those of the annual *Reports,* which included Smithsonian lectures, surveys of progress in various fields, and Henry's commentaries on Smithsonian operations. And so also, beginning in 1862, did the Miscellaneous Collections, devoted to instructions for collectors, reports of explorations, bibliographies, descriptive natural history, physical tables, and the like. All this was useful. In any event, scientists must have taken heart from the opening of new outlets for their findings. Joseph Leidy for one wrote a friend in 1847, "This offers us an excellent opportunity which we shall certainly not neglect." And Leidy did not neglect it.

Despite Henry's experimental predilections, the American natural environment and its weight in American science pushed the Smithsonian inexorably toward natural history, and therefore toward the museum collections that Henry and Bache had dreaded from the start. In 1847 Henry insisted that no natural history collections at Washington be dumped on the Smithsonian. In the Senate his ally Jefferson Davis likened any such incubus to the Siamese king's revenge through the gift of elephants. Yet scientists of stature such as Agassiz, Dana, and Gray urged Henry to accept government collections, especially those of the Wilkes Expedition, and build on them. Dana went further by helping introduce into the heart of Henry's camp an artful and untiring champion of collections, the young Pennsylvania zoologist Spencer F. Baird.

As soon as Dana got Baird to apply for the job of curator, Baird set about marshalling references, not only from leading scientists like Dana, Gray, Agassiz, and Audubon, but also (on Dana's advice) from politicians like James Buchanan and especially George Marsh. Marsh used his influence with certain regents and presently became one himself. An accomplished linguist, he tutored Baird in German and encouraged him to study Danish. Baird promptly edited a four-volume translation of the immense Brockhaus *Bilder-Atlas.* His linguistic facility, according to Henry, became a factor in his eventual Smithsonian appointment.

But Henry stalled the appointment on the grounds that the building should be completed first. It was therefore a triumph for Baird when the regents made him assistant secretary, with Henry's grudging assent, as early as July 1850, at fifteen hundred dollars a year. He was twenty-seven. Henry warned him beforehand that the secretary would demand his assistants' "cordial support and co-operation . . . in carrying out the plans which I deem the best for the interests of the Institution." Baird, moreover, was to assist not only in the collections but also in publications, correspondence, and other business. Baird was unfazed. That fall he settled happily in Washington, bringing with him not only his wife and daughter but also his own rich collections of birds, mammals, reptiles, fishes, and amphibians, comprising thousands of skins, skulls, skeletons, nests, eggs, jars, and barrels. They became the nucleus of the Smithsonian's natural history collections.

In the fall of 1851 Henry wrote Gray, "Baird has proved a very efficient assistant and rather requires the rein than the spur." By then Henry may have begun to suspect that he had another Jewett on his hands. Always amiable, modest, and self-possessed, Baird nevertheless proved to be a prodigious worker. In the winter his workday averaged twelve hours, in the summer fifteen. Henry put him in charge of the exchange program, in which besides the paperwork he did much of the manual labor of making up packages—four tons of them in the first year. But he thrived on it, priding himself on the system's huge expansion thereafter and on the praise won by its efficiency and dispatch. To Marsh he wrote that besides the exchanges and his natural history operations, he had entire charge of the publishing department, "revising memoirs, fighting printers and engravers, correcting proof, distributing copies, &c.," had also taken on the job of permanent secretary of the AAAS, and was hard at work supervising the *Bilder-Atlas* translators, "making out Zoological reports for Army officers, writing 8 or 10 letters every day, attending to dozens of diurnal and nocturnal visitors, etc." He added exuberantly, "I am making every effort, in addition to the reptiles and fishes, to procure the best collection in comparative Osteology in the country."

Nevertheless, during Henry's confrontation with Jewett over the library, Baird's fate, as a fellow naturalist put it, "was suspended on a very fragile thread." Early in the battle, Henry privately lumped Baird with Jewett as trying to usurp power and undermine the secretary for the sake of his own specialty. Baird, wrote Henry, "is I think at heart a good fellow, but . . . from inordinate ambition . . . [adopted] a selfish, and to me, . . . a dishonest course. . . . He is exceedingly plausible and any person who has not had a good opportunity of studying his character would be deceived by him." Still, wrote Henry, he did not despair, once he had "settled with Jewett," of "bringing Baird to a sense of his duty." Baird himself privately sympathized with Jewett but kept mum in public. To a mutual friend, Baird denied any knowledge of a conspiracy to drive Henry out. He loved Henry as much as any man on earth, he wrote, but had been "chilled by his reserve, want of confidence, and not infrequent harsh accusations," especially after the first year of kindness and frankness. If Baird seemed "forward and sometimes presuming," he insisted, it was only the mark of an active mind and a wish to be helpful. Whether this explication helped, or Henry simply recognized Baird's outstanding ability and devotion to science, Baird survived and flourished, and so did natural history at the Smithsonian.

Like Agassiz, Baird rejoiced in every new correspondent who might feed specimens into his insatiable collections. Travellers like the diplomat Marsh, army officers who showed a spark of interest in collecting at their posts, amateur naturalists, all were advised and cultivated by the great collector at Washington. Baird did personal services, executed commissions, forwarded papers and books, and procured supplies for men who would otherwise have been cut off from the right path or uncertain of it. In Mississippi, for example, the planter Benjamin Wailes depended primarily on Baird for help in classifying specimens—shells, insects, or otherwise—sent to Washington; in return Baird sent him Smithsonian publications and helped him get other books. In the 1850s Baird printed up and distributed circulars with detailed instructions for collecting specimens, preserving them, fixing time and locality, recording habits and peculiarities of the living creatures, and so on. Henry notwithstanding, the Smithsonian's collections grew phenomenally. "It has been the work of but three years," wrote Baird in 1853, "to raise this collection from nothing to the front rank among American cabinets." What was more, the collection was no hodgepodge for the casually curious, but a working collection arranged with great skill, care, and expertise for the use of scientists.

In the museum's earlier years, Baird emphasized those fields and areas neglected by other American collections, but even then he surely meant for the Smithsonian's holdings to become definitive for the natural history of the entire continent. And in realizing his imperial dream, Baird had a military strategy. The army and to some extent the navy were to be enlisted.

When Baird became assistant secretary in 1850, he was assigned not only "to take charge of the making of collections" but also "to request of officers of the Army and Navy . . . and of other persons such assistance as might be necessary for the accomplishment of the intended object." This was his marshal's baton, and by mid-1853 he had a good grip on it. "My exploring expeditions are all off," he wrote his ally and confidant George Marsh that July. "I have had in all no less than 19 to equip and fit out. . . . I expect the . . . matter thus collected . . . to have the effect of forcing our government into establishing a National Museum, of which (let me whisper it) *I* hope to be director."

Baird had his way. Completion of Renwick's "Norman cenotaph" ended one of Henry's excuses for delay. American naturalists, including Henry's own cronies, kept up a gentle pressure. Torrents of specimens poured in from the expeditions and from Baird's system of volunteer collectors. Though Henry still feared Congressional appropriations as an entering wedge for political influence, he himself doubtless felt the impact of Baird's zeal and success. At any rate, in 1856 Henry yielded to the proposition that the natural history collections in the desperately crowded Patent Office, notably those of the Wilkes Expedition, should be transferred at government expense to the large, vacant hall of the Smithsonian as a "national museum." This was done in 1858 with the help of seventeen thousand dollars from Congress initially and four thousand a year thereafter (conveyed through the Interior Department to spare Henry the indignity of applying regularly to Congress). By then they represented only a fifth of the total Smithsonian collection, but the principle of a federally funded museum had been established. Baird had succeeded in giving the Institution what proved to be, as Henry had feared, a permanent turn toward natural history.

Even if Henry had been free to spend the Smithsonian's income as he pleased, its meagerness would have limited the scope of the formal "Programme." In the 1850s, for example, the Coast Survey appropriation generally ran to about fifteen times the income from Smithson's endowment. Still, that comparison belies the significance and influence of the Smithsonian. To begin with, for all Bache's resourcefulness, the proportion of the Coast Survey appropriation spent on basic science was necessarily much smaller than was true of the Smithsonian's income. And beyond that were lines of force not defined by dollars.

There was the eminence, the dignity, the wisdom, and the vision of Joseph Henry himself. His aura enveloped the Institution. And like the core of an electromagnet, the Institution gave direction and strength to Henry's influence as a leader of the scientific community. So too with young Spencer Baird among American naturalists, though to the scientific community at large he lived in Henry's shadow until Henry died. The character of the Institution, unique in that period, the publicity attending its conception and birth, its lodgment in the nation's political center, perhaps even Renwick's dramatic setting for it, all amplified the voices of its spokesmen. If Bache and Maury were barons, Henry was at least a prince of the church, though perhaps not a pope.

The Smithsonian set a precedent for endowed foundations that supported scientific research as a primary goal, rather than a by-product as in government agencies or universities. Thus the Institution's very existence was symbolic. To the people and their representatives, Henry's

Reports proclaimed the worth and honor of basic research. Henry's operating policies (despite their concessions to practical work for government) made the same point. In both theory and practice, moreover, Henry and the Smithsonian upheld a high standard of scientific responsibility and quality. Henry's seedbed approach of doing what others would not or could not do, until (as with the weather bureau) the seedlings were transplanted elsewhere or became self-sustaining, served further to maximize the Smithsonian's return on its investment.

These were some reasons why the Smithsonian Institution became a sanctuary for a succession of young scientists, especially naturalists, to whom Henry allotted free lodgings in the otherwise useless towers of the Smithsonian building; and why it became a shrine, or at least an inspiration, to some others. "If I *could once* step within its threshold," wrote young Alexander Winchell from his position in an Alabama "female seminary," "it seems to me my eager eyes and ears and understanding would drink in a world of wonders and knowledge—and that scientific atmosphere! How refreshingly would it be inhaled."

The Smithsonian was significant also because of Henry's insistence that it be more than local or provincial. The mid-forties, when the Institution was born, had brought forth still other antidotes to provincialism in American science. So the Smithsonian reinforced a new note in American science—its national scope and outlook.

Baird's strategy advanced this nationalizing tendency by harnessing the army and navy for scientific research. That policy by itself so multiplied Smithsonian influence as to make meaningless its measurement solely by endowment income.

Chapter 15

Soldiers, Sailors, and Scientists

American earth and life scientists of the 1840s saw at once that the new mood of Manifest Destiny promised not only a magnificent area of study but also the means of exploiting it. In 1842 young Josiah Whitney spoke for them all. Noting War Department plans to survey the resources of the Far West, including Oregon, Whitney wrote home from Paris:

> To this I have been looking forward ever since I began to study the sciences as a profession. I long to be one of those who are to develop the resources of that mighty country. . . . Could we not command influence enough to attach me to any expedition which might be sent out? I would leave Europe willingly.

Even in simpler times not many lone scientists had shown the hardihood of Thomas Nuttall, who scouted much of the Far West on his own before 1820. And by mid-century the rising standards and sophistication of science, even in descriptive natural history, left the field-worker still less time and strength for the demands of physical survival. No longer could loners do the job adequately. Scientific exploration of the Western wilderness now required an expedition, a collective enterprise.

Few American scientists could mount such expeditions out of their own pockets. Asa Gray and others made some use of individual professional collectors like Charles Wright. In 1853 James Hall scraped together fifteen hundred dollars to send two assistants up the Missouri on an American Fur Company steamboat for a summer's fossil collecting in the Dakota Bad Lands. But he did so only "at a great sacrifice." A trip to the upper reaches of the Missouri would have cost anywhere from two to four thousand dollars. Not even the scientific societies could bear such expenses.

In 1843 the botanist George Engelmann mentioned several possibilities for scientific hitchhikers: "One of the collectors has gone with the Oregon Emigrants, the other with a mission of Jesuits. . . . A person who wants to travel with the Santa Fe traders can collect well if he has a servant with him, and a cart or a pack-mule." Five years later, writing about botanical collecting in Mexico, Engelmann touched on an escort with far greater resources and a broader commitment to the general welfare: "The collections . . . would almost always go with return [Army wagon] trains, . . . empty for the most part. . . . As long as our troops are there, science may as well profit by them."

Scientific expansionism did not always imply political expansionism, especially among New Englanders. "I shall offer something on the moon at the meeting of the Association [of geologists]," wrote James Dana in 1846, adding sarcastically, "we may soon expect it to be *Annexed,* and it is important that its resources should be early explored." Nor did scientists suppose that war would do science any good. Besides remembering Franklin's remark that there never was a good war nor a bad peace, Philadelphians could recall how the War of 1812 had cut them off from European scientific publications, and how the infant Academy of Natural Sciences had been weakened by the drawing of its leaders into military service. Even the Virginian Matthew Maury, though a supporter of the Mexican War, complained that it took away some of the Naval Observatory's "most efficient observers." By the spring of 1848 the younger Silliman was writing, "May we not hope that this long and inglorious war is over, to curse us no more with widows & orphans cries?"

Yet "this long and inglorious war" yielded rich scientific booty. So even those scientists who deplored the war lost no time in using it, like John Torrey, who wrote a few years later that the only good to come out of it "was the benefit which it has done to botany." From the Rio Grande in the fateful spring of 1846 a West Point graduate sent a vial of water and mud to the microscopist Professor Bailey at his alma mater. "On the march I collected a few plants & shells," the officer wrote, "but there has been so much confusion & turmoil since the capture of *Thornton* that it is impossible to discover any trace of them." The capture of Thornton was the very casus belli specified in Polk's war message. Science did not merely go where the action was; it sometimes got there first.

Other military officers found science and war compatible, sometimes chillingly so. "As the guerilla's mode of warfare admits of no quarters," wrote one, "we adopted the same, leaving in our track blood & fire, burning & plundering their ranches, sparing none but women & children. . . . You will notice among the Coleoptera I sent home, many not duplicated. They were collected on scouts to parts I had no opportunity of revisiting." The war being popular in the South, volunteers flocked to Charleston on their way to Mexico, and the zoologist John Bachman seized the opportunity to enlist the officers for science. He passed out a long circular listing the specimens wanted, with directions for skinning, stuffing, and shipping, and a price list: fifty cents for small species like bats, rats, and squirrels, a dollar for rabbits, weasels, and others of like size, four dollars for deer, ten for the "Mexican tiger" and the "Lobo."

At St. Louis, "Gateway to the West," through which considerable military traffic was funnelled, the physician and botanist George Engelmann seized even better opportunities. Born and university-educated in Germany, a friend of Agassiz in his Paris days, Engelmann had been the leader of St. Louis science since his arrival in the early thirties and was to be the organizer of the St. Louis Academy of Science in 1856. "We are all in uproar here," Engelmann wrote Asa Gray in May 1846, "and I myself am burning to collect new Cactuses in New Mexico or on the Rio Grande." His medical practice, the largest in town, prevented that; but he found a young German, Augustus Fendler, with teachability and wanderlust. Asa Gray moved fast to make arrangements with the War Department and the Topographical Corps, get advance pay for specimens, and send instructions to Fendler. Engelmann was "delighted to see with what promptness all acted," and began to "hope a little more from this country for science!!" By mid-August, Fendler was off for Santa Fe in company with a regiment of cavalry and despite troubles came back a year later with a rich harvest of plants.

Topographical Corps officers did what they could for science on military expeditions. Ordered to northern Mexico with a Topographical Engineer detachment, Captain George W. Hughes tried hard to get expert civilian help but failed to enlist the entomologist Samuel Haldeman. So Hughes fell back on the less expert but intelligent and zealous Josiah Gregg, author of the classic account of the prewar Santa Fe trade, *Commerce of the Prairies.* Despite danger and hardships, Gregg assembled a good botanical collection, some of it from the battlefields of Monterrey and Buena Vista, and shipped it safely to a Louisville botanist. The latter mailed part to Torrey at Princeton and the rest to Gray at Harvard, but both shipments were lost en route.

More rewarding were the scientific trophies of a party with Kearny's Army of the West in its march to California. Commanded by Lieutenant William H. Emory, a brother-in-law of Bache, it had only one day in which to assemble scientific apparatus. "The object of the Expedition was purely military, not exploration," wrote Emory later, "and what has been done in that way was at moments snatched from military duties and under great disadvantages." Nevertheless, the swashbuckling, red-whiskered Emory had energy, intelligence, and scientific talent. From its adventures in the Southwest, Emory's party brought back fine collections of plants for Torrey to work up, cactuses for Engelmann (already a leader in that field), and even a consignment of geological specimens. Most notable were Emory's geographical and ethnological observations.

These war-borne scientific reconnaissances contributed little of significance. But they kept alive the precedent of letting science ride to far places astride more conventional government operations, whether military, diplomatic, economic, or administrative. And they added to the experience of the Topographical Corps and of the federal government generally in bearing that strange new rider—bearing it at last on an unearthly journey, for James Dana's mocking forecast of lunar annexation had touched a future he may not have foreseen.

Though a mere terrestrial annexation, the huge Mexican Cession at once required an elaborate boundary survey, since the Treaty of Guadalupe-Hidalgo had been negotiated in a dust cloud of geographical ignorance. The Topographical Engineer officers who did the work had railroad routes also in mind, and so had grounds for geological as well as topographical observations. But they went beyond even that to a general scientific reconnaissance of the Southwest. The Mexican Boundary Survey parties took along expert collectors, who made their own reports and also brought back heavy loads of specimens and data for processing by James Hall in geology, John Torrey in botany, and Spencer Baird in zoology. The last of three large volumes of reports did not appear until 1859.

The bonanza of scientific raw material to this point had come in the process of getting title to the Far West. Now the government had to foster the use of that domain, which required knowledge of what it contained. In what has been called the Great Reconnaissance, expedition after expedition, usually conducted by the Corps of Topographical Engineers, crisscrossed the trans-Mississippi West between the end of the Mexican War and the beginning of the Civil War. They searched for wagon-road routes, artesian well sites, navigation possibilities on the Colorado River, and lands for the Indians of Texas. They explored six possible routes for a railroad to the Pacific. And beyond their specific objects, they looked to the whole range of natural resources and environments.

All these expeditions thus yielded grist for the mills of science, and the most prolific were the Pacific Railroad surveys, 1853–55. Science could not give Congress what it had wanted from the railroad surveys, a single transcontinental route so superior as to silence sectional rivalries. Instead, several routes proved feasible, and sectional squabbling continued to delay a choice until war slammed a mailed fist on the Northern side of the balance. But with Bache's erstwhile Senate defender Jefferson Davis as secretary of war, Bache's brother-in-law and Davis's close friend Major William Emory as commander of the bureau in charge, and Bache's friend and former Coast Survey associate Isaac Stevens in command of the northernmost party, it is not surprising that the railroad surveys did a good deal for science. Aside from the data they yielded, they hired more than a hundred scientifically trained men for collecting in the West and classifying in the East, and spent half a million dollars on assembling and publishing the scientists' findings in thirteen massive volumes. Indeed, the *Pacific Railroad Reports* incorporated most of the preceding expeditions' findings as well, and so stood as a compendium of the Great Reconnaissance (though a rather badly organized one).

The Bache influence doubtless did much initially to graft a scientific mission onto the Great Reconnaissance. But the day-by-day guidance and incitement came from other scientists, and especially Spencer Baird. Writing to Captain George B. McClellan, later of Civil War fame, Baird in 1852 promised to "inoculate" a fellow officer "with the Nat. Hist. virus, and if I have as good fortune as with you, shall be well satisfied." Even Isaac Stevens, science-minded as he was, had to restrain Baird's zeal in preparing instructions and apparatus for him. "I want you to understand, Professor Baird," he protested, "that my exploration is something more than a natural history expedition."

In 1853 Baird wrote, "The string of scientific expeditions which I have succeeded in starting is perfectly preposterous." So it was. During the 1850s some thirty or so expeditions brought their findings and collections in huge quantity to the Smithsonian. Henry, for all his misgivings about the lopsided emphasis of American science, could not help boasting in 1857 that the institution's stimulation and scientific direction of such explorations had "done more to develop a knowledge of the peculiar character of the western portions of this continent than all previous researches on the subject."

Baird provided the expeditions with full instructions for observing and collecting. He helped them prepare outfits and procure apparatus, much of the latter being invented or adapted by the Smithsonian for their special purposes. The Institution recommended, recruited, and trained scientific personnel (which meant that Baird had a good deal of scientific patronage to dispense). It received the collections and referred them to experts for study and report. It employed skilled artists to draw the hundreds of new types and species brought in. Baird himself wrote massive and masterful reports, including the classic report on mammals, birds, and reptiles of the Mexican Boundary Survey and the volumes on mammals and reptiles for the *Pacific Railroad Reports.* For the latter series he collaborated with John Cassin and George N. Lawrence on the *Birds* volume, which has been called epoch-making in American ornithology. It and the *Mammals* volume were later republished commercially. Presumably Baird received no royalties. But he had already reaped a reward more precious than money to a nineteenth-century scientist: the attachment of his name to a multitude of new genera and species.

Baird's influence favored zoology, to the occasional discontent of an entomologist or fungologist. But Baird did not have things all to himself. He reassured Joseph Henry's close friend John Torrey in 1853 that "the exploring expeditions this year will do much for botany," and that he had taken care to provide them with botanical supplies such as presses and portfolios. Torrey, who never went on an expedition himself, remained the botanist to whom expedition leaders usually turned for initial advice and final reports, though he referred large parts of the collections to Gray (who complained on one such occasion that "Torrey swamped me"). At Torrey's urging, Henry took over the expeditions' botanical accumulations, enlisted Torrey and Gray in preparing and labelling them, and thus initiated what later became the United States National Herbarium, the nation's foremost. For the geology of the expeditions James Hall assumed a role similar to Torrey's in botany. Something like a scientific-military entente developed between the inner circle of Baird, Torrey, and Hall on the one hand, offering scientific status and credit to Topographical Corps officers, and the officers on the other, offering the scientists financing, job patronage, and specimens.

In some ways, the navy's partnership with science resembled the army's. The two efforts were comparable in magnitude. From 1847 to 1860 the navy sent out or took part in more than a dozen expeditions with some scientific purpose, touching every continent. During the early 1850s a dozen ships of war flew the flag of science, and by 1856 Congress had spent two-thirds of a million dollars on publishing the reports of the Wilkes and Perry expeditions alone. Also, as with the army, the westward surge played a part. The Pacific Coast explorations and the Central American expeditions looked for a better route to California, among other things. And the dream of Asian trade that helped inspire American expansion to the Pacific Coast was also a factor in the navy's Pacific expeditions.

But in other respects, the navy's way with science differed from that of the army. The army scouted the land in the interests of settlement and exploitation and could therefore enlist a variety of sciences on practical grounds. The navy's chief practical goals were to further maritime commerce and whaling, toward which only oceanography could offer much direct help. So the navy made more than did the army of raising the nation's scientific prestige abroad, thus justifying support of other sciences and—not incidentally—keeping up naval appropriations in peacetime. Unfortunately the nationalist motive also inhibited the use of European scientists in navy expeditions. Furthermore, because scientific training for naval officers had come later than for military, many naval officers had little regard for science and scientists.

The nation's first great scientific venture by sea was Captain Charles Wilkes's United States Exploring Expedition of 1838–42. Its initial promoter in the 1820s and 1830s, an Ohio editor named Jeremiah Reynolds, at first had aimed at verifying the weird notion of a hollow earth open at the Poles but soon dropped that fantasy in favor of scientific exploration of the Antarctic and Pacific regions. He won government backing for the project by invoking its potential contributions to the whaling industry, Latin American trade, and better charts for naval and merchant ships. Reynolds could cite the Lewis and Clark Expedition as a precedent, and Congressional jurisdiction over commerce as a constitutional basis for federal support. Another potent argument in the high-spirited 1830s was the glory that government expeditions were winning for other nations. Also persuasive were the rising prestige of science and the glamor of explorations from Cook and Bougainville to Humboldt.

Though Reynolds was shouldered aside as the project took shape, he had planted the idea of using civilian scientists. The secretary of the navy asked and got advice on objectives and personnel from several scientific societies, and a number of civilians applied for scientific posts. Asa Gray dropped out before the expedition left, to the detriment of both the expedition and Gray's later career, but Charles Pickering, James Dana, and other able scientists signed on. The first commander resigned before sailing, partly because the civilian complement was not completely under his authority. His successor, Captain Charles Wilkes, had a good background in mathematics and astronomy, but he likewise pressed for an all-navy operation. As a compromise, the civilian complement was thinned down to seven scientists, two artists, and two technicians, while navy officers were assigned to hydrography, mapping, astronomy, and terrestrial magnetism. "There was a wonderful deal of jealousy of us 'scientifics' when we started," wrote Dana, "and especially on the part of Wilkes lest we should trespass on his departments."

In August 1838, the Wilkes Expedition set sail under orders that made science its primary mission. Wilkes's tendency to suit the expedition's itinerary to his own specialties annoyed the civilian scientists. Dana later called Wilkes "an ignoramus in science." But Dana was kinder in his overall estimate. "Wilkes," he wrote Asa Gray, "although overbearing with his officers, and conceited, exhibited through the cruise a wonderful degree of energy. . . . I much doubt if with any other commander that could have been selected, we should have fared better, or lived together more harmoniously. . . . The Navy does not contain a more daring explorer, or driving officer." The achievement speaks for itself. Six naval vessels set out; one was lost off Cape Horn and another at the mouth of the Columbia River, but the rest came home after voyaging about 85,000 miles. They had explored 280 islands, 800 miles of streams and coast in Oregon, and 1,500 miles of Antarctic coast; cruised the Central Pacific; and brought back hundreds of new species of fish, reptiles, and insects, rich collections in botany, geology, and native artifacts, and a store of charts that served the navy a century later in the Pacific campaigns of World War II. Besides zoology, geology, botany, and hydrography, the expedition took account of ethnology, anthropology, meteorology, and physics.

So far as science was concerned, however, the long voyage was a mere prelude, useless until collections were studied and findings published. The squabbles that delayed and marred that work reveal much about relations between science, government, and the public in the early 1840s. The collections fell first into the clumsy hands of a conglomeration of Washington politicians, amateurs, and outnumbered scientists known as the National Institute for the Promotion of Science. Its officers resented and distrusted the new, specialized, professional elite of science. Before the Joint Library Committee of Congress took over and put Wilkes in charge of the work, many specimens had been damaged, lost, or stripped of their labels.

But Wilkes and the Library Committee also made trouble. Under Congressional pressure, Wilkes tried to force the professionals working on the reports to translate Latin terms into English, in the name of democracy and the public's stake. Only a compromise arranged by Joseph Henry preserved the scientists from such government manipulation of methods and standards. Another challenge came from the very nationalism that had helped launch the expedition, when congressmen and others opposed using European collections or specialists in working up the reports. But the flag-wavers eventually let English scientists help out on algae and fungi, and sent Asa Gray to study botanical collections in Europe.

False starts were made in the execution of the work. The botanist on the expedition turned out to be an incompetent political appointee. Gray reluctantly agreed in 1848 to take over the botched botany report and start fresh, partly for the pay but mostly as a duty to science and the national honor. Titian Peale's volume on birds and mammals was rewritten by John Cassin. Even in the hands of able and industrious men, the immense task seemed endless. John Torrey toiled at intervals on the North American plants from 1848 to his death in 1873, working up about eight thousand specimens. Gray helped him in later years and meanwhile spent a vast amount of time and energy on the Asian and Latin American plant collections—all wasted because his manuscript of more than seventeen hundred pages was never published. The unfinished work haunted Gray for years. Agassiz never finished his two volumes on the fishes. Five volumes of reports were completed but not published.

Thirteen big volumes of scientific reports did make it into print, covering ethnography, philology, zoophytes, mammalogy, ornithology, anthropology, geology, meteorology, mollusca, crustacea, botany, and hydrography. Yet even these fruits of long labor were blighted by Congress's asinine decision to limit the official printing to a hundred copies of each volume, despite the outrage of leading scientists and societies. (Four of the authors used the type and plates to print more copies of their volumes, and others published some findings in scientific journals.)

Theorizing had been discouraged on principle, and so the reports were largely descriptive. Nevertheless, they won respect for American science both at home and abroad. Agassiz saw them as "surpass[ing] in scientific importance the publications of all the exploring expeditions issued by European governments taken separately." They served national pride further by freeing American navigators from dependence on foreign charts in the areas covered. Apart from the reports, American science profited indirectly from the expedition. Though Asa Gray did not join it, his labors on its collections primed him for his fruitful correspondence with Darwin. Navy Lieutenant James Gilliss, who did not go either, used the occasion to "smuggle" the Naval Observatory into the federal establishment. James Dana did sail with Wilkes, and what he saw broadened his perspective on geology to include its overlap with zoology in the island-building coral animals. His report on coral zoophytes was pioneering, yet authoritative. The study of island formation also directed him toward volcanism, and his observations of its work enabled him to make still other notable contributions to geology. The expedition thus did much to win Dana his preeminence in American geology.

A series of navy-sponsored ventures in science followed upon the Wilkes Expedition. The first, Lieutenant James M. Gilliss's four-man astronomical expedition to Chile, 1846–52, was (as we have seen) in the spirit of current science and, despite Lieutenant Matthew Maury's hostility and perhaps deliberate sabotage, relatively productive. But then came several enterprises of Maury's own making and marring. Maury did his best to shut out civilian scientists in favor of scientifically inclined naval officers. He wanted to develop an official scientific corps of navy officers, analogous to the army's Topographical Corps. But the material was lacking. The Naval Academy had not been established until 1845, and even after it was, science languished there. Some staff officers had duties presuming formal education, but line officers, who often had an anti-intellectual bias, looked down on them. Science won few promotions for navy men, and they knew it. Wilkes, Gilliss, and Charles H. Davis were exceptions in having solid scientific training, but the last two were hostile to Maury. Of Maury's small coterie of science-minded officers, several later turned against him. So in trying to confine navy science to navy officers, Maury tended to stifle it.

Most of the Maury projects harked back in methods and aims to eighteenth-century geographical exploration, with little geophysical sophistication. Maury's crony Lieutenant William F. Lynch, for example, visited the Dead Sea for three weeks in 1848 with two officers, ten seamen, and a couple of tourists picked up on the way, but no trained scientists. His immediate motives seem to have been to while away time between assignments and to escape certain family problems. His findings did remain the chief source of technical information on the Dead Sea until the 1930s, but that was not saying much.

In 1850 the wealthy New York merchant Henry Grinnell fitted out two ships for an Arctic expedition. Maury, hoping to verify his own erroneous hypothesis of an open polar sea, got Congress to adopt the project as a navy expedition, and he secured the command for one of his cartographers, Lieutenant Edwin J. DeHaven. But Maury asked no advice of scientific societies, no professional scientist accompanied DeHaven, and so sixteen months of hardship yielded little for science but a meager collection of flora and two minor papers. Soon afterward, with the announced aims of serving science and commerce and the unavowed aim of opening a vast new area for Southern slaveholders, Maury persuaded the navy to finance a trip down the Amazon in 1851–52 by his brother-in-law Lieutenant William L. Herndon and Passed Midshipman Lardner Gibbon. A few specimens came to the Smithsonian, but no report on them appeared.

In July 1852 John Pendleton Kennedy became secretary of the navy. Nothing in his background as a well-known novelist and Maryland congressman explains why he should have been the first navy secretary to accept scientific activity as a legitimate and normal navy function rather than an occasional special duty imposed by Congress. Yet during his eight months in office before the Fillmore administration ended, Kennedy sent out four expeditions and furthered others. Though Kennedy unsuccessfully espoused Maury's pet idea of a navy scientific corps, he had no intention of hamstringing the navy's scientific effort through Maury's policy of excluding civilian scientists. On the contrary, he consulted Bache, Henry, Peirce, Espy, and Agassiz about plans and personnel. In particular, he asked Spencer Baird to work with the navy's expeditions as Baird was doing with the army's. This did not freeze out Maury, since most officers with enough scientific background to command the expeditions were of Maury's circle. But after 1852 Baird and other civilians had a significant influence on the navy's scientific operations—fortunately for science.

At the urging of the New York mercantile community, Kennedy sent a ship to South America in 1853 to explore the Río de la Plata and its tributaries in the interests of both trade and science. A Maury protégé, Lieutenant Thomas J. Page, was given command, and no trained naturalist went along, but Baird assembled scientific equipment and got Asa Gray to write up careful botanical directions. Page's first voyage, 1853–56, and a follow-up, 1859–60, brought the Smithsonian many specimens, though Congress never came through with money for a formal scientific report. Science similarly piggybacked on four expeditions, 1853–60, to look for a possible canal route across the Isthmus of Panama. Not trusting Maury, Baird arranged for two able naturalists to be attached to the most notable of these, 1857–58, resulting in still more specimens for John Torrey and Asa Gray to work up.

As a navy surgeon with the DeHaven expedition, young Elisha Kent Kane had fallen under the spell of Arctic exploration. After the expedition's return, Henry Grinnell consented to furnish him a ship for a second expedition. This time the Washington competitors pulled together. In 1852 Secretary Kennedy enthusiastically detailed Kane and nine others to the expedition on navy pay, Bache and Henry gave advice, Baird assembled natural history equipment, and Maury's Naval Observatory furnished instruments for geophysical observations. A civilian naturalist and an astronomer were added to the party, which set out in May 1853. With heroic determination and energy, Kane brought his little band further north than any previous expedition, before they abandoned their ice-locked ship and retreated in the summer of 1855. Though most of the natural history collection had to be left behind and no official report appeared, Kane brought back useful data on tides, meteorology, magnetism, and geography, most of which found its way into print in one form or another.

The most famous naval expedition of those years, that of Commodore Matthew C. Perry to Japan, 1852–55, made little room for science, though its commander had longstanding scientific interests. Like Maury, Commodore Perry wanted to foster scientific capabilities among naval officers. But he and the navy saw his mission as primarily diplomatic and commercial, one so delicate as to be jeopardized by a contingent not fully subject to naval discipline and bent on other ends. So he fended off civilian scientists, used such scientifically inclined naval officers as he could find, and ordered all officers to make what observations and collections they could in various scientific fields. Incidental and amateurish as they were, the data had some value. Congress spent about four hundred thousand dollars to print thirty-four thousand copies of the three-volume report, for which some leading scientists wrote several chapters on natural history, based on notes by members of the expedition. Far more important, the Perry expedition succeeded magnificently in opening Japan to the world, and vice versa.

To compensate for the low priority of science on the Perry expedition, President Fillmore and Secretary Kennedy stressed it in sending out the Ringgold expedition to the North Pacific. The commercial purpose had carried the $125,000 appropriation through Congress and continued to hold first priority, but Kennedy and American scientists saw to it that science ran a close second. Maury, an early supporter of the project, pressed for an oceanographic emphasis and got his Naval Observatory protégé Lieutenant John M. Brooke assigned as an "officer-scientist" in charge of all astronomical work, the only naval officer aboard who had purely scientific duties. But Maury also solicited and endorsed recommendations from leading civilian scientists as to the scientific program. Spencer Baird later wrote proudly that the Smithsonian had systematized "the whole plan of operations, nominating competent naturalists, making all purchases and settling all accounts, supplying detailed instructions to the corps," and afterward providing facilities for studying the collections. Working closely with Henry, Baird saw to it that Ringgold's was the expedition best supplied with scientific equipment, including daguerreotypy.

This time a corps of civilian scientists went along, and able ones at that. Despite Maury's disapproval, Secretary Kennedy supported that policy, partly because for the honor of the nation he wanted the best men possible, partly because he thought using too many line officers for science might impair navy discipline (he did, however, instruct the officers to handle hydrography, oceanography, astronomy, meteorology, and magnetism). At Baird's urging, Kennedy even held up the sailing for several months until ten properly qualified scientists were rounded up, along with an artist skilled in scientific drawing. On Agassiz's recommendation, his own student William Stimpson went as zoologist; and on Baird's, the noted collector Charles Wright signed on as botanist (to the delight of Asa Gray, Wright's best customer). Baird tutored Ringgold's second in command, Lieutenant John Rodgers, in marine biology; Gray and Bailey furnished advice in their specialties; and Eben Horsford briefed a navy surgeon on chemistry and chemical apparatus (the surgeon was later replaced by the able young chemist Francis Storer). In scale (with five ships), in scope, and in quality of staff and preparation, the North Pacific Expedition eclipsed every navy exploration since that of Wilkes.

From the summer of 1853 to the summer of 1854, the North Pacific Expedition made extensive observations and collections at the Cape of Good Hope, Australia, and Hong Kong, where the ailing Ringgold turned over his command to Rodgers. Thereafter, under Rodgers, explorations went forward along the China coast to Korea and east to the Bonins, the Ryukyus, Japan, and Kamchatka. With his mentor Maury's oceanographic and polar interests in mind, Rodgers also squeezed in a side trip to the Bering Strait and the Arctic Ocean before ending the cruise at San Francisco in October 1855. Lieutenant Brooke used his new deep-sea sounding lead to bring up Pacific bottom specimens, on which Jacob Bailey based a series of important articles. With ex-Secretary Kennedy's help, Baird secured the expedition's large natural history collections for the Smithsonian, after a tussle with Maury, who had wanted them for the Naval Academy.

After much lobbying and petitioning by scientists and scientific societies, Congress came through in 1857 with fifteen thousand dollars toward preparing the natural history reports. With William Stimpson in overall charge, Baird, Cassin, Wright, Gray, William Sullivant, Augustus Gould, and others set to work. But the work was enormous—more than five thousand species of mammals, for example, not to mention birds, reptiles, crustaceans, mollusks, insects, fishes, corals, and plants. In 1861, with work on six of the nine branches completed, the Civil War ended prospects of publication. When Stimpson became director of the Chicago Academy of Sciences, he brought with him nearly all the invertebrate specimens, which were consequently incinerated in the great Chicago

THE SENIOR LAZZARONI

Joseph Henry (1797–1878),
1840s?—Smithsonian Institution Archives

Alexander D. Bache (1806–1867),
c. 1860—Smithsonian Institution Archives

Louis Agassiz (1807–1873),
1857—Harvard University Archives

Benjamin Peirce (1809–1880),
1858—Harvard University Archives

Oliver Wolcott Gibbs (1822–1908), 1870s?—Smithsonian Institution Archives

Benjamin A. Gould (1824–1896), 1860s?—Alice B. Gould Papers, Massachusetts Historical Society

James D. Dana (1813–1895), 1850s?—Smithsonian Institution Archives

Charles H. Davis (1807–1877), c. 1870—Charles H. Davis, *Life of Charles Henry Davis* (Boston, 1899)

FOES OF THE LAZZARONI

William B. Rogers (1804–1882),
1850—MIT Archives

Matthew F. Maury (1806–1873),
1853—Library of Congress

Ormsby M. Mitchel (1809–1862),
1862—Library of Congress

Asa Gray (1810–1888),
c. 1855–60—Harvard University Archives

James Hall (1811–1898), 1856—John M. Clarke, *James Hall of Albany* (Albany, N.Y., 1923)

Spencer F. Baird (1823–1888), 1850s?—Smithsonian Institution Archives

Josiah D. Whitney (1819–1896), 1860s?—Smithsonian Institution Archives

Benjamin Silliman, Jr. (1816–1885), c. 1855–60—Library of Congress

fire of 1871, along with Stimpson's magnificent collection of manuscripts, drawings, and rare books. Broken by the blow, Stimpson died a few months later at forty.

Despite these tragedies, the Ringgold-Rodgers expedition left its mark on science and, for that matter, on commerce too. The Naval Observatory made good use of the hydrographic and astronomical data. Though the war prevented full publication of the expedition's charts, Maury used much of the material in his *Sailing Directions.* After Stimpson's death, a surviving manuscript in which he described many of the lost specimens was found and published in 1907. As in Jacob Bailey's case, other data reached print through individual articles. Most important of all was Asa Gray's recognition of common ancestry among certain Japanese and American plants, an insight that gave strong support to Darwin's theory.

The military and naval expeditions were not free rides. The scientists who went along paid a psychic price. Especially on long voyages, regular navy officers openly resented the civilian interlopers and their troublesome requirements. "The officers," wrote Charles Wright from the North Pacific Expedition, "care not a fig—any of them—for our labors and never put themselves to any trouble to facilitate them. . . . I met with more sympathy among rude teamsters on the plains of Texas." Even among the officers themselves, those with scientific specialties or assignments felt estranged from their brothers.

In the army expeditions, perhaps because they were less physically confining, scientist-soldier relations seem to have run smoother. Still, John S. Newberry complained in 1857 that "a degree of jealousy of the Scientific Corps which I did not expect pervades the military portion of each surveying party, and through them has infected the entire War Dept & especially the Corps Topos Engrs." At Fort Riley, Kansas Territory, in 1855 the army surgeon and amateur entomologist William A. Hammond, depressed by his fellow officers' total lack of sympathy, began considering the resignation that he eventually tendered.

Nevertheless the scientists' uneasy alliance with the military paid off, and in more than data. Astronomical observations for the Wilkes Expedition brought William Bond the first money he had ever earned from his scientific work. James Dana got $1,440 a year for his work on the Wilkes reports, and Asa Gray was elated by his five-year, $120-per-month contract for working up the botany. Though Gray's botanical labors were never published, they freed him from provincialism and gave him a world view in botany; while Dana and others, like their European counterparts Humboldt and Darwin, found in expedition service an early equivalent to graduate training.

The opportunities afforded by the armed services deeply influenced what American science did in those years and how it went about doing it. The government exploring expeditions overshadowed all other scientific enterprises in that day. The scientific leaders who managed to mount such mighty steeds could, within limits, rally American science to their own causes, setting its pace, its tone, its tactics and strategy—as witness Spencer Baird's triumph in shaping the Smithsonian. From the end of the Civil War until the beginning of World War II, American science would not again be so largely influenced by scientists on government horseback.

Part Three

THE GOVERNANCE OF SCIENCE, 1846–1861

Chapter 16

Bache and Company, Architects of American Science

Impersonal factors obviously shape the growth of institutions. Historical momentum, social circumstance, and sheer logic push here and block there. But individuals also make a difference. The development of American scientific institutions in the nineteenth century cannot be fully understood without looking at a small group of men known to themselves as the Florentine Academy, the Lazzaroni, or simply "we," and to outsiders as the Mutual Admiration Society or Bache and Company.

At most, the active core numbered half a dozen, with three or four others (not always the same ones) on the periphery; and their setbacks loomed at least as large as their triumphs. Yet by merely doing battle they left a mark. Even in failure they advanced their cause by proclaiming it. And they were good at proclaiming. Articulate, energetic, and strategically placed, they focussed and projected the needs and aspirations of American scientists like a lens.

During their heyday as a group, from the late forties to the early sixties, they made no fundamental contributions to scientific knowledge. To them, "doing something for science" in those years meant organizing, raising support for, and guiding scientific institutions. "While science is without organization, it is without power," said Bache in a major address of 1851. They did not dominate their sphere as fully as the Founding Fathers in politics or the business tycoons of their own day. Yet they were akin to them in purpose: the organizing of a new era.

Historians have seized upon the playfully self-mocking name the group ultimately adopted for itself: the Lazzaroni. Originally applied to

Neapolitan beggars or idlers, it gained currency in the American press in 1850 with a widely copied story about a secret society of Genoese in New York City. Some contemporaries of the scientific Lazzaroni would have given much for the historian's privilege of reading their private correspondence in all its pungency, playfulness, and pride. But that experience can seduce one into accepting the Lazzaroni's own estimate of their wisdom and power. Words must be measured against deeds. In their running fight with Matthew Maury, for example, their potshots now and then took effect; yet when Maury's hulk of a book, *The Physical Geography of the Sea,* invited a broadside that would have blown it out of the water, the circle merely fulminated in private. There were other such anticlimaxes in the Lazzaroni story.

One historian, Mark Beach, denies that the Lazzaroni constituted "a cabal with coherent and disciplined points of view about the advancement of science." No more than a half-dozen of them ever got together at any one time. They did not always agree on such issues as the proper form of a university, their strong personalities sometimes clashed, and their formally arranged conclaves were infrequent and much given to mere conviviality. The legend of their power as a group arose, Beach suggests, simply because they were scientific leaders who found each other congenial and so inspired jealousy and suspicion among those not included. They were important individually, he implies, rather than as a group.

Beach's disparagement is a useful counterweight. But each of the full-fledged Lazzaroni did see one or another of the group often, they had close professional ties, they corresponded voluminously, and—what is most telling—they consciously saw themselves as a brotherhood, united in promoting the scientific enterprise in America along organized, European lines. They were indeed a "mutual admiration society." They fed each other's egos, rallied each other's spirits, promoted and defended each other's interests, swapped ideas, seconded each other's public pronouncements, and maneuvered en bloc to exalt each other's power. Their operation as a group thus did not depend on the frequency or size of their formal meetings. In any case they remain impossible to assess individually without some reference to their collective relationship.

The Lazzaroni deserve notice also as a natural phenomenon in the development of organized science. In England during the 1830s, when Bache and Henry made their transatlantic pilgrimages, a similar clique was centered in Cambridge University. Much like the later American group, the Cambridge coterie had an active core of five leading scientists with three or four satellite members, and it similarly tried not only to control patronage and advance its members' interests, but also to maintain professional standards and promote knowledge. Its plans for the scientific community were less grandiose than those of the American group, perhaps because the British community of the 1830s had not yet reached the proper stage of development. A generation later, however, there appeared another British clique, styling itself the X-Club, that was even more like the Lazzaroni.

At its inception in 1864, Herbert Spencer described the X-Club as "a small club to dine together occasionally." It turned out to be more than that. "In the course of time," recalled Spencer forty years later, "the existence of the Club became known in the scientific world, and it was, we heard, spoken of with bated breath—was indeed, I believe, supposed to exercise more power than it did." Besides Spencer it included Joseph Hooker, Thomas Huxley, John Tyndall, and five other well-known scientists, all of them friends of long standing and all of one mind on the needs of British science. No others were ever allowed to join, though on occasion the club regaled distinguished dinner guests, including Louis Agassiz and Asa Gray. The X-Club, as such, met only six or eight times a year, its discussions were diluted with gossip and badinage, one or more members were usually absent, and in its latter years political and other disagreements arose between members. Yet its power, though exaggerated, was not altogether mythical. Five of its members became presidents of the British Association, three were successive presidents of the Royal Society, and the members' other scientific honors and offices made a formidable list. Their mutual encouragement and support helped in all this. They also took on a larger mission: to advance British science by directing scientific education, shaping public opinion, defending intellectual freedom, maintaining professional standards, and establishing or reforming institutions that would promote scientific research, dignity, and autonomy. They went beyond the Lazzaroni in trying to inject the scientific spirit into broad cultural and social issues, and they lasted longer as a group. Otherwise, the parallel is striking.

So natural does such a pattern seem, in fact, that sociologists have generalized it for twentieth-century scientists. Small, close-knit, informal groups form, they say, quite consciously about an acknowledged leader, who usually serves as a model for at least the younger members. Such groups tend to generate "tribal folklore," with mock ceremonies. Held together best by the belief that they are advancing a radical new view in science, they may go beyond scientific objectivity in pushing it. And the impression they give of arrogance and exclusiveness often sets outsiders against them, thus making it harder for them to achieve their goals.

A more accurate label for the Lazzaroni would have been the one occasionally applied by their adversaries: Bache and Company. From first to last, the group's kingpin was Alexander Dallas Bache. Enough has been said about Bache's qualities heretofore to suggest why he generated coteries so readily: about his genial, yet commanding personality, his family tradition of politics and public service, his zest for organizational politicking and bureaucratic gamesmanship, his power base in government, and his ambitions for himself, his profession, and his nation. The Lazzaroni customarily referred to him as "the Chief." So we may see the group as having been conceived when Bache first joined with Joseph Henry in a declared resolve to organize and professionalize science in America.

That conception occurred at some time between 1832, when Henry first came to know Bache, and August 1838, when a comment by Henry, writing to Bache in Europe, reveals the living germ of the Lazzaroni program. "I am now more than ever of your opinion," wrote Henry,

> that the real working men in the way of science in this country should make common cause and endeavour by every proper means unitedly to raise their own scientific character. To make science more respected at home [in order] to increase the facilities of scientific investigations and the inducements to scientific labours. There is the disposition on the part of our government to advance the cause if this were properly directed. At present however Charlatanism is much more likely to meet with attention and reward than true unpretending merit.

As for the governance of the scientific enterprise, Henry applauded the "aristocratical" character of the British Association. "Those who have some reputation for science" ran it, he noted approvingly. "The great body of the members have no voice."

The reference to "your opinion" implies that Bache originally articulated the program. In Bache's "opinion," as reflected by Henry, are several elements of the subsequent Lazzaroni agenda: setting scientific standards to separate "charlatans" from "real working men"; raising the status of scientists; increasing support for science; winning recruits to it; enlisting and guiding government in the cause; and organizing scientists themselves, under an elite leadership, to further these ends. Another Lazzaroni goal may be assumed from the central concern of Bache's European tour: the development of scientific education. And still another may be inferred: that since the scientific enterprise must be governed only by "the real working men," not only should scientific "charlatans" be excluded from control but also the public and the government (though their material support would be acceptable if "properly directed"). In short, the autonomy of the scientific enterprise.

If the Lazzaroni began with the two-man nucleus of Bache and Henry, their first triumphs were the appointment of Bache as superintendent of the United States Coast Survey in 1843 (on the strong recommendation of Henry) and the naming of Henry as secretary of the Smithsonian Institution in 1846 (the work chiefly of Bache as a regent). By the latter date a new member had been drawn in. In the spring of 1842 Bache met the short, stocky, whimsical Harvard mathematician and astronomer Benjamin Peirce. Though Peirce made no European pilgrimage, he knew and used European works much more than did most American mathematicians. We have already noted his readiness to take on the famous French astronomer Leverrier in 1847. He also gave occasional evidence of the promoter's instinct, playing a leading part in founding the Harvard Observatory and agitating for a scientific school at Harvard. By the end of 1843, Peirce actively supported Bache's Washington activities, assuring him that "I regard [your success] to be intimately blended with the best interests of science in the country." There spoke a true member of the group. In 1852 Peirce took over direction of longitude observations for the Coast Survey and continued in it, along with his Harvard duties, until Bache's death in 1867, when he succeeded Bache as superintendent.

The advent of Louis Agassiz in 1846 gave the Bache group a prize recruit. Bache put the resources of the Coast Survey at Agassiz's disposal for research in marine biology, a favor for which Agassiz promised to be thankful "for the remainder of my life," and Bache gave Agassiz further "liberal assistance" that helped him decide to stay in America. In Cambridge, Agassiz settled down across the street from Peirce, with whom he developed (according to a friend of both) "a certain subtile tie of affinity" because "Peirce was a transcendentalist in mathematics as Agassiz was in zoology," presumably a reference to the mystical strain in their science. More down-to-earth was the mutual friend's further observation that both men were academic politicians working to advance the interests of science. As for Henry, Asa Gray noted in 1847 how much he and Agassiz "enjoy and admire each other." In 1849 Henry recorded "several long conversations" with Agassiz "on the subject of the means of improving the conditions of science in our country." Bache and Company were clearly in business.

These four—Bache, Henry, Peirce, and Agassiz—constituted the core of the so-called Lazzaroni. But by the early 1850s two lively junior partners had been admitted.

In 1848, at the age of twenty-four, the brilliant, abrasive, headstrong astronomer Benjamin A. Gould returned from Europe, anointed with the friendship of Humboldt, the sponsorship of Gauss, and a Göttingen Ph.D. He also had a mission. "I dedicate my whole efforts," he wrote Humboldt in 1850, "not to the attainment of any reputation for myself, but to serving, to the utmost of my ability, the science of my country—or rather as my friend Mr. Agassiz tells me I must say, science in my country." For all his newfangled credentials, Gould had to scrape by at first by tutoring Harvard students. But he soon found a patron in Bache. Some years later Gould wrote him rather fulsomely, "I came upon a package of your letters from the years 1849 & 50. Chief Bache, to green boy Gould. If it were possible my dear friend, I should say that they made me love you more than ever." In 1852 Gould, like Peirce, began working part-time for Bache's Coast Survey, and he would keep at it until Bache's death.

The other junior Lazzarone, Oliver Wolcott Gibbs, won full acceptance sometime in the early 1850s. Like Gould, his companion in Europe, Gibbs came back full of zeal to "do something solid for science . . . in America." In 1848 Gibbs's friends put him up, without consulting him, for a chemistry appointment at the New York Free Academy. Though John Torrey was also a candidate for the job, Gibbs won out, thanks to his supporters' high-powered campaign. Henry, Bache, and Agassiz knew about the affair and perhaps were impressed by Gibbs's effectiveness as a string puller.

Gibbs also offered the Lazzaroni certain geographical and professional desiderata. The group dovetailed in those respects rather like a president's cabinet. Bache and Henry represented governmental power, the Washington scientific community, established prestige, and the fields of physics and geophysics. Agassiz and Gould represented European doctorates, international reputation in one and youthful brilliance in the other, the scientific community of Cambridge and Boston, and the fields of zoology and astronomy. Peirce reinforced the Cambridge connection and brought in the field of mathematics. Gibbs shared Gould's youthful energy and ambition, had a New York City base, and represented the field of chemistry.

The roster of peripheral or associate Lazzaroni also suggests a pattern of conscious balancing and link-forging. The chemist John F. Frazer was no star in science, but he had been Bache's lab assistant, succeeded him at the University of Pennsylvania in 1844, edited the *Journal of the Franklin Institute* from 1850 to 1866, and helped Bache maintain his ties with the scientific "Clique" in Philadelphia. Frazer also offered his Philadelphia home as a pleasant stopover between Boston and Washington. So, without really admitting him to their inner councils (as late as 1858, Gibbs did not know him by sight), the Lazzaroni allowed Frazer to claim comradeship. In James D. Dana, on the other hand, the group actively sought a three-way link with Yale, Silliman's *Journal,* and geology. Dana humored them in this and did make the *Journal* hospitable to their interests. But somehow he never managed to attend their formal powwows. And finally, for particular undertakings (as will be seen), certain strategically placed auxiliaries joined the group, notably the New York City civic leader and Columbia trustee Samuel B. Ruggles in the struggle for an American university and the naval scientist Charles H. Davis in lobbying to establish a national academy of sciences.

The roll of Lazzaroni omitted several American scientists of stature or promise. Asa Gray and James Hall had their own fiefdoms; and each in his way was something of a loner, or at least uncongenial to cliquism, Gray quietly so, Hall irascibly (though the latter kept in touch with the "Boston circle" during the late forties and early fifties). The Rogers brothers, William, Henry, Robert, and James, formed their own independent closed circle. During the early fifties Gibbs's old crony Josiah Whitney shared a house with Gould and soon afterward married Gould's cousin. But Whitney had a prickly temperament and also spent much of his time on geological field trips. Anyway, in Dana the group could at least nominally claim a geologist to match Hall, William Rogers, or Whitney. And though Joseph Henry's old friend John Torrey could have given it a botanist in Asa Gray's league, Torrey was too much a scientist of the old school to fit in—not to mention his uncomfortable relationship with Gibbs and his closeness to Gray.

Group consciousness had developed to the point of an accepted name by May 1852, when Peirce addressed Bache as "president of the Florentine Academy." In the summer of 1853 Peirce suggested an annual group dinner and went on to such group concerns as a national academy of sciences and the presidency of the AAAS. In the spring of 1855 Fairman Rogers, writing to another Philadelphia scientist, John L. LeConte, referred to the Bache group as "the Scientific Lazzaroni" with an emphasis suggesting recent coinage.

In 1856 Gould revived Peirce's suggestion of a regular annual get-together, enthusiastically proposing "one outrageously good dinner" every winter and drawing up a dinner list that has stuck in historians' minds as the definitive Lazzaroni roster. Besides the six full-fledged members, it included Dana, Frazer, and Cornelius C. Felton—the last-named not even a scientist but a Harvard professor of Greek, put on the list perhaps because he was convivial, influential in Harvard affairs, and married to a sister of Mrs. Agassiz. But Felton's presence on the dinner list does not establish him as having been an operating member of the Scientific Lazzaroni. Indeed, Gould himself wrote that for the formal dinner there would be "no name, no officers, no club, no society." Furthermore, there seem to have been only two such dinners held before the Civil War, in 1857 and 1858. Though they have attracted historians' attention, they cannot be taken as the touchstone of the group. The real test of membership was in combat, not cookery.

As in the case of the British X-Club later, the existence of the Lazzaroni became hazily known to the general public and their power romantically exaggerated. In November 1858 the *New York Times* alluded to "a notorious society which boasts of its control over every scientific appointment of value in the country." But the Lazzaroni never really wielded such control, and furthermore, internal strains were by then beginning to sap such power as they did possess. Peirce was vainly striving to stave off a confrontation between Agassiz and Dana over the merits of an Agassiz protégé. Bache had already dropped some mildly slighting remarks about Henry, and Gibbs was much harsher. Peirce had complained that Gould "has been at the bottom of all the difficulties of my life." One Harvard undergraduate, 1859–62, later remembered Peirce and Agassiz as "enemies, with occasional intermissions of loving friendship." Setbacks on more than one front had dampened Lazzaroni spirits. Gibbs tried to organize another annual dinner in 1859, but Agassiz seemed indifferent. Bache talked of a final effort to rally the group, then in March 1860 flatly declared "the Lazzaroni defunct." In 1860, 1861, and

1862, Frazer, Peirce, and Bache himself wrote hopefully of another possible dinner, but without result.

Lazarus rose for one last victory. Early in 1863, Bache, Gould, Peirce, and Agassiz, and the adjunct Lazzarone Charles H. Davis met in Washington and contrived, without Henry's knowledge, to slip a bill unobtrusively through the Congressional mill establishing the National Academy of Sciences. In August 1864, however, Bache suffered the stroke that would carry him off two years later. With his incapacitation the Lazzaroni as a group were indeed defunct.

So much for composition and chronology. The group's mere existence and life cycle hold interest for sociologists of science. But the Lazzaroni's significance for the history of science lies in what they actually did—and failed to do. Even this, of course, is only part of the story that now follows, the story of how the scientific enterprise in America was organized and directed in its formative years.

Chapter 17

Support without Strings

In dubbing themselves Lazzaroni—beggars—the Bache group may have been venting their frustration at what they had to do for support. Gibbs complained about the drudgery of teaching undergraduates, Henry about pressure to demonstrate practical utility, and Agassiz about the incomprehension and "blunders" of "legislative bodies and governments." What they wanted was self-rule for science, support without strings, the time and money to do research without having to account to laymen for its direction or consequences.

But how could he who paid the piper be kept from calling the tune? Alexander Dallas Bache canvassed the possibilities in an 1844 address. As head of the Coast Survey, he could scarcely reject government support of science for legitimate government purposes. But congressmen were too ignorant of science to aid pure research wisely through direct grants or stipends. Perhaps such funds could be disbursed safely and soundly through an otherwise independent government bureau of experts. Aid from state or local governments might be filtered through local scientific societies.

The societies could also channel private philanthropy and perhaps raise funds from the general community by expanding their membership. Bache praised the British Association and the Royal Institution of London for underwriting research, notably that of Faraday. The pending Smithsonian funds might be used in that way. Perhaps a wealthy benefactor could set up another foundation like that of the Lowell Institute in Boston, except that the latter "offers a bounty for good lectures; we want a bounty for research."

Most revealing was the half-heartedness of Bache's hopes for today's best-known haven for unfettered research, the university. His speech placed German universities at the center of German science. But American colleges, he said, "by overloading the professor with work, by stinting him in his means of research, by affording few facilities of intercourse with his fellow laborers, by undervaluing the results of his labors in extending his science, . . . sometimes are far from aiding science." His guess was "that our colleges might be improved, but not . . . *at the present day.*" American colleges had developed to meet the wants of American society, and those wants would have to change before colleges could.

But times and the needs of society changed fast in the "Young America" of the 1840s, and as the decade ended in a surge of economic growth Bache and his liegemen dared to hope that the day of the true university in America was at hand, with all that would mean for free-ranging research.

In Germany a main element of the university ideal was *Bildung,* the cultivation of the student's soul, personality, and intellect. Another element, somewhat akin, was Idealism, a quasi-mystical sense of some inner spirit or relationship among all things, distinct from their physical properties or external semblance. American scientists seemed to ignore all this as incomprehensible or immaterial. On the other hand, American universities, when they finally appeared in more than name, would be more strongly marked by democratic openness and eclectic pluralism than in Germany. The main concern of American scientists, however, was for what the German university exemplified in extended and specialized training of students, peer review and recognition, exchange of ideas, and support for unconfined research. Lazzaroni exhortation in particular dwelt less on what the university could do for students than on what it could do for scientific faculty.

In the winter of 1850–51 Wolcott Gibbs plotted to escape from his time-devouring duties at the New York Free Academy by starting a "polytechnic school" in New York City. Josiah Whitney heartily seconded the plan out of "utter disgust" with the way things were going at Harvard, and they sounded out Benjamin Gould and James Hall as possible colleagues. But they had no financial support; and when Hall himself launched a more ambitious project, they climbed aboard and let their own plan drift into limbo.

Gibbs and Whitney had seen in New York City not only a mercantile population needing scientific instruction, but also the new Astor Library, good communications with the rest of the world, and opportunities for outside income. In 1851 Henry P. Tappan's book *University Education* suggested that the city's wealth could give solid footing to a true university. But Albany had strong claims, too. It had the state legislature, already generous in support of James Hall's work. It lay on the main route between the seaboard and the West and was close to New England. With its medical school, preparatory school, agricultural society, and museums, and with nearby Rensselaer Polytechnic Institute and Union College, Albany already had a respectable scientific and technological community. And above all it had James Hall.

"I feel too isolated here," wrote Hall in 1846. But he already had an emotional stake in his New York survey, relished his triumphs over legislative appropriation committees, and so decided to stay put and try moving the mountain to himself. By 1850 he had rallied his influential Albany friends—politicians and businessmen as well as scientists—to the idea of a university at Albany. In April 1851 the legislature incorporated the Albany University. Its first circular promised support for "professional and profound research."

Not only Gibbs and Whitney, but also Peirce, Agassiz, and Gould offered the enterprise their advice and services, provided that the faculty control appointments and courses and that research be guaranteed support without strings. Anxious to enlist the scientific community generally, James Hall contrived to bring the American Association for the Advancement of Science to Albany in August 1851 for its annual meeting, and Albany's leading citizens set an antebellum record for lavish hospitality. In his presidential address, Bache pointedly asserted the priority of university research over teaching. The assembled scientists heard much of the project in private conversation. One of them, John P. Norton of Yale, shortly announced a course of lectures in the new University's School of Theoretical and Applied Science. "I have never felt more interested or more hopeful," Norton wrote that fall. But his fellow scientists waited more cautiously for a legislative appropriation, and the only savants on hand for the winter term were Hall and Norton. That meager session turned out to be both the university's birth cry and its death rattle.

Norton gave his lectures and also his life. That winter he exhausted himself trying to sustain his Yale and his Albany courses concurrently by commuting weekly between Albany and New Haven. The railroads of 1852 were not good enough for that. Spitting up blood, he had to quit both jobs in March; and in the following September, barely thirty years old, he died, apparently of tuberculosis. The tragedy epitomized events at Albany. In February 1852 the university's supporters opened a campaign for state scholarships. But disunity weakened it. The New York Agricultural Society planned a school of its own. Rensselaer Polytechnic Institute and Union College drew back from what they took to be a rival. Bache and his group may have too publicly ranked research above teaching, whereas the citizens of Albany wanted practical aid for economic development. Opponents of the scholarship bill killed it in the spring of 1852, whereupon Peirce and Bache shifted their hopes to the University of Pennsylvania. By the end of 1853 the Albany university movement had, as Whitney put it, "completely fizzled out."

As early as the winter of 1851–52, Bache had talked with Bishop Alonzo Potter, a trustee of the University of Pennsylvania, about raising that school to the European level. In April 1852 Bache suggested that it "gather in the harvest of students" that the now-doomed Albany scheme might have attracted. Potter pressed the university idea on his fellow trustees, and that winter they debated the proposal heatedly. The Lazzaroni, including their Pennsylvania adjunct John Frazer, supported it. But its opponents argued that not enough American college graduates would forego an early start at making money. Potter suggested fellowships, but no money seemed in prospect for that purpose. And so the Pennsylvania vision in turn "fizzled out."

Now the Bache circle turned to Columbia College. With a mere 156 students and 14 faculty, with science represented in its archaic curriculum only by a professor of mathematics and another of natural philosophy and chemistry, and with a timid, church-bound board of trustees, Columbia might have seemed a poor prospect for a university. But in addition to the general advantages of New York City, the college had a potential bonanza in its downtown property, especially if it should move to cheaper land uptown. And it had a dynamic and forward-looking trustee in Samuel B. Ruggles, a lawyer who had retired young to throw his energies into public service. As an active worker for the Albany project, Ruggles had become an acknowledged though temporary member of the Lazzaroni, who in 1852 still called themselves the Florentine Academy. That fall Ruggles tried to move his fellow Columbia trustees in the university direction. The conservatives dragged their heels, but in November 1853 Ruggles won approval for relocation uptown and, rather tentatively, for a graduate program.

At this point Wolcott Gibbs became the hapless center of a storm that blew the whole venture off course. He had known Ruggles for years, and in the summer of 1852 Ruggles had warmly though unsuccessfully recommended him for an appointment in chemistry at Pennsylvania. Gibbs then angled for an appointment at Yale. But that prospect vanished, according to Whitney, when Gibbs "remarked that he wished the Faculty to take into consideration that he often worked on Sunday, and rarely went to church." In 1853, however, Ruggles and his son-in-law George T. Strong proposed Gibbs for a chemistry opening at Columbia as a step toward upgrading the college. Though Ruggles and Strong, Episcopalians like most of the other Columbia trustees, knew that Gibbs was a Unitarian, they considered the fact irrelevant to teaching and research in chemistry. Gibbs's brother George had a stronger sense of impending trouble. "How the devil," he asked Wolcott, "does Trinity Church permit a heretic to enter the sacred walls?"

An indicator of wind direction was a letter from one trustee asking Gibbs whether he thought the Bible should take precedence over physical science where they conflicted, to which Gibbs replied that in such a case one of the two must be misunderstood. That cost him some trustee votes at the outset. Gibbs's fellow Florentines began to take a hand—rather too heavy a hand for Gibbs's good. "We are hard at work for

Wolcott Gibbs, and want your aid," wrote Gould to Peirce. Glowing testimonials were showered on President Charles King, not only from the Florentines but also from other leading scientists. A refrain ran through the accolades that the trustees may have found less moving than intended: the appointment of Gibbs was important "for the sake of science in America" (Gould), "American science" (Peirce), "the cause of science" (Dana), "science in our country" (Bache). The trustees might have been moved in the opposite direction had they read Gould's note to James Hall: "Wolcott will galvanize the possums & Col. Coll. yet expand into a university but for the Lord's sake don't say anything about *that.*"

The impetuous Ruggles met the sectarian challenge head-on, pointing out that both the charter and state law prohibited any religious qualification for office in the college. His opponents resorted to the casuistry that their individual religious objections to Gibbs did not constitute an official religious qualification. Poor Gibbs's struggling candidacy began now to be weighted down with an array of Larger Issues. Most obvious was that of religious freedom, implying a contest between bigots and infidels, as each side saw the other. One trustee believed that the charge of bigotry merely turned fence-sitters against Gibbs. Ruggles himself became, in the words of a nontrustee, "almost delirious on the Gibbs question." Strong had Gibbs's recommendations printed and distributed; this also offended some. The New York press took up the subject, all pro-Gibbs except the *Herald.* And Ruggles and Strong stirred up the alumni, from whom the college in its century of existence had not yet received a single gift of money. Now at least they gave Columbia their attention and Gibbs (himself an alumnus) their overwhelming endorsement.

There were other issues too. The Florentine Academy had pressed too hard and too obviously. The formidable leader of the opposition among the trustees, Gouverneur M. Ogden, was prejudiced not so much against Gibbs personally or against Gibbs's faith as against change of any kind, and he feared the worst in this case. In February 1854 he wrote: "I have a suspicion very strong that the most favorable recommendations [of Gibbs] . . . , extravagant as they are, have proceeded from interested motives, from men who expect places in a grand scheme for education expected to be carried into execution by us; of which his elevation is to be the entering wedge." Ogden's suspicion was shrewd indeed. As Ruggles expressed it to Peirce that very day, "It is not the question of electing Wolcott Gibbs or not—but whether Science is to be cabinned, cribbed & confined by the Church. . . . The War will be long and bitter—but a University will spring out of it." To the Florentines, religious and academic freedom were but corollaries to the grand issue of scientific independence. To Ogden, the issue was precisely the same: whether the laymen who paid for science should have a say in its doings.

In April 1854, by a margin of one vote, the trustees chose Richard S. McCulloh, a second-rate Southerner currently at Princeton. Refusing offers from President Horace Mann of Antioch College and President Frederick Barnard of the University of Mississippi, Gibbs stayed in New York City on grounds of sentiment and its facilities for scientific study. One chemist wrote another in 1856 that "Gibbs is engaged six or seven hours every day in teaching little boys in the Free Academy." Meanwhile, at Columbia, Ruggles's prediction showed signs of coming true, though in a painfully slow and halting fashion. The controversy had inspired Ruggles and Strong to write a pamphlet, *The Duty of Columbia College to the Community,* powerfully arguing for "a great national University" at Columbia that would allow students to pursue their studies as far as possible and "above all" would support "original research and discovery by the ablest men the world can furnish." "It must," wrote Ruggles, "be pre-eminent in Physical Science, which . . . plainly characterizes the present age."

From time to time during the rest of the fifties Ruggles pushed the proposal. The college moved to Forty-ninth Street in 1857, acquiring more than half a million dollars thereby. Better faculty were hired, including Charles Joy in chemistry (Gibbs in 1857 flatly refused to be considered). In 1858 schools of science, letters, and law were organized. The Law School survived and grew. The others were dropped in 1861 for lack of students. European competition in graduate study, and lack of American incentives for it, remained major obstacles. But the alumni had been permanently roused by the Gibbs affair, the Ruggles-Strong pamphlet had pointed the way, and historians have generally agreed with Strong's own retrospective judgment in 1868: that at a time when "the institution seemed about to decompose and perish," the "prodigious row about Wolcott Gibbs" had initiated "the slow and interrupted progress of the College toward convalescence." Richard Hofstadter, a mid-twentieth-century historian at Columbia University, put it more strongly: "Columbia University arose out of the case."

The Lazzaroni, as they called themselves by 1855, kept out of Columbia's post-Gibbsian affairs. Bache, for one, recognized that they had overplayed their hand there. Instead they backed Henry Tappan's renewed call in the spring of 1855 for a great university in New York City. That August, Bache's presidential address to the American Association for the Advancement of Education proclaimed such a university to be "the want of our country." Young Americans should not have to go abroad for the advanced preparation required by the modern world, he argued. He stressed the need for a large endowment, not to put up elaborate buildings but to support faculty salaries and research.

In July 1856 Gould followed this up with a similar address, touting faculty research as well as deeper and more specialized education. Gould, however, questioned the adequacy of private gifts. He thought state or federal aid would be needed. This, he commented, might be a good thing despite the risk of political meddling, because "the principle of power without immediate responsibility is too much at variance with the whole tenor of American republicanism, to escape distrust and animadversion." Peirce chimed in soon afterward with a printed plan for a university, emphasizing well-supported research facilities fully controlled by their users. And in October 1856 Bache returned to the subject with another published address.

The drum beatings in New York reverberated in other quarters. Yale's existing Department of Philosophy and the Arts granted no advanced degrees, but in 1856 James D. Dana asked the Yale alumni at Commencement, "Why not have here, THE AMERICAN UNIVERSITY!" The foundation was in place, he pointed out. Only more faculty, another building, an astronomical observatory, and a farm for the agricultural department were needed. Then the tide of students going to Germany would turn. The Yale Corporation spent four years looking before it leaped, but at last in 1860 it adopted a report, signed by John A. Porter but probably owing much to Dana and Daniel C. Gilman, that expanded and systematized the department's offerings and instituted the degree of Doctor of Philosophy "in accordance with the usage of German Universities." In 1861 Yale awarded the first three American Ph.D. degrees, one in physics, the others nonscientific.

Harvard meanwhile inched along a different route, that of faculty quality and outlook rather than institutional structure. One historian has neatly characterized the Harvard of that period as "an undistinguished institution embracing distinguished men." Peirce complained to Bache in 1855 that President James Walker's "ideas of education are but just worldly wise without any prospective enlargement."

Not all the faculty were distinguished, at least not for research and scholarship. Those who assessed Joseph Lovering after his death seemed most impressed by his effective teaching and the fact that he had held the Hollis Professorship of Mathematics and Natural Philosophy for a record fifty years (1838–88). "Professor Lovering's life," said President Eliot, "seems to me to be better characterized by the word fidelity than by any other," and he laid most stress on Lovering's "capacity for assiduous routine labor." Eben Horsford had come from Liebig's lab with the hope of replicating it, but he had to struggle with insufficient funds, ill-prepared and often weakly motivated students, and a gamut of tasks from elementary instruction to research duties, tasks that had engaged half a dozen men in Giessen. Worn and disheartened, he recovered from an illness in 1852 (the year young John P. Norton died) fearful of destitution for his family in case of his death. His series of papers on the chemistry of corals, of which he knew little, drew critical fire from fellow scientists. So he turned away from Liebig's example and acted instead on one of Liebig's admonitions: to make science useful—and lucrative. After 1852 his energies went to developing a miscellany of commercially promising patents such as safety lamps, india rubber, gas burners, lead pipes, grindstones, soap, ink, and yeast powder. From then until he finally resigned in 1863, he took less and less interest in his college duties.

Nevertheless, Harvard had Asa Gray, Louis Agassiz, and Benjamin Peirce. Moreover, as professor of anatomy the medical school had Jeffries Wyman, a tall, slender, alert Yankee, whom an ardent disciple of Agassiz credited with a wider and more accurate knowledge of natural history than the master's, though he published much less. "In some ways," recalled the student, "he was the most perfect naturalist I have ever known." He was also the foremost American comparative anatomist of his day. These men and some outside the college but within the Boston community, such as William B. Rogers and the young scientists of the Nautical Almanac, were gradually creating a professional, cosmopolitan, research-oriented spirit—and therefore a university spirit—at Harvard.

Josiah P. Cooke bridged the old Harvard way and the new. In 1850 the twenty-three-year-old Cooke succeeded the lately hanged John W. Webster as chemistry professor in the college (Horsford taught in the Lawrence Scientific School). The appointment raised faculty hackles. As a chemist, Cooke was self-taught, except for some lectures he had attended during a year in Europe. In keeping with the older way, he came of Boston gentry, had graduated from Harvard, and was a friend of President Jared Sparks. On the other hand, he foreshadowed the new way in being committed to chemistry as a lifetime career and, as Jeffries Wyman conceded, was "industrious, zealous, & . . . a conscientious labourer." While filling in during Webster's trial for murder, he had wangled the use of a cellar room without gas or running water and, at his own expense for chemicals and apparatus, offered the first teaching laboratory in Harvard College proper. For several years the students got no credit for lab work. By 1854 five lab courses were offered. Though handicapped, as one student remembered, by "a nervous trembling which affected his voice as well as his hands," Cooke proved to be an able teacher. Indeed, Joseph Henry in 1859 called him "one of the very best lecturers on chemistry I have ever heard." Desperate by 1856 to get out of the cellar, he won the corporation's consent to tap "the wealthy gentlemen of Boston" (in a safer and seemlier way than had his late predecessor) and by 1858 he had a large new chemistry laboratory.

But it was Agassiz who struck a gusher of philanthropy. He had all the necessary drilling tools: an incomparably winning enthusiasm, an international scientific reputation known to all literate laymen, an extraordinary capacity for enthralling them with the wonder of science, and a secure position in a prosperous community distinguished by the "liberality of the wealthy gentlemen" (he "only associates with 'upper crust' nowadays," wrote Gould half-jokingly in 1860). Agassiz's passion for collecting specimens had long consumed much of his own money without being satiated. In 1848 he had got Harvard to buy him an unused bathhouse on the Charles as a storehouse. That was only the beginning.

In 1855, fortified with $10,500 from rich Bostonians and $1,800 from the college, he took over the original Lawrence Scientific School building for a "zoological hall." He had already sent out more than six thousand circulars to scientists, natural history teachers, and scientific societies begging specimens for a projected *Natural History of the Fishes of the United States.* Now he announced instead a series of ten volumes covering all American natural history and launched a great campaign for subscriptions, appealing to national pride.

The publisher of Agassiz's projected *Contributions to the Natural History of the United States of America* insisted on a guarantee of five hundred subscribers for each of the ten volumes at twelve dollars a volume. Five times as many subscriptions came in. Two volumes on American turtles appeared in the fall of 1857. Turtle fanciers got their money's worth: fine bindings, excellent drawings, a learned, thorough, and many-faceted coverage—all in all, a major contribution to zoological knowledge. And for those who looked beyond turtles, Agassiz included an "Essay on Classification," heralded as a conspectus of all zoological creation and in fact a heroic effort to update Cuvier. Unfortunately Darwin topped it and eventually buried it with *The Origin of Species* in 1859. In 1857 scientists like James D. Dana and Asa Gray felt disappointment and uneasiness about the immutability of Agassiz's long-held ideas, despite the scope and skill of his fact gathering. The third volume in 1860 ignored Darwinism and concentrated on jellyfish. Even specialists found it of limited value because of its sterile Cuvierianism. Agassiz never finished the series.

Agassiz's most important contribution to American science during those years was not so much what he did *in* science as what he did *for* science. Despite his grandiose publication plans, Agassiz kept expanding the scale of his collecting. By 1854 he was thinking in terms of a great teaching museum, a separate institution within the university, excelling the Jardin des Plantes in Paris. To this goal he won over Francis C. Gray, scholar, philanthropist, industrialist, who at his death in 1856 left $50,000 to endow research exclusively. Agassiz got the executor to use the designation "The Museum of Comparative Zoology at Harvard College" (the public would always think of it as the "Agassiz Museum"). Now began a fund-raising drive for the building. By 1859, $50,000 in private donations was assured. Again pleading the nation's scientific honor, Agassiz beguiled the Commonwealth into earmarking proceeds from state sales of new-made Back Bay land as matching grants against private gifts to five educational institutions, including up to $100,000 for his museum. With the state grant, private gifts, and the Gray bequest, Agassiz thus piled up $220,000.

Dana was reserved about the "Essay on Classification," but not about this. "I do rejoice with you—$220,000!" he wrote Agassiz in April 1859. "I believe you are doing your greatest work for American Science in securing this endowment." Along with subscriptions to the *Contribu-*

tions and further gifts from private benefactors, Agassiz raised a grand total of $591,000, not counting his college salary, from 1853 to 1860. Agassiz's triumphs of grantsmanship presaged the era of colossal private benefactions and the scientific research foundations that would arise from them.

In one respect there was a difference. Agassiz's enormous "bounty for research" belonged to Agassiz rather than zoology in general. The two-story brick and granite building that opened in 1860 became a temple to Cuvier, an anti-Darwinian fortress—in short, a symbol of one man's increasingly granitic opinions. The "personal equation" in Agassiz's fund raising, as the historian Howard S. Miller has pointed out, becomes even more obvious when contrasted with the relative modesty of the $23,000 raised in 1863–64 for the herbarium of Agassiz's colleague Asa Gray. By then Gray commanded more respect than Agassiz among scientists, but he had less celebrity, personal charm, social influence, and promotional genius. In that sense, the ideal of research support free of nonscientific considerations was not fully realized in Agassiz's triumph.

The most bizarre manifestation of the drive for support without strings grew out of the Albany university project. One of that project's advocates, the astronomer Ormsby Mitchel, had pleaded so eloquently for a university observatory that the concept floated free of the general wreck and was taken in tow by two leading Albany citizens, the banker Thomas Olcott and the physician James Armsby. Both men were cultivated and enterprising, leaders in charity and education, patrons of art and science, models of civic-mindedness. They were also men of power and pride, taking special pride in their city's high cultural and intellectual standing and in what they had done to help achieve it. In that spirit they took charge of a fund-raising campaign, to which the largest contributor was a wealthy widow, Blandina Dudley. Her late husband was to be memorialized in the projected Dudley Observatory.

In the fall of 1852 Mitchel, pleading other obligations, recommended Benjamin Gould as director. But Gould, who was working for Bache's Coast Survey in Cambridge, hung back, and meanwhile the observatory project languished. At last, in the summer of 1855, the desperate trustees struck a deal with the Lazzaroni. To oversee the scientific work of the observatory, they named a "Scientific Council" consisting of Gould, Bache, Peirce, and Henry. Upon the Scientific Council's insistence, the trustees promised to raise an endowment for an annual operating income of at least $10,000 (eventually Mrs. Dudley came through with $50,000 toward that end). At the urging of Peirce, they wheedled another $14,500 from Mrs. Dudley to buy a heliometer, a telescope designed to measure small angular distances between stars. This would develop a more precise base for Coast Survey measurements, for which Bache wanted maximum precision, and it would put American astronomers on a par with

European in addressing the questions then uppermost in astronomers' minds.

In return Bache stationed a Coast Survey man at the observatory for Coast Survey observations and for other work as an unpaid volunteer, and Gould agreed to supervise the observatory without pay so far as he could from Cambridge. The trustees were overjoyed, not least by the prestige these luminaries shed on them and the observatory. But the involvement of Gould spelled trouble. Though brilliant, hard-working, and full of ambition for himself and American science, Gould was also arrogant, tactless, and touchy. He was a driven man, highly competitive, whose all too evident self-esteem contended with spells of self-doubt and depression. "He has more personal enemies than any person with whom I am acquainted," Joseph Henry would comment in the midst of the protracted turmoil that accompanied Gould's association with the Dudley Observatory.

From the start Gould did his acerbic best to instruct Armsby, Olcott, and the other trustees in the significance of both his own appointment and their commitment to a heliometer. Physical astronomy, the study of the physical nature of heavenly bodies, was not to be the observatory's concern, despite public interest in it. *Practical* astronomy, the study of position and motion, and *theoretical* astronomy, its mathematical elaboration, were to be the Dudley's work. That was what the heliometer and meridian circle were for, not for gawking at Saturn's rings or the mountains of the moon.

The building already put up on Mitchel's specifications was to be partially but expensively reconstructed to accommodate the new instruments. The piers, the instruments, and other arrangements must be up to Gould's high standards of accuracy, stability, and dependability, no matter how long or how much it took to meet them. Meanwhile Armsby and Olcott, on the scene as Gould was not, busied themselves with procurement and building in ways that, in Gould's view, wasted money and compromised standards. Gould was soon convinced "that the empty dazzle of temporary show was, in the wishes of the managing Trustees, paramount to any ideas of scientific usefulness or dignity." The trustees in turn saw Gould as infringing on their fiscal responsibilities and delaying the commencement of working operations through inefficiency, extravagant perfectionism, and perhaps lack of confidence in his own ability to use the instruments to good purpose. Gould's insistence on directing the work from Cambridge certainly impaired both the efficiency of the work and his communication with the trustees. In the spring of 1857 Gould suffered something like a nervous breakdown. "No power to act, to think, or to contemplate," he wrote a friend from Cambridge. "No vigor nor energy nor hopefulness nor vital force. Trying to work too hard, something snapped. . . ."

Gould recovered presently, but in the fall of 1857 his Coast Survey assistant at Albany detonated what may have been an inevitable explosion. The Danish astronomer Christian H. F. Peters had, like Gould, studied with Gauss and Encke and earned a German Ph.D. Gould had known him briefly in Europe and so took him on as a Coast Survey assistant in Cambridge, after which Bache, on Gould's recommendation, assigned him to Albany. There Peters curried favor with Olcott, notably by giving Olcott's name—contrary to astronomical usage—to a comet Peters discovered in July. Gould, as editor of the *Astronomical Journal,* reported it simply as "Fourth Comet of 1857." Olcott, who had been happy at his "immortalizing . . . in this world" by association with the Scientific Council, may have been disgruntled at having his name rudely stripped from an eternal wanderer among other worlds. He was certainly enraged when in November Bache reassigned Peters to Cambridge on grounds of financial stringency, Peters thereupon resigned from the Coast Survey, and Gould refused to hire Peters as an observatory employee.

Chided by Olcott, Gould threatened to quit as unpaid director. Bache talked him into agreeing to move to Albany and take full charge of the observatory while continuing his Coast Survey duties. But in January 1858, before the new arrangement could be put into effect, the trustees suddenly appointed Peters director. Henry later charged that Peters had wooed the trustees with "a chimerical scheme of making the Observatory a democratic establishment," but the trustees had probably turned to Peters in the hope that he would speed completion of the observatory and thereby reassure the prospective donors on whom its future desperately depended, especially after the Panic of 1857. The Scientific Council in all its majesty descended on the trustees and got Gould reinstated on the basis of the new arrangement, while Olcott's influence obtained a place for Peters as director of the Hamilton College observatory, where he made a distinguished career for himself over the next thirty years.

But irreparable damage had been done by the confrontation. From afar Gould had already annoyed the trustees beyond endurance by his very absence, his seemingly unreasonable perfectionism (which he did not deign to justify to laymen), his sharp-tongued spurning of their well-meant efforts to help out, and what they saw as his shabby treatment of a respected fellow professional, Peters. In person Gould enraged them still more. During the crucial meeting of January 1858 he had stood before them and disavowed responsibility for delays and cost overruns, on the grounds of Armsby's meddling. This compounded injury with insult. Each side in the escalating conflict thenceforward imputed ignoble or sinister motives to nearly everything said and done by the other side, brushing off explanations as evasions or lies. The Lazzaroni, in the guise of the Scientific Council, leaped to the defense of their junior member.

What followed calls to mind Matthew Arnold's "darkling plain swept with confused alarms of struggle and flight, where ignorant armies clash by night." Fortunately the "confused alarms" need not be rehearsed in detail here, since Mary Ann James, in her 592-page dissertation of 1980, has traced the convolutions of the Dudley Observatory affair with admirable thoroughness and understanding. In the course of the struggle, both sides multiplied petty harassments, the Scientific Council was dismissed but refused to surrender the observatory, Gould on occasion had the doors locked against the trustees, and a war of thick pamphlets raged, spilling far more ink than the Lincoln-Douglas debates of that year. The Scientific Council fired off a 91-page blast, which even a friend of Bache thought "hideously long." It was only an opening shot. Fifty-one printed pages of Gould-trustee correspondence followed. The trustees responded with fifteen thousand copies of a 173-page statement, which drew a 126-page published rebuttal from a Gould supporter. Finally, in January 1859, Gould himself published a 366-page reply to the trustees. Its very length and stridency persuaded at least one scientist that "there must be something wrong with the man." Newspapers and magazines filled the intervals with long editorials and commentaries, other partisans joined the pamphlet war, and impassioned public meetings now and then took up the question. Cabinet members and senators were drawn in at Washington, as were leading political figures of New York State. Longstanding Albany animosities and rivalries brought recruits to both sides who knew and cared little about astronomy, while resentment of Lazzaroni power contributed to arousing and dividing the national scientific community over the issue.

But what was the issue? Though Gould charged the trustees with pandering to the public by sacrificing true science to empty show, later events indicate otherwise. The trustees' favorite, Christian Peters, made his reputation at Hamilton College by the same sort of "practical astronomy" (i.e., star catalogues and positional computations) for which Gould had wanted the heliometer (which was never acquired) and on which Gould rebuilt his own scientific career in later life. After the trustees regained possession of the observatory by hiring strong-arm men to evict Gould forcibly, they devoted it (first under Mitchel and later under Lewis Boss) to precisely the same sort of work, not to "empty dazzle." The Albany elite, after all, had supported science long before the Dudley was born, and they readily gave basic research precedence over popular enlightenment.

Actually there were several issues. Personality played a part, especially that of Benjamin Gould. So did the injured pride of eminent scientists on one side and civic leaders on the other, all of whom had to some degree invested their reputations in the Dudley enterprise. Rising emotions had bred mutual distrust and misunderstanding. The several combatants' links with scientific circles like the Lazzaroni and with social and political circles in Albany and beyond had widened the conflict. One scientist believed in 1865 that the affair "had its origin . . . perhaps before the revolution of the colonies against Great Britain; in other words, that it was a social feud long pent up."

On a higher level of principle, the trustees effectively pleaded the rights of property and the rule of law. The Lazzaroni assured the public that the trustees had given them full and irrevocable power over scientific operations, and that they therefore had a "moral right in equity." But among themselves, even they admitted the trustees' stronger legal case, and public opinion swung to that view. Americans were historically and currently hypersensitive on the issue of property rights.

The welfare of science, in short, was far from the sole issue. It may not even have been the most important issue. But it surely loomed large in the minds of Gould and his allies, along with personal considerations. "In the success or failure of the Dudley Observatory," wrote Gould, "I believed that the success or failure of future institutions of science in America was largely involved." And why should Gould and his fellow Lazzaroni have believed that? Because the setting up of the Scientific Council had in their view been a crucial experiment in scientific freedom, support without strings, or as their adversaries saw it, power without responsibility. Save for the accident of Gould's personality, the pattern might have been set their way at a time when the whole structure of science was crystallizing. Or so the Lazzaroni probably believed.

The Dudley debacle thus shows how passionately the Lazzaroni, especially Bache, pursued the ideal of scientific self-rule. But it also shows something more: the social and political limits of that ideal, the suspicion and hostility latent in the layman's obeisance to the witch doctor, at least in America. Not the least piquant irony in the affair is the fact that Gould himself had, as earlier noted, unwittingly stated the moral before the tale began: "The principle of power without immediate responsibility is too much at variance with the whole tenor of American republicanism, to escape distrust and animadversion." Yet neither he nor his comrades saw the applicability of that remark to the Dudley affair.

One consequence of the Dudley affair is plain enough. It was the rock on which the Lazzaroni split. Wolcott Gibbs had danced about the combatants, whooping them on, offering advice, collecting material for pamphlets. Bache was less frenzied but still staunch for Gould. Peirce, however, was more equivocal, rating Gould high as a scientist but low as an administrator. Peirce did his duty as a member of the Scientific Council and the Lazzaroni, but grudgingly. Agassiz shared Peirce's estimation of Gould, but kept out of the affair altogether and tended to his collections. Dana took a coolly detached view of the "Observatory comic tragedy" and declined to publish anything on it in his *American Journal of Science.*

Henry suffered most and broke most sharply with his comrades. "I do not like controversy," he complained. "The longer I live the more I am convinced of the meanness of the great majority of mankind." Samuel Morse was just then attacking him on the question of priority in the telegraph. Smithsonian problems also bedevilled him. During much of the hectic summer of 1858 he was troubled by "a nervous affection of the head" that made him feel constantly "on the verge of breaking down." He feared that the whole "dirty business" of the Dudley would put the Smithsonian and Coast Survey in Congressional jeopardy. After the Scientific Council's published *Defence of Dr. Gould,* he vetoed a full and speedy rejoinder to the trustees' lengthy rebuttal. This infuriated Gibbs, who denounced Henry to Bache as a "cowardly selfish & ungrateful sneak." Peirce sympathized with Henry and urged against sacrificing him uselessly for Gould. Bache bemoaned the division "in our ranks" and disagreed with Henry even in retrospect. Henry's worst fears were not realized. The Nautical Almanac lost its 1859 appropriation on the mistaken supposition that Gould was connected with it, but the Smithsonian and Coast Survey somehow got all they asked for from Congress. The Lazzaroni breach was never fully healed, however. Bache and Henry remained close, but no Lazzaroni gathering could be organized until the last one in 1863—from which Henry was omitted.

At first glance, the Lazzaroni's other crusades for support without strings in the 1850s seem also to have been turned back short of Jerusalem, especially if Agassiz's fund-raising triumph is counted as his alone. But viewed more closely, significant influences can be traced to them. The Lazzaroni lost the battle for Gibbs's appointment to Columbia but did much to win the war for the making of Columbia into a true university. James Dana's closeness to the Lazzaroni must have reinforced and enriched, perhaps even inspired, his commitment to the university ideal; he in turn helped move Yale in that direction; and the Yale movement involved Daniel Gilman, who was to realize the ideal at Johns Hopkins, from which it would spread through the American academic world. Lazzaroni preachings during the university movements at Albany and the University of Pennsylvania, and their general public exhortations in 1856, spread the gospel widely. Frederick Barnard took up the chorus at Mississippi in 1857, and as a Lazzaroni protégé he would eventually preside over Columbia's final metamorphosis.

Chapter 18

Communication and Conflict

To grow, science must know, trust, and build on what has been done before. And so the means of recording, communicating, and evaluating its discoveries must grow with it.

In simpler times much of that work was done by learned societies. Despite such names as American Philosophical Society and American Academy of Arts and Sciences, they remained local well into the time of the republic. But they also remained useful, as Agassiz and Bache attested in the mid-1840s. A few of them prospered in the fifties. Agassiz's museum drive in the late fifties probably did more to help the Boston Society of Natural History by stimulating popular interest than to hamper it by competing for money. In the mid-fifties it had stagnated, and its museum had suffered when the young men who kept it in order became absorbed in professional work. But in 1860–61, major gifts enabled the Society to build fine new quarters on reclaimed Back Bay land given it by the legislature. The rise of science at Harvard revitalized the American Academy of Arts and Sciences during the fifties. With Gray, Wyman, and Agassiz on hand, it shifted more toward natural history than before. On the eve of the Civil War, the Academy and the Society together won a place in the history of science with their notable debates on Darwinism. In Philadelphia the Academy of Natural Sciences raised enough money to add a story to its building. And the Cleveland Academy of Natural Sciences grew more active in the middle and late fifties.

But such gains were not the rule. Even in Boston the rise of Harvard, which helped the societies in the short run, overshadowed and so diminished them in the long run. In New York a visitor in 1860 assessed the New York Lyceum of Natural History as "a small one and not possessed of much enthusiasm or vigor—but respectable for position age &c." In Philadelphia the drain of leading scientists to Washington meant the decline of the American Philosophical Society during the fifties—a sad falling off in attendance, communications, and publication. The Panic of 1857 hit the Chicago Academy of Natural Sciences, as it did others, since they relied on member contributions. The Western Academy of Sciences wasted away in Cincinnati. The number of nonspecialized scientific societies in the nation rose only slightly in the fifties; and in New England, the bellwether of American science, it actually declined.

Local societies would survive, of course. Some of their functions had a place even in the dawning age of big, national science. Through museums and popular programs they could bring science to the public, especially to the young. They could carry on relatively modest but still worthy publication. They could provide informal and easily accessible meeting places for those with shared intellectual interests. But they could not support research and publication on the scale science increasingly required. Specialization and professionalization left behind the amateur "cultivators" of science who constituted much of their membership. Other institutions—government, business, universities, professional journals, and national societies—outdid them in size, financing, status, and reach. And though Boston led other cities in science, its lead was not so great that one of its local societies could, as in London or Paris, assume the national authority that an increasingly national science looked to.

In the 1840s a revolution in communication and transportation began making the scientific community more fully national. In 1843 it had cost only twelve cents to send a barrel of flour from New York City to Troy, but eighteen cents to send a letter. In 1847 the first American postage stamps appeared, and by 1851 a three-cent stamp would take a letter three thousand miles. Between 1845 and 1854 the number of letters sent through the United States Post Office tripled. In 1853 Spencer Baird in Washington could answer a letter mailed by Agassiz from Cambridge two days before (though an Alabama geologist complained a year later that "our mails have lately been so deranged that here we have lost all confidence in them").

Scientists like the botanists John Torrey and William Sullivant hailed the promise of faster and cheaper scientific exchange. And for such workers in natural history, the exchange could be of more than words, thanks to concurrent advances in freight transportation. Zoological and botanical specimens could be exchanged far more quickly and cheaply than before. In 1817 freight had taken at least seven weeks to reach Cincinnati from New York; by the early fifties it made the trip by rail in a week. On the average, railroads and steamboats carried long-distance freight five times as fast as wagons and canal boats had. Private express companies sprang up everywhere.

As with mail, the freight network was not perfect. In the summer of

1849 a South Carolinian had to wait for vessels to start running to New Orleans in the fall before sending a box of specimens to a Mississippi correspondent. Another South Carolinian complained in 1848, "There appears to hang over the matter of transporting natural history specimens, a kind of fatality, they either never reach their destination or reach only in a broken condition." But service improved. In Cambridge by the mid-fifties, Asa Gray and Louis Agassiz could summon their hordes of plant and fish specimens from throughout the continent by private correspondence and shipment, even without Spencer Baird's military and naval auxiliaries.

In developing a national scientific community, the printed word remained indispensable. Commercially published scientific books were not the principal medium, partly because of expense and poor sales. In natural history, good illustrations were essential; but competent artists were scarce in the United States, and lithography ran to fifteen dollars a plate. In the late forties, press runs of scientific books seldom went over two hundred fifty copies, and a naturalist was lucky to sell two hundred of an illustrated work. The author usually took a heavy loss. Scientists therefore had to depend largely on government publications, society journals and proceedings, and independent periodicals.

Government publications had serious drawbacks. Contracts were let out to private printing firms. The quality of craftsmanship accordingly ranged from the outstanding printing and engraving of David Dale Owen's geological report on Wisconsin, Iowa, and Minnesota in 1852 to the outrageously shoddy workmanship and paper of John W. Foster's and Josiah Whitney's reports on the Lake Superior district in the same year, so bad that Congress had them done over (to little avail). Of the latter reports Whitney wrote that the printer was "notoriously defrauding the Government," but "there are so many who have their fingers in the spoils," that nothing could be done about it. At about that time Congress made some reforms in contracting for publication of its own documents, and so they improved in paper, presswork, and proofreading, though not in binding. Nevertheless, John Torrey was embittered by the slipshod printing in 1859 of his important botanical report for the Mexican Boundary Survey. "You know how it generally is with public printing, especially that done for Congress," he wrote a fellow botanist. "The lucky fellow who gets the fat job cares for nothing but to receive his money in the shortest possible time & so he runs the work through with railroad speed. . . . The best that can be done is to get up a copious list of errata if they will allow this."

Workmanship aside, Congress was capricious in the quantities provided—for example, only a hundred copies each of the Wilkes Expedition reports as against thirty-four thousand copies of the far inferior reports of the Perry Expedition. The distribution of government scientific publications made no sense either, most of them going to politicians and influential constituents who cared nothing about them. "Valuable documents which are reported to applicants as all exhausted, do wholesale duty as wrapping paper for Washington grocers and market men, at a standard price of four cents a pound, maps and plates included," complained a scientist in 1854.

But the establishment of the Government Printing Office in 1861 ended the corrupt farming out of printing jobs; and meanwhile, with all their shortcomings, government publications had carried a large share of American scientific output. Natural history predominated; scientists were drawn to it by abundance of opportunities, government by concern for economic development. State and federal geological surveys, army and navy exploring expeditions turned out volumes by the score on all aspects of natural history. Bache's Coast Survey and Maury's Naval Observatory gave geophysical sciences their place in print, with a nod to astronomy, which also had the Nautical Almanac. And the Smithsonian reports and contributions touched all fields.

Scientists have always tended to publish articles rather than books, and antebellum America offered a variety of outlets, though often transitory, for their papers. In addition, immigrant scientists published about a third of their articles in Europe, native scientists about a tenth of theirs. At home, some learned societies continued to publish good work, notably the two leading societies in Boston and the two in Philadelphia. But the American Philosophical Society cut back from one volume every two years to one in seven, while the Academy of Natural Sciences leaned heavily on one man, Joseph Leidy. And all four had a natural history emphasis, which meant that government publications increasingly overshadowed them.

Two efforts in astronomy suggest the problems of specialized journals in science. During the first half of the century more than a quarter of American astronomical papers appeared abroad, almost all in one German and one British journal. The rest made their way into society proceedings, Silliman's *Journal,* and the *Journal of the Franklin Institute.* In July 1846 Ormsby Mitchel launched the eight-page monthly *Sidereal Messenger,* the first exclusively astronomical journal in the United States, on the shaky basis of three or four hundred subscriptions charmed out of the people of New Orleans during a popular lecture series. His prospectus promised a record of the Cincinnati Observatory's work, the latest astronomical intelligence from foreign journals, and—characteristically—a special department "to excite an interest among the people in the elevating study of astronomy—and to give a permanent support" to the observatory. Unfortunately Mitchel's courting of the populace put off serious astronomers, touchy like all American scientists in that decade about the taint of amateurism and superficiality. He got more encouragement from leading Europeans, but though he pleaded desperately for articles, few came.

Mitchel himself wrote half the material, despite a multitude of other demands on his time, the rest being mostly copied from other sources.

His writing was not only sprightly but also accurate. The discovery of Neptune and subsequent wrangling over it gave him a good running story, which took the side of Peirce against Leverrier. During the first year, circulation rose to almost a thousand. But Mitchel's dramatic accounts of his financial problems raised more doubts of permanence than they did money. As a reporter of astronomical news, the *Sidereal Messenger* lacked thoroughness and regularity. Circulation levelled off. The *Messenger* grew duller and more perfunctory in its second year and expired in October 1848.

The nation's second astronomical journal likewise bore the image of its maker, but it was an image almost the opposite of Mitchel's. Benjamin Gould arrived home from Europe the year the *Sidereal Messenger* winked out, intent on establishing "something like the Astronomische Nachrichten," though he could find only two American astronomers who thought such a journal could be "intellectually supported" by Americans. He won enough financial support from donors (including Bache) to set it going, though with less than a hundred paying subscribers. Its announced aim was advancement of knowledge rather than diffusion, to which, Gould suggested, Americans already paid ample attention. The *Astronomical Journal* was "designed . . . rather to aid and serve astronomers, than to interest lovers of astronomy." In private Gould wrote more pungently, "Worlds on fire, comet-collisions, poetic inferences & all species of astronomy for the million are exceedingly good in their proper place, but there ought to be something besides."

Gould and Mitchel had one thing in common, however: the patriotic purpose. Gould proposed to distribute at least sixty copies to European observatories and academies, paid for or not, so that American astronomers could "take their true position in reference to transatlantic scientists." The *Journal* should "not be merely creditable, but a glory to our country." It would have no reprints or popular articles, only original scientific work. And as an earnest of this, the first issue consisted chiefly of a formidably mathematical paper by Benjamin Peirce. The engineer and lover of astronomy Uriah Boyden guaranteed the second volume against a deficit. After that, every year seemed likely to be the last. By 1856 Gould himself was making up the deficit. Expenses incurred in his Dudley Observatory ordeal forced him to publish fewer issues and eliminate diagrams and illustrations. In the judgment of a later generation of astronomers, Gould's *Astronomical Journal* quickened research and raised standards. But the outbreak of civil war brought a twenty-five-year interruption of the struggling *Journal,* until Gould himself revived it in 1886.

Among general scientific journals, the *American Journal of Science,* more commonly known until the 1850s as "Silliman's *Journal,*" held its lead. In 1846 Joseph Henry considered using Smithsonian funds to publish a journal with more concern for physics than Silliman's. But in that same year Silliman's *Journal* increased its issues from four to six per year, and James D. Dana, Silliman's son-in-law, became an associate editor.

Henry conceded an improvement. From the start Dana did his best to restrain the elder Silliman from accepting unsound articles as indulgences to old friends or as bids for popular readership, and in 1851 Silliman yielded control entirely. Dana called on Baird and other specialists to evaluate articles submitted in their fields. He solicited articles from established scientists. He added Gibbs, Gray, and Agassiz as associate editors. He got Gray and others to make regular contributions. He employed a knowledgeable Paris correspondent to report on European developments. And he recruited better book reviewers, cautioning them to criticize content, not personalities.

Dana struggled with the problem of increasingly specialized articles in a journal dependent on a general clientele. He had to compromise, admitting some material of general appeal and yet getting by with no more than a thousand subscribers during the fifties. But Henry abandoned his journal project, partly because of limited funds and partly to avoid undercutting Silliman's *Journal* and others.

Bache's 1844 address called for a journal to keep Americans better informed of European scientific work and also to review American work more critically and fully. In 1850 David A. Wells, a twenty-one-year-old Agassiz student, began trying to meet the former need with his *Annual of Scientific Discovery,* a 392-page compendium of short reports on new work in technology and science from European and American sources (including Silliman's *Journal*), as well as a classified index of articles in scientific journals and reports. A leading twentieth-century bibliographer of science has called it "the earliest successful attempt [in the United States] at maintaining a kind of 'Science Abstracts'." Wells skillfully balanced scientific soundness with popular appeal, working his Harvard connections to get articles by Agassiz, Wyman, and Horsford, while sweetening the dose with more popular material. It went down well with at least one Illinois layman: William Herndon remembered years later that his partner Abraham Lincoln told him in the fifties, "I have wanted such a book for years." Silliman's *Journal* generously praised the early volumes. Perhaps as gratifying to Wells was the *Annual*'s financial success over the next twenty years.

Communication is sometimes equated with understanding, and understanding with peace. But in science, communication may breed conflict on two counts. It requires validation, which implies the possibility of rejection. And even when a finding is validated, recognition for it depends on priority. "Most scientific disputes arise from questions of priority," remarked Silliman's *Journal* in 1861.

Disputes have always enlivened the history of science. In Europe, science in the 1840s rang with the clashes of Liebig and Mulder in chemistry, Leverrier and Adams in astronomy, Murchison and Sedgwick in geology. Yet mid-century American scientists considered themselves (or more precisely, each other) as uncommonly contentious. This view did not derive from ignorance of Europe. The Swiss-born botanist Leo Lesquereux wrote in 1852, "The longer I live in America, the more I am offended about the jealousy of all your scientific men. Every one of them, with a few exceptions, is looking to his brother in science like to a foe whom he has to crush before he can ascend a higher step." In 1853 the German-born botanist George Engelmann was "deeply pained to witness the spirit of rapacity, envy and sickly emulation evinced" by rival paleontological parties in the Bad Lands. Josiah Whitney, who knew European science firsthand, complained that same year about other squabbles: "There seems to be nothing but quarreling among the scientific men in this country."

In 1862 the geologist Benjamin Shumard thought it too bad that American paleontologists and geologists "find so much fault with each other" in public. James Dana in 1865 alluded to "American hate & jealousy" in science. Joseph Henry in 1858 was "sad and depressed at the state of feeling among the cultivators of science as it is exhibited in this country. The geologists appear to be arrayed against each other and look apparently upon different strata of rocks . . . as . . . personal property. . . . We should seek [science] for its own sake, and not for the emolument or reputation which may result from it." Henry himself was a model of forbearance under provocation, but sometimes even his feelings broke through. "Henry sticks it into Morse," reads an 1846 entry in the diary of one of his Princeton students. "Says Morse's assistant Vail has lately published a book purporting to be a history of the telegraph and hasn't mentioned him at all in it."

Disagreement has several gradations. A discussion may become a debate, then a controversy, then a quarrel, and finally a feud. The issues can likewise be graded in descending order of dignity: theories, modes and methods, priority, competence, and mere personalities. Mid-century American scientific encounters ran both gamuts.

The Taconic Controversy touched every level of conflict. It began in 1841 when Ebenezer Emmons formally declared that a system of strata in the Taconic Mountains, towering over the Hudson lowlands in the Hudson-Champlain Valley, represented a distinct geological period, the first to have identifiable life forms. The controversy over this hypothesis lasted, in the words of its modern historian, "for the remainder of the century and affected the lives and careers of every prominent geologist in the United States and Canada."

In the New York Geological Survey reports of 1842, Emmons not only set forth his Taconic system but also rejected the hypothesis of the German geologist Abraham Werner that rock strata had been deposited uniformly throughout the world from a primordial sea. Whereas the Wernerians or Neptunians classified strata solely on the basis of mineralogy, Emmons also considered structure and fossil content, and he named the strata thus determined after New York localities. His method of classification (as distinct from the Taconic system) was accepted with little dissent. Yet it meant more than mere shuffling of nomenclature. It expressed the basic ideas and approach of modern geology. And since the New York survey became a standard for the rest of the nation, Emmons may be credited more than anyone else with this transformation of American geology.

Yet in the Taconic battle, Emmons suffered from being labelled "an old fogy." He did hang on to some outmoded ideas, and he failed to keep up with new work in the American West. But his background and career also may have told against him. Born in western Massachusetts—in the Taconic area—in 1799, and thus being classifiable with an earlier generation of geologists, he studied at Williams College, Rensselaer Institute, and the Berkshire Medical School. Along with his geological work, he also practiced medicine, taught obstetrics at Albany (though unwillingly), and even wrote state reports on New York horticulture and entomology. Such a conglomeration of callings struck the new generation of specialists as ludicrous—and worse still, unprofessional.

Besides that, Emmons lacked personal charm. Though grim in appearance, he was shy and easily rattled. He was also stubborn and graceless. More particularly he incurred the hostility of the redoubtable James Hall, as did so many others. Hall had borrowed four hundred dollars from him and failed to repay it for many years (we sometimes forgive our debtors more easily than our creditors). Hall also had been Emmons's assistant in the New York survey, and afterward had succeeded in getting assigned to do the report on paleontology, which he then parlayed into his great lifework of fifty years. This left Emmons to cope with the agricultural reports for which he came to be mocked.

The merits of Emmons's Taconic hypothesis even now resist a simple judgment. The strata in question, like the controversy itself, are extraordinarily disturbed and confused, a fact that goes far to extenuate Emmons's self-reversals and rearrangements in defending his views. But apart from the validity of the system itself, the controversy over it is a classic example of scientific communication and evaluation as they operated in mid-nineteenth-century American science.

Besides the New York state report of 1842, Emmons used other channels to present his Taconic system. Having first broached it to the New York Board of Geologists in 1839, he tried it out in a paper before the new American Association of Geologists and Naturalists in 1841, a paper upon which most comments were unfavorable. In 1842 Henry Rogers rejected the system in a paper before the American Philosophical Society. Then in 1844 Emmons seized upon two new-found trilobite fossils as definitive for the system and as the first of all living organisms, from which view Hall dissented. In 1846 a committee of the AAGN found for Emmons as against Hall. But that year the new finds led Emmons, in his agricultural report, to postulate a complete, convulsive flip-flop of the whole series of strata. This contortion, along with his unprepossessing personality, tended to alienate the new establishment in geology. Charles Lyell, the world's leading geologist, offhandedly pronounced against the system. Still more ominously, James Dana joined the opposition. Though Dana's *American Journal* published the AAGN endorsement of Emmons, it also printed Hall's rebuttal, and the *Journal* thereupon began to snipe at and then bombard Emmons and the Taconic system. Emmons turned to the new American Association for the Advancement of Science with a paper in 1848, but T. Sterry Hunt of Canada (later converted) attacked the system in another AAAS paper in 1850.

At this point Emmons made a fatal error in tactics. A small-town schoolteacher named James T. Foster (not the respected geologist John W. Foster) had prepared a geological chart for New York school use. When Louis Agassiz and James Hall (who had planned to make money with a chart of his own) publicly heaped scorn on the Foster chart, Foster sued both men for libel. Meanwhile Foster enlisted Emmons to revise the chart, putting in both the New York nomenclature and the Taconic system, and to appear as the prosecution's sole expert witness. This made the case in effect a trial of Emmons's competence and the validity of the Taconic system. A formidable phalanx of scientific notables—Henry, Whitney, Dana, Horsford, John W. Foster, and Agassiz—appeared in court to bear down mercilessly on Emmons's reputation and Taconic system, while Emmons stammered and stumbled under cross-examination. The bewildered and overawed jury of course found for Agassiz and Hall. Hall got his own chart adopted by New York, with complete scientific and commercial success. Virtually ostracized by his peers, Emmons retreated to North Carolina as head of the state geological survey and there finished out his life.

In the mid-fifties, Emmons published the most elaborate survey of American geology till then, including a long defense of the Taconic system, only to see the book pilloried (with some justice) by Hall in an *American Journal* review. At an AAAS meeting in the 1850s William B. Rogers pointed dramatically at Emmons in the audience and pronounced the Taconic system "dead, dead, dead." That report was exaggerated. The noted European geologist Joachim Barrande came out powerfully for the Taconic system in 1860, when only Emmons still supported it. Barrande's conversion made a strong impression in America. Emmons was heartened, and even Hall made a partial concession in the *American Journal.* Jules Marcou, who had no love for Dana or Hall, became the system's most vocal advocate over the next thirty years, upholding it in a debate with William B. Rogers before the Boston Society in 1860. The conversion of Sir William Logan in Canada encouraged Emmons to write privately in 1861 that "it may be regarded as being adopted both in this country & Europe." Other geologists agreed at least that its day was at hand.

That report too was exaggerated. In 1862 Barrande dropped his advocacy of Emmons's Taconic nomenclature. Dana's 1863 *Manual of Geology,* the most influential text on American geology, studiously omitted the Taconic system. This in itself may have doomed it. Though Emmons died that year, the controversy did not. Supporters came forward in the late sixties and early seventies, stirring Dana to write fifteen papers on the subject from 1872 to 1888. Marcou kept up the fight, and international geological congresses in 1885 and 1888 came near adopting the Taconic system. But the votes fell short, and so the Emmons system passed into history, except as the "Taconic Sequence," a local term. Even now the tangled geology of the Taconic range challenges and intrigues geologists.

Thus the Taconic Controversy brought into play a complete assortment of mechanisms for scientific communication and debate: journal articles, reviews, open letters, and editorials; society proceedings, government reports, textbooks, monographs, and the popular press; government boards and special committees of societies; papers, discussions, and formal debates before local societies, specialized societies, national societies, and international congresses; and even court hearings.

While the Taconic Controversy displayed all the gradations of reason and passion, principle and prejudice, it was essentially a debate about theory. A less lofty battleground was that of scientific modes and methods. The Dudley Observatory donnybrook, for example, incidentally touched on the issue of "practical" as against "physical" astronomy. Bache and Maury clashed over differing modes of science, the Arago approach as contrasted with that of Humboldt. Further down the scale were priority disputes, the obsession of Charles T. Jackson in telegraphy, anesthesia, and guncotton. During the late 1850s James Hall, who had an insatiable hunger for priority, and George C. Swallow, the Missouri state geologist, quarrelled publicly with Fielding Meek over priority in identifying Permian fossils in the United States. (Meek won out.) Still less high-minded were challenges to individual competence, as in the case of James T. Foster and his chart, or James Dana's devastating critique of the work of Agassiz's protégé Jules Marcou in the late fifties (though Dana tried to downplay personalities).

At the lowest level were purely personal rivalries and antipathies. These, of course, developed sooner or later in most controversies, like the Dudley and Taconic, as well as in Hall's periodic brawls with Meek and other former assistants. Mere rivalry for office and influence disrupted more than one learned society. The American Philosophical Society suffered from a persistent struggle between amateurs and professionals. In the American Academy the professionals battled each other, Peirce and Agassiz against William Bond and Asa Gray. The Peirce-Bond feud grew partly out of differing modes, Peirce favoring practical astronomy and Bond emphasizing physical, but jockeying between Peirce as astronomy professor and Bond as observatory director at Harvard played a part. In 1851 Peirce rushed in to corner a subject Bond's son and co-worker George had been developing, and in 1859 George riposted by winning appointment as his late father's successor, despite Peirce's desire for the job. When Peirce found George Bond on the same Academy committee with him in 1861, he wrote, "The dirty dog! I cannot and will not stand it."

Which were the bloodiest fields of science? Joseph Henry in 1859 gave the palm to meteorology: "No part of science has given rise to more angry personal discussion the object of which appeared to be the support of adopted opinions rather than the discovery of new truths." If meteorology was stormy, geology was seismic, and astronomy had its star wars. But even if the formal apparatus of debate, criticism, and judgment did not always bring instant peace, without it the conflicts would have been far more protracted and inconclusive than they were. The arena of science needed all the rules, referees, and judges it could muster.

Chapter 19

Liberty and Union: The American Association

The growing scope and complexity of the scientific enterprise in the 1840s made national organization imperative. And American circumstances, material and psychological, increasingly favored it—except that in America the logic of centralized authority would grate against the ideology of individual freedom.

In organization as in other things, European science set the style. Founded in 1831, the British Association for the Advancement of Science offered an appealing model. By mid-decade dozens of Americans had attended its meetings, and many more had read about them in Silliman's *Journal.* The BAAS, they learned, brought scientists together from all over the nation and from all fields, yet had a place for specialization in its sections. It channelled funds to research. It spoke for science to the public and government. But Silliman's *Journal* said less about its courting, or at least its toleration, of amateurs. Joseph Henry knew about that weakness firsthand and feared that in an American version it would be fatal. In America the 1830s were the Age of Jackson, exalting mass self-culture over elite expertise; and with its gangplank unguarded, an American society on the BAAS model might well be swamped by a rush of confidently ignorant lubbers. So when the American Academy of Arts and Sciences asked the American Philosophical Society to join it in organizing such an association in 1838, Henry opposed the idea as premature, and his fellow members of the APS agreed with him.

To have built on existing general societies, shot through as they were with localism and amateurism, would indeed have been building on sand. But a single field of science, already mature enough to have a national and professional outlook, could safely organize a national specialized society and, once it was solidly established, enlarge it to include other fields. If not the author of this strategy, the geologist-theologian Edward Hitchcock of Amherst College seems to have initiated it.

The geologists had organized a national society in 1819 with members in every state. But the difficulty of travel in that day kept attendance at its New Haven meetings largely local, and it expired in 1826. Hitchcock had belonged, and he missed the contacts it had given him. "I am tired," he wrote in 1838, "of running my nose every now and then against a post which my neighbor might have told me to avoid." By then it had become "a sort of hobby" with him to advocate an annual meeting of American scientists. The proliferation of state geological surveys during the 1830s suggested to him that geology could now take the lead. The surveys had created a solid cadre of professionals and bolstered the self-confidence of the profession. Hitchcock himself had directed the Massachusetts survey. He and his counterparts "had long felt the need of meeting to compare notes and try to reduce American geology to some uniform system." Now the need had become pressing.

Hitchcock began with the New York Board of Geologists and Henry D. Rogers, director of the New Jersey and Pennsylvania surveys. In 1838 the New York board fell in with Hitchcock's proposal for a meeting of the geologists "and other scientific men," a phrase revealing his larger objective. Hitchcock played skillfully on Rogers's vanity—how skillfully is measured by Rogers's description of him as "an engaging, unpretending, and guileless man." After a year or so of correspondence among geologists and a persuasive unsigned article in the *New York Review,* Rogers and the New Yorkers called on state survey geologists to meet in Philadelphia in April 1840.

With Hitchcock presiding, the eighteen who showed up from as far away as Michigan constituted themselves the American Association of Geologists. Yet from the start Hitchcock's broader vision beckoned them, that of expanding the new society "gradually and quietly . . . to embrace all the sciences." Accordingly they modelled their proceedings after the European associations and named the chemist Benjamin Silliman, Sr., as chairman of the 1841 meeting. At the resoundingly successful 1842 meeting in Boston, excited by a notable mountain-building hypothesis presented by Henry Rogers and his brother William, graced by the participation of the great Charles Lyell himself, and including several nongeologists among the forty in attendance, the society broadened its name to the Association of American Geologists and Naturalists and adopted a constitution with some resemblance to that of the BAAS.

In Washington, meanwhile, there loomed an ominous rival of precisely the stripe dreaded by the professionals. In 1840 Secretary of War Joel R. Poinsett, a dabbler in science, led in organizing the National Institute for the Promotion of Science. His conception betrayed two grave faults of the old order in American science: amateurism and political involvement. "Few of [the] members can bestow their whole time to the purpose of the Institution," he remarked, "but all may devote some portion. . . . The mind requires relaxation from the labors of a trade, or profession, or the cares of the state." When the Whigs took over the government in 1841, Poinsett withdrew, leaving the Institute in the hands of a State Department clerk named Francis Markoe.

At first the professional scientists showed some sympathy for the new society. But Charles Wilkes took umbrage when his expedition's collections were scooped up and mishandled by the Institute, and he humiliated it by pulling political strings to have them rescued. Moreover, in the hope of winning the Smithson bequest for the Institute, Markoe recruited politically influential amateurs as nominal members and claimed to speak for the national scientific community. His political ploy repelled the new breed of professional scientists, and his claim of national leadership for the Institute challenged those who were grooming the American Association for that role.

The rivals collided in April 1844 when the Institute held an elaborate "Scientific Convention" in Washington, inviting the Association and "all other scientific and learned societies," though the Association had set that place and time for its own meeting. This struck most Association leaders as an attempt to overshadow or swallow up their organization. They responded by postponing their meeting and organizing a scientific boycott of the Institute's carnival of amateurs. A few reputable scientists did attend the Institute extravaganza (complete with the Marine Band and a galaxy of politicians from President Tyler on down), but these defectors were mostly nongeologists not firmly attached to the Association, some of them seeking federal support. The Association's May meeting stood in pointed contrast: quietly professional, well attended, and showing signs of a broader base with some chemists and naturalists among the geologists. The political opposition to "Markoe & Company" needed little more than this professional snub to block Congressional funding for the Institute.

The Institute was not solidly based. Most of its regular meetings had been attended by a score or less, largely from the Washington area. Encumbered by a hodgepodge of donated curiosities without the funds to care for them, it soon fell moribund. Efforts at resuscitation were made in 1847 and 1848, but three years later it had only twenty-eight paying members. Its charter lapsed in 1862, but by then no one cared.

The threat thus beaten back may not have been as real as the professionals feared. But the episode had symbolic and psychological significance. It displayed both the fears and the aspirations of working scientists. It probably also heightened them. Whether or not it hastened the development of a more lasting organization would be hard to determine. There was no instant emergence from the AAGN chrysalis. Though the AAGN spoke with growing assurance for all of American science and renewed its welcome to all fields, geology still hogged the proceedings. After the 1846 meeting a Southern scientist complained that "naturalists and their pursuits" were "a mere circumstance in the Assoc."

Still the AAGN had shown vitality and staying power. It was dominated by the Northeast, but so was American science in those days, and the technology of travel was encouraging a wider reach. In 1840 the two outlanders at the Association's first meeting had taken a week to come from Michigan to Philadelphia, travelling day and night by the most direct route, more than once piling out with the other passengers to help pull their stage from the mud. By 1846 railroads were well along toward cutting travel time by two-thirds and fares by more than half, while steamboats reduced the river run between New York and Albany from several days to eight hours and the fare from two dollars to fifty cents. When in 1851, after its final metamorphosis, the Association met in Cincinnati, its president would express surprise at the large attendance in a city so far from "the Atlantic slope" and credit it to "the great facilities for personal communication which our times present." The Erie Railroad had by then shortened travel time from New York City to forty hours, and Cincinnati would be twenty-eight hours from Philadelphia in another year.

Psychologically as well as physically, the national character of American science grew palpably stronger during the mid-forties. In this bracing climate the AAGN scheduled its 1847 meeting for Boston, by then the leading center of American science. Like Lyell five years before, Agassiz gave the Boston meeting a special distinction. Joseph Henry presented his plans for the new Smithsonian Institution. Benjamin Peirce was on hand from Cambridge. Coast Survey duties kept Bache away, but the Lazzaroni-to-be were well represented. So was American science generally. Slightly more than half of the papers were in fields other than geology and paleontology. Almost unanimously the Association voted at Boston to transform itself into the American Association for the Advancement of Science. Like the other members, Henry, Peirce, and Agassiz were all for the change. Agassiz's enthusiasm alone would have insured it.

Early in 1848 a three-man committee—Agassiz, Peirce, and Henry Rogers—set about framing a new constitution. Rogers privately called it "my Constitution," and perhaps he did provide the working draft. But the temperaments and experience of his two colleagues strongly suggest that they had their say. The resultant draft, duly circularized, used the BAAS as a model. Despite the forebodings of the 1830s, membership was to require only current membership in some other reputable society or nomination by the standing committee and a majority of members present—both tests likely to be little more than pro forma. Rogers characterized the proposed constitution as "democratic," in that the right to vote for officers and committees belonged to all members rather than a limited category as in the BAAS. Readiness now to run a risk unacceptable ten years earlier showed confidence rather than weakness. Agassiz, a philosophical democrat, doubtless approved all this, perhaps even instigated it. Major powers of decision during and between annual meetings, however, were given to a standing committee made up of the president, secretary, treasurer, and their outgoing predecessors, plus the two section chairmen and six members elected at large. The general scheme was deliberately left "flexible" or "pliant" to avoid wrangling and delay, and the constitution was indeed unanimously adopted at the Philadelphia meeting that September.

The City of Brotherly Love lived up to its name. Even the Philadelphians, characteristically aloof at first, came round by the end of the meeting, which William Rogers applauded for its "general harmony." Agassiz was the lion of the occasion with his twelve separate addresses, though the three Rogerses in various combinations also figured prominently. Meanwhile the "small fry," as one member called them, were kept in their proper place. Joseph Henry praised the meeting as the best he had yet seen. "The association bids fair," he wrote, "to exert an important influence on the progress of science in our country."

Hopes, in short, ran high. But what exactly were those hopes? What did American scientists suppose the new Association could and should do for their world? Alexander Dallas Bache missed the 1848 meeting as he had that of 1847. But he spoke more authoritatively for antebellum science than anyone else, and in an address in 1844, he had thoroughly canvassed the opportunities and dangers of a national scientific society.

Through periodic meetings in different cities, Bache said, such an association could spread the gospel of science and help recruit new workers. By providing mutual encouragement and raising social status it could retain and stimulate them. Through informal colloquy and formal papers it could augment communication. As a spokesman for science it could enlist the aid of government and private philanthropy and channel it to individual researchers. Most important of all, Bache seemed to imply, it could govern an autonomous, national community, provided it were given continuity between meetings by a cadre of strong officers. In that role it could insure that support would come without strings, and that meddling politicians would be elbowed aside. A major object would be to "repress charlatanism," while uncovering new talent. Strict scientific criteria should be set for membership; and those who somehow slipped through the net, as well as the mere small fry, should be given little or no voice in the association's affairs.

At its first annual meeting, the Association adopted its own statement of objects:

> By periodical and migratory meetings, to promote intercourse between those who are cultivating science in different parts of the United States; to give a stronger and more general impulse, and a more systematic direction to scientific research in our country; and to procure for the labours of scientific men, increased facilities and a wider usefulness.

Presumably to forestall bickering, the statement kept to broad generalities. Most significantly, it said nothing of just how "more systematic direction" would be imposed on American science, by whom, and for what ends. Its silence on this question attested to the difficulty of answering it. Home rule for science was one matter; the question of who was to rule at home was quite another. Who would distinguish between charlatans and pathbreakers? Who would anoint new talent? Who would judge the judges? Where might American science find its balance point between law and liberty, union and democracy?

The American Association for the Advancement of Science began with 461 members in 1848, peaked at 1,004 in 1854, and stood at 644 in 1860. From year to year membership fluctuated, in part because occasional meetings outside the Northeast drew in local citizens who soon dropped out and because in 1851 and 1854 the rolls were purged of nonpaying members. As this suggests, entry could be easy and participation casual. "No scientific eminence or attainments are requisite for membership," conceded an unsigned article in 1853 (written by either Benjamin Gould or Wolcott Gibbs, probably Gould); the Association embraces "all those who love science and are disposed . . . to advance and extend it in the United States." Among those of known occupation who joined before the Civil War, more than half were in medicine, theology, business, law, or journalism. The register of those attending the 1860 meeting included lawyers, journalists, clergymen, farmers, teachers, librarians, physicians, a "mechanic," a "deacon," and a "Sanskritist."

And yet the founders' longstanding fear of domination by incompetent amateurs proved groundless. The very transience and passivity of such members rendered them innocuous. From 1848 to 1860, some two thousand persons joined—more than twice as many as belonged at any one time. Almost half left within three years of joining. Scarcely one in four members in good standing were likely to show up at any one meeting. This gave a sufficient advantage to the serious and faithful, since only those present could vote.

The 1857 meeting did provide for a separate category of "associate members," and the next two meetings considered proposals to restrict voting and office-holding to full members, defined as those "devoting themselves to scientific pursuits." But backers dared not insist that all such devotees earn their livings primarily from science. Amateurs might still make useful contributions to science. Furthermore, to stigmatize them might be to forfeit the support of the general public. One nonpracticing member, a *New York Times* reporter, insisted "that it was no use to legislate to make oil rise to the top of water," that "the scientific members would naturally be above the others without any rule." The real leaders, who already understood that fact, gave the voting proposal little support, and Bache himself tacitly dropped his 1844 recommendation. So the "associate member" category was never put to use.

During the 1859 meeting the *Springfield* (Mass.) *Republican* informed the visiting savants that "scientific investigation . . . should aim at popular ministry and service," rather than science for its own sake. Joseph Henry bluntly retorted that it was not "the object of the association to diffuse knowledge to the public . . . but to assist each other—to obtain new views, to criticize, to receive suggestions." Those objects were not questioned by those who ran the Association. Confident of their control, they had accepted Bache's suggestion in 1850 that some extra meetings be frankly aimed at laymen. Evening sessions, offering free public lectures by such stars as Agassiz and Silliman, enhanced the Association's public image without debasing the working sessions. But popular diffusion, as distinct from scientific communication, stopped there.

Nevertheless, the AAAS stood unchallenged as the voice of the national scientific community. Its eminent members gave it authority, and the press made it heard. During the 1840s and 1850s, science made good copy. The newspapers of each host city ran stories and editorials about Association sessions every day for a week or more. At Charleston, South Carolina, in 1850, the *Courier* at first gave ten times as much space to the cotton market, but at last a debate on the unity of the human race roused the paper to full coverage. At Cincinnati, the *Gazette* gave the 1851 special meeting detailed running treatment, with considerable editorial comment. The *Albany Journal* reported the regular meeting of that same year even more fully. By the end of the fifties, correspondents were showing up from out-of-town papers—four from Boston and three from New York for the Springfield meeting of 1859. As early as 1850 Benjamin Gould had observed sourly that "every temptation is held forward to the scientist to 'make capital' . . . by having himself puffed in the newspapers," and by 1859 Joseph Henry was privately remarking that "much good would be done if the discussions were conducted with closed doors." But most scientists enjoyed the spotlight.

Civic groups sought the honor of the Association's presence. Railroads and hotels offered special rates. Leading citizens opened their homes to the more distinguished members. Receptions, banquets, and organized tours regaled the savants. At Providence in 1855 Maria Mitchell summed it up in her diary: "For a few days Science reigns supreme —we are feted and complimented to the top of our bent, and . . . one does enjoy acting the part of greatness for a while!" That Quaker lady was intoxicated by compliments. Her male colleagues went further. "We had a capital meeting," wrote Spencer Baird after Cincinnati in 1851, "and were treated like princes, invited to revel in wine cellars . . . tea'd, dined, and otherwise eaten and drunken." Albany dazzled them in both 1851 and 1856. After the Baltimore meeting of 1858, one member wrote that "a New York delegation, who were on hand to invite the Association to that city next year, backed out when they saw how things were going on; they couldn't venture to put themselves in competition." To minimize "such carryings on," the Association's leaders planned the next year's meeting for Springfield, Massachusetts, which had not invited it. But after that meeting, Bache more than half-seriously complained to the local committee that "Springfield would not let [the Association] inaugurate the policy which they had intended."

As Bache's remark suggested, surfeit sometimes dulled the pleasure. Maria Mitchell wrote, "I was tired after three days of it, and glad to take the cars and run away." "I am glad," wrote Wolcott Gibbs after Baltimore, that "we are going next year to a quiet place where we shall not be eaten & drunk up but have time for science & scientific converse." Robert Hare objected publicly to what he called "gadding about." Not all the host cities worked so hard at hospitality; Cleveland seemed disgruntled by the previous postponement of a meeting there because of a cholera scare, while at Montreal the hotels were jammed and few private homes were opened to visiting members. Nevertheless, the honors and good will more often accorded them must have done much to make scientists feel at home and respected in American society.

But in carrying the gospel of science to the people and raising the prestige of scientists, the Association had also led the public to expect more than it could be given. "There is a real hungering and thirsting after science in the popular mind; but the savans are silent or write only for their brethren," complained John Swinton of the *New York Times,* a brilliant pioneer in the reporting of science for a general newspaper. "Frequently, during the sessions of the Association," he observed, "I have thought what an almost hopeless gulf separates . . . present . . . scientific treatment from . . . the great popular heart." The "two cultures" were diverging. The popular evening lectures did less to bridge the gap than to mark it by their separation from working sessions. And when the public looked to the Association for scientific guidance in social, economic, and philosophical issues, the Association naturally and no doubt wisely drew back from the abyss.

The question of racial differences, for example, gripped that whole generation and ultimately plunged it into bloody civil war. Papers at the Charleston meeting in 1850 ventured to address the issue, and Agassiz freely displayed his racial prejudices in discussion. But the Reverend John Bachman, though he chaired one such session, urged that the complex and explosive subject be kept out of Association proceedings thereafter; and so it was, at least until after Darwin's *Origin of Species* gave it new scientific interest.

The Association found it harder to choke back religious pronouncements for public consumption. Indeed, the risk seemed to lie in not being pious loudly enough. Bache in 1850 warned the Association of a wave of religious hostility to science he saw sweeping the South. Agassiz saw it too; notwithstanding his God-ridden cosmology, he had been pounded by preachers for denying the brotherhood of all races through descent from

Adam. Thus defensiveness gave urgency to many scientists' public avowals of their faith in both science and God. "Religion and Science [are] co-ordinate branches of human inquiry," Joseph Henry told the Association at Cleveland in 1853, without recorded objection.

Some scientists were privately scornful. "Agassiz treated us to an embryological demonstration of the existence of a personal God," wrote Josiah Whitney in 1856. "The Association came near voting to print all the sermons published in Albany during the session. Also the first thing on the programme was to assemble at a church in the city and have religious exercises!! Oh, *Potzdonnerwetter!* How pious we are getting in this nation of filibusters, slaveholders, and speculators (i.e. swindlers)." Religious asides seem to have been meant and taken as rhetorical embellishments, unrelated to serious scientific matter. And yet they grew more common as the fifties went on.

For all the Association's public piety, by the end of the fifties the press was showing some impatience with its internal squabbling and external aloofness. Swinton of the *New York Times* charged the Association with "either cowardice or incompetence" in ignoring "the great Darwin question" at its 1860 meeting. "Even Agassiz told me he would wholly avoid the question," Swinton reported. Apparently the question did get a vigorous informal airing by Agassiz and William B. Rogers. But it did not figure in the official program.

These were minor criticisms. As a mediator between science and society, the Association surely did more good than harm. The annual meetings and attendant publicity must have attracted more young men to careers in science than were repelled. Those casual members who wandered in and out of the Association, even nonmembers who merely read press reports of its doings, must at least have formed a clearer idea of what science and scientists were up to; and if a consequent realization of their own scientific ineptitude drove some mere dabblers away altogether, so much the better for science.

The Association's attitude toward the public was more apprehensive than amorous, however. It worried more about warding off political meddling, religious hostility, and public ridicule than it did about tapping the public's purse or otherwise actively exploiting public favor. In 1853 the Association formally resolved against even trying to build up a permanent endowment. Not until 1873 would the AAAS get its first modest endowment for research—a thousand dollars. Bache suggested in 1851 that the AAAS might stimulate research by advising or directing scientific projects undertaken by government or philanthropy. But even this hope came to little more than an occasional memorial to a state legislature, expressions of support for Henry's vision of the Smithsonian, and committee reports lauding Bache's Coast Survey or urging more scientists for government exploring expeditions. The Association's main concern was not what it could get the public to do for scientists, but what it could help scientists do for themselves.

. . .

In the Association's 1848 statement of objectives, the first on the list turned out to be first in importance: "to promote intercourse between those who are cultivating science in different parts of the United States."

One way to do that was through publication. Despite the Association's slender means, it managed to publish a substantial volume of *Proceedings* every year. The South Carolina physician and amateur naturalist Robert W. Gibbes made his most noteworthy contribution to American science in seeing the first volume through the press—hounding laggard members about their papers, editing the papers when received, and even putting up a hundred dollars of his own money. The indefatigable Spencer Baird, who became the first permanent secretary in 1850, complained while struggling to decipher and organize manuscripts and correct galleys, "Of all unpleasant jobs, this is the worst I ever knew." But experience and system lightened the burden in later years.

Much as scientists valued the chance to read, hear, and publish formal papers, they cherished still more the opportunity for personal contact, mutual encouragement, and scientific cross-pollination. "Much of the value of these meetings," said Joseph Henry, "is derived from the personal conversation of the members." "This is like water in the atmosphere," wrote a back-country Carolinian, "it keeps you from drying up." The amateur geologist and botanist Increase Lapham of Milwaukee, modest, quiet, self-taught, had won the respect of many eminent scientists and corresponded frequently with them. But the AAAS meeting in Cleveland gave him more than he could get from letters and print: "renewing old acquaintances, 'arguing points' in science, exhibition of specimens, instruments, &c., newly brought to light." In 1857 an unsigned article asserted that "all our men of science look forward" to the annual AAAS meeting and "have more or less regard to it in their studies and researches during the whole preceding year." The meetings also, as another article put it, brought "studies the most widely remote into intimate connection." The tendency toward specialization was running strong, and the AAAS yielded to it in organizing specialized "sections." Yet this made scientists all the more anxious to keep lines of communication open between specialties.

As a marketplace of ideas, however, the AAAS was not entirely free. Secure in their control, the professionals promptly set about defining and enforcing standards.

The professional scientific community had policed itself before the AAAS came along. It forced the erratic naturalist Constantine Rafinesque to present his research in privately printed pamphlets, little known to other scientists, for twenty years before his death in 1840. It knocked down and stamped on Robert Chambers's *Vestiges of Creation* despite the book's popularity among laymen. AAAS censure was only one of the cudgels brought down on Ebenezer Emmons. But the Association spoke with special force as the national embodiment of American science, the court of last resort now that the complexity of science made an appeal to the general public pointless.

Leadership control of programs tightened gradually. At first the planners asked only to be informed of paper titles in advance. Then they required titles and abstracts to be sent to the permanent secretary (Spencer Baird to 1854, Joseph Lovering thereafter), who in practice rejected some or returned them for revision. The leadership, fearful of European ridicule, suppressed the locally published proceedings of the 1853 Cleveland meeting and substituted an expurgated Cambridge edition. When the author of one deleted paper appealed to the general membership, the members endorsed a subcommittee report frankly asserting the right to throw out bad papers for the sake of the Association's reputation. Thus the professionals' control won official sanction. Two years later the Association resolved that "no notice of articles not approved shall be taken in the published proceedings."

At the 1855 meeting the professional temper showed itself when the reading of a paper was (according to the local press) "summarily interrupted by those in authority, on the ground that the members of the Scientific Association did not meet as mere school boys to listen to elementary truths." ("Those in authority" presumably included William B. Rogers, chairman of the section meeting.) Such high-handedness enraged those members put down and offended others on the grounds of simple fair play. In European scientific societies, similar displays of elite power raised similar objections. But a special factor figured in the American reaction: the ideal of democracy.

To Americans, democracy meant more than merely rule by the people. It also meant free speech, equal opportunity, and equal justice under law. All those principles would be involved in the running debate over whether the American Association, and by extension the American scientific community, was as democratic as it could and should have been.

The question was a valid one. American society itself fell far short of those ideals, even in the Jacksonian era, which some historians have called "the age of the common man." Aristocracy in the form of an elite of wealth and family influence flourished in Jacksonian and post-Jacksonian America. The professional scientific community could be viewed as an offshoot of that aristocracy. Most professional scientists came from professional or upper-middle-class families. Three-fifths of the scientists listed in the *Dictionary of American Biography* were sons of professionals, as against one in seventy-five of the general population. Nearly another fifth were sons of entrepreneurs. Obviously it helped to be born to a family that could stimulate, encourage, and finance a scientific career. Among the Lazzaroni, Bache could claim the most distinguished lineage, but Gibbs was not far behind, and all of them moved easily in elite social circles, as marked, for example, by membership in Boston's Saturday Club, New York's Century Club, or Philadelphia's Wistar Party. Such men were not apt to be doctrinaire democrats.

The only *DAB* scientist of the time listed as the son of an unskilled laborer was Joseph Henry, yet he was the most hostile toward rule by the unsifted majority. Before Henry had been in Washington a year, he was privately disparaging congressmen in general and praising the Supreme Court justices because their life tenure shielded them from the electorate. The European revolutions of 1848 did inspire him to affirm that "man has a right to self government as soon as he is prepared by moral and intellectual culture for so doing." But his reservation was crucial, and like Europe he soon relapsed into authoritarianism. In 1861 he judged "our present form of government" too weak to long resist "demagogues and infuriated mobs." In 1864 he privately favored high educational, "moral," and property qualifications for voting, as well as life tenure for the president. In 1868 he still favored restricting suffrage, and in 1872 he remarked that "there is much in hereditary descent which determines our character." Less than two years before his death he wrote, "I fear our system of general election will prove a failure," and in recording an offhand remark by General Sherman that the nation might be obliged to turn to monarchy, he asked only "when and how?"

Henry's political philosophy was consistent with his views on the governance of science. One of his favorite maxims was that scientific opinions should be weighed, not counted. When he attended the 1838 meeting of the British Association, he thought it a good thing that "the great body of the members have no voice in the management." The fear of majority rule led him to oppose an American association at that time.

All this suggests that the antebellum scientific community in America was indeed undemocratic. But something may be said on the other side. In science, at least, Joseph Henry clearly meant that merit, not inherited privilege, should be weighed. Scientific careers were not entirely a closed preserve of the upper class. As in American society generally, the expansion of science made entry easier, and the expansion of state colleges and universities would soon do likewise.

Scientific eminence cannot be handed down intact from father to son like a family fortune. Family background might point the way, and family money and influence clear the path. Fifteen per cent of scientists active 1800–63 were related to other scientists, and a third of the whole came from families with members already actively interested in science. Father-son patterns did appear, as in the Peirce, Dana, Silliman, and Agassiz families. But the chain of generations did not stretch as far in science as did the lineage of Medicis, Rothschilds, or Rockefellers in wealth. Family ties among leading American scientists tended rather to be within a single generation, such as brothers like the Rogerses or cousins like the LeContes, or by scientists' marriages to relatives of other scientists, as with Dana and Silliman, Whitney and Gould, or Henry and his brother-in-law Stephen Alexander.

Furthermore, the scientific community had an ethos not conspicuous in the community of wealth. At least in theory, scientists placed the whole enterprise above the individual. They also set store by pure reason. Those two precepts, taken together, ruled out power derived from anything but merit. The dilemma is, of course, that power once acquired is likely to outlast merit. But so long as such power is not heritable, death or debility at last put an end to it.

Thus the ideals of science, the nature of its work, and the demand for new workers, all supported at least two elements of democracy as defined by Americans: the ideal of equal opportunity, of careers and positions open to talents; and the ideal of free speech, of untrammelled expression of ideas and opinions. Even Joseph Henry did not quarrel with the first ideal, though he had reservations about the second. He and his fellow Lazzaroni contended more doggedly against the ideal of equality in power, of counting votes rather than weighing them, and against the ideal of equal justice under law, of equal treatment in the judging of disputes. But the linking of these ideals with the others in the political creed of the larger society pulled all of them into the politics of the AAAS.

The public had long suspected that science was somehow out of tune with democracy. In 1846 a congressman charged that the Smithsonian would be "one of the most withering and deadly corporations, carrying with it all the features of an aristocracy the most offensive that could be established in any country under heaven." In defense, scientists had long argued that science would benefit all the people and that national pride required it. After Jackson they began to claim that their professional community was itself democratic. Ormsby Mitchel not only promoted his observatory as a democratic rebuttal to Russian autocracy, but also proclaimed himself "the first democratically elected astronomer," since the directorship was elective. In 1856 the Yale astronomer and physicist Denison Olmsted, in an article entitled "On the Democratic Tendencies of Science," insisted on the compatibility of science with democracy, pointing out that poor boys could enter science by working their way through college, and rehashing the argument that science contributed to the material welfare of all the people.

The argument for democracy was thus bound to crop up in Association politics—democracy, that is, for professionals, not amateurs. William B. Rogers, who peremptorily silenced an amateur in mid-paper and who favored excluding amateurs from voting and office, at the same time led the struggle for democratic procedure among the professionals.

Bache and his Lazzaroni fought for paternal authoritarianism. Henry was elected president for 1849 and was succeeded in order by

Bache, Agassiz, Peirce, and Dana. It was not so much the presidency, however, as the fourteen-member standing committee that gave the Lazzaroni their power. Though the committee usually included only three or four Lazzaroni, they dominated it through prestige, common purpose, faithful adherents, and easily overawed local members. The committee conducted general business year round, rejected or suppressed papers—the power of scientific life and death—and even assumed the exclusive right to nominate association officers and its own elected members. Here, it began to seem, was power without check or accountability.

By the summer of 1850 Henry was sounding out Bache's views on "the final organization of the Association." As outgoing president in 1851, Bache delivered a magisterial discourse on the state and needs of American science and the Association's proper role therein. It confirmed him as the leading spokesman and mentor of the American scientific community. At that meeting the standing committee, on grounds of efficiency, took it upon itself to appoint the two section chairmen, who were ex officio members of the committee itself.

Grumbling ensued, notably by Henry D. and William B. Rogers and by Charles Wilkes, a longstanding proponent of democracy in science. Resolutions were passed to preserve some section autonomy. The Lazzaroni bridled at even so mild a show of recalcitrance. Looking back on the meeting, Henry recognized "a spirit . . . not easy to control." Bache admitted that "every powerful engine is hard to guide right." Dana thought the grumblers merely a few "small minds" goaded by "petty ambition"—"yet in our democratic land . . . we may expect to find inflated stuff abundant." Never mind, he told Bache, "those who know what true Science is should strive to keep the Association in the right path." When discontent flared up again in 1853 and 1854, Dana's annoyance deepened. "Discordant elements seem to grow more discordant with each year," he complained, "& I do not look for much real good from the Association."

By 1855 William B. Rogers had emerged as the anti-establishment champion. His scientific credentials yielded nothing to those of the Lazzaroni. He and his brother Henry were among the dozen most frequent contributors to Association programs (the Lazzaroni, including Dana, accounted for seven of the others). Rogers and Bache even saw eye to eye on many issues: higher education in science, enforcement of standards, enlistment of government support, professionalism, and the regrettable inevitability of specialization. Rogers's opposition clearly sprang from democratic principle, not from disagreement on scientific goals nor from "petty ambition."

Having placated the malcontents by setting up a committee to revise the constitution, the Lazzaroni packed it with their supporters. At the 1855 meeting it concocted a last-minute draft reinforcing the power of the standing committee. Rogers was ill, but he pulled himself together and gratified the eager press with "a controversial debate of the most exciting and animated character." Shoulder to shoulder with Rogers stood

Ormsby Mitchel, the living symbol of democracy in science. They were backed by Charles Wilkes, William's brother Robert, and others. The light of self-interest seemed at last to break on the hitherto uncomprehending and docile majority, which (according to a Providence reporter) now manifested "a strong determination . . . to resist anything which might look like the encroachments of 'despotic power'. . . ." By a standing vote a large majority postponed the question to the 1856 meeting. Rogers rejoiced at thus "foiling the well laid plan."

At Albany in 1856 the committee on revision again submitted a draft strengthening the standing committee. But William Rogers, put on the revision committee as a token of fairness, presented his own minority report requiring open nominations and election by ballot. Debate once more erupted, Mitchel again supporting Rogers. At last, "after much cunning evasion and manoeuvre," as Rogers later put it, the Lazzaroni were compelled to let the membership choose, and the Rogers version passed by a wide margin.

The "constitutional" or "democratic" party (as it was now variously styled) was jubilant. Rogers considered the other side "entirely routed." Josiah Whitney expressed delight that "the democratic party carried the day against Bache, Henry, and the Cambridge clique." "The Cliques put down & all right," noted Charles Wilkes in his diary, "huzza huzza after 3 years fight." The press, and through it the public, by now saw the issue as clear-cut. Of course they took the side of democracy and the unhampered right to rise—which was, after all, the central idea of the new Republican Party's philosophy. The outcome was good news for "young scientific aspirants," in the view of the *New York Times.*

In fact, neither issue nor outcome was quite that clear-cut. Like the American colonists, the "democratic party" had rebelled more against the prospect of tyranny than the actual exercise of it. The Lazzaroni had not—at least not yet—used their power to suffocate science. On the other hand, Bache observed privately afterward that "the Lazzaroni after being whipped remained masters of the field after Zachary [Taylor]'s Buena Vista fashion." So it seemed. No more full-fledged Lazzaroni held the presidency before the Civil War, but they continued to figure prominently in the standing committee and the programs. Ormsby Mitchel still complained about the clique in 1858, and Matthew Maury quit the Association that year in protest against it. In 1859 the Lazzaroni got Benjamin Gould elected vice president, whereupon an amateur member commented sourly that "the Mutual Admiration Society have . . . had things their own way." All this suggests that the Lazzaroni had scarcely needed to trifle with the constitution in the first place.

Yet the Lazzaroni had by then lost much of their spirit. Agassiz stayed away from the 1858 meeting. So did Gould and Frazer, despite Bache's pleas. Gibbs, who did come, found Bache as cross as "a grizzly bear with a sore head." The Dudley Observatory debacle had just left the Lazzaroni divided and exhausted, and they had no taste for renewed combat. Dana, for one, was "gratified . . . to find things moving on more smoothly than heretofore."

William B. Rogers contented himself with the constitutional victory of 1856. He resigned himself to any influence his foes might gain on their merits (though he privately characterized many of their Association papers as stale scraps, trite and trivial). By 1860 Gibbs and Agassiz talked of making Rogers president. Bache, however, thought it not enough that "after being a very naughty boy for many years, he held up for a little while," and so Bache "went for further probation."

Some years later Joseph Henry looked back and lamented the "time, mental activity, and bodily strength . . . expended among us in personal altercations which might have been devoted to . . . the advancement of science." The wrangling in the AAAS had distracted and wearied those involved, encroached on the limited time of the meetings, and tarnished the public image of scientists. Still, it did spotlight the perennial dilemma of science: how to strike the best balance between order and flexibility, experience and originality, the old and the new, union and liberty.

Like so many American dilemmas, this one would be mitigated, if never finally solved, not by revolution nor pure reason but by growth, change, and compromise. After the 1850s the Association would never again hold the power and bear the responsibility so nearly alone. There would be other institutions and organs to share them. Furthermore, the size and diversity of the national scientific community would preclude domination by a single clique.

"Ever since 1853," wrote Gould in 1860, "both Gibbs & I have felt that the Assocn. did as much harm in one way, as good in others; and have earnestly hoped for its speedy dissolution." Not only the issue of who should govern but other issues besides remained unsettled. Objective, systematic peer review had not been achieved. Personal prejudices had worked injustice. Incompetent, even preposterous papers had got as far as oral presentation. Trivialities had clogged the programs. Important scientific questions had been dodged.

And yet the good seems to have outweighed the bad. Before the AAAS, American scientists had little to orient and steady themselves by except the patronizing comments of European scientists or the increasingly meaningless approval of the public. Now the Association served as a pole star in the spinning world of science. It had promoted a sense of community and common cause (despite parliamentary squabbling), a professional spirit, concern for standards, a definition of goals, scientific intercourse, cross-pollination of fields, and public, foreign, and self-respect. By its continued existence it left open the way to further advances.

But just ahead lay a jolting break in the road.

. . .

In early August 1860 the AAAS met at Newport, Rhode Island. Peace at last prevailed. One member had seen no other meeting so quiet and pleasant. "The weather is perfect," he wrote, "& I should have seen the whole of the Scorpion's tail last night, if it had not been for the full moon." Yet one cannot help projecting back upon that quiet gathering in an elegant seaside resort the poignancy of other doomed summers, those of 1914 and 1939. So it was with Joseph LeConte. From the far side of the American Armageddon he remembered sensing in everyone at Newport "a deep suppressed uneasiness. . . . It was like the stifling air before a storm."

The fateful presidential election of 1860 was running its course. Abraham Lincoln kept mum in Illinois, John Bell measured out platitudes in the border states, John C. Breckenridge in the South professed his desire to save the Union—on Southern terms. Of the four candidates, only Stephen Douglas, having given up hope of election, spoke passionate truths in both North and South, trying desperately to draw them together and back from the abyss. As the Association met, Douglas paused at Newport for a breather.

The malaise at Newport was scientific as well as political. Attendance had dropped to 140, and several prominent members were absent. A solar eclipse expedition to Labrador kept at least ten members away. Gray, Torrey, Wyman, and Leidy were absent also. Peirce and others were in Europe. "The papers read were no great shakes," Bache noted sadly. "Even Agassiz had taken no pains & only jawed over again his types of structure slightly modified. . . . Henry seemed quite out of sorts . . . & delivered a dispirited lecture" to a small audience. The younger Silliman saw "a decadence in the scientific character of the Association. . . . Signs of concession to scientific charlatanism were visible." By the end of the meeting, he wrote, many felt "that unless the old order of proceedings could be restored, the American Association must come to a speedy and disastrous end."

In Association politics, the South swept the board, even though only eighteen attended from slave states, and those mostly from Baltimore and St. Louis. When the Labrador eclipse party showed up near the end of the meeting, one of the party, Frederick Barnard of Mississippi, found to his surprise that the Lazzaroni had made him the next president. They had picked John Newberry for vice president, but John Gibbon of North Carolina outfoxed them and slipped in Robert W. Gibbes of South Carolina. John W. Mallet of Mississippi became general secretary. In 1859 the Association had voted to hold the 1861 meeting "in some Southern city," and Gibbon now got Nashville, Tennessee, designated as the place and April 1861 as the time.

The Southern sweep was a fluke of internal politicking, however, not a token of sectional reconciliation nor a clue to the future. Hindsight reveals more accurate omens at Newport. On August 4, Captain Edward B. Hunt of the Army Corps of Engineers gave a well-attended paper on the technology of war. "War," he said, "is the applied science of destructive projectiles." He saw many applications ahead: chemical substitutes for imported saltpeter in gunpowder, purer iron, rifled cannon, asphyxiating shells, telegraphic reconnaissance balloons, the Drummond light for night operations, and more. On the same day, Bache was reading a paper when a buzz of excitement ran through the audience and left him droning on unattended—Stephen Douglas had entered the hall. Science had lost the stage to political crisis and war.

Part Four

WAR AND RECONSTRUCTION, 1861–1876

Chapter 20
Science and the Shock of War

Does the mind of the scientist throw a brighter, purer, steadier light on political issues than that of the layman? It evidently did not during the 1850s and 1860s. In those turbulent years the scientific method resolved no political questions, not even to the satisfaction of the scientists. American scientists held opinions on the clash of North and South as various as those of their nonscientific countrymen, and not perceptibly different from them.

Scientists, like the general public, were divided on the fundamental issue of slavery even within their geographic regions. In the South the botanist Henry W. Ravenel of South Carolina reported that "our people here" (presumably whites) had studied the enslavement of blacks thoroughly in its "moral, religious & political" aspects and found it good "in the sight of God." But another Southern botanist, Moses Curtis of North Carolina, wrote that slavery's evils "are not unknown & unfelt by us," though he thought remedies were best left to Southern whites. Such Northerners as John Torrey and Robert Hare had no quarrel with slavery, at least as of 1850; and in 1860 the upstate New Yorker Joseph Henry judged slavery to be "in accordance with the general tendency of modern civilization founded as it is on the application of scientific principles [i. e., racism] to the arts of life." Even the purebred Massachusetts Yankee Benjamin Peirce was reported in 1856 as thinking slavery "a blessing rather than a bane."

Peirce's views were known in antislavery Boston, but he suffered no ostracism thereby. Not so with another dissenter from local opinion, the North Carolina chemist and mathematician Benjamin S. Hedrick. Born and raised in an antislavery section of his state and sharing its views, Hedrick graduated from the state university and then studied mathematics briefly under Peirce at Harvard. In 1854, at twenty-seven, Hedrick became professor of mathematics and chemistry and head of an embryo scientific school at his alma mater, the University of North Carolina. Hedrick kept his antislavery opinions to himself until a student in 1856 asked which presidential candidate he preferred, whereupon Hedrick replied that he liked Frémont, the Republican. This alarming news reached the leading proslavery newspaper editor, whose public fulminations Hedrick met with a powerful open letter against slavery. The resultant outcry frightened the university administration into dismissing him, with private regret (the president gave him credit for having "the courage of a lion and the obstinacy of a mule" as well as being unsurpassed in his profession). Hedrick left his native state, not to return until Reconstruction days. Peirce refused to help him find a job; indeed, a friend assured Hedrick that "the whole of the Cambridge clique is decidedly averse to you." But after a series of short-lived positions, Hedrick at last settled down to spend the war in Washington as a Patent Office examiner.

More often, of course, scientists went along with their sections. Benjamin Silliman, Sr., moved in the vanguard of his, denouncing slavery publicly as early as 1805 and a half-century later drawing wide public attention with an open letter in support of arming free-soil migrants to Kansas. The younger Silliman in 1847 refused a chair at the University of Alabama, "such is my abomination of negro slavery." Still others, like Asa Gray, the Rogers brothers, and the Pennsylvania meteorologist James H. Coffin, an organizer of his local Republican party, were staunchly antislavery, though not radical abolitionists.

But the private correspondence of antebellum scientists suggests by omission that most of them regarded political controversy as an annoying distraction from their ruling passion. Some believed that scientific institutions, such as the Smithsonian or the AAAS or a great national university, might help bind the Union together and thus abate the nuisance. Even when secession came, they hoped the movement would blow over soon without war. "It will all come right in a few months," wrote John Torrey four days after South Carolina seceded. "There is a considerable amount of gas to be let off first." On the same day, a Northern zoologist wrote a wholly nonpolitical letter to a fellow scientist at Charleston, promising (without any evident sense of double meaning) to send a "fine collection of Northern shells."

As other states followed South Carolina, some Southern scientists, like Moses Curtis and the LeConte brothers, reacted with sorrow and dread. "Hang politics," wrote Curtis, "both sides . . . are wrong, grievously, madly wrong." But even these doubters soon yielded to the secession madness, the Massachusetts-born John L. Riddell of New Orleans being one of the few exceptions. A noted Charleston zoologist, the Reverend John Bachman, left a full record of his more gradual conversion to disunion. Born and raised in New York State and not an ardent defender of slavery, he had "venerated the union" and condemned nullification. In

June 1851 he even passed along secret secessionist plans to a New York political leader. A few months later he still saw secession as entailing "long years of poverty & misery," but felt a daily weakening of his own attachment to the Union and a resolve to go with his state if the "evil day" of secession arrived. In 1852 he deserted the Whigs for the Democrats. By 1857 he considered the Union all but doomed by the "black republicans." In December 1860 he gave the opening prayer of the session at which the South Carolina Convention passed its Ordinance of Secession, and two months later he wrote, "I am one of those who rejoice that this union is overthrown."

When the guns began to roar, Southern scientists closed ranks. The Union capture of New Orleans in April 1862 relieved the Confederacy of the stubborn Unionist John Riddell. After the University of Mississippi shut down in October 1861, its president, the Massachusetts-born Frederick Barnard, made his way to the North; but though he was a covert Unionist, his political views were not the only nor even, probably, the chief reason for his move.

A few Northern scientists, even in wartime, dissented more boldly than did their Southern counterparts. Richard McCulloh, who had got the Columbia University appointment for which Wolcott Gibbs and the Lazzaroni fought so fiercely in 1854, resigned it in the fall of 1863 and became an engineer officer in the Confederate army; Joseph Henry thought he must have been "under the influence of an attack of monomania." But McCulloh was, after all, a Southerner. Henry's diagnosis might have been more applicable to James Hall, who though born in Massachusetts railed against the war, the Negro, and meddlesome "New England propagandists" so intemperately that the younger Silliman questioned his loyalty and Wolcott Gibbs saw in him "a clear case of moral insanity." Hall, for his part, wailed that "people are *mad! mad!* some day we shall become sane."

Mad or not, however, most Northern scientists rallied to the Union cause when war began. Once indifferent to slavery, John Torrey of New York came to support Lincoln and oppose concessions before Fort Sumter, and soon afterward he proclaimed himself "ardent for the Union" though "the war will probably last till next winter." William B. Rogers, formerly of Virginia, held identical views. From Iowa in February 1861, the English-born botanist Charles C. Parry wrote Torrey: "We of the West are strong union men, & don't much like having the Miss. blocked up by a foreign country."

The gentle botanist Asa Gray, already an antislavery Republican, fiercely declared, "God save the Union, and confusion to all traitors." Gray wrote his old friend and correspondent George Engelmann in agonizingly divided Missouri that "the life of a rebel is duly and justly forfeit," to which Engelmann replied sorrowfully, "I did not think you so bloody minded, harsh and onesided." Gray was fifty and had accidentally lopped off part of his left thumb shortly before war began, but he drilled with a home guard in Cambridge, scraped up what he could for government bonds, and regretted his childlessness only because "I have no son to send to the war." Through most of the war Gray carried on a tense political debate by mail with Charles Darwin, who hated slavery but also disliked the crass, money-grubbing, Anglophobic North and therefore rooted for Confederate independence. Only the amiable temperament of both men preserved their long friendship.

"If you think me belligerent," wrote Gray to Engelmann, "I am nothing to Agassiz." Peirce complained in January 1861 that Louis Agassiz had become "violently republican"; and on the day after war broke out, one of Agassiz's students found the master "in the gray dawn walking in Divinity Avenue weeping, almost raving in his misery." Agassiz had not overcome his racism, but he loved his adopted nation and its ideal of democracy for whites. Among Agassiz's fellow Lazzaroni, Benjamin Gould regretted on the morning after Sumter that "hemp in December & January" had not stopped the "lowlived wretches at Charleston"; Wolcott Gibbs was reported as feeling "a grim satisfaction in the idea of the South going to ruin" and by midsummer 1861 was complaining that the Lincoln administration lacked zeal; and James Dana, though less frenetic, heartily supported the war from first to last.

Before the war Alexander Bache, the Lazzaroni chief, had been a pro-Southern Democrat and a great admirer of Jefferson Davis. But Bache had prudently kept out of active politics himself and had fought successfully to prevent the politicizing of his Coast Survey, most notably in 1857 when his Southern superior tried to get the names of Republican sympathizers in order to purge them from it. Secession turned Bache into an uncompromising Unionist. He denounced "the breaking up of a great nation in which we have a birthright!" And such would be the burden of Bache's wartime labors for the Union that he may be said to have ultimately given his life for it. He was, after all, the great-grandson of Benjamin Franklin.

Not all the Lazzaroni were wholehearted in the struggle. As to the proslavery Benjamin Peirce, Joseph Henry testified after the war that though loyal, Peirce "had always been a Democrat, and that even in the midst of the war, [he] had not given countenance to the measures of the North." Henry meant that as praise.

Henry's own case is complex. In adjusting to the war, he labored under several handicaps. As we have seen earlier, he had long considered popular government to be the nation's fatal weakness. One of Agassiz's young assistants recorded in his diary on the day of Lincoln's first inauguration that Henry remarked "this government cannot exist 25 years longer even if the present crisis is safely passed." In a corner of his mind Henry may have harbored a grim sense of vindication by the apparent wreck of the Union. In any case, the new president's vision of the war as one to save government of, by, and for the people could not have moved Henry very deeply. Neither could the later goal of emancipation.

Henry never shook off his irrational, which is to say unscientific, racism —shared, it must be said, by most whites, North and South. "I would not permit the lecture of the coloured man to be given in the room of the Institution," he wrote Bache self-righteously in 1862, the "coloured man" being Frederick Douglass. Later Henry remarked that "to liberate the Negro *ever* in this country [would be] certain death to the race." Finally, Henry had long admired the South and white Southerners, perhaps as exemplars of aristocracy. In February 1861—before war had begun, to be sure—he could actually write, "I am pleased that Jefferson Davis [a long-time Washington friend] has been appointed President of the Southern Confederacy for I put confidence in his talents and integrity." And in 1862 John Torrey found him "rather inclined to let the Cotton States go."

During the war Henry tried to keep the Smithsonian out of what he called "partisan politics," only to drag it in when he resisted the use of the Smithsonian hall for antislavery lectures, the "peculiar doctrines," as he put it, of "an avalanche of strong minded *women* and weak minded *men* . . . from the north." This put the Smithsonian in bad odor with a number of congressmen for some years. Not surprisingly, intermittent nighttime flashes of light from the Smithsonian towers led to accusations of treasonous signalling to Confederate forces across the Potomac. Actually the flashes were experimental signals to the Coast Survey office, as well as light from the lanterns of resident students climbing to their tower rooms and assistants reading meteorological instruments on the roof.

Early in 1862 Wolcott Gibbs alerted Bache to "extraordinary stories in circulation" about Henry's supposed disloyalty. Gibbs had said harsh things about Henry during the Dudley Observatory affair, but he scoffed at these tales. "I should as soon expect to be called disloyal myself," he wrote. Henry's wartime record would vindicate Gibbs's faith. Notwithstanding his pro-Southern and anti-Negro proclivities, Henry labored as strenuously in behalf of the Union cause as did Bache himself.

Henry evidently never subscribed to Abraham Lincoln's political views. As late as October 1864, Henry's daughter Mary, who probably reflected her father's opinions, called Lincoln "a bear and totally incompetent," one whose reelection might bring on "a rebellion at the North." Lincoln and his Cabinet never attended a meeting as Smithsonian regents ex officio; Lincoln's reply to Henry's invitations was always, "I guess that you can run the machine better than we can." Yet Lincoln and Henry came to know each other well during the war, and the quondam doter on Jefferson Davis felt the attraction of Lincoln's character and personality.

Bitterly opposed to Lincoln at the start, Henry called on him a few weeks after the inauguration and found him "already improved in manner & appearance." Lincoln took to Henry at once. By one account he called Henry "one of the pleasantest men I have ever met; so unassuming, simple, and sincere." Actively interested in science and technology,

Lincoln more than once drew on Henry's knowledge and wisdom in such matters, and in others also. Those who occasionally reported secesh signals from the Smithsonian did not know that Lincoln himself came to at least one of the signalling experiments. At Lincoln's request, Henry checked up on alleged coal deposits put forward as the mainstay of a proposed freedmen's colony in Central America, and he helped prevent a tragedy by reporting their worthlessness. He also helped Lincoln by unmasking the fakery of a spiritualist who was preying on Mrs. Lincoln's grief for the Lincolns' dead son Willie. A federal official years later recalled Henry's wartime appraisal of Lincoln as a man of unusual honesty, strong intellect, quick comprehension, and greatness of character. That recollection may have been somewhat heightened for effect, but Henry himself wrote his wife after Lincoln's death in praise of Lincoln's "honesty of purpose . . . kindness of heart and . . . moderation and prudence of action."

At the AAAS meeting of 1869 Benjamin Gould remarked that war had stimulated "physical science" more than it had "any other art." To a modern reader this might seem a mere commonplace, obviously alluding to the recent Civil War. But Gould was talking about France, not the United States, and by "physical science" he meant only those aspects "useful to the [army] engineer, the topographer and the artillerist," a rather limited range. His listeners would certainly not have regarded the American Civil War as a boon to science generally. Neither, presumably, would Gould himself, since he had complained in 1864 that his war work "very essentially delayed" his astronomical research. They all knew the Civil War had laid low the struggling science of the South and staggered that of the North.

What the Civil War did to science came as no surprise to the scientists. The War of 1812 and the Mexican War (as distinct from its territorial spoils) had hurt more than helped American science. In 1861 scientists expected the worst. Science must "go to the wall," mourned both Benjamin Silliman, Jr., and James Hall. The rising young geologist Ferdinand Hayden in Washington and the ill-starred old geologist Ebenezer Emmons in his North Carolina exile took it for granted that science would be muted if not silenced by the war—as indeed it was for both of them. In October the entomologist John L. LeConte in Philadelphia quite matter-of-factly reported the growing "paralysis" in scientific pursuits.

Fortunately for American science as a whole, the armies tussled and trampled where science had least to lose—in the South. Only there did scientists suffer the physical destruction of their work. But every man is his own universe, and weighing the aggregate could not have much lightened the burden of seeing one's own lifelong hopes and labors go for nothing.

Luck spared some, like Eugene Hilgard, whose geological collections survived repeated military clashes in the neighborhood of Oxford, Mississippi. Enlightened intercession saved others. John L. Riddell, standing on his record of loyalty to the Union, managed to preserve the New Orleans Academy of Science building from occupation and probable pillage by Union troops. In November 1862, when William Fry, a *New York Tribune* assistant editor, learned that the federal government was about to auction off the contents of public and college libraries seized in Beaufort, South Carolina, he raised an editorial cry for civilization, with allusions to the burning of the Alexandrian Library. Fry also appealed privately to President Lincoln, and Samuel F. B. Morse seconded him by telegraph in the name of science. Lincoln stopped the sale and, with Joseph Henry's ready assent, had the Beaufort books stored at the Smithsonian for postwar restitution (unfortunately they all burned in the Smithsonian fire of January 1865).

Malice, greed, ignorance, and idle mischief more often had their way. That champion of slavery and secession Henry W. Ravenel lost twenty years of botanical records, manuscripts, and correspondence when Wheeler's Confederate cavalry sacked his house; and Confederate soldiers also looted the Louisiana State Seminary (later Louisiana State University) of its scientific books, models, equipment, and specimens. But Union troops, as was to be expected, wreaked most of the scientific havoc. In Tennessee the eighty-three-year-old engineer, scientific writer, and college professor John Millington saw not only all his property swept away but also "all my beautiful museum at the [Memphis] Medical College destroyed." On the eve of war the Virginia Military Institute had dreamed of becoming a major scientific school, but Hunter's troops wrecked its buildings, apparatus, and dream in their Valley raid of 1864. A like fate befell the books and collections of the University of Alabama and those of the Medical College at Charleston, where William Hume of the faculty later found only a bagful of brass and copper fragments from valuable instruments broken up for their metal. "Mineralogically and chemically I am ruined, and I must also add pecuniarily," wrote Hume. At the start of the war the Elliott Society at Charleston nursed hopes of postwar greatness; it survived the war but Sherman's troops destroyed its natural history papers and collections just before the end. In the state capitol of Mississippi, shelves and cases of geological specimens were apparently swept with the butts of muskets, the floor strewn with broken specimens and shattered jars, and nearly all the labels lost.

The heaviest blow came with the burning of Columbia, South Carolina, in February 1865 (whether Union or Confederate troops or mere accident were chiefly to blame remains in dispute). The library of the South Carolina College was saved, but John McCrady of Charleston lost his own scientific library, manuscripts, and journal. "This," he wrote soon after, "forbids me to hope that I shall ever be able again to devote my life to science." Robert W. Gibbes lost his rich scientific collections and everything else; a few months later he died of a fever, perhaps without reluctance. John LeConte (not to be confused with his Philadelphia cousin John L. LeConte) entrusted all his scientific manuscripts and notes, "the labors of my *life,*" to a Catholic priest, whose house was then burned along with the manuscripts. John's brother Joseph took his own manuscripts out of the city in a couple of wagons, which were overtaken, seized, and burned by Union troops. The scientific apparatus of the Citadel military college at Charleston had been sent to Columbia for safekeeping; it was destroyed there. John Bachman's natural history library, including many presentation copies from European societies, his correspondence with eminent European and American scientists, and his extensive herbarium had all been sent inland from Charleston, but they were detained at Columbia, where they perished. "The stealing and burning of books," wrote Bachman with understandable bitterness, "appear to be one of the programmes on which [Sherman's] army acted."

Perhaps as damaging to Southern science was its wartime isolation. Northern scientists regretted losing the chance to do field work in the South or to get data and advice from their Southern counterparts. Jeffries Wyman, investigating the effects of snake venom, missed his supply of Virginia rattlesnakes—missed the rattlesnakes, he remarked, a good deal more than he missed the Virginians. But while Northern scientists were cut off from the South, Southern scientists were cut off from the whole outside world by battle lines and the blockade. "I have been so *blockaded,*" wrote a Tennessean in June 1862, "that I have lost quite a year in scientific matters." An Alabamian in 1863 confessed ignorance of all botanical progress since the start of the war; "scientific pursuits are pretty much suspended in the South now," he reported. At the end of the war Joseph LeConte wrote Spencer Baird that he had seen no scientific papers during the war, and he asked Baird for a list of significant articles published in those years. Moreover, Southern scientists depended largely on Europe and the North for scientific instruments. Lewis Gibbes, at the College of Charleston, had cannily stocked up on apparatus from Boston and New York during the fall of 1860, but few others were so forehanded.

Besides being the cockpit of the war, the South suffered far more than did the North from the war's impact on higher education. Though Southern colleges had not produced many scientists, at least they had hired some. But when war came, a much greater proportion of Southern college students than Northern took up arms. Upperclass Southern whites had a long tradition of violence, military zeal, and hair-trigger sensitivity about "honor." Southern college students were especially volatile, unruly, and romantic. Hyperenthusiastic for the Southern cause, certain of a quick triumph, they rushed to seize glory while it still glittered. In 1861 entire graduating classes and in one instance an entire student body enlisted en masse. A heavy proportion of undergraduates also joined the colors, despite faculty efforts to sober and hold them. Many never returned. Among those who had been enrolled at the University of Virginia in 1861 alone, war deaths amounted to about half the toll of Harvard alumni and students of all years. Entering students during the war consisted mostly of disabled veterans and others unfitted for military service by youth or physical disability.

For lack of students many Southern colleges shut down altogether during the war. Some never revived, including half of those in Alabama and four-fifths of those in Texas. Some that remained nominally open became little more than preparatory schools or military academies. On the eve of conflict, grandiose plans were afoot for a great "University of the South" in Tennessee. But it did not open until after the war, and then on a much humbler and poorer scale, its buildings burned, its rich endowment lost, its chief promoter killed in action. By then the financial collapse of the Confederacy had wiped out most college endowments and had ruined or impoverished potential benefactors. Faculty salaries during the war fell in some cases to mere token levels, and in others ceased entirely. Some faculty members who survived the war found other occupations or moved to other regions, like the LeConte brothers, who rebuilt their scientific careers at the University of California. Still others were drained of intellectual vitality by their ordeals in war and reconstruction. The war left the South even further behind the North in science than it had been before.

The Civil War hurt American science in ways that touched Northerners as well as Southerners. On both sides it distracted scientists from their work, reduced their income and support, diverted them—sometimes for good—to other activities, and even killed them. Since most American scientists were Northerners, the story now turns largely to them.

"You may see me soon in Washington in a New York regiment," Wolcott Gibbs wrote Alexander Bache a few days after Sumter. "Here all classes are volunteering." Gibbs, who was thirty-nine, presently had second thoughts. So did others once the first flush of patriotism cooled. Younger scientists in Washington began research on ways to avoid the draft. At the Smithsonian, Baird fretted but somehow escaped. William Rhees contemplated taking the title of "Professor," "as I believe they are exempt," but in the end got off for physical disability. So did Fielding Meek, who was deaf. At the Coast Survey, Julius Hilgard was saved by a rupture, to his great relief, while Charles Schott hired a black substitute. They were evidently in no hurry to court the fate of Archimedes at Syracuse. Even some who had gone willingly to war lost their early zeal. "I long to escape from the Army and be at work again in Natural History," wrote Ferdinand Hayden after more than two years of it. "Patriotism is moonshine." But that attitude, as the war settled into what began to seem endless and perhaps futile bloodletting, did not set the scientists much apart from their fellow citizens.

Among those scientists eventually listed in the *Dictionary of American Biography* who were between fourteen and forty in 1861, only about a quarter saw military or naval service in the Civil War, as compared with something approaching half of the same age group of men in the general population and a third of *DAB* technologists. This is not an entirely fair comparison, however. There were surely potential or fledgling scientists in the service who failed to make the *DAB* because they were killed or turned to other careers by the war—for example, by its interruption of their higher education. So the disparity need not imply a peculiar coolness of young scientists to the call of country. Evidence for this view appears in the comparative numbers of *DAB* scientists who entered their fields during successive five-year periods. In the last half of the 1850s, sixty-two entered. But in the war years, only forty-one entered. Even in the last half of the 1860s, only fifty-two entered. And the first half of the 1870s brought in only sixty-one, still short of the prewar number. The *DAB* roster, in other words, is probably skewed toward those young scientists of 1861–65 who were somehow spared military service and so were more likely to go on to scientific eminence.

Comparisons aside, many scientists did forsake their work, at least temporarily, for military service. Ninety-eight members of the Boston Society of Natural History, a "large proportion," served in the war. Louis Agassiz complained in 1864 that his assistants at the Museum of Comparative Zoology kept going off to war, some dying in battle or in the hospitals, "so that I have had great difficulties in maintaining the proper activity in the Museum." One, Albert Ordway, had been doing notable work on trilobites when he enlisted, but rose to brigadier general and never came back to science. Death claimed other promising young men, including Sydney Coolidge, a Harvard Observatory associate killed at Chickamauga; Newton Manross, a Wöhler Ph.D. who died leading a charge at Antietam; and Theodore Parkman, another rising Ph.D. in chemistry, torn apart at twenty-five by a shellburst in North Carolina.

The best-known scientist to die in military service was Ormsby Mitchel, the people's astronomer. A West Point graduate, he accepted a commission as brigadier general in the summer of 1861. While his cherished Cincinnati Observatory lay dormant, its steps rotting, its roof leaking, the great equatorial dulled with tarnish and missing its eyepieces, Mitchel led his troops in a brilliant campaign in Tennessee, gaining control of a crucial railroad and launching the storied exploit that came to be known as "the Great Locomotive Chase." For this, "Old Stars," as his troops were said to call him, got a new star. But friction with an inept commanding general led to his transfer to South Carolina, where he died of yellow fever at Beaufort in October 1862—a loss to Northern science that more than balanced Beaufort's books.

Now and then naturalists in uniform snatched moments for science on the side. A Pennsylvania soldier in Maryland busied himself with botanizing and resolved to become a naturalist after the war. A nineteen-year-old Wisconsin soldier, Edward L. Greene, carried a botanical handbook in his bedroll and botanized in Kentucky, Tennessee, and Alabama between battles. He saw no prospect of becoming an eminent naturalist, what with "nothing but my own hands and nature before me," but "as I love the study I shall give all my leisure to it." After the war he did become a noted botanist. Some were favored by location and duties. As a surgeon on the Carolina coast, Ferdinand Hayden saw "the way opened for splendid collecting," and he took it. By the fall of 1863 he had only half his time occupied with medical duties and so had made "some good collections"—though he still panted for a return to full-time science.

Higher-ranking scientist-officers had more power to indulge their scientific cravings. Brigadier General Richard Owen followed geological as well as rebel formations through Mississippi. Colonel Elisha Andrews did the same through several states. Others encouraged science-minded subordinates. General William T. Sherman obliged Frederick Barnard by inquiring after the geologist Eugene Hilgard in Mississippi with a promise of safe conduct and assistance—something the scientists of South Carolina would have welcomed from him later on. In California, Colonel Richard C. Drum assured a soldier-collector that off-duty science would speed, not slow, his advancement in the army. John L. LeConte got a short leave to finish a scientific paper. Albert Bickmore in North Carolina got a leave and then light duties in order to collect for Agassiz's museum. And Fielding Meek, though a shy civilian, met a navy surgeon with an interest in paleontology who wangled a navy steamer and crew for a month's collecting by Meek along the Potomac, with the navy's blessing, in the critical month of June 1863.

A few such brief science breaks counted for little, however, against the toll of time, energy, commitment, and life itself exacted from scientists by military and naval service. The Illinois entomologist Benjamin D. Walsh put the case more callously than would most in telling of a friend who "was just beginning to take hold of ornithology, when the war broke out & off he went as a Regimental surgeon!" Walsh commented, "Just as if the saving of the lives of a few thousand soldiers, who will die anyhow in a few years more or less, was of any importance whatever compared with the discovery of great scientific truths which never die!" But Walsh, who was addressing the military surgeon John L. LeConte, probably meant that as grim raillery, for which he had a taste.

In both North and South the war competed with science for the minds and energies even of the civilians. Beginning in June 1861 the work of the United States Sanitary Commission made heavy demands on Alexander Bache, Wolcott Gibbs, and especially the geologist John S. Newberry, who took charge of its operations in the Mississippi Valley. "I am in for the War," wrote Newberry in July 1861, and he meant it. In 1865 he wrote, "as my head has been for the past four years full of the arts of *war,* those of peace have had little space left them." In July 1864 the Sanitary Commission enlisted Benjamin Gould to direct an elaborate program of physical measurements of soldiers. This was doubly unfortunate, since it took Gould away from his genuinely scientific work for many months, and since its seriously flawed data became a pseudoscientific prop for generations of American racism.

On the other side of the battle lines, John Bachman in Charleston turned himself into a one-man equivalent of the Sanitary Commission. During much of the war, though in his seventies, he spent most of his days tending the sick and wounded, raising money for the hospitals, and receiving and distributing funds and food. As the Confederacy drew near its end, his wife died, and Bachman found his only relief from despair "in constant occupation day & night. I wander everywhere collecting for the hospitals—preaching every Sunday." Whatever he may have done for relief and religion, American science was the poorer for it.

Less tangible but no less real was the psychological impact of the war. It hit Southern scientists hardest, their would-be nation being in the more desperate case. "The absorption of the mind in the war and its possible results made abstract thinking and writing seem an absurdity, if not a crime," recalled Joseph LeConte afterward. His brother John finished an important paper on sound velocity in the summer of 1861 but did nothing more in pure science while the war lasted. Both Moses Curtis and Henry Ravenel confessed in 1865 that they had done nothing in botany during the war. "Other matters were too absorbing," explained Ravenel; and Curtis blamed "the great topic of all overlaying nearly every other."

In the North, war excitement put science out of mind for many at the start. William B. Rogers in Boston, Ferdinand Hayden in Washington, John Torrey in New York, and the conchologist John G. Anthony in Cincinnati, all acknowledged as much in the summer of 1861. Intermittently and in lesser degree, the distraction persisted as the war dragged on. In Massachusetts during the summer of 1862 the entomologist Asa Fitch put it vividly:

> We are having strange and stirring times. . . . The getting up of our regiment—its rendezvous here in Salem—war meetings in every neighborhood—calls and consultations over the thousand arrangements to be made—squads of recruits passing by with drums beating and flags flying—all together, it's keeping my head in such a whirl that it's impossible to fix my thoughts steadfastly upon my own work, so as to make progress therewith. . . .

That fall Joseph Henry reported that "all thoughts are absorbed in that of the War," and William B. Rogers found that the war, as well as family anxieties, had "greatly encroached upon my time for scientific reading & work." "The state of the country," wrote the geologist J. Peter Lesley in 1863, "renders all steady thinking on other subjects impossible." At Cleveland early in 1865 the zoologist Jared Kirtland was "reading four daily papers and every obtainable publication relating to this great Contest." Consequently his researches were "without any system or order—of course without important results." Indeed, wrote Kirtland, "this War appears to have obliterated all taste for cultivation of the Natural Sciences in the West."

To all these distractions the war added money problems for many scientists. The first year or two brought hard times for Northern business and thus for scientists who depended on it for part or all of their livelihood. In Cleveland the paleobotanist Leo Lesquereux's business ruin in 1861 took all his property, home included, and left him heavily in debt. The turmoil in Missouri unsettled the amateur zoologist Hiram Prout's business and so kept him from his Bryozoa for months. Industrial chemists had little work; and even college chemists, who had got used to supplementary income from paid analyses and other odd jobs, were straitened when that source suddenly dried up.

The economic boom that began early in 1863 relieved this class of scientists, but the accompanying inflation hurt the far more numerous class that depended on fixed incomes from teaching or government work. "My salary of 3,500 has not been sufficient by nearly 400 dollars during the last 12 months," wrote Joseph Henry in September 1864. Wolcott Gibbs at Harvard and the young astronomers Asaph Hall and Simon Newcomb at the Naval Observatory echoed his complaint.

The boom, and perhaps inflation also, turned some scientists' thoughts toward making money, science or no science. They had known the yearning before, as Benjamin Silliman, Jr., had abundantly demonstrated in the 1850s. When Eben Horsford gave up his Rumford Chair at Harvard in 1863 for full-time business, he cited "the pressure of the times," the needs of his five daughters, the state of his health, and the demands of his prospering Rumford Chemical Works; but he added that he had "longed and prayed" for financial security for years past. Still, the speculative urge that had run through American life since Columbus now reached new levels of frenzy. Gold and silver bonanzas, the oil fever, the stock market, trading in gold and greenbacks, fat government contracts, all dazzled Americans, and some scientists eyed the lure.

"Men of science," wrote Joseph Henry in 1866, "are not generally among the number of those who have made fortunes by the war." Lack of capital to invest did limit the temptation mostly to those whose expertise had direct monetary value, especially the geologists. Ferdinand Hayden, though still in military service, and James Hall both sniffed hungrily at mining and oil ventures. So, of course, did the younger Silliman. Hall was offered half of any profits in return for advice on oil leases in Pennsylvania, and he probably accepted. If so, the profits apparently never materialized. Robert E. Rogers fared worse, losing heavily in an oil venture that had rested on his geological judgment and in which he himself had been the largest investor. Peter Lesley had been similarly burned in a coal-desulphurizing venture just before the war, and thereafter, according to his biographer, stayed out of business undertakings and "confined his energies to purely scientific and literary work." How "purely" he did so is questionable, however. Near the end of the war Lesley informed James Hall that his fees for examining coal, iron, and oil properties were "established at a high figure now. . . . Moneyed men *must* have opinions & will pay *anything* for them." He pleaded with Hall not to spoil things by working cheap, and Hall seems to have granted his prayer.

The applied scientists' cousins in technology rode the same economic roller coaster. For the first two or three years, the war depression hit the civil engineering profession hard. In 1861 an engineer in Pennsylvania bemoaned "the almost total cessation in the construction of public works." Secession cut off Charles Ellet, Jr., from his Southern connections and thwarted his plans for river improvement. Then in 1863 came the war boom. Du Pont black powder sales for mines, quarries, and construction projects had dropped sharply through 1863 but nearly tripled in 1864. Railroad track construction showed the same pattern. The civil engineering instrument sales of a leading Northern maker had fallen off to almost nothing in the summer of 1861, but were rising strongly by the spring of 1863 and reached record heights in 1864. Mining engineers had fared somewhat better from the start, since production of most metals held up close to or a little above the 1860 levels. The wartime Nevada silver boom created fewer jobs than it might have, thanks to the unusually tractable ores and also to disillusionment with pseudoscientific "process-peddlers," but metals prices generally soared in 1863 and so did prospects for mining engineers.

All these wartime problems and distractions did not mean that Northern science shut down for the duration. "I do not do so much scientific work as before the war," Asa Gray wrote Charles Darwin in 1862, "but still I keep pottering away." Gray's remark applied to most Northern scientists. When Darwin wondered that they could think of science at all in the midst of war, Gray explained, "Well, first, we get used to it. Second, we need something to turn to."

It helped to be out of touch with war news. At a cabin in the Rockies on the day after First Bull Run, the botanist Charles C. Parry, blissfully ignorant of the disaster, was exulting over the most productive six weeks of collecting he had ever known. The geologist William H. Brewer spent most of the war in the California wilderness for Josiah Whitney's state survey. "The national troubles . . . are in my mind by night and day," Brewer (then thirty-two) noted in his diary in May 1861, but his references to the war thereafter were rare and perfunctory. Some scientists in the troubled East tried to shut their minds to the war. Aside from increased printing costs and the loss of Southern contacts, James Hall's geological work did not seem to be much impeded; perhaps his opposition to the war served him as a means of disengagement from it. At the Nautical Almanac Office in Cambridge, John D. Runkle retreated into work and a perennial conviction that the war would soon end. A fellow worker remarked with wonder that Runkle would never stop work to talk about even the most exciting war news.

The brilliant astronomer Simon Newcomb, aged twenty-six, was also working at the Nautical Almanac when war began. He joined a Cambridge drill club; but the Lazzaroni circle got him a navy commission as "Professor of Mathematics" at the Naval Observatory in Washington, and he snapped up the offer. In 1862 he angled half-heartedly for a military commission, but in the end his only taste of army life was emergency duty during Jubal Early's brief raid on the outskirts of Washington in 1864. Newcomb did not miss the field of Mars, being well into the orbit of Neptune. He seemed to feel—correctly, as it turned out—that more important work than soldiering lay before him in science. By 1864 he was breathing down Benjamin Peirce's scientific neck in working out a theory of Neptune's motion. Peirce got his fellow Lazzarone Rear Admiral Charles H. Davis to restrain Newcomb from doing it on navy time (in a suggestive slip, Peirce at one point spelled his young competitor's name "Newcome"). But Newcomb went ahead on his own time. His diary entry for April 10, 1865, reads in full: "Surrender of Gen. R. E. Lee, and probable end of the war, announced by 500 cannon at daylight. Too bad a cold to go out, so finished Neptune, getting the last of the theory ready for the press."

Wolcott Gibbs likewise carried on with his scientific work, despite the war, his labors for the Sanitary Commission, and a move to Harvard in 1863 that filled his time for two years with organizing the scientific school anew and teaching his courses. In 1864 he managed somehow to make what has been called his most important contribution to chemistry: electrogravimetry, the use of electrolysis in quantitative analysis. In geology, notwithstanding the war and a nervous affliction that limited him to two or three hours of work a day, James Dana in 1862 finished his *Manual of Geology,* the most important American geological publication of the war years. Though Dana had not yet read *The Origin of Species,* his *Manual* dissented from Darwinism in passing; and it derived virtually all geological history from the contraction of the cooling earth, a view now discarded. Nevertheless, the *Manual* in its successive editions constituted the most authoritative compendium of American geology for many years. In it, American geology came of age.

Dana would have been more in fashion, as well as on firmer ground, had he said nothing at all about Darwinism in his 1862 edition. James Hall, the leading American paleontologist, stubbornly refused to express an opinion on it or even discuss it to the day of his death in 1898. "It was ever conditions not theories that confronted him," says his biographer, "and he went on heaping up new facts to the end." During the Civil War —whether or not because of it—American scientists said surprisingly little about Darwinism. In meetings of scientific societies, only two serious commentaries were offered, one against it in 1861 and the other for it in 1863. Scientific publications carried almost nothing on the subject. Only Asa Gray, through book reviews, kept Darwinism alive in the *American Journal of Science,* defending its compatibility with religion and pointing out the growing acceptance of it by European scientists. Louis Agassiz combatted it in three popular lecture courses and a score of articles in the *Atlantic Monthly,* but his professional compeers did not consider any of that to be serious science. In 1863 Asa Gray commented privately on Agassiz: "This man, who might have been so useful to science and promised so much here has been for years a delusion, a snare, and a humbug, and is doing us far more harm than he can ever do us good."

On the day war began, Agassiz had told his excited students that "those who sought Nature always found relief if their minds were distressed." What the students wanted, however, was more news, not relief. When he caught Albert Ordway reading a newspaper in the Museum of Comparative Zoology, Agassiz threatened to ban all newspapers from the building. He did not do so, perhaps realizing the futility of it. Instead, as his assistants (including Ordway) slipped off to war, Agassiz tried to turn the excitement to scientific account, as had Prussia in the time of Napoleon. "In these days of trial of the nation," he wrote President Lincoln in March 1862, "everything which manifests intellectual activity and life is important, especially in reference to the estimation in which we are to be held abroad." He saw himself throughout the war as thus making his own chief contribution to the cause through well-publicized perseverance in the scientific enterprise.

But no ingenious rationalization, no physical or mental disengagement, no high resolve to carry on, could disguise the fact that wherever the war touched individual scientists it hurt them. One must look beyond individuals to the institutions of science to detect even an indirect benefit among the multiple wounds of war.

Chapter 21

War and the Structure of Science

For scientific institutions, the Civil War was not a total disaster. It broke up old patterns and let new ones form. Lincoln's first Congress in 1862, freed of Southern obstructionism and nerved for bold action by the crisis, ranks among the foremost half-dozen in American history for the sweep and significance of its innovations. Among them were the Morrill Land Grant Act for the support of colleges and the acts establishing the Department of Agriculture and the National Academy of Sciences.

Those acts were not war measures, however. Secession and war merely blasted a way through longstanding barriers. The sponsors of the college act and the Agriculture Department never pleaded war needs. Advocates of the National Academy cited the government's need for wartime advice, but on that score both debate in Congress and actual performance were perfunctory. Apart from those collateral legislative openings, the war gave scientific institutions little aid or comfort. On the contrary, it dealt some of them bruising blows.

The business slump of 1861–62 discouraged giving to scientific institutions, and war-related philanthropies like the Sanitary Commission competed for what funds there were. Ormsby Mitchel's Cincinnati Observatory might have fared poorly even if its promoter had stayed out of the army. That of Washington University in St. Louis had to be left half-finished. By the end of 1862 Christian Peters of the Hamilton College Observatory had not been paid for many months. The Harvard College Observatory, which got only $200 a year from the college, could barely scrape by. For its salaries, instruments, publications, and other costs, the observatory had an annual endowment income of only $5,600, aside from the college's mite. Its director, George Bond, gave up his coffee, borrowed his brother's newspapers, and wore his overcoat in the library to save fuel. The worry and privation may have hastened his death from tuberculosis in 1865 at the age of thirty-nine. Only some timely private gifts allowed Bond to publish his great monograph on Donati's comet in 1862.

Bond doubtless envied Louis Agassiz's magic touch. In the first month of the war, Agassiz cajoled the Massachusetts legislature into granting his museum $20,000. Agassiz's wooing of the public, in which he was abetted by a claque of distinguished literati, did him no good with his professional peers, but it kept the money rolling in. During the first two years of the war, the Museum of Comparative Zoology got nearly $40,000 in state and private contributions. In 1864 Agassiz raised another $6,200 by popular subscription. At the end of the war, the museum's endowed income exceeded $12,000 a year, but Agassiz went on getting and spending.

By then the boom of 1863–65 had helped reverse the decline in philanthropy. Not only was money more plentiful, but also, said one fund raiser, people "learned to levy on their own pockets, to pay voluntary as well as involuntary taxes." Even Asa Gray got a modest endowment and a building for the herbarium and library he had gathered and hitherto maintained at his own expense. On a visit to Chicago in 1864, Agassiz so fired a number of wealthy men with zeal for natural history that the Chicago Museum of Natural History was established soon after. In 1863 Dr. William J. Walker, who had retired and made a fortune in railroad stocks, gave large sums to the Boston Society of Natural History and sixty thousand dollars to the projected Massachusetts Institute of Technology. At his death, near the end of the war, he left each about a quarter-million. "I think," wrote a young entomologist on hearing the news, "that science and scientific men are to be fostered in this country as they never have been before."

Though the ups and downs of wartime giving may have roughly balanced out, the war's net impact on scientific societies was decidedly negative. Most societies had always relied more on the time, talent, and labor of their members than on philanthropy. The Boston Society of Natural History built handsome new Back Bay quarters with its financial windfall, but lost ninety-eight members to military service. The drain and distraction of war killed the Cleveland Academy of Natural Sciences, the Indiana Association for the Advancement of Science, and smaller groups like the once sprightly Pottsville (Pa.) Scientific Association.

Even before the fighting started, secession forced the American Association for the Advancement of Science into suspended animation. The historical accident of a Mississippi president and secretary, a South Carolina vice president, a decision to meet in Tennessee on April 17, 1861, and an overwhelmingly Northern membership posed an obvious problem. By mid-February 1861 only a few leading members, North or South, entertained even the possibility of such a meeting. Wolcott Gibbs thought it a fit occasion to "let the monster die a natural death," though no one else seems to have avowed such a purpose. In the end, the permanent secretary, Joseph Lovering of Harvard, with the nearly unanimous consent of both outgoing and incoming standing committees, belatedly sent out a circular postponing the meeting "for one year, at least." Early in 1862 Rochester, New York, offered to play host and Lovering favored the idea, but the war dragged on and nothing was done. By then James Dana, for one, professed to be glad the Association had "died out." Soon after the war Julius Hilgard implied that the neglect had been deliberate, in reaction against the Association's excess of democracy. But there is no hard evidence for that view, other than Gibbs's offhand remark.

Scientific publication suffered also. In 1863 Agassiz's Museum of Comparative Zoology began publishing its *Bulletin,* the first strictly zoological periodical in the United States, but elsewhere scientific journals languished or died. Benjamin Gould gave up on his *Astronomical Journal* early in 1861 because of insufficient funds and the mental stress of "the horrid crisis," and John D. Runkle's *Mathematical Monthly* ceased publication "on account of the present disturbed state of public affairs," leaving the nation with no mathematical journal until 1874. The leading scientific periodical, *The American Journal of Science,* lost all its Southern subscribers, a fifth of the total, and some of its Northern subscribers as well. In December 1862 it had only nine subscribers in the whole state of New Jersey, and by the end of the war it was losing six hundred dollars on each issue. Somehow it survived.

The war also interfered with book production. Demand for war-related books crowded out scientific material at some presses, at least in 1861. Military matter preempted government presses, interrupting the publication of scientific reports.

During the first two years of war, Northern colleges felt the impact of financial stringency and the call to arms. From Washington University in St. Louis, William Chauvenet wrote in May 1862: "We are among the very few of the western colleges that are not either closed altogether or in pecuniary difficulty." Money problems, lack of students, and the occupation of its main building by a military garrison shut down the University of Missouri during most of 1862 and aborted a planned scientific course. In 1861 the war frustrated a desperate endowment campaign by President Thomas Hill of Antioch College. When Antioch closed early in 1862, Hill accepted the presidency of Harvard.

Many college students were too young for military service. Of the ten colleges that listed a minimum age for enrollment, seven of them set it at fourteen. At the University of Pennsylvania in 1861, most students entered at fourteen or fifteen and graduated at eighteen or nineteen. But enough college students were of military age to reduce enrollments significantly. At Indiana State University, Daniel Kirkwood reported a substantial drop in enrollment as of June 1862. "Some have gone to the war," he wrote, "others home to take the place of brothers in the army."

As in years past, some students went abroad. The young astronomer Cleveland Abbe, having volunteered in 1861 and been rejected—paradoxically—for nearsightedness, was "overwhelmed" in 1864 when Otto Struve accepted him as a sort of postgraduate fellow at the great Pulkovo Observatory. The only nineteenth-century American known to have studied science in Russia, Abbe would look back on his two years there as the high point of his life. Coincidentally with the Northern draft act of 1863, Americans at Heidelberg "increased in number in a very extraordinary fashion," and Göttingen had ten Americans at the end of that year, five of them chemists.

But the Northern groves of academe were not levelled like those of the South. The older and sturdier colleges survived; and some that closed were soon reopened, like Missouri and Antioch. Northern superiority in manpower told here as on the battlefields. In the last year of the war, the hundred or so leading Northern colleges enrolled about ten thousand students, fourteen hundred of them in scientific courses. The business revival of 1863 also helped. In 1864 enrollment at Lafayette College in Pennsylvania had dropped to nineteen, but its faculty responded by projecting a scientific course. When the coal and iron magnate Ario Pardee challenged Lafayette's president to justify that quixotic enterprise, the president did it so convincingly that Pardee wrote him a check for $20,000 on the spot. Pardee eventually gave Lafayette a total of half a million. In 1863 and 1864 Princeton ran its most successful endowment campaign to that time, raising $100,000, though not enough to open its contemplated school of applied science. Meanwhile its rival Rutgers was raising $145,000 and converting its science course into a separate department, with an eye to Morrill Act support.

In the hard times of 1861–62 even the technological colleges suffered. The University of Pennsylvania's feeble School of Mines, Arts, and Manufactures lapsed into a suspension that would last a dozen years. In May 1861 William B. Rogers postponed the opening of his Massachusetts Institute of Technology for the duration of the war, though the school's hard-won state land grant depended on his raising a hundred thousand dollars within a year. Pleading war conditions, he won a one-year extension, but only at the last minute did a large gift clinch the grant.

The Rensselaer Polytechnic Institute had languished for lack of money and leadership since Benjamin Greene's departure in 1859. Enrollment fell below a hundred all through the war, and only twelve graduated in 1865. A great fire in Troy consumed the school's buildings in 1862. But RPI survived. The war boom of 1863–65 helped the trustees to find a better site and put up new buildings by 1864, and tuition went up from $100 to $150 that year without discouraging enrollment. So the boom brought some relief to technological as well as scientific schools, while the land-grant college act of 1862 promised still more to the lucky or nimble. But the boom merely balanced out the preceding war depression.

Although one historian has called the war's effect on chemical education "an overwhelming catastrophe" that turned "the tide of progress . . . backward for nearly twenty years," there is much contrary evidence. In 1861 the students of Union College organized a chemical society under the direction of Charles F. Chandler. It died when Chandler moved to the Columbia School of Mines in late 1864, but meanwhile it had enlisted 138 members, held 96 meetings, and heard at least 41 papers. (A dozen years later, perhaps encouraged by that experience, Chandler took a leading part in founding the American Chemical Society.) And at Yale's Sheffield Scientific School in February 1862, Samuel W. Johnson counted more chemical laboratory students than ever before, though the business recession had shut off his own outside work.

The Sheffield School easily weathered the prevailing shortages of money and students in 1861–62. It was, recalled one of its faculty, "a small affair, barely tolerated by some of the authorities and ridiculed by some of the members of the Classical Department." But part of its strength lay in its care to live within its modest means. For fifty years it never tolerated a permanent deficit. Its able faculty settled for lower salaries than were offered elsewhere. It built on narrow but solid foundations of applied science and engineering rather than trying to be an instant Giessen or Göttingen. So the lower enrollment of wartime did not drag it under. As Daniel Webster once said of Dartmouth, it was a small school but there were those who loved it; and more important, one of those was rich. On the eve of the war, the wealthy philanthropist Joseph Sheffield had given the Yale Scientific School a building, a fifty-thousand-dollar endowment, and its new name. He continued to aid it. Fortune favored it also with designation as Connecticut's sole beneficiary of the Morrill Act, and all concerned moved so fast that it became the first school actually to receive money under the act, two-thirds of its total income for the 1864–65 school year. This windfall enabled it to raise salaries to two thousand and hire William H. Brewer away from the California geological survey as professor of agriculture. The new strength of the Sheffield School, added to the growing scientific strength of Yale College, made science at Yale equal or superior to that at any other American college, at least until the Johns Hopkins University came on the scene in 1876.

For the leading college in the most militant Northern state, Harvard seemed strangely untouched by the war. "College life went on much as usual, and with scarcely diminished attendance," comments its best-known historian. Not until the fall of 1864 did freshman registration drop off as much as it had in the depression of 1857. But Harvard and its patchwork appendage, the Lawrence Scientific School, generated their own turmoil in those years—or perhaps it should be said that Louis Agassiz generated it.

Asa Gray's besting of Agassiz in the Darwinism debates of 1860 had lowered Agassiz's standing in professional eyes, and it may have shaken

Agassiz himself. Though to the public he had never stood higher, he now reached for still more acclaim and authority. His spirited young acolytes at the museum, however, began to strain at the leash by which Agassiz held them back from Darwinism. They saw themselves as underpaid, undervalued, intellectually and professionally trammelled. Agassiz felt the tension and checked them all the harder. Early in 1864 five of the brightest broke away in a body because, as one put it, "exceedingly arbitrary and tyrannical rules . . . have been imposed . . . by which all intellectual independence is taken away." Others had already left. All went on to scientific distinction. Agassiz replaced them with older, established naturalists, able and more docile, but without the élan of the departed student rebels.

While secession brewed in his own domain, Agassiz set out to shape the destiny of Harvard. When his brother-in-law Cornelius Felton died early in 1862 after little more than a year as president of Harvard, Agassiz and his fellow Lazzarone Benjamin Peirce succeeded in replacing Felton with Thomas Hill, a Unitarian minister and Harvard alumnus from Massachusetts who had briefly been president of Antioch. Hill was known to be science-minded, and his two Lazzaroni sponsors looked forward to "exciting discussions about modes and plans of education, and . . . changes favorable to science." Agassiz promptly urged a grandiose scheme on Governor John Andrew whereby Massachusetts would put up two million dollars to create a great university on the European model, using Harvard as a nucleus and taking advantage of the Morrill Act. Of course, nothing came of that fantasy. Nor did anything come of Agassiz's move to give the disjointed, drifting Lawrence Scientific School a new form and mission, since Agassiz himself turned against the movement when his archrival Asa Gray began to take the lead in it.

Early in 1863 Eben Horsford told Agassiz privately of his decision to resign the Rumford Chair and recommended Wolcott Gibbs as his successor. Both Agassiz and Gibbs leaped at the idea. Almost ten years earlier the Lazzaroni had lost their noisy battle to place Gibbs at Columbia as the first step toward making that dormouse of a college into a university lion. Now Gibbs stood higher than ever in his field, and he pined more than ever to escape "the fatigue of 3 lectures or recitations a day" at the New York Free Academy, which was, he said, "eating me alive." The vigorous backing of Agassiz and Peirce elated Gibbs, who promised "to work with you all in establishing an American University" at Harvard.

It was unfortunate that young Charles W. Eliot stood in the way. Though only an assistant professor of chemistry and mathematics, Eliot had taken over the chemical laboratory of the Lawrence Scientific School from Horsford nearly two years before and was also serving as the school's acting dean. Unlike Gibbs, he was a Harvard graduate and came of an old and prominent Massachusetts family. In an earlier day he would therefore have succeeded Horsford as a matter of course. But

Harvard was indeed entering a new era in which professional merit came first. Eliot had the support of Asa Gray and most of the faculty. But he had no European training and no flair for research. Gibbs had both, along with the favor of Agassiz and Peirce and hence of their protégé President Hill. Spurning Hill's half-hearted efforts to find a modest alternative place for him, Eliot angrily resigned and went off to study educational institutions in Europe. After returning he taught chemistry at MIT. Gibbs became Rumford Professor and dean of the Lawrence School. "I felt," Gibbs wrote, "as if a heavy burden had been taken off my shoulders." The feeling would last only a half-dozen years. In defeat at Columbia, Gibbs and the Lazzaroni had set a train of events in motion that would ultimately gain their main object; in victory at Harvard, time would yield a similar result—but in a way totally unforeseen by them and bitterly ironic for Gibbs himself.

Soon after the Gibbs victory at Harvard came another Lazzaroni success. This too had its elements of irony, among them its apparent reversal of the earlier Lazzaroni defeat at Columbia in 1854. Their Columbia standard-bearer this time was Frederick Augustus Porter Barnard.

In 1837 a series of adversities had led the twenty-eight-year-old Barnard, a Massachusetts-born graduate of Yale, to accept a teaching position in mathematics and natural history at the University of Alabama. In 1854 he moved to the University of Mississippi, where he taught a melange of sciences—chemistry, math, physics, astronomy, and civil engineering. His own interests shifted in the fifties from chemistry and photography to physics and astronomy. Despite severe deafness he was an effective teacher, and his Yankee energy, which had merely irritated the Alabama president, so impressed the Mississippi trustees that they made him president in 1856. In that role he came out for electives and graduate courses, with true university status as an ultimate goal. The dream did not come true in his time there, but he did tighten student discipline, build up enrollment, add new buildings, and develop an astronomical observatory, taking special pride in an order to Alvan Clark in Cambridge for an 18½-inch telescope lens that would outdo Harvard's 15-incher.

Meanwhile Barnard kept up his own scientific standing through articles and AAAS meetings. His espousal of the university movement and his laudatory AAAS report in 1857 on the Coast Survey endeared him to the Lazzaroni and their chief. Hence Bache's choice of Barnard as AAAS president for 1861.

Despite all this, Barnard's Mississippi years were the most miserable of his life. In 1858 he wrote that he would rather be a day laborer than a Southern college president, and he bewailed the "meanness, and coarseness, and ignorance and brutality" pervading "this whole community." He emphasized science too much for local tastes, and his Yankee ways stirred suspicion. When a student raped and beat one of Barnard's female slaves, Barnard made trouble and was therefore tried by the board of trustees on a charge of being "unsound on the slavery question." He was acquitted, but he kept his eyes open for a job in the North. When secession came, Barnard was fifty-one and loath to follow Benjamin Hedrick into jobless exile. So he played down his longstanding Unionism, except to Northern friends, expressed concern for "the cause of our injured section," and even tried to get Clark's great lens through the blockade (to George Bond's chagrin it ended up at the Dearborn Observatory in Chicago). When the University of Mississippi shut down for lack of students in the fall of 1861, Barnard set out on a university mission that eventually brought him to Norfolk, Virginia, where he was liberated by Union forces in the spring of 1862.

Bache tided Barnard over with a Coast Survey job. But Bache had larger plans. First he suggested Barnard for the Harvard presidency, but Agassiz and Peirce put forward Thomas Hill for that position. Then Bache, Dana, and Henry pushed Barnard for the Columbia professorship vacated by the renegade Richard McCulloh—the very position Gibbs had been denied in 1854. The physicist Ogden Rood prevailed, partly because of Gibbs's support for him, partly because of Barnard's deafness. But when Charles King resigned as president of Columbia early in 1864, Gibbs joined Bache and Henry in getting Barnard chosen as King's successor.

Hampered by the same trustee obstructionism that had blocked Gibbs in 1854, Barnard did not become one of the great postwar leaders in the university movement. Still, he served Columbia well for a quarter century. And as fast as the reluctant trustees permitted, he piloted Columbia toward university status, beginning with his strong support of the Columbia School of Mines, which opened on November 15, 1864.

Recalling World War II and developments since, one might suppose that government science flourished during the Civil War. It did not.

On the state level, antebellum science had owed most to geological surveys. All those in the Confederacy stopped short when war began, and in several the records of work already done were lost or ruined through neglect. Mississippi, while suspending its survey, did find money for its state geologist, who analyzed saline waters because of the salt shortage, poked around in caves (unsuccessfully) for niter deposits, and did other odd jobs. Some Northern state surveys stopped for lack of funding. Maine established one and then failed to appropriate money for it. In 1864, however, George Cook's adroit lobbying and cut-rate cost estimate induced the New Jersey legislature to revive a survey discontinued in 1857.

Declining yields in the gold fields had led California in 1860 to authorize a survey under Josiah D. Whitney, whose national reputation enabled him to recruit a corps of outstanding young scientists. Whitney's plans were ambitious and his hopes high. Scientifically important work was done while the war raged in the East. But Whitney's arrogance, sharp tongue, and emphasis on pure science rather than practical data got him into trouble. "It is not the business of a geological surveying corps to act . . . as a prospecting party," he told the legislature. His survey outlasted the war, but as a lobbyist Whitney was no James Hall. Unlike their New York counterparts, the California legislators finally concluded that it was not their business to advance paleontology.

Some scientists assumed at the start that in wartime the federal government would spend no money on scientific research. Increase Lapham confirmed that hypothesis when he sought a three-thousand-dollar appropriation for the Patent Office to support his proposed study of North American grasses. He argued that the study would help the war effort by helping the farmers. But two years of writing congressmen brought only the reply that immediate war needs came before any such frills. "Particularly inimical" to the idea was Justin Morrill of Vermont, even then pressing his land-grant college bill—which like the National Academy involved no appropriation.

Far from expanding scientific research in the armed services, the war took some army and navy officers away from their scientific assignments. Since 1857 Captain George G. Meade of the Topographical Engineers had been doing an outstanding job in charge of the Great Lakes Survey. In 1861 his friend Joseph Henry begged him not to give up a brilliant future in science to become, as Henry put it, mere food for powder. But Meade pulled strings to get the command of a Pennsylvania brigade and went on to win his place in history as commander of the Union forces at Gettysburg. He never returned to science.

Just before the war began, another Topographical Engineer officer, Captain Andrew A. Humphreys, with the assistance of Lieutenant Henry L. Abbot, made a name for himself in science with his *Report upon the Physics and Hydraulics of the Mississippi River,* the fruit of nearly a decade's work. After studying the theory and practice of river control in Europe, 1853–54, Humphreys had systematically tested what he learned abroad against elaborate and comprehensive data gathered on the Mississippi with exemplary skill and ingenuity. When foreign theory and home data failed to match well, Humphreys had revised or discarded the former and had ambitiously developed what he hoped would serve as a new general formula to measure water flow in rivers everywhere. The final report, to be sure, did not explicitly claim any fundamental breakthrough in theoretical analysis. Nevertheless, the *American Journal of Science* pronounced it "one of the most profoundly scientific publications ever published by the U.S. government," and Benjamin Peirce hailed it as "the finest discovery in hydraulics of modern times," one that called for "rewriting the whole theory of hydraulics." Though much of the new theory in the report was later rejected, Humphreys's formula did stimulate later research that led to important advances in theory, and meanwhile the report won Humphreys numerous international honors.

But the war cut short his promising career as a scientist, calling him and Abbot into the field, where Humphreys acquired the unscientific nickname of "the fighting fool of Gettysburg." After the war Humphreys became chief of the Army Engineers Corps. But somehow he could not manage to pick up where he had left off in science. On the contrary, he became unscientifically dogmatic in defense of his report as gospel for the Engineers' river work, to the eventual detriment of the corps' influence and standing.

The Corps of Topographical Engineers, for more than a generation the chief governmental instrument of natural history exploration in the Far West, had come close to the end of its scientific mission, functionally and geographically, with the Northwest Boundary Survey of 1857–61. Still, it was the Civil War that prevented the issuance of that survey's final report. More than that, the war put an end to the corps' very existence. Not only Humphreys, Abbot, and Meade, but other officers as well left the corps, some to join the Confederate army, others to assume field commands with the Union army. The corps' scientific work had no relevance to a war fought mainly east of the Great Plains. Most of its remaining officers served as staff officers in field commands, and early in 1863 the corps itself was officially and finally reincorporated into the Corps of Engineers.

Science in the navy lost its two best-known exemplars. Since his great United States Exploring Expedition of 1838–42, Captain Charles Wilkes had been on special service in Washington superintending publication of the expedition's massive reports. By 1859 Congress had appropriated more than a quarter million dollars for the purpose, and more was needed. But the war suspended further publication, and Wilkes himself was ordered to sea duty in April 1861. That November he seized two Confederate diplomats from the British mail steamer *Trent,* precipitated an Anglo-American diplomatic crisis, and thereby won more celebrity than science had ever brought him.

On April 20, 1861, after the secession of Virginia, Commander Matthew F. Maury resigned from the navy, walked out of the United States Naval Observatory in a civilian suit, and proceeded to Richmond. He had disliked slavery, but accepted it as legal and moral; he had dreaded secession, and indeed had tried to arrange a mediation to end it, but made up his mind in advance to go with his state. This decision virtually ended his career as a scientist, though he would do significant work in naval technology for his new government. Bache and his Lazzaroni rejoiced at the "absquatulation" of the "Southern humbug."

Maury's absquatulation did, in fact, lead to a revival of the observatory's astronomical work. Captain James M. Gilliss, who had done more than anyone else to create the observatory and had then been passed over in favor of Maury as its superintendent in 1844, found belated justice as Maury's successor in 1861. Behind the observatory lay malarial flats and marshes; before it, during the war, were kept a vast number of army horses, whose tributes to the navy made access "very disagreeable." But to Gilliss his new command was sweet indeed. For the sake of national prestige as well as wartime utility, Congress was generous with appropriations. Gilliss equipped the establishment with up-to-date instruments. In 1863 he ordered a great transit circle from a noted Berlin firm, though he did not live to see it. He added three more civilian "aids" to the astronomical staff to help make and reduce observations, and work began on reducing and publishing a mass of crude observations accumulated over the preceding decade. The recruitment of young Simon Newcomb presaged further progress. Other astronomical institutions began cooperating with the reborn observatory.

But the war brought problems along with opportunities. "The equipment of our fleets taxes me almost exclusively," wrote Gilliss in November 1861. The observatory had to procure, care for, and distribute all of the navy's charts, chronometers, spyglasses, compasses, and other instruments. The wartime shortage of manpower probably contributed to Gilliss's difficulty in finding good civilian assistants. And the war postponed a permanent organization of the observatory. Gilliss's sudden death from apoplexy in February 1865 left that task to his successor, Charles Henry Davis.

In April 1861 Alexander Bache played a major part in bringing Captain Davis from Cambridge to Washington, counting on him, as Peirce's brother-in-law and an associate Lazzarone, to help brace Bache's Coast Survey against the shock of war. While running his Nautical Almanac by mail, Davis took control of the navy's powerful new Bureau of Detail, which assigned officers and purchased ships. In September 1861 Davis relinquished his Almanac and Washington duties for a year's service on the North Carolina coast and the Mississippi, but he retained enough influence to bring about the creation of a still newer and more powerful bureau, the Bureau of Navigation, intended by Davis to gather in the navy's principal scientific agencies—the Nautical Almanac, Naval Observatory, Hydrographic Office, and Naval Academy. To Davis's chagrin, however, Congress left the academy out and put the wholly nonscientific Bureau of Detail in. This eventually undid the dream of coordinating navy science. Davis headed the new bureau from the fall of 1862 to the end of the war. But he was not succeeded by a scientific officer, and so the Bureau of Navigation became little more than the older Bureau of Detail after all—though its chief would be the most powerful officer in the navy until after the war with Spain.

Of the two leading civilian institutions in scientific Washington, the Civil War harnessed one, the Coast Survey, and hobbled the other, the Smithsonian Institution.

Without the bureaucratic wizardry of its chief, Alexander Bache, the Coast Survey might even have ceased to exist, like the army's Corps of

Topographical Engineers. Despite the Survey's remarkable growth under Bache, it had remained officially temporary, assigned to a finite, nonrecurring task. Despite Bache's continental aspirations, it also remained a coastal operation, and much of the coast now lay in enemy hands. And despite its civilian status under the Treasury Department, it needed navy help in personnel and facilities. Indeed, it ran the risk of losing part or all of its functions to the navy's Bureau of Hydrography, much as the Topographical Corps was absorbed by the Corps of Engineers. It was no wonder that Bache's enemies cherished the belief that his agency would be an early casualty of the war. Bache himself feared as early as January 3, 1861, that "the terrible disruption of our country . . . will sweep our organization away entirely, or sadly cripple it."

But Bache saved the day by promptly making the Survey both influential and indispensable in carrying on the war. "Every scrap which we know about the coast is eagerly sought by the authorities," he wrote happily in the spring of 1861. The Survey met the needs not only of the navy but also of the army. General McClellan testified in January 1862 that "the only reliable topographical information we have" of the war areas "is derived from the Coast Survey." And Bache did not rest with distributing what the Survey already had. He also arranged to have calls made upon it for reconnaissances, and he made special surveys. The routine, peacetime work of the Survey necessarily slowed, but as early as June 1861 Bache had actually sent some of his assistants to serve with the army in the role of topographical engineers.

More than that, Bache worked his way into the councils of war. In June 1861 he managed to get a board set up to plan coastal and blockade operations. The board consisted of Bache himself, Major John G. Barnard of the Army Engineers (Frederick Barnard's brother), Captain Charles H. Davis, and a good friend of all, Captain Samuel Du Pont of the navy. Besides establishing the Survey's wartime importance, this congenial foursome cemented the Survey's relations with both the War and Navy Departments. When in December 1861 Bache learned of a move by some congressmen to suspend the Coast Survey for the duration of the war, he was prepared. Agassiz rallied his political friends, McClellan expressed his "great concern," the *New York Times* published an encomium on the Survey's war contributions, and the House voted down the pernicious amendment three to one. A couple of weeks later, a young assistant joined the Coast Survey and noted the impressive scale on which it operated (except for salaries, his being four hundred dollars a year): "an entire block of buildings besides several detached houses . . . divided off into a number of different departments . . . I believe, about a hundred assistants in the different offices in this city, and several hundred in field service."

So the Coast Survey survived and prospered. Yet the expansion of its practical services meant a curtailment of its more purely scientific work. And the increase in its remarkable superintendent's work and worry probably hastened his incapacitation and death. "Every day brings me some new call or responsibility," Bache wrote in May 1861; and four days later he wrote a friend, "I am trying to work up to a maximum without going so far as to break down." That summer and fall he did begin to break down, complaining of "sick headaches" from "brain wear." The young Survey recruit of January 1862 had seen an engraving of his new chief but did not recognize him from it, "he is so much altered; he has a long white beard, and . . . looked very badly."

Outdoor field work later that year did Bache good, but the decline resumed. He worsened rapidly after working eighteen hours a day to plan and supervise the strengthening of Philadelphia's defenses during Lee's campaign of June and July 1863. Other extraofficial war labors, especially the Sanitary Commission and a navy commission to evaluate inventions, further drained his strength. "This War has played the mischief with me," Bache wrote Gibbs that September, "interesting me to such an extent that I would rather die than not do all the opportunity gives me to do & that my education makes me feel that I can do." Frederick Barnard's appointment as president of Columbia in February 1864 was Bache's last notable coup in behalf of American science. That spring, failing in mind and body, he gave up active work. Three years later, at sixty, Bache died—from softening of the brain, said the physicians.

Meanwhile, life in wartime Washington had not been easy for Bache's old friend Joseph Henry, nor for the Smithsonian Institution. "The greater part of our intimate acquaintance have left," Henry wrote sadly in 1863, "and the city is now overflowing with strangers from the north & west." The "Norman Castle" in which he and his family lived continued to gall him as a memorial to misspent endowment. Through the long, hot, Washington summers, he wrote, "the thick walls . . . absorb heat from day to day until . . . they become like the sides of an oven." All during the war a legion of fleas, brought in with Spencer Baird's endless cortege of bird and animal cadavers, murdered sleep for Henry and his family. Henry thought the vermin came from Mexico and the Far West, but he apparently sought no entomological determination. In his sweltering, itching wakefulness he may have brooded instead over Baird's diversion of funds and attention to the fleas' former hosts. Far worse than any of this was the sudden death of Henry's only son in October 1862. A few hours before the end came, the young man "started suddenly upright with a wild look in his eyes crying out—'Oh they are chaining him! They are chaining Father!' " "Father has grown touchingly gentle since Will's death," wrote Henry's daughter Mary. Henry himself confessed that the blow "has lessened my desire to live."

But he went on, and so did the Smithsonian. In the first days after Fort Sumter, the Norman Castle had seemed likely to be tested under siege. Baird prepared for rebel assault by packing up rare and unique eggs and small birds. Henry ordered that no flag be flown, so that the Institution might plead its stated mission to all mankind and trust to "the sense of propriety of the besiegers." (The no-flag policy continued through the war, despite public criticism, and on the day of final victory, when Washington blossomed with bunting, not one flag could be bought or borrowed for the Smithsonian.) After troops arrived from the North on April 25, 1861, the Smithsonian got back to normal, except for hundreds of soldiers in the Hall, decorously viewing the specimens.

The Smithsonian's endowment gave it more independence from Congressional caprice than the Coast Survey enjoyed, but less adaptability to new conditions. Wartime inflation meant short rations for Smithsonian programs, since the Smithsonian's fixed income was mostly paid in depreciated greenbacks. Interest on some Indiana bonds, making up about a tenth of total income, was paid in gold and thus rose as greenbacks fell; but nearly as much had been invested in bonds of seceded states (most of it as late as 1860), and so yielded nothing at all during the war. "Terribly gloomy here," wrote Baird at the end of 1861; "I don't know what is to become of Smithsonian matters." Henry's annual reports complained publicly that inflation cramped operations, and he wrote privately in 1864 that "the efficiency of the establishment is materially diminished" in consequence.

Despite constraints on spending and wartime pressure on printers, Henry managed in 1862 to begin an additional series of publications, the Smithsonian Miscellaneous Collections. But the daily weather reports had to be suspended until the spring of 1862 and then only partially resumed, because the South was cut off, army observers were withdrawn from the West, and telegraph lines were jammed with war business.

The heaviest blow in those years, however, had nothing whatever to do with the war. To save money, the wings and towers of the building had been finished inside with wood instead of fireproof material. "Always anxious about fire," as he wrote in 1851, Henry strictly banned smoking and combustible litter, but disaster struck in an unforeseeable way when workmen installing a stove in January 1865 mistook a brick-lined furring space in one wall for a chimney flue. The hot air came out in a loft just under the roof, drying and heating the roof and rafters for eight days before they burst into a mass of flame. Firemen seized the opportunity to swill some whiskey they found in the building, "and would have died had they not vomited freely," since it was laced with copper sulfate for preserving specimens. Wind direction and the fireproof sections, however, saved most of the building.

The losses were heavy enough: the manuscripts and effects of James Smithson, the books being held for postwar restoration to the Beaufort (S.C.) libraries, a large stock of Smithsonian publications, $10,000 worth of instruments and apparatus, and other valuables. Flames consumed the product of a decade's arduous work by the artist John M. Stanley, nearly a hundred and fifty precisely detailed and finely executed portraits of Indians from forty-three Far Western tribes, "an irreparable loss to students of history and ethnology," a modern student has written. An equally irreparable loss to history was that of the Smithsonian's official, scientific, and miscellaneous files, including eighty-five thousand pages of correspondence, four manuscripts accepted for publication, and many of Henry's own scientific manuscripts. It took twelve years and $125,000 merely to repair the building and correct faulty construction discovered during the work.

In the case of the only lasting wartime gain won for scientific institutions by the scientists themselves, Henry was deliberately kept in the dark by his own Lazzaroni brethren while they concocted it. This was in creating the National Academy of Sciences.

Since the mid-1840s, Bache had argued privately and publicly for an American equivalent of the French Academy, with its government subsidies for research, its publications, its spur to scientific work through public honor, its advice to government, and its perpetual secretary Arago (Bache's personal model) as "the real dictator." The democratic unruliness of the AAAS in the mid-1850s made such an elite, authoritarian institution all the more appealing to the Lazzaroni. Bache and Peirce sounded the call publicly. Agassiz drew up a plan for an academy to start with a dozen appointed members who would elect more members, join with them in electing still more, and so on till a specified limit was reached. But the very strength of the democratic spirit discouraged the Lazzaroni from pushing any such plan—until the war came and with it the activist Thirty-seventh Congress.

In May 1862 that Congress transformed the agricultural subdivision of the Patent Office into an independent department, authorized to hire "chemists, botanists, entomologists, and other persons skilled in the natural sciences pertaining to agriculture" and to make "practical and scientific experiments." Its first commissioner carried over the entomologist Townend Glover from Patent Office days and as chemist took on a Liebig-trained Bache protégé, Charles M. Wetherill, who soon gave way to the Swiss-born Henri Erni. Wetherill and Erni have been called the first chemists ever employed by the federal government. But the farmers, not the scientists, had been behind the creation of the new department, and its scientific efforts were fumbling and insignificant. Scientists generally paid little heed to it, overlooking its potential for the support of research.

A few weeks later the farmers had their way again. In agricultural chemistry and entomology, and in the educational programs and public service of schools like that at Yale, farmers had begun to see the usefulness of applied science. During the fifties, agricultural societies and journals increasingly advocated government aid for such schools. In 1857 Representative Justin S. Morrill of Vermont introduced a bill to foster them through land grants. Morrill's bill passed both houses in 1859, only to be vetoed by the pro-Southern President Buchanan. But secession ended Southern obstruction, and the Morrill Act became law in July 1862, offering each state thirty thousand acres of public land per senator and representative for endowment of "colleges for the benefit of agriculture and the mechanic arts . . . without excluding other scientific or classical studies." Except for Evan Pugh, a Göttingen Ph.D. in chemistry and head of a state-supported agricultural college in Pennsylvania, scientists had done little to promote the bill. Nevertheless, Yale, Rutgers, and other colleges wasted no time bringing in the sheaves. Even scientists like Louis Agassiz, till then not much concerned with the farmer, suddenly awoke to the possibilities.

By late 1862 the innovative spirit of that Congress had thus been manifested, while changes in the technology of war suggested a need for expert advice to the government. The time now seemed ripe to realize the Academy dream. When Bache and Commodore Charles Davis urged action, however, Joseph Henry opposed it. The cry of aristocracy would doom the proposal in the House, Henry thought; and if not, the Academy would be attacked by those excluded, would probably get no funds from Congress, and might be politicized. Then Agassiz heard about the discussion, probably through Peirce. Hankering after Morrill Act money for the Lawrence Scientific School and hobnobbing with the Massachusetts congressional delegation, Agassiz was alive to the new legislative opportunities. By February 1863 he had broached the scheme to Senator Henry Wilson of Massachusetts, who thought he could get such a bill through Congress.

Bache happened to have arranged a Lazzaroni dinner, the first since 1858, for February 21, 1863. But Bache and Davis decided to go ahead on the Academy scheme without the reluctant Henry. So they advanced the meeting to February 19 without informing him. That evening, Bache, Davis (now a rear admiral), Agassiz, Peirce, and Benjamin Gould met with Senator Wilson at Bache's house. Bache and Davis presented drafts, and the final bill was worked out. By its terms, the National Academy of Sciences was to be incorporated, empowered to organize and govern itself, and required to report on scientific questions as requested by any government department, without compensation except for expenses incurred thereby. Wilson introduced the bill on February 21.

Henry found out about all this some days after the event, but he expected the bill to fail and so made no public fuss. He underestimated Wilson's parliamentary skill. Late on the harried last day of the session, March 3, 1863, Wilson suavely asked the Senate's leave "to take up a bill, which, I think, will consume no time, and to which I hope there will be no opposition. . . . It will take but a moment, I think." Requiring no appropriation, it passed both houses by voice vote without comment, and President Lincoln signed it by midnight, doubtless also without comment.

When comment finally came from an astounded scientific community, it ranged from amusement through scorn to outrage, for the act limited the Academy to fifty members and appointed all fifty by name for life. It has been suggested that naming all fifty at once made the bill easier to pass without debate. But Davis's draft, like Agassiz's plan of 1858, had provided for a dozen or so named incorporators empowered to elect the rest. It is not clear why such a provision would have been considered more likely to incite debate. Within the week, Davis expressed keen regret that his plan had not been followed, while Bache defended the Wilson version. Neither man alluded to parliamentary expediency as a consideration. Another hypothesis rings truer, to wit, that the Lazzaroni simply felt they knew best who should be in.

Henry, however, denied the clique's "moral right to choose . . . the members," and even Charles Davis agreed. Furthermore, the clique had, as both Henry and James Dana remarked, put many on the list who did not belong there and left out others who did. George Bond, Spencer Baird, Elias Loomis, and James H. Coffin were all excluded, as was John W. Draper, the noted chemist, physicist, physiologist, and historian. Yet John F. Frazer, the Lazzaroni's undistinguished man in Philadelphia, was on the list, and so were several small fry from Bache's Coast Survey and Davis's Nautical Almanac. Agassiz's hostility toward Baird (who had resisted sending him more specimens from the Smithsonian), and Peirce's fierce hatred of Bond, made those exclusions all too obviously malicious. Bache himself privately explained that Bond and Baird were simply "too mean to bring into our [!] Academy." And aside from personalities, the distribution of fields was skewed. "No end of mathematicians & as(s)tronomers with small show of chemists geologists &c!" complained Josiah Whitney (who was on the list). Thirty-two were in the physical sciences (mainly geophysics) and only eighteen in natural history. This reflected the composition of the Academy's founding fathers —three astronomers, one zoologist, and a geophysicist chief—but not that of the scientific community.

Besides Henry, Dana, and Whitney, others named to the Academy condemned its origins. Among them were William B. Rogers, Asa Gray, John Torrey, George Engelmann, Jeffries Wyman, Joseph Leidy, and both the Sillimans—a formidable band of malcontents. If Henry had declined membership, most of them might well have followed suit, and the National Academy would likely have gone the way of the National Institute, Congressional charter and all. But Henry accepted the Academy as an accomplished fact, privately urging members to give it "a proper direction" and "remedy as far as possible the evils which may have been done." Sooner or later the dissidents accepted their appointments, though some held aloof from Academy proceedings for a time.

So when the organizing meeting convened in New York City on April 22, 1863, thirty-two incorporators showed up, a respectable turnout in the circumstances and better than Bache himself had apparently hoped for.

Frazer had drawn up the proposed constitution, with modifications by Bache. It specified life terms for the officers. But democracy's watchdog, William B. Rogers, had decided to attend, after some soul-searching, and he rose at the last minute with a solemn warning of disaster that startled the membership into cutting the terms to six years. Even so, as one member remarked, "the President is made absolute." Bache, of course, was elected president, Dana vice president, Agassiz "Foreign Secretary," and Gibbs "Home Secretary." Henry declined to accept any office. In the meeting, Rogers roundly denounced the injustice of the membership list. The press was still more censorious. Even Dana's *American Journal of Science* commented pointedly that "so far as any honor may attach to membership, it will be shared much more largely by those" elected to vacancies by the full body.

Born under that cloud, the infant Academy seemed unwanted by its own progenitor. No one in the government but Bache, Davis, and Henry sought its advice. Those three induced or trumped up the half-dozen requests that came in 1863, all of them inconsequential; and only four more, equally trivial, came during the remainder of the war. In 1863 both the Treasury and Navy Departments resisted paying even the expenses of the Academy committees. At the start of 1864 Henry still nursed a hope that the government would find the Academy worth funding or endowing. But Congress did not see fit to do so. In 1865 Senator Wilson tried to sneak a small appropriation through Congress on the last day of the session, but this time the tactic failed. The Academy would have to wait until 1941 for the first Congressional appropriation even to defray the expense of its services to the government.

To encourage scientists, therefore, the National Academy would have to depend on the mere honor of election. This meant that the meeting of August 1864 in New Haven would be a crucial one, for there were three vacancies to fill. Would election by the whole Academy now indeed confer more honor? Would it be untainted by favoritism or prejudice? Or would the heavy hand of the Lazzaroni show itself once again, perhaps fatally?

Draper was apparently not seriously considered. The implacable Peirce may have somehow blocked nomination of Bond, now mortally ill with tuberculosis (he died six months later). But the nomination of Spencer Baird threw down the gage of battle. Agassiz, heady with triumph from a grand Midwestern lecture tour, insisted that no merely descriptive naturalist should be elected. Most of the other naturalists would have been excluded by Agassiz's new test, and they took it personally. Henry thought them ready to resign in a body over the issue. Dana, presiding in Bache's absence, was all for Baird. Henry, who had reason to resent Baird and often took him to task privately, nevertheless recognized his zeal, industry, and knowledge, and therefore supported his nomination. The Lazzaroni had evidently disintegrated. Asa Gray, who had "kept wholly aloof before," came to this meeting in behalf of justice. Agassiz stood alone in his own natural history section, and his tortuous parliamentary dodges availed him nothing. Baird was overwhelmingly elected. Blaming Gray, Agassiz quarrelled noisily with him on the train back to Boston, threatening bodily harm if Gray ever dared speak to him again, and five months later he was still characterizing Gray as a "nasty skunk."

Frederick Barnard afterward deplored the strong language and weak science of the New Haven meeting. If discord persisted and apathy increased, he expected "a great defection from the ranks—in other words, that the Academy is going to blow up." But it was the Lazzaroni that had blown up. The Washington meeting of January 1865 was not, as Barnard had feared it would be, "the last meeting of the National Academy of Sciences." Attendance was poor—only twelve at the opening session—but harmony prevailed. Despite its failure to get money from Congress, the Academy lived on.

Respect for the Academy had somehow survived the shame of its founding and the bickering of its early years. When Ogden Rood of Columbia was elected that January, even his hostile colleague Charles Joy called it "a great honor to the College." Elected at the preceding meeting, Leo Lesquereux in a private letter had made it still plainer why the Academy would live: "As I have given to science most of my life it is now pleasant indeed . . . to see my meagre attempts acknowledged. . . . And the position gives me some new impulse to scientific work and also some more of working power because now I am sure that I may find help and sympathy."

The end of the Civil War thus coincided with the dawn of a new era in the power structure of American science. The failure of the Academy to get federal funds left it significant as a touchstone of recognition, but not as a center of power and patronage. Nor would the postwar revival of the AAAS restore the centrality of that institution. The downfall of the Lazzaroni ended the domination of the national scientific community by a single clique. The Civil War had exalted the nation over the states in politics, but at its end the scientific community was moving from central authority toward a federalism of specialized fields. American scientific achievements in the century since, compared with those of more centralized science in other nations, suggest that the movement was not necessarily for the worse.

Chapter 22

"Small Potatoes": Science and Technology in Arms

In 1863 Joseph Henry argued the importance of science to winning the war. "The art of destroying life, as well as that of preserving it, calls for the application of scientific principles, and the institution of scientific experiments on a scale of magnitude which would never be attempted in time of peace," he wrote in his annual Smithsonian report that year. "New investigations as to the strength of materials, the laws of projectiles, the resistance of fluids, the applications of electricity, light, heat, and chemical action, as well as of aerostation [ballooning], are all required." Little of that was done by anyone during the Civil War, however, let alone by the Smithsonian.

The American Civil War, to be sure, has been called the first great modern war largely because of its technology. Railroads, telegraphy, breechloading and repeating rifles, ironclad warships, machine guns, rifled cannon, land and sea mines, and wire entanglements were all used effectively for the first time in a major conflict. But all those innovations had been developed before the war. Some types of Civil War weapons that we think of as modern—aerial reconnaissance, submarine vessels, incendiary weapons, and war rockets—had actually been used in combat generations earlier.

Even on the level of mechanical invention, neither Southern desperation nor Yankee ingenuity gave birth to any fundamental *wartime* breakthroughs. The Confederacy had both opportunity and need to think anew, but lacked a solid foundation of science, technology, and industry. The Union had the foundation, but its adversary's limitations blunted the spur of challenge. From first to last, each side expected the war to end soon and therefore saw no point in long-range government research and development. The same reasoning, reinforced by the certainty of immediate profits from government orders, discouraged research and even innovation by private industry.

The Union's choleric chief of army ordnance, Brigadier General James W. Ripley, fiercely—and with some justification—resisted all innovations, for the sake of speedy production and efficient supply. Anyway, trained ordnance officers were desperately few and overworked, many of their ablest brother officers having contrived to get field commands. The Union army's foremost weapons developer, Major Thomas J. Rodman, had his hands full as commander of the Watertown (Mass.) Arsenal. A Congressional committee asked him in 1864 if ordnance officers had worked at improving weapons since the war began. "No," he replied, "not to any great or practical extent. Their duties have been so much increased." Secretary Gideon Welles said much the same of the Union navy.

In the Confederate navy, Matthew Maury developed contact-detonated underwater mines, and John M. Brooke (one of Maury's Naval Observatory protégés in the early 1850s) not only designed projectiles, fuses, and a rifled cannon but also became chief of the navy's ordnance bureau. The Confederacy's chief of army ordnance, Pennsylvania-born Brigadier General Josiah Gorgas, worked wonders in mobilizing arms production. But he instituted no coordinated, ongoing research program, though his subordinates, notably George W. Rains and John W. Mallet (a Göttingen Ph.D. in chemistry), conducted occasional experiments on their own—more, in fact, than their Union counterparts.

In the North, General Ripley's blunt obduracy made his name a byword among the numerous civilian inventors and incurred the repeated anathemas of the *Scientific American,* which periodically agitated for an independent research-and-development board unconstrained by responsibility for production. But the closest effective approach to such an agency was in the person of President Abraham Lincoln, who had a longstanding interest in science and technology. Lincoln screened many weapons proposals by civilians and arranged tests of those he deemed promising, often attending himself. He gave the first known order for machine guns by any government, the only Union order for breechloading cannon, and the war's largest order for breechloading rifles, and he promoted a variety of other military gadgets, including incendiary weapons. But until late 1863 he could find no officer both competent and willing to replace Ripley, who had proved tireless and resourceful in frustrating Lincoln's development efforts. By then Lincoln himself apparently believed the war's end too close for new weapons projects.

In February 1863 the Union's Navy Department set up a "Permanent Scientific Commission" to evaluate inventors' proposals. It had been conceived by the Washington Lazzaroni, Bache, Henry, and Davis, who with two of their close associates composed its membership. The navy gave it neither funds nor pay, so the commission confined itself merely to passing judgment on what came before it. By the end of the war it had met more than a hundred times and reported on 257 proposals, but nothing significant emerged. What the navy wanted anyway was simply relief from the importunities of inventors, and that was about all it got. What Bache, Henry, and Davis had really wanted was to make a place for science in the war effort. But the Permanent Commission's work did not bring science to bear in any significant way. Nor did any other government agency, despite Henry's hopeful urgings as secretary of the Smithsonian.

Astronomy, higher mathematics, and the earth and life sciences could offer no help to the art of war in that day, though Thomas S. Ridgway tantalizingly described himself in 1863 as "late geologist to Maj. Gen. John C. Frémont in his military campaign in Western Virginia." Nothing came of Henry's suggestion that physicists work on ballistics, strength of materials, and fluid mechanics. More promising, however, was Henry's recommendation of "chemical action" as a field of war research.

So, at any rate, some chemists believed. Charles Wetherill speculated in 1861 that by taking up "the chemistry of warfare I might possibly be of some little use to our country in its present trial." Benjamin Silliman, Jr., characteristically asked Frederick Genth that spring, "Can't you make use of some of your abundant technical knowledge now for Government & so get some of the flood of money which this war pours out?" Eben Horsford considered seeking some government position that could use his chemical knowledge, then decided that he "would not have such a place with its annoyances on any account," so he tinkered on his own with a cannon primer, a submarine, and a "marching ration" of compressed meat and grain, none of them successful.

In 1861 Professor Robert O. Doremus of the New York Free Academy began experimenting with gunpowder, using government equipment at West Point and the Washington Navy Yard. In 1862, however, he accepted Napoleon III's invitation to Paris, reportedly selling his cannon cartridge to the French for twenty thousand dollars by year's end and remaining for another year to perfect a compressed granulated gunpowder. In 1863 Henry Wurtz, a well-known chemist and geologist, developed a gunpowder additive intended to generate greater pressure as the projectile moved along the bore. Lincoln took an interest in it and ordered a trial in December. But the gun burst, Lincoln gave up, and Wurtz turned his talents to more peaceable uses.

Incendiary compounds, rocket propellants, and poison gases might have been developed further by trained chemists. A British chemist came close to smokeless powder in 1865. But the incendiary mixtures used by both sides seem to have been concocted through trial and error by unschooled tinkerers, and such weapons achieved little except to light the fires of righteous indignation and set off charges of barbarism. The rockets that were used drew on no chemical expertise and had little military effect—"small potatoes and few in a hill," remarked Lincoln of one rocket test. In 1862 Robert Doremus designed apparatus for generating chlorine gas to fumigate a cholera-tainted steamer in New York Harbor. This may have inspired a letter to Lincoln that year from a New York schoolteacher named John W. Doughty, who suggested that heavy shells be filled with liquid chlorine. The chlorine, he explained, would upon release expand into a choking gas of many times its original volume, which being denser than air would sink irresistibly into trenches and bombproofs. But there is no evidence that Lincoln saw the letter. Someone, presumably a White House secretary, routed it to General Ripley, who ignored it. On writing Ripley directly, Doughty was told that the Ordnance Department was too busy to test the idea. Two years later Doughty wrote again to no avail. By then Fort Sumter had been under siege for months, twelve thousand men had been killed or wounded in a few hours before the Confederate entrenchments at Cold Harbor, and protracted trench warfare had set in around Petersburg. Half a century later the Germans, using a cruder technique than Doughty's, produced an astonishing collapse in the Allied lines at Ypres.

Chemists might have been enlisted on still another technological front. For years before and during the war, the War Department had been recommending the establishment of a national foundry, which could among other things "serve as a great laboratory." By the time of the Civil War, moreover, the Bessemer process had opened the age of steel. As its chief American promoter, Alexander Holley, remarked in 1864, it was "a chemical process . . . conducted on chemical principles," and Holley chided the United States government for not having encouraged efforts to improve it. Steel guns had already shown promise. The German firm of Krupp, which had exhibited one at the Crystal Palace in 1851, had since then sold a large number to a dozen governments. While not much of the steel used thus far had been Bessemer, Holley pointed out that encouraging trials of Bessemer-steel guns were well along.

The Army Ordnance Department felt no yearning for steel guns, however, Bessemer or otherwise. Krupp's New York City agent Thomas Prosser offered some to General Ripley in August 1861, but Ripley replied that enough field artillery was already on order. The governor of Connecticut sent a Krupp steel gun to the War Department in the following spring, but its fate is obscure. In July 1864 Ripley's successor, Brigadier General George Ramsay, discouraged a Pittsburgh firm's proposal to make cast-steel guns, on the grounds that bronze and wrought-iron fieldpieces were just as good and cheaper.

The navy proved to be more open-minded. Prosser sent a tracing of Krupp's largest steel cannon to Captain Dahlgren, Chief of Navy Ordnance, in January 1863, though confessing that the Krupp works were too busy to start on an order for at least three months. Lincoln, who had passed along a proposal for steel cannon to the indifferent General Ripley more than a year earlier, may have been apprised of Prosser's letter. At any rate, Lincoln's confidant, Assistant Secretary of the Navy Gustavus Fox, soon afterward called John Ericsson's attention to the pressing need for a Bessemer plant in the United States. Ericsson thereupon recommended Alexander Holley to a Troy, New York, firm, which in turn dispatched Holley to England to acquire American rights to the Bessemer process. It was this mission that made Holley the major prophet of Bessemer in America, and it is remembered as a first giant step in the rise of the American steel industry. By war's end, Bessemer steel was being made at Troy. But no national foundry came along to heed Holley's call for government encouragement of the process; and even if one had, it could hardly have yielded significant results in time for use in battle.

Chemistry did no more for the Confederate cause than for the Union. The Confederacy's closest approach to enlisting scientists as such came when it set up an independent Nitre and Mining Bureau. This used a few chemists to find and work sources of niter (potassium nitrate) for gunpowder, chiefly artificial beds and long-time accumulations of organic wastes in caves and on farms. It also employed a few geologists to locate and work deposits of iron, copper, lead, and coal. But the bureau's dozen or so scientists used their special knowledge in routine operations, rather than adding to it through research or even applying it to improve technology. Other men with no scientific training did the same work for the bureau just as effectively, whereas two of the South's ablest scientists, Joseph and John LeConte, had little success in trying to produce niter from artificial beds in South Carolina.

A study of nine Alabama chemists in Confederate service shows three in the Nitre and Mining Bureau, one running an army pharmaceutical laboratory, one (John W. Mallet) in the Ordnance Department, and four on service quite unrelated to science. Of the three in niter and mining, the only one with an assignment beyond the merely routine was sent to England in 1864 for information about the manufacture of acid and ammunition. Whatever he may have learned presumably came too late to be useful in the war.

With all that has been said here in disparagement of scientific and technological research as factors in the Civil War, there were glints of things to come. In the summer of 1861, for example, Matthew Maury got the University of Virginia to establish a laboratory for weapons development, staffed by faculty but under government authority. The arrangement was too little productive and too soon forgotten to be called a precedent, but it resonates with our own time.

Likewise not a precedent, yet even more suggestive, was President

Lincoln's Timber Creek project. It developed from uneasiness at the government's dependence on niter from British India as an essential ingredient of gunpowder. Since war with Great Britain could not be ruled out in that period, Joseph Henry in 1857 had urged Secretary of War Jefferson Davis to mount a search for alternative sources or substances. Nothing was done, but the outbreak of civil war and the worsening of relations with Great Britain led Dahlgren to raise the subject with Lincoln in the spring of 1861. That fall the growing fear of a niter shortage roused the Navy Department to send Lammot Du Pont to England, where he bought up all the niter available. After Captain Charles Wilkes took two Confederate diplomats from a British steamer in November, however, the British forbade the shipment of Du Pont's purchases, which fact probably weighed in Lincoln's decision to release the diplomats. With the niter embargo thereupon lifted, General Ripley reported ample supplies on hand in the spring of 1862.

Nevertheless, Union defeats in the summer of 1862 foreboded a long struggle with the Confederacy, and once again war with England loomed. So when Isaac R. Diller, an old Springfield friend, approached Lincoln that August as agent for a German chemist's chlorate-based powder, Lincoln asked the new Agriculture Department chemist Charles Wetherill to make up a small batch for trial. The results were so promising that in mid-December Lincoln formally agreed to buy at least a thousand pounds of the new powder for more extensive trials and to recommend buying the American rights for $150,000 if those trials were successful. Lincoln stood alone in the government as a backer of Diller's project. Secretary Welles thought the deal "well-intentioned but irregular" and regarded Diller with suspicion. Nevertheless, taking an oath of secrecy, Diller, Wetherill, and two helpers set to work in a ramshackle rented building on Timber Creek in New Jersey, about six miles from Philadelphia. The work went hard. Sickness caused delays—Timber Creek was "a most unhealthy location"—and the machinery broke down more or less regularly. But the four workers shared confidence and zeal, and by June 1863 twenty-five hundred pounds were ready to test.

A navy ordnance officer wrote privately that the powder "has proved a decided success. . . . If it preserves its properties after graining it is everything that is to be desired, and will probably supersede common nitre powder." In that reservation came the rub. Though Wetherill worked on until October 1863, he could not come up with a way to grain the powder cheaply for safe storage and transport. By then, prodded by Dahlgren, American powder firms had managed to convert plentiful Chilean sodium nitrate to potassium nitrate. The War Department had eight million pounds of niter stockpiled, enough for two or three years. And the threat of war with England had passed away. So Lincoln let the Timber Creek project drop, to the distress of Diller, who lost heavily, and of Wetherill, who found himself replaced at the Agriculture Department (Henry took him on at the Smithsonian for a couple of years, after which

Wetherill served as professor of chemistry at Lehigh until his death in 1871).

The parallels between Lincoln's Timber Creek project and Franklin Roosevelt's Oak Ridge project are striking. But of course the historical consequences of the two projects are not in any degree comparable. Much the same can be said of science and scientific technology generally as factors in the Civil War. Their historical interest lies in their portent, not their performance.

Though science and scientific technology had little effect on the war, the war (as we have seen) had a substantial effect on them, and it was not much more agreeable than what happened to Diller and Wetherill. In science and technology, as in the nation at large, the postwar period was therefore a time for reconstruction.

Chapter 23

Many Mansions: The House of Postwar Science

Most scientists had assumed in 1863 that the National Academy of Sciences would permanently supplant the seemingly defunct American Association for the Advancement of Science. But by war's end sentiment had changed. The Academy's limit of fifty members, Lazzaroni high-handedness in arbitrarily naming all fifty for life at the start, and the government's spurning of the Academy's advice turned many, even such erstwhile Lazzaroni as Benjamin Gould and James Dana, to thoughts of resurrecting the AAAS after all. James Hall missed, among other things, "the opportunity of rapping the knuckles of egotists & asses without the waste of time to write long articles." So the last AAAS president, Frederick Barnard, and a surviving majority of the old standing committee cheerfully accepted an invitation from Buffalo in 1866 to meet there that August.

Bitterness and poverty kept all the Southern members home. But the wartime hiatus had at least healed the Association's internal wounds. Bache was hors de combat, and his former Lazzaroni henchmen Henry, Dana, Peirce, and Agassiz were kept away by health, personal affairs, or business. So was their old archfoe William B. Rogers. When a local dignitary praised the Association for a past "remarkably free from . . . bickerings and jealousies," and its members for having been "so absorbed in the search for truth as to forget self," some listeners must have smiled inwardly. But he was a better prophet than historian. Though some National Academy members had come mainly to keep the rabble under control, their very presence scotched the notion that the two organizations could not coexist. One Academy man, Frederick Barnard, presided with "dignity, precision, and dispatch," according to William Whitney, and paid tribute at the end to the meeting's "unbroken harmony." Attendance was not large, and as Whitney put it, there was a lot of "trash read & talked." But there were good papers too. The weather was cool and pleasant, Buffalo hospitality was lavish, and bonhomie prevailed. That counted most in the trial reunion, and it worked. Annual meetings followed in unbroken succession, each with its published volume of proceedings and papers. Permanence was ratified in 1874 by incorporation under a Massachusetts charter.

The 1850s proposal to restrict voting and officeholding to full-time professionals resurfaced in 1874 and touched off a last flare-up of anti-elitism, led by William Rogers's brother Robert. But the standing committee put the measure through by granting the privileged status of "fellow" to all who were members that year and requested it. At the next annual meeting the president rejoiced in the time now saved by letting the standing committee run things. When the AAAS held its long-delayed Nashville meeting in 1877, President Simon Newcomb remarked that "this is not an Association over which it is a very difficult matter to preside. We are not torn by conflicting influences nor by internal dissensions of any kind." And the *Nashville American* hailed the Association as "a happy blending of the aristocracy and democracy of research."

The new provision of 1874 did not cut down membership as its opponents had predicted. On the contrary, membership began to rise steadily, topping the prewar record of a thousand in 1875 and reaching the two thousand mark by 1885. There it hovered until the turn of the century, when it rose to new heights. Yet the AAAS did not loom as large in American science as it had before the war. Local and state societies and the National Academy survived, though narrowly, and found their new niches. Specialized societies sprang up and became national. Specialized journals joined the AAAS *Proceedings* and the *American Journal of Science* as outlets. Industry began to foster research, and philanthropic endowment expanded. Universities, technological institutions, and the federal government all enlarged their roles. So, like other American institutions, those of science found both strength and free play in federal pluralism.

The development of New York City as the nation's business center helped some of its specialized societies, like the Torrey Botanical Club, become national in scope and standing. One such was first proposed in 1874 at a national convention of chemists in Pennsylvania to celebrate the centennial of Priestley's discovery of oxygen. The idea was shelved in favor of an AAAS chemical subsection. But early in 1876 New York City chemists, led by Charles F. Chandler of Columbia, called for a citywide society, then, after an unexpectedly strong response, raised their sights and organized the American Chemical Society, electing John W. Draper president.

The AAAS had already moved to accommodate specialization in new subsections, and in 1882 it increased its two full-fledged sections to nine. Its migratory meetings gave its chemical section a more national character than the New York-centered ACS, which lost ground to it. But the AAAS met only once a year. In 1891 the ACS reorganized as a federation of local sections, meeting monthly, and by 1900 it counted seventeen hundred members. About then the AAAS formally recognized "affiliated societies," and by 1910 some thirty were meeting with it. Once again federal pluralism had resolved a dilemma.

Far from claiming the life of the AAAS, the National Academy of Sciences hung for some time on the verge of losing its own. Eminent members resigned or were dropped for nonattendance. The Academy's twice yearly meetings rarely drew twenty, and some, like James Hall, did not come because so few came. With annual dues of five dollars and no government support at all, the Academy could afford to publish only four reports and one volume of memoirs during its first fifteen years. And the federal government remained disinclined to seek the Academy's counsel or even receive it when offered, thus negating the Academy's official reason for being.

Joseph Henry saved the National Academy by reluctantly taking on its presidency after the death of Alexander Dallas Bache in 1867, mainly out of loyalty to Bache's memory. He proposed to resign in 1873 but was prevailed on not to. So he bore the cross to the day of his own death in 1878. Had he not, the Academy would have gone under. The challenge, however, as he wrote in 1868, was not just to keep the Academy alive but also to "render it useful." Asa Gray and James Dana urged that it be made into a local learned society, but that would have betrayed Bache's dream. Henry still hoped that the government would ultimately deign to accept free advice on a regular basis. Perhaps, he suggested, the Academy could also defend standards, champion the cause of basic research, and serve science generally as an advocate before the public. In none of those roles, however, was the Academy to be conspicuously effective during its first half-century.

Josiah Whitney tested the Academy's limits as a tribunal in 1874, when he demanded that it expel Benjamin Silliman, Jr., not for his science so much as for his business ethics. Whitney was sincere in his concern for ethics; and the Academy itself, according to Wolcott Gibbs, twice voted against admitting Charles F. Chandler "on the ground that he uses science only as a means to make money." Silliman had certainly earned a reputation for the rollicking pursuit of profit. But quiet exclusion was one thing, noisy expulsion quite another. Besides, the main reason for Whitney's vendetta was his rage at Silliman for trumpeting the promise of certain oil and gold deposits in California after Whitney had loftily ignored such things in his California state geological survey. "If Silliman's reports are correct," Whitney wrote, "I am an idiot and should be hung when I get back to California." Thenceforward he blamed Silliman for legislative reluctance to appropriate funds for the survey, though it was obvious then and is now that Whitney's own arrogance, sharp tongue, indifference to economic payoffs, and repeated failure to meet deadlines were to blame.

Whitney persuaded scientists at Yale and elsewhere that Silliman had brought shame to science. Cruelly snubbed at Yale, Silliman relinquished or was denied all his courses except in the medical school. Whitney also presented the National Academy's council with a savage indictment of Silliman, charging him with falsifying geological reports for gain, and demanding his expulsion. But Whitney was strangely dilatory in backing up his charges, though Silliman had to endure newspaper slurs for more than a year. So at last the council declined jurisdiction and dismissed the case. Whitney and his brother William quit the Academy in a huff, but Silliman remained a member and lived to see his gold and oil reports spectacularly borne out by events.

Though the Academy council dodged the Silliman issue, it is suggestive that Whitney and Silliman considered Academy membership the supreme trophy of battle. Ineffective as a court of moral judgment, an information center, a funding agency, a representative body, or an adjunct of government, the National Academy nevertheless remained potent as a certifier of scientific status. This was another function Henry had seen for it. And he had also recognized that for that function the number and qualifications of its members were crucial.

The initial coup of the Lazzaroni had frozen membership with some fields scanted and others overrepresented and with some arbitrary selections obviously based on politics and personalities rather than abilities and achievements. Since new choices had to wait on deaths or resignations, the remedy seemed glacially slow. Almost at once demands were made to raise the ceiling on membership. Henry supported the idea in order to "rectify the mistakes of the original selection." But others feared cheapening membership and risking other changes by Congress. In 1869 Alexis Caswell resigned on principle expressly to make room for some younger man, while John W. Draper, stung by his omission from the original list, organized "The American Union Academy of Literature, Science and Art" in New York City and proclaimed its "all-embracing character." Few submitted to its embrace, and it soon perished. But 1869 also brought a petition by some National Academy members to dissolve their own academy on the grounds that "its existence . . . is unnecessary." So at last, with Henry's backing, Congress was persuaded in 1870 to end the limit on membership. This appeased the would-be disbanders.

The National Academy elected thirty-three new members in 1872 and 1873, thereby bringing in younger men and fresher views, redressing old injustices, and making progress toward a fairer balance of fields. At the same time Henry helped preserve the distinction of membership by insisting that original research be a prerequisite, and by supporting a subsequent limit of five new members in any given year. John W. Draper was finally admitted in 1877, along with his son Henry. By then membership had reached a hundred, remaining at or below that level until the early twentieth century.

In proposing a series of science textbooks the publisher Henry Holt wrote in 1874: "There will be no one concerned in it whose standing is not above all question. I shall try to find all the authors in the National Academy." That sort of respect, strengthened by the reforms, insured that the Academy would survive into the next century, when it would at last live up to the vision of Bache and Henry, government counsel and all.

Government turned its back on the National Academy but not on science generally. Before the war, government science had edged warily around constitutional inhibitions; afterwards it grew bolder in a new era of governmental activism. The *New York Times* argued in 1872 that government ought to help finance basic scientific research because it was not self-supporting and because "it is *truth,* to be loved and sought after for its own sake."

States were even more activist than the federal government. In Connecticut, with Western farmers competing for the new markets opened by urbanization, the legislature at last heeded Samuel W. Johnson's longstanding pleas for agricultural experiment stations on the German model. When Johnson's former Yale student Wilbur O. Atwater raised the cry again and the wealthy farm editor Orange Judd put up a thousand dollars, the state joined in funding the nation's first such station, under Atwater at Wesleyan University in 1875. Two years later Johnson got it moved to suburban New Haven as a permanent establishment with full state support and himself as director. "We *hope* to do some good *scientific* work," Atwater wrote in 1875. There and at later stations in other states, pressure for practical services squeezed out most basic research. But in 1887 Congress began annual subsidies to such stations for research, the first significant federal research aid to states and universities, and thereafter station scientists exploited breakthroughs in bacteriology, virology, genetics, and botany, with spectacular results.

On the other hand, though state geological surveys resumed after the war, the war's interruption of education led to a shortage of trained geologists, other publishing outlets lessened the need for state publications, and the federal government took over much of the theoretical research. So state surveys settled down to practical, local work.

In the federal government the first dozen years of postwar science were transitional. The new Department of Agriculture marked time. Its first commissioner, a Pennsylvania dairy farmer whose chief scientific acquirement was his name, Isaac Newton, simply appointed a chemist, an entomologist, and a botanist-nurseryman, who handled whatever problems happened to come their way. He and his successors were political appointees with short terms, and appropriations on the average did not rise for twenty years. Department scientists had no security in office, no voice in policy, and no leeway in research. Not until the eighties did the Department evolve long-term, problem-oriented programs with dedicated, secure workers, legislative and outside support, and regulatory responsibilities. Then it began to show what science could do for farm production, and appropriations grew accordingly.

When Alexander Bache died in 1867, Joseph Henry and others prevailed on Benjamin Peirce to take over Bache's Coast Survey. To the amazement of those who knew him, Peirce turned out to be a highly effective administrator, in part because of his zeal for national greatness through science and in part because of his dramatic personality—a newspaper reporter that year saw him as the popular idea of a genius, with long, iron-gray hair and beard, and in his eyes "a look of fine frenzy such as belongs to the poet." During his seven years as superintendent, Peirce maintained Bache's high standards of precision and finally realized Bache's old dream by commencing triangulation along the thirty-ninth parallel to connect the Atlantic and Pacific surveys, thus giving the Coast Survey a continental mission. In 1878, four years after Peirce went back to full-time teaching at Harvard, the bureau became the "Coast and Geodetic Survey" in recognition of its new scope.

The Reconstruction years were transitional also for science in the armed services. A war-weary nation, caught up in moneymaking and confident of its oceanic moat, trimmed its army and navy to near-token size. Furthermore, the proud new professional scientists considered science too demanding to be a mere sideline, and so they resented uniformed part-timers.

After Matthew Maury's defection in 1861, Captain James Gilliss had begun the Naval Observatory's scientific renascence, and after Gilliss's death in 1866 Rear Admiral Charles H. Davis, a scientist as well as a naval officer, had succeeded him. But when Davis left for sea duty in 1867, a nonscientist, Commodore Benjamin F. Sands, took over. Though Sands tried to give his scientists leeway and credit, discontent simmered. Cleveland Abbe campaigned without success for a civilian superintendent, and after Sands resigned for health reasons Simon Newcomb wrote privately, "No one but ourselves know what a strain upon . . . the observatory professors especially myself the 7 years of Sands' administration were." Years later Abbe still grumbled that army and navy officers in charge of scientists always took more credit for the work than interest in it, knowing that their reassignments were frequent and unpredictable.

Newcomb himself became head of the Nautical Almanac in 1877. His brilliant work on planetary motions during his years with the observatory and the Almanac gave him a worldwide reputation and reflected glory on those agencies as well. His retirement in 1897 spelled the end of the Almanac's great days, however, and left it to routine work with small appropriations.

In 1868 Cleveland Abbe left the Naval Observatory to head the Cincinnati Observatory, where he began a weather forecasting service. This inspired Increase Lapham of Milwaukee, a veteran weather observer for the Smithsonian, to propose a federal storm warning system. And getting wind of this, Colonel Albert J. Myer, chief of the Army Signal Service, persuaded Congress in 1870 to assign his neglected command the duties of a nationwide weather bureau.

Joseph Henry, in line with his seedbed principle, cheerfully ceded his volunteer network to Myer, whereupon most of the volunteers dropped out. Though Abbe had fought against military control of the new service, he joined it as a civilian in 1871, thus beginning a forty-five-year career as a government meteorologist. Myer put practical service before research, forbidding Abbe to publish without permission and seldom giving permission. "I can't help pining for freedom & a telescope," wrote Abbe privately in 1877. In 1880 Myer was succeeded by a more research-oriented chief. But friction developed between civilians and officers, and so Congress in 1891 at last shifted the whole operation, along with Abbe, to the Agriculture Department.

By then the most ambitious of the army and navy scientific operations, the exploring expeditions, had long since faded away. In the years 1869–74 several navy expeditions to Central America once again checked out possible routes for an interoceanic canal, but the eventual choice rested on diplomacy, economics, and technology. Otherwise, navy exploration gave way to routine chart making. The army held on more doggedly. Indian resistance prolonged its Western mission and made data gathering in advance of settlement too risky for civilian scientists alone. For a few years after the war, engineer officers in the field still collected information on geography and mineral resources. By 1876 Congress had sharply cut funds for such ad hoc reconnaissances. Nevertheless, in its last campaign as a standard-bearer of science, the army sponsored a noteworthy tandem of surveys, those of Clarence King and Lieutenant George M. Wheeler.

The civilian Clarence King, an alumnus of the Sheffield School and the Whitney survey, conceived, promoted, planned, and directed his undertaking. Now he lives chiefly in the memoirs of his devoted friend Henry Adams as a paragon of charm, wit, intellect, and daring, or in the phrase of John Hay as "the best and brightest man of his generation." In his own time, at the zenith of his ill-fated trajectory, King was a star shell of science, lighting up the Western landscape. His eloquence and economic arguments charmed the redoubtable Secretary of War Edwin M. Stanton into sponsoring, supplying, and providing escorts for a hundred-mile-wide survey along the fortieth parallel, embracing the construction route of the Pacific railroad, from the Rockies to the Sierras. The Geological Survey of the Fortieth Parallel began work in the summer of 1867. Its dashing civilian director, aged twenty-five, enlisted the ablest of young civilian scientists. He avoided Josiah Whitney's fatal error by starting work in the mining regions and publishing his and James D. Hague's solid report on them in 1870, years before the other survey volumes. Congress was appropriately generous in response.

Bent on matching European standards, King refused to "rush into print" with the other scientific reports. After the field work ended in 1872, six years of painstaking desk work began. All the authors were at the beginning or in the course of notable careers. Sereno Watson's volume on botany was an almost instant classic. From Leipzig, King imported Ferdinand Zirkel, whose study of rock sections brought the new technique of microscopical petrography squarely into American view. The volumes on geography, paleontology, and ornithology ranked high. King's own eight-hundred-page geological history was richly dramatic in subject and style, though (consistent with King's dynamic temperament) it leaned far too strongly toward catastrophism. Finally, in 1880, came Othniel Marsh's volume on extinct birds, which Darwin thought the strongest evidence yet for evolution. The King survey tried to cover too much territory too fast to avoid some weaknesses. But it brought modern professional science to the study of the West, setting a pattern and a standard for subsequent work.

The expedition of the young West Pointer Lieutenant George M. Wheeler was more decidedly an army show. Wheeler's initial proposal emphasized ethnological studies and topographical maps for the army in Apache country, though with such attention to other scientific matters as time might allow. Soon he expanded his vision to include mapping all of the nation west of the hundredth meridian over a fifteen-year period. Wheeler's surveying parties began work in 1872 and grew year by year in size, quality of personnel, and sophistication of techniques. At various times they included young Grove Karl Gilbert, eventually one of the most distinguished American geologists, the eminent paleontologist Edward Cope, and still other civilians who later became leaders in their fields. Bright young West Point engineer officers also vitalized the survey, though they kept getting transferred just as their experience ripened. By 1879 Wheeler's survey had mapped most of the nation's southwest quarter. Seven thick volumes of final reports, 1875–89, included excellent work in geology, paleontology, zoology, botany, archaeology, and ethnology. Yet, though Wheeler strove to accommodate them, the civilian scientists chafed under military restrictions. In the showdown of 1878–79 that feeling worked in favor of the two major civilian surveys led by Ferdinand V. Hayden and John Wesley Powell, respectively, both under the Interior Department.

The short, slight, energetic Hayden had studied under John S. Newberry, been one of James Hall's assistants, and worked in paleontology for seven years in the Bad Lands and Northern Plains with Fielding Meek, the Topographical Engineers, or on his own. As a Civil War army surgeon he dreamed of leading a great postwar government survey in the West. In 1867 he won a federal assignment to survey Nebraska resources and parlayed it into the Geological Survey of the Territories, under the Interior Department. By 1879 it had covered all of Nebraska and large parts of Wyoming, Idaho, and Colorado.

Hayden cultivated the public, especially businessmen, by looking first to economic resources, even including tourism. He and his brilliant photographer, William Henry Jackson, did much to promote the establishment of Yellowstone National Park in 1872. All this helped make his survey the best funded of all four. But he did not shortchange science. He lived up to the Indians' name for him, "man who picks up stones running," by covering as much as possible as fast as possible and reporting it as soon as possible. Though breathlessly discursive and miscellaneous, his own reports teemed with new data, abounded with insights, and raised fruitful new questions. Moreover, he recruited such eminent scientists as Newberry, Meek, Leidy, Lesquereux, and especially Edward Cope, as well as young men like John M. Coulter and C. Hart Merriam who later rose to eminence. And in its voluminous publications, the Hayden survey welcomed papers by still others who had not actually travelled with it. This pleased the scientific community.

John Wesley Powell had been an Illinois schoolteacher with a passion for natural history. The Civil War made him a major and cost him his right arm. In 1867, having left the army, he scraped up funds from a variety of sources for a collecting excursion to the Rockies by a party of students and amateur naturalists. A year later he took a larger party to the Colorado River country and then in 1869 made his famous descent of the turbulent Colorado for almost nine hundred miles, including the Grand Canyon. On this last exploit he took no scientists along but he kept geology in mind. More important, he emerged a popular hero and so was able at last to get a Congressional appropriation for a survey of the region. From year to year during the seventies he managed to get further appropriations. As a scientist Powell was largely self-educated, but a fortunate moment of Congressional befuddlement placed him under the aegis of the Smithsonian, thus raising his stock among the professionals. In 1874 the Interior Department assumed jurisdiction, and in 1876 Powell's enterprise was officially designated the "United States Geological and Geographical Survey of the Rocky Mountain Region," thereby assuring its independence of the Hayden survey.

The Powell survey might seem to have been the runt of the litter. It was the newest, smallest, and by far the most skimpily funded. Its years of romantic adventure, 1869–72, were followed by a half-dozen years of mundane mapping, with few trained scientists along and Powell himself usually in Washington. Its agenda was narrower, primarily topography and geology, with a later turn toward Indian philology, ethnology, and human adaptation to the arid environment. And yet its scientific weight exceeded that of any other survey.

In large part this was due to the brilliant insights of Powell himself, inspired by a region that blazoned its origins with rare clarity and emphasis. Powell did not have to dig for his discoveries. He began to find them as early as his 1869 expedition in the visible forms and patterns of the landscape. His eye was less on stratigraphy and fossils, as with the other surveys, than on physiographic geology, for which the modern term is "geomorphology," study of the shape of the land. The slow rising of whole regions, not merely through horizontal compression and folding but also through vertical uplift of huge blocks, and the removal of great masses by imperceptible degrees over millions of years by the commonplace workings of wind and water—these dominated Powell's thinking and, through him, that of what came to be known as the "American school of geology." Powell grasped the mechanics of erosion and coined terms to express them, terms that embodied new concepts or sharpened earlier ones: "base level of erosion," "recession of cliffs," and "antecedent," "consequent," and "superimposed" valleys. Henry Adams notwithstanding, modern geologists see Powell's uniformitarian ideas as far outshining the catastrophism of the coruscating Clarence King.

Fortune blessed Powell, moreover, in two of his otherwise undistinguished personnel choices. Captain Clarence E. Dutton of the Army Ordnance Department was, like Clarence King, a Yale man of cultivation, charm, and eloquence. He joined the Powell survey in 1875 and remained with it and its successor for fifteen years, making fruitful use of Powell's ideas and on his own hook reviving and establishing the theory of isostasy, which held that as transported sediment weighted down and depressed one area the eroded land rose in compensation.

Grove Karl Gilbert came over from the Wheeler survey in 1874 out of discontent with military constraints. Drawn to science at Rochester, New York, by work in Henry A. Ward's remarkable supply house of natural history specimens for teachers and museums, young Gilbert had entered geology through John S. Newberry's Ohio survey. Gilbert refined and developed ideas that Powell tossed off loosely, but like Dutton he had ideas of his own, which Powell encouraged by giving him a free hand. In Utah's Henry Mountains (named by Powell for Joseph Henry) Gilbert recognized and authoritatively analyzed the blisterlike domes called laccoliths, formed by lava intrusion between strata, and his report on the Henry Mountains also included a classic essay on geomorphology. Gilbert remained with the Powell survey and its permanent successor for forty-four years.

Though the four major surveys emphasized different scientific fields, they tripped over each other territorially. The Grant era scandals sensitized the public to waste, and army-civilian animosities burst into the open in 1873 when the Hayden and Wheeler surveys overlapped in

Colorado. Congress began to ask questions. Disconcerted by the solidarity of civilian scientists behind Hayden, the army shortened its lines by conceding geology while hanging on to topography. But overlapping continued. By 1878 fifty thousand square miles, mostly in Colorado, had been surveyed by both Wheeler and Powell, and one section of Utah had been favored with the attention of all four surveys. And the army's very success in subjugating the Indians had by then undercut its raison d'être in science.

In 1878 Clarence King suggested that the National Academy of Sciences serve as a referee. The Academy's acting president, Othniel Marsh, stacked the committee against the army, and so its report, following Powell's suggestions, called for replacement of all the surveys by a permanent agency, the United States Geological Survey, under the Interior Department. In 1879, despite complex and frantic maneuvering by Hayden, who feared that King would get the directorship, the deed was done. Supported by Powell, King did become director, and when he quit in 1880 to pursue his luckless business ventures, Powell took over. But Hayden swallowed his chagrin and joined Gilbert and Dutton as an underling in the new bureau.

The outcome held more than one meaning for American science. Powell had remarked in 1874 that there "is now left within . . . the United States no great unexplored region, and exploring expeditions are no longer needed for general purposes." In Utah, Powell's own survey had already found and named the last unknown river (the Escalante) and the last unknown mountain range (the Henry Mountains) in the United States, excluding Alaska. The new United States Geological Survey saw its principal mission not in geographical discovery but in the ongoing professional study and economic application of geology. The frontiers of American science were now not so much in exploration and description as in analysis and interpretation, and so were shared with scientists everywhere.

The new agency also exemplified a general change in the terms of government science. Before the Civil War, government science had tended to be ancillary and ad hoc, uncertain in both constitutional footing and bureaucratic tenure. By 1880 Congress had established permanent scientific bureaus, secure in their legitimacy and missions.

Where did the Smithsonian fit in the taxonomy of scientific institutions? The question sometimes arose in Congress but with no clear answer. The Smithsonian was neither fish nor fowl, though it embraced numerous specimens of each. Part private, part governmental, it was an improbable hybrid—or to borrow one of Joseph Henry's favorite words, a chimera.

In origin and law the Smithsonian was a private institution. Yet Congress had established it and now was regularly funding its National

Museum. Its board of regents had a statutory government majority. Its Washington domicile tended to involve its secretary in government boards and advice to government leaders. It planned and equipped government expeditions, took charge of their collections, and even had nominal jurisdiction over the Powell survey for a time.

Having noted a sharp decline in Atlantic Coast commercial fish catches, Spencer Baird added another governmental link in 1871 by getting Congress to create the Commission of Fish and Fisheries with himself as unsalaried head. Its stated purpose was to study the problem and seek remedies. Baird's real interest, however, was in the study of marine ecology, which he knew would be required.

Putting in six hours a day over and above his Smithsonian duties, Baird led the Fish Commission in helping save the fisheries, including those in the Great Lakes and the Pacific, as well as establishing carp culture in thousands of ponds. Dearer still to his heart was the field center he set up at Woods Hole, Massachusetts, unusually rich in marine life. There he welcomed scientific researchers, many of whom came at their own expense. Of course, shoals of specimens went to Baird's National Museum. Baird induced colleges, universities, and private individuals—notably Louis Agassiz's son Alexander, a copper-mining tycoon as well as a scientist—to buy land at Woods Hole and support research facilities there. In 1888 the Woods Hole station became the private Marine Biological Laboratory, now flourishing as the Woods Hole Oceanographic Institution. The commission itself evolved into the present Fish and Wildlife Service of the Interior Department.

The bill establishing the Fish Commission had required the commissioner to be chosen "from among the civil officers or employees of the Government," and yet no one had challenged Baird's eligibility. That fact may have given pause to Joseph Henry, who steadfastly denied that the Smithsonian was or ought to be a government agency. He considered dependence on government to be incompatible with the Institution's first duty, that of giving scientists support without strings. Nothing troubled Henry so much as the draining of the Smithsonian's endowment (about seven hundred thousand dollars in the mid-seventies) or its revenues (about fifty thousand dollars a year) for purposes other than support of research. After the war he managed to unload the library on Congress, the herbarium and insect collections on the Agriculture Department, the art collection on the Corcoran Gallery, the weather bureau on the Signal Service, and the human skulls and bones on the Army Medical Museum. But these divestitures fell far short of making up for the uncontrollable growth of Spencer Baird's National Museum.

Baird could not be checked by withholding money for acquisitions. Free of charge, they tumbled in like a confluence of avalanches, from private donations, exchanges of duplicates, Smithsonian explorations, and especially (by Congressional decree) from the federal expeditions and surveys. At Baird's urging, seventy-eight freight cars full of Centennial Exhibition specimens and artifacts, foreign and domestic, many of them samples of industrial products, arrived in 1876 for permanent deposit. The Institution's four-hundred-foot-long basement was already jammed with "very many skeletons, beautifully prepared by Professor Ward, some 40,000 jars of alcoholic specimens, together with several hundred large copper tanks, likewise filled; over 50,000 skins of birds . . . several thousand skins of mammals; large numbers of fossil vertebrates, minerals, ethnological objects, &c." So the Centennial knickknacks, still crated, went to a four-story armory building for storage, filling its rooms from wall to wall and floor to ceiling. In 1877, without benefit of Centennial accessions, 11,398 more specimens flooded in.

Baird pointed out that federal money covered the cost of the museum's operations. But Henry brooded over the endowment tied up unproductively in the museum building, the cost of its upkeep, and its demands on the Institution's staff. He kept begging the government to buy it or take it as a gift, and also to detach the National Museum from the Institution. The museum's function was to collect and display objects, he said, whereas the Smithsonian's solemn obligation was to make discoveries and diffuse them. Besides, the museum forced the Institution to crawl to Congress for support. The Smithsonian should stand alone and free, disbursing its support for research from modest quarters.

In the face of Henry's public reproaches, Baird remained calm—and collected. In an unsent letter of 1868 he declared that Henry's "is the governing and controlling brain, while I am only assisting to carry out his plans." This makes one think of Uriah Heep, but that would be unjust to Baird. Despite their policy differences, a strong personal bond had formed by then between Baird and Henry. Baird seems to have been deeply grieved when Henry died of nephritis at eighty in May 1878, and genuinely embarrassed when chosen by the regents as secretary the day after Henry's funeral, though Henry himself had recommended the choice in advance. In his first report as secretary, Baird pledged to carry out Henry's principles: the seedbed principle; the confining of work to science; a balanced budget; and the extension of benefits "to the whole world." This appeared to leave room for the National Museum, private contributions to which, Baird noted happily, continued "to increase in value and magnitude year by year." In 1879 Congress appropriated $250,000 for a new museum building on the Smithsonian grounds. Baird was delighted.

By then the policies of the Smithsonian were becoming less crucial to American science. Another estate of science, the academic, was on its way to overshadowing the Smithsonian in support of research.

Chapter 24

The New Education

Postwar observers of higher education in America perceived what Ralph Waldo Emerson called in 1867 "a cleavage . . . in the hitherto firm granite of the past." They came to know it as "the new education." In cleaving the monolith of the old education, science was the wedge and technology the maul. Science-trained college presidents took the lead in the movement, while science widened and deepened its penetration of the curriculum. Technology, as the supposed offspring and actual partner of science, gave that movement the weight and force of economic incentive.

Many small colleges lacked the wherewithal for the new education, even when they had the will. Others resisted it on grounds of principle. The proud and long-established small colleges of New England in particular stood up for the traditional curriculum in the name of liberal culture and thereby heartened their daughter institutions further west, even to the Pacific. When the bright young chemistry professor Ira Remsen in 1872 asked Williams College for a small room to do research in, the president admonished him, "You will please keep in mind that this is a college and not a technical school. . . . The object aimed at is culture, not practical knowledge."

But the tide ran strong against them. Like Francis Wayland of Brown in 1850, Frederick Barnard of Columbia in 1870 blamed a decline in college enrollment per capita on college slowness to respond to science and technology. Wayland's diagnosis had proved to be premature. Postwar academic leaders sensed that Barnard's was not. Williams College gave Ira Remsen his little research lab before the year was out. At Princeton the conservative President James McCosh likewise came to terms with science, organizing a school of science in 1873, adding scientific equipment, and instituting graduate work. As a professor of philosophy

McCosh had preached Scottish realism, and in academic policy he practiced it.

One of the ways in which science expanded its academic presence was through the system of elective courses. Harvard had tried it in the early 1840s and then drawn back for lack of resources. But by the late 1860s Harvard had money enough to revive it for half of the upperclassmen's courses. Cool toward it previously, the new president, Charles W. Eliot, embraced it in his inaugural address of 1869 and thereafter preached it so ardently that the public came to believe he had invented it. Electives freed the Harvard seniors first, then the juniors and sophomores, and by 1884 reached down to much of the freshman year. From Harvard the system spread throughout higher education. Conservatism limited it in the small New England colleges, poverty in the Southern state universities, but by the 1880s nearly all except the neediest colleges had adopted it to some extent.

The elective system decreased the proportion of students taking science courses, but those who did take them by choice (or at least by preference) were more apt to bring out the best in their teachers. They could also take more science and on a higher level. Specializing furthermore quickened the research spirit in professors. In the mid-1880s the historian Francis Parkman privately sounded out Harvard faculty opinion on the elective system. The scientists generally thought it had improved the quality of their students. Wolcott Gibbs, noted Parkman, "wants to abolish the old 'American college' and turn Harvard into a university [with] untrammeled election." As Gibbs implied, undergraduate electives paved the way for true universities. Frederick Barnard observed in 1879 that graduate courses were "a natural and necessary consequence" of the system.

Some who favored more science feared nevertheless that without some restraints students would choose foolishly, or dodge hard courses, or wall up their minds with overspecialization. For such doubters Yale's Sheffield School offered the "group system," under which students could choose a field such as chemistry or civil engineering, but had to take a "group" of courses tailored to it. Cornell and Johns Hopkins followed suit. Somewhat akin were "scientific departments," granting the B.S. degree for three years of study. More than twenty-five colleges created such departments during the sixties; and Rutgers, which had set up one in 1863 to edge out Princeton for New Jersey's land grant, lengthened its curriculum to four years in 1871.

Still more ambitious, at least in name, were the "scientific schools" affiliated with but not part of regular colleges. In 1873 the AAAS president commented sarcastically that "nowadays every college must have a scientific school attached, else it is not thought complete," though the supply of competent professors was far from adequate. According to the federal commissioner of education, the number of such schools increased from seventeen in 1870 to seventy in 1873 and the number of their instructors from 144 to 749. If he was right, the bottom of the barrel must indeed have been scraped hard for those instructors. Some leaders urged birth control for such schools, but the movement crested in 1873 of its own accord. Some of the weak schools expired during the ensuing depression, others in the relatively prosperous eighties.

As before the war, the scientific schools leaned heavily toward engineering and applied science, especially chemistry. Perhaps for that reason, perhaps also because of their detachment, inchoateness, shorter curriculum, and more transient students, they were still regarded by the colleges as inferior appendages and their students as of low caste.

At Yale, however, Sheffield students and faculty cheerfully acquiesced in the conservative President Noah Porter's salutary neglect. Indeed they would resist integration with the college until long after Porter had faded into the past he loved. Sheffield was thus free, for example, to organize a multidisciplinary course especially for two students in 1870. By 1871 Yale College proper had given up offering any classes for Sheffield students. That was the college's loss. Though Sheffield had its spells of short rations, its benefactor and namesake Joseph Sheffield enlarged its quarters in 1866 to make what George Brush called "an exceedingly well furnished building with convenient laboratories, lecture rooms, museums, etc." and provided a second building in 1873. Furthermore, segregation from the college did not stunt the scientific students' cultural growth. Their courses in the late 1860s included modern languages, literature, history, and political economy, all within Sheffield. In the early 1870s Professor Thomas Lounsbury of Sheffield revolutionized college teaching of English by turning it away from arid rhetoric and grammar to modern literature. Sheffield's enrollment rose from 120 in 1868 to 247 in 1873.

An alumnus of Harvard's Lawrence Scientific School, the chemist Frank Clarke, conceded in 1876 that Sheffield surpassed his alma mater in making scientists. A former Lawrence faculty member and acting dean, Charles W. Eliot, in two widely noticed *Atlantic Monthly* articles entitled "The New Education," charged in 1869 that Lawrence was a disjointed congeries of professorial fiefdoms. Unlike Sheffield, Eliot wrote, Lawrence required no meaningful entrance examinations, no courses common to all, and indeed no knowledge of any kind first and last other than that from "inconceivably narrow" study mostly under one professor. And it set a minimum residence requirement of only one year for the degree of Bachelor of Science (though most students in fact remained from eighteen to thirty months).

Eliot's flings at the Lawrence School infuriated Wolcott Gibbs, who had beaten out Eliot for the Rumford Professorship in 1863 and since then had been dean of the school and also in charge of chemistry. By this time Gibbs had become, in a professional judgment rendered forty years later, "the most commanding figure in American chemistry." He looked the part. A newspaper account some months earlier described him as "tall, well built, with the heavy neck, the full features, the curly hair which used to make up the type of the New York politician in the days when to be a politician was to be a man of ability." More of an experimentalist than a theorist, Gibbs was nevertheless a fount of ideas, many of them left to others to develop. He was not unmindful of his professional eminence and dignity. His own mother told Joseph Henry in 1866 that Harvard "was rendering him so conceited that she feared he would be spoiled." Gibbs himself wrote that year, "I like my professorship here extremely as no one could have a more independent position."

His swelling pride injured, Gibbs took Eliot's charges personally and published a tart rebuttal. Just as it came out, however, its author wrote privately in anguished apprehension:

> An event has occurred which may change the whole course of my future life. Mr. Charles W. Eliot whom I have just publicly posted as a liar has been nominated for the presidency of Harvard University and will probably be elected! . . . Knowing Eliot as I do I see clearly that he will make my place as uncomfortable for me as possible and I am looking around for another position.

Though Gibbs and Josiah Cooke urged that the Lawrence School drop technology and concentrate on advanced science as part of a true university, Eliot took the opposite course, replacing Gibbs as dean with Henry Eustis, the engineering professor, stressing technology, and lengthening the curriculum to four years. But MIT proved to be unbeatable, and Lawrence's enrollment remained stuck at about thirty through the 1880s. Not until 1906 did it become the Graduate School of Applied Science. Meanwhile Eliot took chemistry away from Gibbs, leaving him to teach physics, not his chief interest. As a grudging nod to Gibbs's standing as a chemist, Eliot let him have a room for his chemical research, provided he paid all other expenses. Gibbs could have retired to independent research on an inheritance in 1871, and he was offered a chair at Johns Hopkins in 1876. Yet he chose to stay at Harvard until his retirement in 1887 at sixty-five. Meanwhile he invented the ring burner, his most important contribution to chemical apparatus, and brilliantly carried through a long, notable series of delicate and immensely difficult researches in analytical chemistry, involving the complex acids of tungsten and molybdenum. By then Johns Hopkins had taught Eliot the advantages of winning prestige through research; and when Gibbs retired, Eliot tendered a private half-apology for his treatment.

Elsewhere philanthropic businessmen gave vital support to academic science and technology. Cornell University, born in 1868, owed its financial health to the telegraph tycoon Ezra Cornell's benefactions and his management of New York's college land grant, and the school dutifully lived up to his much-quoted call for "an institution where any person can find instruction in any study" on equal terms. Ario Pardee's coal-mining money transformed Lafayette from a dying classical college to a flourishing, science-oriented school in which liberal arts and engineering students were integrated on an equal footing. At Lehigh, founded and supported by Asa Packer's railroad fortune, the classicists were actually the underdogs.

Still more supportive of academic science, at least in the long run, was the Morrill Land Grant College Act of 1862. It required that grant recipients teach "such branches of learning as are related to agriculture and the mechanic arts" without excluding "other scientific and classical studies," thereby affirming the academic parity of science and technology with other fields. By the end of the seventies its land scrip had nourished more than forty institutions, including private and state colleges and universities, agricultural colleges, and technological schools. Some of them antedated the act, but others sprang up or were augmented in direct response to it. And its stimulus reached beyond its actual grantees. The existing Agricultural College of Pennsylvania (now Pennsylvania State) got Pennsylvania's grant. Nevertheless the agricultural and technological college organized in 1864 by the University of Pennsylvania in the vain hope of a share was, though dormant for a time, the germ of the Towne Scientific School, opened by the university in 1872 on the strength of a private benefaction.

Agriculture, to be sure, got off to a shaky start for lack of qualified teachers and a well-developed body of knowledge. Florida State, for example, fell back on a professor of "Agriculture, Horticulture, and Greek." Agricultural science was Greek to many farmers anyway, and even the education-minded Granger movement of the late sixties and early seventies badgered land-grant institutions with demands for more down-to-earth training.

Debate persisted as to whether the land-grant schools' emphasis should be on science or vocational training. Donations and appropriations certainly came easier for the latter. The public mind seemed unclear on the difference. Progress toward higher academic levels was often dishearteningly slow. Yet it proceeded, and so, therefore, did academic acceptance of basic science and scientific technology. And by democratizing access to higher education, the public land-grant institutions deepened the pool of talent available to both.

As business boomed, 1865–73, so did the demand for trained engineers. The number of engineering schools rose from seventeen to eighty-five in the 1870s, and they turned out 2,259 engineers as compared with 866 in all the years before. By 1880 the majority of American engineers were college-trained. Prejudice against them persisted among working mechanics and some employers, but it had visibly waned since prewar days. Leading technologists did not share it, even those without college training themselves.

Science was accepted as an essential element in engineering. In 1872 a Pennsylvania steelmaster declared that his industry needed "thoroughly prepared scientific men . . . the physicist, the geologist and mineralogist, the chemist, the engineer," and the eminent mechanical engineer Robert Thurston insisted that "Science and Art must always work hand in hand." A geologist-engineer defined mining engineering in 1874 as "the application of science to the discovery and working of mines, and to the subsequent treatment of ores." Major construction projects were now entrusted only to college-trained engineers drawing on scientific theory and sophisticated experiment. An English engineering educator appraised some American civil engineers he met in 1873 as "men of considerable scientific attainments." Professional pride encouraged this image of the engineer as scientist. Engineers coveted intellectual honor just as scientists claimed credit for economic boons. Each aspired to a share of the other's chief glory.

So the Sheffield School in 1869 officially spelled out its aim of teaching "the principles of science, the laws of its application, the right methods of research, the exact habits of computation, analysis, and observation, and a fundamental reverence for truth." And at the first national conference on engineering education, organized by Alexander Holley and held at the Franklin Institute in 1876, no one questioned the importance of scientific theory, methods, and standards in the engineering curriculum.

Engineering curricula otherwise varied widely in length (one to four years), entrance requirements, and emphasis, both from field to field and from school to school.

The Rensselaer Polytechnic Institute continued to drift without strong leadership or solid endowment. In 1870 it jettisoned all but its civil engineering course. But that four-year course, broadened to include such subjects as metallurgy and thermodynamics, was enough to carry the school through the ensuing depression and toward its twentieth-century vigor.

Until 1863 American mining engineers had all been trained on the job or in Europe, except for the few graduates of a shortlived program at the Polytechnic College of Pennsylvania. Then Thomas Egleston, a young graduate of Yale and the Ecole des Mines in Paris, persuaded the trustees of Columbia to let him open a school of mines there for men who would carry on the nation's mining and metallurgical work "on thoroughly scientific principles." In November 1864 the Columbia School of Mines opened in a former broom factory with Egleston and two other professors. The school did not have to stay there long. The timing of its debut had been perfect. In American mining, the lone prospector had lately run out of easy pickings and given way to an engineered industry, one of deep-shaft, hard-rock mining with expensive equipment and corporate financing. Western mining now looked not only to gold but also to silver, copper, and lead, with their special problems. Trained engineers and managers therefore commanded dazzling salaries. President Barnard of Columbia believed that such allurements had something to do with the instant success of the new mining school.

Despite high admission standards and an unusually demanding program (initially, forty hours a week of lectures, lab, and drawing, with no time for lunch), enrollment kept growing. From 1873 to 1888 it actually exceeded that of Columbia College itself. Buildings, faculty, and courses were added. The first Engineer of Mines degree was awarded in 1867, and in 1868 the curriculum was lengthened from three years to four. Not until the 1890s, however, did the school become a balanced general school of engineering. Of 871 mining degrees granted in the United States up to 1892, 402 came from Columbia and 126 from MIT, with no other school accounting for as many as 60.

The new scale, tempo, and technological sophistication of American industry and transportation required growing numbers of more highly trained mechanical engineers. Engineering scientists in Europe were laying down a theoretical base in thermodynamics, strength of materials, and elastic theory with a consequent need for higher mathematics, a range of expertise that traditional shop culture could not well encompass. Further incentive came from the Morrill Act with its land-grant subsidy explicitly for "mechanic arts." So after the Civil War and especially in the years 1868–72, the first formal mechanical engineering programs in America sprang up.

That of the Sheffield School stressed "theoretical or pure mechanics mathematically treated," including "theory of mechanism" and "application of the principles of dynamics." At the other end of the spectrum was the privately endowed Worcester County (Mass.) Free Institute of Industrial Science (eventually Worcester Polytechnic Institute), which first offered its three-year courses in 1868 for mechanics, farmers, businessmen, and schoolteachers, with ten hours a week of vocational shop practice. But pressure from its faculty and the outside world raised it gradually over the years to a full-fledged four-year engineering school. Somewhere between Sheffield and Worcester stood Illinois Industrial University (later the University of Illinois), a land-grant institution whose mechanical engineering course was one of the first to include a machine shop for engineering education rather than vocational training. Its unusually able first professor, Stillman W. Robinson, gave the course a strong start. Most land-grant mechanical engineering programs, however, suffered from administration neglect, mediocre instructors, and uncoordinated courses until the late 1880s.

The best such program in the country took shape at the independent, privately endowed Stevens Institute of Technology in Hoboken, New

Jersey, which opened in 1871. Under its brilliant first president, Henry Morton, Stevens Institute concentrated on mechanical engineering, attracting a strong, research-minded faculty. One of them, Alfred M. Mayer, became the leading American authority on acoustics. Another, Robert H. Thurston, whom Morton recruited from the Naval Academy faculty, drew up an ambitious, detailed four-year program, including basic sciences as well as an elaborate series of courses in engineering theory and practice. At Stevens, Thurston set up the first permanent mechanical engineering laboratory in the New World, complete with director, professional staff, research program, and facilities praised as "very complete and extensive" by an English engineering educator in 1873. It trained both engineers and teachers of engineering in the research approach and helped raise the prestige of the mechanical engineering profession. The Stevens program as a whole influenced such programs elsewhere, as at Purdue University in 1874.

Other fields of engineering education had yet to emerge, but their time was coming. Charles E. Munroe, a young Harvard assistant in chemistry and future leader in the field, prefigured twentieth-century chemical engineering with his course in "Chemical Technology," 1872–73. During the seventies industrialists saw no need for college graduates in electricity. There were no courses in electrical engineering in 1880. But the eighties inaugurated the age of electric light and power, and attitudes then began to change.

Unlike Stevens, Columbia, and RPI, the Massachusetts Institute of Technology in Boston was an all-round engineering school. In the fall of 1866 it moved from its temporary quarters to a big new building in the Back Bay, embellished with Corinthian columns and a broad flight of stone steps, and staffed with an outstanding faculty. By the time it sent forth its first graduates in 1868, MIT had a national reputation and an enrollment of 172. Failing health led its founder, William Barton Rogers, to resign as president in 1870, but by 1873 enrollment had climbed to 348.

The depression shrank enrollment and forced painful belt-tightening. Staff and salaries were cut, equipment ran short. President John D. Runkle wrote privately in 1878 that he wanted to hold on to his part-time Nautical Almanac job in case the Institute collapsed under him. Worn out by work and worry, he turned back the presidency that year to the reluctant Rogers, who filled in until Francis A. Walker came up from Sheffield in 1881 and took over. By then the economic tide had turned, and thenceforward MIT prospered.

From the start MIT had captured national attention with its effective use of teaching laboratories. The concept had been stressed in Rogers's published plan of 1864. Francis H. Storer and Charles W. Eliot (both of whom had previously assisted Josiah Cooke in his Harvard College teaching lab) planned and conducted MIT chemistry lab instruction, and also produced the world's first inorganic chemistry textbook written with lab work in mind. Rogers's 1864 plan had included a physics teaching lab, for which funds finally became available in 1869. Edward Pickering, a Lawrence graduate in his early twenties, built the physics lab up to an impressive scale by 1877, when he left to become director of the Harvard Observatory. Like chemistry students in Liebig's lab during the 1840s, many of Pickering's boys stayed past closing and asked permission to work in off hours. A separate lab was soon opened for mechanical engineering students, and in 1871 a young chemistry instructor, Robert H. Richards, for the first time anywhere organized mining and metallurgical labs large enough to demonstrate actual operations, yet not too large for the school to finance or the boys to work.

The MIT example, and that of Stevens Institute, spurred lab teaching generally. Physics professors from the West came during summers to study Pickering's methods. By 1880 lab teaching was reported in physics at 35 colleges, universities, and technical schools, and in chemistry at 148.

So it was that the academic junction of science and technology grew more solid and serviceable, whether in college-affiliated schools like those at Yale, Columbia, Cornell, and Illinois, or independent schools like RPI, Stevens, Worcester, and MIT. Graduates went out into the world of technological enterprise with knowledge of science, respect for it, and minds attuned to its standards and ways. One writer in 1882 complained that calling such institutions "schools of science" demeaned "those who pursue science for its own sake." However that may have been, the "pure" scientists, in being thus associated with the technological cornucopia, could more effectively press for support in pursuing basic research and rearing their pure-minded successors. Private giving to higher education averaged six million dollars a year during the 1870s. It was not entirely by coincidence that in those years the day of the true university dawned in America.

For most American scientists, research and graduate study were the essence of the "true university." Frederick Barnard, for example, dated "university instruction" at Columbia from the establishment of a "graduate department" in 1880. But in both research support and graduate offerings, the seventies were years of beginnings only.

Harvard's Eliot and Cornell's Andrew White initially focussed on reshaping undergraduate education. Though Cornell has been called "the first American university" by a leading historian of education, and though it became a model for other institutions, only 10 per cent of its first entering class stayed long enough even to get bachelor's degrees. In 1872 American universities could count less than two hundred graduate students in all fields, and those in science were not always well served.

Barnard claimed in 1879 that "our young men have no longer need to go abroad in order to fit themselves for the scientific professions." But the number of young men who evidently did not share that opinion continued to increase for several years thereafter. In 1867 young Ira Remsen, like his antebellum forerunners, felt that advanced study in chemistry required a trip to Germany. Remsen and thirteen other Americans got chemistry Ph.D.'s at Göttingen in the seventies. Though Wöhler retired in 1880, the numbers of American chemistry students at Göttingen peaked in the eighties and remained high until the eve of World War I. In the last third of the nineteenth century American zoology students likewise flocked to Germany, whose influence dominated their field. In that and other fields, American professors often spent leaves and summer vacations in German study. Rutgers in the seventies gave leaves to a large part of its science faculty expressly to encourage such pilgrimages.

As in former years, American students in Germany during the seventies enjoyed a reputation for hard work and good conduct, and in America a German Ph.D. degree still carried weight in academic hiring. Yet there were signs of change. Ira Remsen himself commented scornfully in 1878 that "very few" German-trained American chemists "accomplish much in the way of original work." As a professor at Johns Hopkins, with its aspiration to supplant German study, Remsen no doubt spoke ex parte. But that very aspiration and its growing credibility were themselves portents.

More and more alternatives to German study appeared in America. In 1871 the University of Pennsylvania followed Yale's lead and became the second American school to grant the Ph.D. A year later, Harvard established its "Graduate Department," which in 1873 produced two Ph.D.'s and an Sc.D. In 1876 James Dana's son Edward, back from chemical study at Heidelberg and Vienna, took his Ph.D. at Yale after all. By then twenty-five American institutions had awarded forty-four Ph.D.'s, though some may have been gratuitously bestowed on faculty members to tone up the catalogue. In that memorable year the Johns Hopkins University made its bow, bent on creating a Ph.D. program to beat the Dutch: no superficial examination and brief dissertation as in Germany, but two years of specialized study beyond the baccalaureate and an "elaborate thesis" requiring most of a year.

To dedicated scientists the primary mission of a true university was not teaching but research. The twenty-two-year-old RPI graduate Henry A. Rowland wrote in 1871: "I have only *commenced* to learn and so I must have time for study and experiment; this I can never have if I am an engineer and so I have chosen teaching." But to academic administrators and the public, the subordination of teaching to research was less acceptable.

President Eliot's inaugural address at Harvard in 1869 bluntly declared that "the prime business of American professors in this generation must be regular and assiduous class teaching." Eliot was not hostile to faculty research. Now and then he praised it. But he took no pains to foster it, and in Wolcott Gibbs's case he disdained it. Among Eliot's hiring criteria during his early years at Harvard, prowess in research ranked low. Even after Johns Hopkins's competition for faculty stars made him change his tune, his concern for faculty research often seemed more expedient than principled. Similarly, Andrew White's inaugural address at Cornell in 1868 rated teaching above research. In its early years Cornell made small provision for support of faculty research. The school's considerable influence as a model for other institutions, especially state universities and land-grant colleges, tended more toward social service than advancement of knowledge for its own sake.

Neither Eliot nor White set much store by the German model. Pressure in that direction came from the professoriate. The frequency of admiring articles on German universities rose in the seventies and peaked in the eighties. As in antebellum times, American scientists viewed the German university through their own prism, ignoring its principles of *Bildung* and *Idealismus* in favor of its specialization and meticulously detailed research, which had in fact become more pronounced in Germany since the fifties.

The chance to realize the American conception of the German system presented itself in 1873, when the Baltimore merchant and railroad investor Johns Hopkins died, leaving about three and a half million dollars to found a university. This, the largest bequest yet made to an American college or university, gave the trustees a free hand in designing the new institution, and they used their freedom with epochal results. Their enlightened vision of a true university, dedicated to research and graduate education, transcended the humdrum recommendations given them by Harvard's Eliot and Michigan's James B. Angell. The third of the academic Magi consulted, Andrew White, encouraged the trustees in that higher course. All three, and even Noah Porter of Yale, independently proposed the same man as president: Daniel C. Gilman, formerly a policy-making professor at Sheffield, currently president of the new University of California. Gilman accepted the Baltimore job early in 1875.

As a graduate student at Yale and Harvard, 1852–53, Gilman had discovered how badly such students needed positive administration encouragement and support. A tour of Europe in 1854–55 formed his ideas of the German university system, ideas subsequently modified and colored by twenty years of American academic life. Gilman's remarkable energy, charm, organizing talent, wide-ranging interests, and grasp of human nature were put to the test in California by sectarians, labor leaders, and Grangers, all convinced that the university's high-flown intellectual aims were subverting religion, democracy, and the utilitarian intent of the Morrill Act. This trial may have pushed him in reaction toward the Johns Hopkins vision of research and graduate training.

Until the mid-seventies Gilman had stressed utility and culture as much as pure research. Even in his inaugural address at Hopkins in 1876,

Gilman spoke of "freedom" to investigate and "obligation" to teach; and in his first annual report he wrote that for a university "teaching is essential, research important." His public pronouncements played on crowd-pleasing themes of character, religious faith, morals, and optimism. In public he justified research mostly as an aid to good teaching. But all this was probably in tactical furtherance of his and the trustees' fundamental objectives.

In hiring for Hopkins, Gilman renounced "sectarian and partisan preferences." His tests were scientific or scholarly achievement first, and teaching ability second. Though the surviving Lazzaroni welcomed the long-awaited materialization of their university dream, none of them, including Wolcott Gibbs, chose to join in it, even at a higher salary. Other established scientists also hung back. On the recommendation of Joseph Henry and Benjamin Peirce, however, Gilman did manage to enlist the eminent though amiably eccentric English mathematician James Joseph Sylvester, who brought to Hopkins not only a towering reputation but also a vivifying enthusiasm. And in science otherwise, Gilman spotted and recruited two young men just beginning brilliant careers, Henry Rowland in physics and Ira Remsen in chemistry. Of the six departments first established, three were scientific: chemistry and physics, mathematics, and "natural science" (geology, mineralogy, botany, and zoology).

Of the fifty-four original graduate students at Hopkins, thirty-six were in the sciences and two in engineering. The quality and success of those who came owed much to Gilman's pioneering graduate fellowship program. Previously a few schools had assisted their own graduates with fellowships, usually for study abroad. The Hopkins fellowships, first announced in 1876, inverted that policy by awarding more fellowships with larger stipends to graduates of other schools for full-time study at Hopkins, emphasizing specialization and requiring periodic evidence of achievement. Nine of the first twenty-one fellows were in science. Many of the fellows eventually joined the faculties of other institutions, spreading the influence not only of the fellowship system but also of Johns Hopkins generally.

The Johns Hopkins influence did not immediately sweep everything before it. Despite Gilman's unsurpassed skill in public relations, his university glorified the elite scholar too much and heeded the lowly undergraduate too little for the taste of the American public. The research ideal made its way slowly. It had little effect on the land-grant colleges until the 1890s, with a few exceptions like Thomas J. Burrill of the University of Illinois, a founder of plant pathology in the early seventies. Henry Rowland complained in 1883 that no American institutions yet supplied research assistants and few even employed teaching assistants.

But the other side of the coin was bright. The American university system was pragmatic and flexible enough to modulate the research-teaching balance as the public mind became more receptive to research. By the 1890s research had widely established itself as a major university function. By 1900 most universities required the Ph.D. of their teachers and made publication of research a weighty factor in their promotion. The role of Johns Hopkins as prototype and propagator of that change helped make 1876, the year of its birth, an epochal date in the history of the scientific enterprise in America. So did certain other events and aspects of the Centennial Year.

Chapter 25
Taking Stock

The decimal system decreed that 1876, as the hundredth year of the Republic, would be a time of taking stock, of looking backward, forward, and inward. In the national self-scrutiny one lens was the great Centennial Exhibition at Philadelphia. There, as in the Crystal Palace exhibitions of the early fifties, works of technology stole the show. In their stronghold, Machinery Hall, they were powered and dominated by George Corliss's linked pair of four-story-tall steam engines, towering over the awe-struck crowd like shining Buddhas endowed with muscle and motion. Those icons of the new age were, as their maker had intended, the centerpiece of the Exhibition. Yet technologically they signified less than certain other exhibits little noticed by the press and public.

In an obscure corner of the French section, a small steam engine drove the dynamo of the Belgian inventor Zénobe Gramme, which in turn lit up a tiny arc in a glass globe and powered an electric motor that operated a little pump. Elsewhere a newer, self-excited dynamo designed by the American Moses Farmer also kept an arc lamp glowing. And yet another, built from magazine descriptions of the "Gramme machine," whirred away under steam power in a machine shop exhibit by mechanical engineering students of Cornell University. But the crowds were neither excited nor enlightened. To them, electricity still meant telegraphs and electroplating, not light and power, and the former were represented in profusion.

Neither did the crowds take note of an electrical device incongruously displayed for a couple of weeks in the Massachusetts education exhibit: the newly patented "speaking telephone" of a Boston University professor of speech, Alexander Graham Bell. On June 25 the young inventor got it tried by a party of scientists and also by Dom Pedro, the

Emperor of Brazil (who did *not* exclaim "My God, it talks!" but simply "I hear, I hear!"). Even so, the only press notice in the next two months was a tepid comment in a Boston paper, perhaps inserted at the instance of the inventor. The Centennial throngs passed the exhibit by. But Sir William Thomson (later Lord Kelvin), who had been among those testing the telephone, grasped its significance, sounded its praises in due course before the British Association for the Advancement of Science, and so helped wake up the world to it.

Sir William was also captivated at the Exhibition by the automatic telegraph of another young inventor, Thomas Edison, who won an award for that and his still more remarkable quadruplex telegraph. But the Centennial visitors who came by train from New York City could have seen an Edison invention far more significant than either, had they glanced in passing at a hill near the hamlet of Menlo Park, New Jersey. They would not have guessed, of course, that the white-clapboarded, two-story structure just erected in an open pasture was no country meetinghouse but an independent, commercial research and development center—or "invention factory," as its inventor called it.

The crowds could scarcely be blamed for missing these portents. The *Journal of the Franklin Institute* observed in June 1876 that anyone with a smattering of science could understand the specimens of technology at the Crystal Palace of 1851, but that such matters had since become too complex for the average fairgoer to fathom. The wonder of it all moved visitors nonetheless, and inspired some. Fifteen-year-old Elmer Sperry, an upstate New Yorker who had helped the Cornell students construct their dynamo, revelled in Machinery Hall and settled on the course that would eventually make him a great figure in American technology. Foreign engineers flocked to the Exhibition and marvelled at American achievements. In Machinery Hall, reported a German, "the diligence, energy and inventive gift of the North Americans celebrates its triumph over all that had ever been achieved by other nations in the invention and construction of machines."

Amid hosannas to technology in the seventies, a few faint catcalls could be detected. In literature Herman Melville's long poem *Clarel* (1876) cast science and technology as menaces to God, nature, and social order. In theology the clergyman (and amateur geologist) George F. Wright asked in 1876 "whether in our religion there is moral power enough left to control and keep in harness the [technological] giant we have awakened." An *Atlantic Monthly* writer declared three years later that "we have already lost power over it." J. Lawrence Smith's presidential address told the AAAS in 1873 that "while mankind has not changed, Galvani's experiment has, and instead of a frog, it is now a world that is convulsed by the electric force then discovered, . . . battles are fought, victories announced, commerce controlled, and, I am sorry to say, tyranny abetted, by that wonderful agent." The Franco-Prussian War of 1870 was popularly supposed to have been detonated by a telegram, the "Ems

Dispatch." German militarism, as demonstrated in that conflict, now cast doubt on the old dream that technology would make war too terrible to wage, a dream explicitly ridiculed in September 1870 by a writer for a new journal, the *Technologist.*

The technologists themselves did not share the public's misgivings. They looked to a glorious future, though in some respects it would be more distant than they thought. Robert Thurston's 1875 vision of a privately financed "mechanical laboratory" enlisting "the chemist and the physicist as well as . . . the mechanic" would not be fully realized in modern industrial research for another generation. Though Edison's Menlo Park establishment foreshadowed such research, it nevertheless hired scientists for the knowledge they had, rather than for what they might discover if let off the leash. For all the distinction of its chemistry program, Johns Hopkins got few calls for its graduates from manufacturers in its early years. Andrew Carnegie, Procter & Gamble, and the Pennsylvania Railroad employed full-time chemists by 1876, but for quality control rather than research. The petroleum industry shunned them until the 1880s. Yet though the liaison with science ripened slowly, the scientizing of technology went on in higher education. And in several fields, professionalization culminated by 1880 in the establishment of national engineering societies.

In 1867 ten civil engineers met in New York City and revived the American Society of Civil Engineers (ASCE), comatose since 1855. This time it acquired permanent quarters, held regular meetings, heard papers, assembled a library, and in 1873 began publishing transactions. When recruiting lagged, it went after a national membership with peripatetic annual meetings and a mail ballot. It formally incorporated in 1877 and ended the century with more than two thousand members.

The ASCE identified itself with the engineering profession. In contrast, the American Institute of Mining Engineers (AIME), organized at Wilkes-Barre in 1871, looked to the mining industry, admitting anyone "practically engaged" in mining or metallurgy, as well as businessmen "associates." Its stated goals were more efficient operations and greater safety, whereas the ASCE aimed at "the advancement of science and practice," the development of "experimental knowledge," and the reporting of professional experience.

The mechanical engineers steered a middle course. The examples of ASCE and AIME and the heightening of professional self-awareness by the Centennial Exhibition and schools like Stevens Institute encouraged them in 1880 to follow the leadership of Alexander Holley and form the American Society of Mechanical Engineers (ASME) in New York City. Regular members had to meet strict professional standards as in the ASCE, but nonprofessional managers and businessmen could be "associates" as in the AIME. Like the other two societies, the ASME held regular peripatetic meetings and published transactions. Also like them, it took firm root (despite the untimely death of Holley in 1882) and still flourishes.

Thus, with its prototypical societies, journals, and colleges, its precursors of industrial research, and its emerging partnership with science, the technological enterprise in Centennial America laid out the pattern of its modern form.

Displaying tangible objects, some in motion, technology easily upstaged science at the Centennial Exhibition. Even the exhibits of American scientific apparatus fell flat, though the Smithsonian's serried ranks of mounted animals, Indian artifacts, minerals and other specimens scored with the public. American science expressed itself more forcefully outside the gates in a Centennial spasm of public self-criticism.

Outgoing presidents of the AAAS had more than once scolded their colleagues. Benjamin Gould in 1869 called American science "far inferior" to European, J. Lawrence Smith in 1873 charged it with haste and impatience, and Joseph Lovering in 1874 read a thirty-six-page disquisition on the century's advances in instrumentation, mathematics, astronomy, and physics without citing a single American contribution. In an 1874 article on "exact" as distinct from natural science in America, Simon Newcomb took up the refrain, and at the AAAS meeting in 1875 the astronomer Hubert A. Newton joined in the chorus. In the Centennial Year, the chemist Frank W. Clarke complained in print that "America, when compared with other first-class nations, occupies a low position," especially in physics and chemistry, though he admitted that zoology and geology were "pretty fair." Henry Adams, as editor of the *North American Review,* commissioned a reprise by Newcomb for a Centennial symposium, asking Newcomb to pull no punches. Newcomb complied.

Historians have suggested that this public self-abasement was disingenuous, meant simply to spur on American scientists and shame the public into generosity. Certainly that motive entered. But Benjamin Silliman, Jr., John L. LeConte, and James D. Dana wrote Newcomb in private agreement with his assessment. Moreover, Europeans also looked down on American science. A Finnish correspondent of Leo Lesquereux, for example, alluded casually to "North America, where it is so few men of science." Dana, Edward Cope, and others all felt and resented European condescension and disparagement. The fact that contemporary evaluations have been called into question, however, suggests that we should look for ourselves at the state of science in Centennial America.

The Civil War had left its mark on the geography of American science. Weak to begin with and totally prostrated by the war, Southern science made little progress toward recovery during Reconstruction. Scientists born in the South Atlantic and Gulf states accounted for 9 per cent of leading scientists active in 1861 but only 2 per cent of those beginning their careers between 1861 and 1876. Some Southern scientists tried to pick up the pieces and renew old ties with Northern correspondents. But the obstacles were daunting. Grinding poverty left the Southerners little time for nonpaying science. Books had perished and outside journals had stopped coming during the war, but replacements were hard to find and money for them harder still.

The dream of building a new South inspired calls for higher education in applied science. The University of Virginia opened schools of applied mathematics, applied chemistry, and agriculture in the late sixties and a school of geology at the end of the seventies. But apart from Vanderbilt University, other Southern colleges lacked Virginia's philanthropic support. Indeed, for want of funds the College of Charleston abolished its chairs of geology and natural history in 1869. Frederick Barnard, who had bitter memories, advised Northern scientists not to teach in the anti-Yankee South. The South's antebellum intellectual leanings persisted. And so the "new education" made little headway in that region.

On the other hand, by expanding the federal government, the Civil War helped vitalize Washington, D.C., as a scientific center. This was a delayed effect. In July 1861 Joseph Henry called Washington the worst place in the country for a scientist because of official drudgery and political constraints. Henry may have been embittered by his own political maladjustment; but after six months in Washington young Simon Newcomb remarked in 1862 that he preferred Cambridge, and six months later he complained that no one in town took any interest in science apart from official duties. Yet by 1868 Henry noted that Washington had more men in scientific employment than any other American city, and in 1872 he agreed with Charles Peirce that "all places in this country, even Cambridge, appear dull and uninteresting in comparison with Washington."

One mark of Washington's scientific invigoration was the superseding of its informal and largely conversational "Scientific Club" in 1871 by the more formal Philosophical Society of Washington. Julius Hilgard claimed to have initiated the move, but Joseph Henry presided over and dominated the new society until his death. It retained some of the social and nonprofessional character of the old club, but it served as a more stable nucleus for the city's scientific community. In 1876 that community had still not outstripped the Boston-Cambridge area, which had forty-five AAAS fellows to Washington's twenty-seven. But it had pulled ahead of New York City's twenty-three, New Haven's seventeen, and Philadelphia's fourteen, the only other cities with more than seven.

As that ranking suggests, the geographical distribution of scientists had not fundamentally shifted since prewar days. Not until the end of the century would state universities help make the Midwest a major scientific region. In the Far West the California Academy of Sciences flourished during the mid-sixties under the leadership of Josiah Whitney, then declined. Cut off from the world during the late war, John and Joseph LeConte had fled black ascendency in South Carolina for the University of California in 1868. Now they bewailed the isolation of West Coast scientists. By the measure of AAAS fellows, the Boston-Washington corridor remained the backbone of American science in 1876, the ranking states being Massachusetts with sixty-four, New York with fifty-six, Connecticut with thirty-four, and Pennsylvania with twenty-five, followed by Ohio's seventeen.

In 1876 perhaps 3,000 Americans used scientific training in their work, but only about 370 of them were distinguished enough to make the *Dictionary of American Biography.* Of those notables, about 20 per cent were in the earth sciences. The proportion of *DAB* scientists entering that field had dropped from 24 per cent in the late forties to 16 per cent in the early seventies, a relative decline that would extend to 11 per cent in the next two decades and still further in the twentieth century. This, no doubt, was the mark of a maturing field with rising competitors.

American geology gained in world stature especially from the great postwar surveys of the West, from which came the outstanding contributions of John Wesley Powell, Grove Karl Gilbert, Clarence E. Dutton, and others, as we have seen. James Dana's notable articles of 1873 on the forming of mountains by horizontal compression due to the contraction of the earth brought a telling critique in 1874 by Dutton, the chief proponent of the isostasy doctrine of vertical uplift. Louis Agassiz's Ice Age theory, long resisted by those who preferred Noah's flood on grounds of theology and tradition, at last won general acceptance in the late sixties. In the seventies Gilbert in Ohio, Thomas C. Chamberlin in Wisconsin, and George F. Wright in New England all clearly identified terminal moraines, and Chamberlin showed that the ice had advanced and retreated more than once. Ironically the limits thus determined for glaciation undid Agassiz's argument that a universal Ice Age had ruled out evolution. Agassiz trumpeted a claim, after a high-spirited expedition in 1865–66, to having found signs of glaciation throughout the Amazon Valley, but he persuaded no scientists. Platform eloquence and popular articles meant nothing to the new breed of American geologists who in the seventies were making their field a quantitative science, concerned with the solid geometry and mechanics of earth structures and not with the hypothetical fiats of divinity.

Paleontology, straddling the earth and life sciences, captivated Centennial America with its colossal dinosaurs. The roundup of those Western giants preoccupied the two greatest American paleontologists of that day, the jut-jawed Philadelphia Quaker Edward D. Cope and the stocky, balding Yankee Othniel C. Marsh, who took each other on in the most savage scientific feud of the postwar era, not over theory so much as personal immortality through priority. Both men were brilliant, arrogant, and independently wealthy by inheritance. Cope was incorrigibly sarcastic. He once crushed an inoffensive colleague writing a monograph on apes by recommending a portrait of the rather simian author as frontispiece; the monograph never appeared. Marsh was secretive, suspicious, and insatiably acquisitive in collecting fossils. A somewhat edgy acquaintanceship turned sour in 1871, when Cope poached on what Marsh considered his own dinosaur-hunting preserves in Wyoming. Each man accused the other of unscrupulous tactics, and both were right. Their escalating war was marked by cloak-and-dagger secrecy and spying, false trails, subverting of each other's hired collectors, and public accusations of lying, theft of specimens, and false dating of papers. The struggle raged for two decades, past the limits of this narrative. Its climax and denouement in the nineties might be called the survival of the richest, namely Marsh, or so it seemed at the time. In longer perspective, Cope is generally rated the greater scientist.

But their fierce rivalry also spurred both men on to labors and achievements as massive as the bones they snarled and scuffled over. Joseph Leidy had given them a solid foundation with his dinosaur studies of the late sixties. Though Cope had been one of Leidy's students, Leidy privately called him "overzealous in science" and not above taking "unfair advantage of his opportunities." Outclassed in both ferocity and finances, the gentle Leidy presently surrendered vertebrate paleontology to his brawling juniors. Cope's long suit was an astounding productivity in publications, 1,395 in his lifetime, many of first importance, especially his monumental *The Vertebrata of the Tertiary Formations of the West,* known to generations as "Cope's Bible." But in his haste Cope made frequent errors of interpretation and nomenclature. Leidy once remarked in discussion that Cope had either reconstructed one of his dinosaurs wrong end to or the beast had a head at each end. In contrast, Marsh's writings were concise and judicious. But he had the perfectionist's imperfection of delay, and so his published output fell far short of Cope's. Marsh's strong point was the scope, completeness, and meticulous ordering of his immense collections. Bones by the boxcar came to his Yale fastness. His techniques of field work and reconstruction were models for his successors. Both men were surely in the mind of Alpheus Packard when in a Centennial survey of American zoology he remarked, "We may congratulate ourselves on the high position of our paleontologists in the scientific world."

By 1876 the larger significance of the battling bonehunters had become clear. In their ravening after specimens they were filling major gaps in the fossil record of species development and thereby making Darwinism more than a mere hypothesis. Though Marsh's celebrated monograph on toothed birds of the Cretaceous period did not appear until 1880, his first discoveries of them in 1872–73 were promptly recognized as steppingstones across the crucial divide between reptiles and birds. And his carefully organized sequence of fossil horses astonished and delighted Thomas Huxley, Darwin's foremost disciple, when Huxley visited America in 1876. Not a link was missing in the chain of descent from the little four-toed Eohippus to the modern horse. Papers and presidential addresses before the AAAS similarly trace the developing attitude of scientists generally toward evolution as evidence for it mounted, not only from paleontology but also from ornithology and botany. The last gasp of resistance came in a section debate in 1873. At Nashville in 1877 Marsh's vice presidential address began: "To doubt evolution to-day is to doubt science, and science is only another name for truth."

Marsh's fossils had given Louis Agassiz pause as early as 1868. Agassiz's objections to Darwinism thereafter made less of metaphysical dogma and more of scientific evidence than before, and he freely expressed his admiration for Darwin as a man, a scientist, and a friend. Yet Agassiz never yielded the main point. To a former student, Burt Wilder, he wrote in 1871, "I have read both volumes of Darwin's Descent of Man, which he sent himself with a few very pleasant words. . . . We are truly friends, much as we differ in views. . . . Time will sweep [away] the delusion . . . I feel daily less and less interest in the matter." A few weeks later Agassiz's Darwinian colleague Jeffries Wyman wrote Wilder that Agassiz could not be badgered into offering any countertheory "except the preposterous one of immediate creation of each species. . . . He was just the man who ought to have taken up the evolution theory and worked it into a good shape. . . . He has lost a golden opportunity."

It turned out to be Edward Cope and a brilliant Agassiz student, Alpheus Hyatt, who by the early seventies independently arrived at and jointly promoted the countertheory that for years appealed to many American scientists—perhaps, for a time at the end of the seventies, even a majority. Indeed, it was often thought of as the American school of evolution. It rejected not evolution in general but the Darwinian mechanism of natural selection from among small, random, unexplained variations. Instead it postulated a vital principle in living creatures that moved them to develop individually in a progressive direction and pass on their acquired characteristics to their offspring. This comforting hypothesis left room for divine purpose, a progressive goal, individual choice, and evolution of species by sudden leaps (rather like Agassiz's "special creations"), thus consisting with the relatively brief age then ascribed to the earth, tens of millions rather than billions of years. Though not drawn directly from the ideas of Jean de Lamarck, the Hyatt-Cope scheme fitted them so well that it was christened neo-Lamarckism. Cope, Hyatt, and Alpheus Packard edited the *American Naturalist* as its principal organ. The vast extension of the earth's estimated age through radiometric dating eventually made ample room for Darwinian evolution after all, and the inheritance of acquired characteristics did not stand up under investigation. But meanwhile the vogue of neo-Lamarckism served as a way station to Darwinism for stubborn theists like James Dana.

Lamarckian or Darwinian, the evolutionary concord affected other aspects of zoology. Now that species were seen as mutable, the zoologists of 1876 tended to demote many species to variations shaped by local environments. Spencer Baird and his protégé Robert Ridgway generalized on geographical distribution and climate. All this, of course, revolutionized nomenclature. The new era in zoology was also marked by specialization. In 1870 a Harvard professor began teaching only entomology, and by 1881 four exclusively entomological journals were current. A scientist commented in 1865 that "the tendency to specialities, Conchology generally being such a wide field, is evidently increasing." Alpheus Packard called for more professorships confined at least to zoology alone.

Recruitment of *DAB* scientists to the life sciences dipped somewhat in the early seventies but afterwards returned to a steady share of about 30 per cent. Agassiz students led among the new generation. Thanks in part to Huxley's influence, laboratory teaching spread, thereby stimulating demand for teachers. Packard in 1876 claimed that in zoology Americans now equalled the Scandinavians and Dutch, were creeping up on the French and English, and might even someday rival the Germans.

Like the life sciences, chemistry dipped in *DAB* recruitment during the early seventies and then recovered to its accustomed 25 per cent. As in other fields, specialization quickened, especially in organic chemistry. Ira Remsen accused American colleges of sadly neglecting that subject, in which only three schools offered lab instruction. Most American-trained chemists, he said, did not know enough of it to appreciate their own ignorance. Yet the rate of American publication in organic chemistry increased more than 500 per cent between 1865 and 1880. Frank Clarke offered a key to the paradox when he pointed out that most such articles were descriptive studies of closely related compounds. The chief aim of most American chemists, he charged, was merely "to discover immense numbers of new compounds, and to theorize upon their constitution." This was true even of Remsen. One compound that came out of Remsen's Johns Hopkins laboratory in 1880, however, was saccharin. (Remsen disputed the credit with his research fellow, Constantine Fahlberg, who nevertheless patented it and got rich.)

A few Americans in the postwar years did solid work in analytical and inorganic chemistry, notably Wolcott Gibbs, Frank Clarke, and Josiah Cooke. Gibbs and Cooke also made good determinations of several atomic weights in the seventies. (A pupil of Cooke's would win a Nobel Prize in 1914 for doing likewise.) Clarke was outstanding in physical chemistry, but the great work of his long life was in calculating, compiling, and publishing authoritative tables of chemical and physical constants. Yet despite his own distinction Clarke himself in 1876 deplored the low standing of American chemistry and physics.

In another Centennial Year assessment, Charles A. Young informed the AAAS that "American Astronomy has not yet passed its infancy." Most of the thirty-odd American observatories were fairly well equipped, but all of them needed more manpower and endowed income. As before the war, many were college observatories used for instruction but not research, other than setting the chapel clock. Only the Naval Observatory and that of Harvard were comparable to Greenwich or Pulkovo. Simon Newcomb having turned down the job, thirty-year-old Edward Pickering was just about to begin leading the Harvard Observatory to the twentieth century. Half a dozen other observatories were moderately respectable and productive. Lewis Boss took over and reactivated the mothballed Dudley Observatory that July and found it in better shape than he had expected; he liked the trustees, thought them committed to the "true scientific prosperity" of the Dudley, and believed that their interest had been "shamefully abused and misrepresented" in the old time of troubles under Benjamin Gould. James Lick's promise of a major observatory gift in California encouraged Young in 1876 to hope that a new day of adequate funding had dawned; but, Lick and promise notwithstanding, the old lament sounded through most of the 1880s.

In America, astronomy was still generally regarded as a tool for practical ends, for determining longitude and latitude, time and tide. In Europe the government supported most astronomical work, but in America it maintained only the Naval Observatory and Nautical Almanac. The latter two agencies produced Simon Newcomb's masterful studies of planetary motion and star positions, but some of the most important American astronomical research was carried out in spectroscopy and photography at the personal observatories of Lewis Rutherfurd and Henry Draper (son of John W. Draper) in New York. The old astronomy had focussed on *where* the stars were, not *what* they were. Draper opened a new window in astrophysics with his 1872 photograph of absorption lines in a stellar spectrum, four years ahead of the English pioneer William Huggins. Meanwhile Charles Young at Dartmouth and Samuel P. Langley at the Allegheny Observatory were studying sunspots and the solar spectrum, and in 1876 Young made the first good measurement of the sun's rate of rotation.

Astronomy's share of *DAB* recruits held remarkably steady at about 10 per cent from 1856 to 1876, then dropped sharply in the 1880s and after. The mathematics contingent, on the other hand, after dropping from about 10 per cent in the late fifties and early sixties to 4 per cent in the late sixties, then shot up to about 15 per cent for the rest of the century. Many American mathematicians had, like Benjamin Peirce, been astronomers also; and as the two fields began to separate, mathematics may have had the stronger pull.

To the AAAS in 1875 the astronomer Hubert A. Newton pleaded the growing importance of mathematics in the physical sciences and its probable future importance in life sciences and social sciences. Such appeals doubtless left American undergraduates cold. College math teaching remained uninspired and unappealing even in colleges embracing the "new education." The "wild, grotesque hilarity" of the old student ritual known as "the Burial of Euclid" seems to have died out at Yale after 1863, only to appear again in a revised but unexpurgated version at Worcester Polytech in the seventies as the cremation of Chauvenet (the latest and best freshman math textbook), complete with funeral pyre and oration, volunteer band, and supper downtown.

Yet historians of mathematics in America have called the changes that came after 1876 "little less than revolutionary." Mathematics asserted its independence from astronomy, and students back from Europe gave it a strong turn toward pure analysis and geometry. Further impetus came from James J. Sylvester and his students at Johns Hopkins. In 1870 Benjamin Peirce produced his most original and important work, *Linear Associative Algebra,* which few Americans read and still fewer understood. Even Europeans were put off by its vague proofs and lack of a clear statement of purpose. Peirce justified it not by any practical or even mathematical applicability but as a way of communing with the mind of God. An eminent English mathematician called it "valuable" but "outside of ordinary mathematics." Now, however, it can be seen as a pioneering work in modern abstract algebra. Also fundamental, and more immediately influential, was the work of the New Yorker George W. Hill beginning in the seventies, work which the great French mathematician Henri Poincaré said contained the germ of all subsequent progress in celestial mechanics.

Though the physicist Joseph Henry was still the dean of American science, American physics in 1876 ranked low against that of other nations. Few colleges adequately prepared their students for graduate study in that field. Teaching labs in physics remained a novelty. At Harvard that year, no physics course drew more than twelve students, though a sophomore chemistry course attracted over a hundred. Fewer than seventy-five Americans called themselves physicists, and only about twenty published regularly. What work they did was mostly in the Baconian tradition, experimental rather than theoretical, with much emphasis on meteorology and geophysics. Weakness in mathematics made that bias difficult to right. American physicists furthermore shared with the British an inclination to explain things on the basis of mechanical models. Some American physicists, in fact, went in for technology as a sideline. Industry, however, had no jobs for them as physicists.

And yet something was stirring. Like mathematics, physics surged upward in its share of *DAB* entrants into science, rising from 4 per cent in the late sixties to 12 per cent in the early seventies and remaining at that level or higher until the eve of World War I. In 1876 Henry Rowland took up his duties at Johns Hopkins, after a year's study with Helmholtz in Berlin, where he had made a discovery of some significance for later electron theory. At Hopkins, Rowland distinguished himself in experimental physics, notably with his diffraction gratings for spectrum analysis and his measurements of physical and electrical quantities. It was also in 1876 that another American physicist, Josiah Willard Gibbs, published the first installment of one of the greatest works of theoretical physics to appear anywhere in the nineteenth century.

Born in 1839 in the house later used as the first home of the Yale Scientific School, Willard Gibbs came of an intellectual and academic family quite unrelated to that of Wolcott Gibbs. Willard's father was a distinguished philologist and Yale professor of sacred literature. The boy's interests, however, ran to technology and mathematics. As a Yale College student, 1854–58, young Gibbs was influenced by Hubert Newton, the mathematical astronomer, and perhaps also by Elias Loomis. But the science courses otherwise tended to be practical and descriptive, and so Gibbs entered the Scientific School in 1858, earning the nation's second Ph.D. in science (or first in engineering) in 1863 with a thesis on gear design. After patenting a railroad-car brake, however, Gibbs spent the years 1866–69 studying mathematics and physics at Paris, Berlin, and Heidelberg. Thanks largely to Hubert Newton, Gibbs in 1871 was appointed professor of mathematical physics in Yale College. For nine years the position carried no salary, but Gibbs, a lifelong bachelor with simple needs who lived with his sister's family, got by somehow. In 1873 he turned down a paying job at Bowdoin College. What mattered to him was his complete freedom at Yale to teach what and when he wanted in three or four lectures a week to one or two students. His interest in engineering problems turned him toward thermodynamics in 1872 but he went far beyond the merely practical. The half-dozen years that followed were the most brilliantly and intensely creative of his illustrious career.

His great work in thermodynamics began with two relatively short papers on plane and solid graphical representations, published in the Connecticut Academy of Arts and Sciences *Transactions* in 1873. They were masterful and illuminating. The great Scottish physicist James Clerk Maxwell was entranced by them, and his enthusiastic public notice initiated Gibbs's rise to fame in international scientific circles (though not among the public at large). Gibbs's truly epochal paper, "On the Equilibrium of Heterogeneous Substances," appeared in the *Transactions* in two parts, one of 141 pages in 1876 and another of 182 pages in 1878. No one on the Academy's publication committee could understand them, but the committee accepted them anyway on faith in Gibbs. Maxwell immediately grasped their importance and spread the word. Gibbs himself sent copies to nearly all of the world's mathematicians, astronomers, physicists, and chemists likely to appreciate them, 507 in all, including some in India, Brazil, China, and Japan. The Connecticut Academy, moreover, regularly sent its transactions to 30 American and 140 foreign learned societies. So the papers were not physically inaccessible. Their intellectual accessibility was more limited, but not impossibly so. "I had been dimly groping," an English chemist wrote Gibbs, "but you have told me what my gropings meant." In America, Henry Rowland and Wolcott Gibbs quickly recognized the importance of the work. On its strength, Willard Gibbs was elected to the National Academy of Sciences in 1879 and offered a Johns Hopkins professorship at three thousand dollars in 1880. Yale held him, however, on the basis of sentiment, loyalty, and—at last—a two-thousand-dollar salary.

The famous "Equilibrium" paper, unlike the first two, was almost entirely analytical rather than graphical in its approach. Fundamental yet strikingly original, powerfully and flawlessly logical, it formed a completed whole from which most subsequent developments in the broad fields it covered derived naturally and inevitably. From its prime equation, which produced the theory of chemical equilibrium, could be deduced all the thermal, mechanical, and chemical properties of a complex system. It virtually engendered the field of physical chemistry. Its celebrated "phase rule" yielded manifold applications in metallurgy, mineralogy, and petrology, as well as in theoretical chemistry. Another segment of the paper played a large role in the electrochemical industry. The paper's contributions to understanding of entropy have shaped ideas of the universe and its future. Gibbs's very methods were so fundamental and so logically irrefutable that they survived the revolution in twentieth-century physics and continued to generate ideas and solutions even in quantum mechanics. Some discoveries were made independently by others and then recognized as implicit in Gibbs's monograph. In a 1936 symposium on its scientific implications and consequences, both pure and applied, a number of leading mathematicians, physicists, and chemists filled some eight hundred pages.

Given all this, one may still ask what Gibbs signified in the history of the American scientific community. Was he above and beyond it, sui generis, a towering irrelevancy, a special creation à la Agassiz? In discussing the "Equilibrium" paper, his most scientifically authoritative biographer remarks: "A genius that requires stimulus from outside . . . could scarcely have achieved it." Contemplating Willard Gibbs, one thinks also of Benjamin Franklin and Joseph Henry. But Franklin and Henry, anomalous and unforeseeable as their careers may have seemed, were not pure cases of scientific self-levitation. Franklin got a boost from Boston and Philadelphia, Henry from Albany. Even Newton stood on the shoulders of giants, and Einstein acknowledged the usefulness of Simon Newcomb's planetary studies. Conversely, the villages of past ages must have harbored ignorant Einsteins along with mute, inglorious Miltons.

If Willard Gibbs had been born female, black, or a child of the nineteenth-century American South or frontier West, it seems unlikely that his genius would have been heard. His social, economic, cultural, geographical, and institutional circumstances did not make him a genius, but they gave him a chance. If he came out of the blue, it was Yale blue.

It must be granted that Gibbs's theoretical science, like that of Benjamin Peirce and George Hill, was not in the prevailing American mode. American scientists of the seventies, as earlier, were best at data gathering, measuring, experimenting, and instrumentation. This suggests scientific immaturity, as does the unevenness of quality among major fields. In some other aspects also, Americans had only begun to develop. In 1876 the American Chemical Society set an example of specialized national societies, but other fields were slow to follow. Though Joseph Henry claimed in 1869 that existing scientific publications were enough for the amount of good work Americans produced, that double-edged remark was at odds with the general view by then. At any rate, Americans often sent their best work abroad.

The Government Printing Office turned out hundreds of volumes of scientific reports in the seventies and eighties, but few scientific books came from commercial presses. Some scientists found an outlet in observatory bulletins or, as in the case of Willard Gibbs, society transactions, but any one of a dozen European societies spent more on publications than all the American societies put together. In 1876 Simon Newcomb pointed ruefully to the *American Journal of Science* as "the solitary standard journal of pure science published in the country." Yet in that year the *Journal*'s subscribers numbered a hundred or two hundred fewer than forty years earlier. The depression had hurt. More fundamentally, the *Journal* faced the longstanding and now worsening difficulty of publishing papers in one specialty without losing readers in other specialties. The *Popular Science Monthly* after 1872 lured some subscribers away. Specialized journals increasingly posed a still graver threat.

Journals in conchology and applied entomology appeared in the late sixties, then faded out; and Benjamin Gould gave up an attempt to revive his *Astronomical Journal* in 1869. But in 1867 four former Agassiz students established the enduring *American Naturalist.* In 1874 a more viable entomological journal commenced in Cambridge, and the first mathematical journal since 1861 appeared under the unlikely but effective editorship of a self-trained mathematician in Des Moines, carrying serious work for nine years there, and subsequently at successive universities. In 1876 Johns Hopkins began moving toward the first American university sponsorship of scholarly journals. It launched the *American Journal of Mathematics* in 1878 and Ira Remsen's *American Chemical Journal* in 1879. The latter emphasized basic research and far outclassed the American Chemical Society's weak *Journal,* born the same year.

All the key elements of the modern American scientific establishment had thus made at least an initial appearance by the end of the seventies. The image of American science called up by this Centennial scan is that of a newborn foal, callow, ungainly, yet fully formed and swiftly gaining strength. It still had a long way to go. But it had some definite ideas of how to get there.

One of those ideas had been current for decades: that American scientists needed to rise above mere description and practical application to basic theory and experimental research. So had its corollary: that they should go beyond the short-range, one-step goal, the target of opportunity, to patient, long-term research and thought. The astronomer Simon Newcomb, the chemist J. Lawrence Smith, and the entomologist Samuel Scudder all sounded those familiar themes.

There was a newer idea, in the spirit of what Mark Twain had just labelled the "Gilded Age." "It is characteristic of the present time," Benjamin Gould told the AAAS in 1869, "that it is a period requiring cooperation and associated effort in scientific research, not merely for the sake of needful distribution of labor, but because combinations of resources and acquirements are requisite, to which no individual can attain." Thus Frank Clarke declared in 1879 that chemistry needed "a body of workers under an efficient head . . . endowed laboratories devoted to pure research." And in 1877 Edward Pickering described in wistful detail the design, equipment, and proper location of a dream lab, a sort of science factory, a Menlo Park for pure research, with a corps of researchers and a legion of assistants and technicians, where "research should be rendered as systematic as are the processes of the mechanic arts."

All these prescriptions required money for stipends, publications, staff, buildings, and apparatus (which as Henry Rowland pointed out in 1883 was growing increasingly elaborate and expensive). "Who Should Pay for Scientific Research?" asked a *New York Times* editorial in 1873. The *Times* thought federal support appropriate for expeditions or large-scale team projects. But "private enterprise and munificence," such as university endowments, seemed more in keeping with American ways. In either case the citizen would have to be cultivated, as always in America.

Frank Clarke in 1872 accordingly summoned his fellow scientists to the lecture platform. But Ormsby Mitchel was dead and Agassiz in decline. Furthermore, the *Nation* in 1867 had scathingly cited Agassiz's case as evidence that performing for the public meant sacrificing the character of true scientist. A few postwar scientists risked it, in John Wesley Powell's words, "to carry on our fieldwork," or in John LeConte's, to buy "some winter-clothing for my family." But they did it grudgingly. So in 1872 twenty-five American scientists invited the English physicist

John Tyndall, who gloried in his virtuosity as a lecturer, to tour America as a pitchman for science. Tyndall called upon the large crowds who heard him to support pure science. But something, whether Tyndall's notorious agnosticism or the approach of hard times, blighted response to that appeal.

The AAAS no longer stood on its dignity. The press now and then praised it for striving to popularize science, and in 1875 its president freely avowed the purpose. Newspapers and general periodicals did a more effective job themselves, however. In 1876 the *New York Times* ran a weekly column on scientific progress. The nationally read *New York Tribune* and the *Chicago Inter-Ocean* made a specialty of scientific articles in their semiweekly editions, circulated over a wide area. Among magazines, *Lippincott's, Scribner's,* and the *International Review* published frequent articles on science. In 1869 Spencer Baird began editing a regular science department in *Harper's Monthly,* and in 1872 John Fiske followed suit in the *Atlantic.* The fascination of Darwinism kept science in the limelight and had much to do with the success of Edward L. Youmans's *Popular Science Monthly,* begun in 1872. Youmans dedicated his magazine to enhancing the status and independence of scientists, but he also made it a spokesman for evolution and the philosophy of Herbert Spencer. It had twelve thousand subscribers by the end of its first year.

In their pitch for science, its hucksters credited it with all material well-being as extravagantly as before the war and even more convincingly, thanks to new technological wonders. "Modern civilization depends upon science," Joseph Henry assured a British government commission two weeks before the Ems Dispatch, and "every discovery is connected with good." "Ten thousand illustrations have taught the plainest and most practical men the value of science," said Daniel Gilman, thereupon citing many such boons; he later added Number 10,001 by calling science "the mother of California." The *Scientific American* in 1876 applauded a suggestion that the government call scientists together to "evolve, discover, or create" a science of monetary policy; and Simon Newcomb did indeed later write extensively on economics and psychic phenomena. To the shame of Americans, an Englishman went them one better in 1872. As science spreads "over the planet and upward to the skies," he promised, "Earth will be made a paradise. . . . Hunger and starvation will . . . be unknown. . . . Disease will be extirpated . . . immortality will be invented. And then, the earth being small, mankind will migrate into space." He set no date.

A somewhat less high-flown argument for science was its indispensability to genteel culture. "To be indifferent upon that point," advised the *Buffalo Commercial Advertiser* in 1876, "is to be 'behind the times.' " Peter Lesley wished the Philadelphia gentry would love science as much as did "the educated merchants' class of Boston." Benefactions to science by self-made millionaires, who were regarded by Social Darwinists as the end products of evolution, confirmed the cultural prestige of science in the eyes of the fitter sort, those who flocked to hear Tyndall.

Some scientists were not satisfied with all this, however. They scorned the utilitarian argument. "Many a time has somebody asked me, 'But what is the use of all your work?' " the young astronomer Cleveland Abbe wrote his mother in 1865. "I have nothing to do with that. It is not my business." Even the cultural, religious, or nationalistic justifications annoyed them. Increasingly they advocated science for the sake of science alone. Logically that position may have strengthened their claim to support without strings. Strategically it had its dangers. It smacked of arrogance and self-conceit. The laity, after all, held the purse strings, directly or indirectly. And the laity was not as tractable as some scientists may have assumed.

After a lull in the late sixties, religious hostility to Darwinism flared up in the early and middle seventies. The Darwinian doctrine seemed of a piece with other rising threats to organized religion in the Gilded Age, including secularism, skepticism, and urban growing pains. The increasingly complete capitulation of professional scientists like James Dana and Joseph LeConte to evolution of some sort, Darwinian or Lamarckian, threatened to isolate theologians. Charles Hodge of the Princeton Theological Seminary, the most influential Presbyterian theologian of his time, had accepted the new astronomy, but he gagged on Darwinism. With two books in 1872 and 1874 he sounded the bugle for a counterattack. Measured by volume of published polemics, it climaxed in or about the Centennial Year, but its subsidence still left uneasiness among scientists and suspicion among some of the public.

Thus, the trustees of Howard University sought assurance from Joseph Henry in 1873 that Frank Clarke, if hired, would not mix atheism with his chemistry like his predecessor; since Clarke got the job, they must have been satisfied. Thomas Huxley, the high priest of Darwinism but otherwise a barefaced agnostic, raised a storm in 1876 when he gave a public lecture at Johns Hopkins unsanctified by an opening prayer. At Vanderbilt University, controlled by the Methodists, Alexander Winchell was summarily fired in 1878 for having published a tract inconsistent with Biblical doctrine (it suggested that blacks were not descended from Adam). Winchell, like some others in such cases, gained more than he lost, being appointed to a chair at the University of Michigan. And the older Eastern colleges hired more and more Darwinists. Yet inhibitions persisted elsewhere, and at the end of the century Daniel Gilman reported that candidates for appointments in biology or natural history at some colleges were still being quizzed about their soundness on Biblical cosmogony.

Still worse in the long run, the public showed signs of being bored or baffled by science. In the mid-eighties science coverage by general newspapers and magazines was noticeably less, perhaps because Darwinism was no longer a hot issue. A knowledgeable observer in 1887 judged the most effective disseminators of popular science by then to be religious journals, aided "to some extent" by agricultural journals. Appleton and Company had led in publishing popular science, but in 1879 it informed Simon Newcomb that "the life of Professor [Joseph] Henry, even from your pen, would fail to cover expenses." In 1866 William Dean Howells had called science "the coldest element in our civilization," and presumably others felt that way too. Benjamin Gould in 1869 guessed that science's concern for hard, precise data repelled some people of a more romantically vague temperament. Alpheus Hyatt in 1884 saw more concrete forms of hostility to science, in the antivivisection movement, for example.

In science even more than in technology, the advance of specialization and complexity was leaving the public behind. Press comments on AAAS meetings repeatedly made that point. At Hartford in 1874 the secretary of the local committee, a clergyman, playfully remarked, "We have looked upon you with mystification. Our daily papers since you came have not been intelligible. . . . I had in mind, indeed, to preach a sermon upon the text: 'such knowledge is too wonderful for us.' " A Detroit reporter in 1875 indulged in more biting humor. The AAAS, he wrote, had been "pouring out acre after acre of learned papers, whose weird interest and thrilling nature are finely illustrated by the fact that the morning papers which have printed them nearly in full, have lost a large portion of their already small circulations this week." Of the same meeting, the *New York Times* correspondent remarked that the younger men were more technical and less intelligible than the older.

Most ominous of all was the darker side of the public's identification of science with technology. To the public mind, science begat technology. Technology begat power. And power begat both hope and fear.

Still, misgivings notwithstanding, modern American science and technology had now been launched beyond recall on their trajectory into the unknown.

Chapter 26

Epilogue: The Last of the Lazzaroni

By 1876 Louis Agassiz had passed from the scene he had done so much to shape. Depression, self-doubt, and weariness had plagued him since the war. Some younger scientists in the postwar years spoke of him contemptuously in private, and their public criticism of his inaccuracies, stale ideas, and appeals to the crowd grew too sharp for him to shrug off. He stood almost alone among scientists in his rejection of Darwinism. His best students had gone over to the enemy or at least to neo-Lamarckism. In 1868 he still contended that mankind was divided into distinct species, but J. Lawrence Smith reported to the AAAS in 1873 that "the unity of the human race is insisted on by nearly all the leading naturalists, who teach [that] . . . all men are brothers; all ought to be treated as such, whatever the origin, the blood, the color, the race."

Agassiz vowed in 1866 to give up popular lecturing, museum administration, even his professorship, and devote himself to rigorous, undiluted research, but he could not bring himself to do it. On the contrary, he clung desperately to popular acclaim, eagerly playing the lion of the occasion at the opening of Cornell in 1868 and throwing himself into exhausting campaigns to raise funds for his museum, though a cerebral hemorrhage laid him low for months in 1869–70. As if sensing that his time was running out, he made his personal peace with Asa Gray, Charles Darwin, and even his neo-Lamarckian "rebel assistants," for whom he gave a dinner of reconciliation in 1869 and all of whom remembered him later with admiration and affection. For all his weariness and disappointments, Agassiz's charm and commanding presence remained evident in his last undertaking, a summer school for teachers and students of natural history in 1873. Late that November, however, he wrote his former student Burt Wilder, "I cannot attempt another move with reference to science. If scientific men are ever to be placed on a proper footing of independence in this country, it is for the younger men to work for it." Three weeks later he was dead of a cerebral hemorrhage at sixty-six.

Time was running out for others of the older generation also. Matthew Maury, the stormy petrel of American government science, had come to rest at last in 1868 as professor of physics at Virginia Military Institute, turning down an assortment of Southern college presidencies to serve the state for which he had once deserted the Union. Early in 1873 he died there of a gastric ulcer. Rear Admiral Charles Henry Davis, Maury's old adversary in navy science and civil war, resumed command of the Naval Observatory in 1874. His health had been sapped by malaria contracted in his wartime Mississippi River duty; he overtaxed it with a multiplicity of responsibilities, including the observatory's Centennial exhibit, in the blazing hot summer of 1876; and early in 1877 he died at the observatory. Benjamin Peirce, Davis's fellow Lazzarone and cofounder of the National Academy, had given up the superintendency of the Coast Survey in 1874 but continued to fascinate and mystify math students at Harvard until his death at seventy-one in 1880. Two years later, across the river in Boston, William B. Rogers met an end that would have been more appropriate for Agassiz. On Commencement Day 1882, Rogers formally turned over the presidency of MIT to Francis A. Walker before the assembled graduates and guests in the great hall of the Institute. Walker's eloquent tribute to the seventy-seven-year-old Rogers as leader of a new era in technological education stirred strong emotions in Rogers, who began to reply, then suddenly collapsed on the platform and died instantly at the moment of his final triumph.

The survivors of the Lazzaroni clique against which Rogers had contended a generation before were now reduced to three: Wolcott Gibbs, Benjamin Gould, and James Dana (who had never been wholeheartedly one of them). All three still did good work against odds.

Being thrust by President Eliot into a backwater at Harvard may after all have been a boon to Gibbs, as a similar status was to his unrelated namesake at Yale. With his inheritance he was able to hire a skilled assistant and concentrate his mind and energies on researches in analytical chemistry "so abstruse and difficult," a colleague wrote, "that most chemists would have shuddered at the idea of attacking them." In 1873, on the motion of President Barnard, Columbia conferred an honorary LL.D. on its distinguished alumnus, whom it had barred from a job on sectarian grounds in 1854. In 1877 Gibbs's Columbia friend Ogden Rood tried to get a chair for him there, but nothing came of it. Perhaps Gibbs himself was disinclined, though his Harvard salary never rose above thirty-five hundred dollars. In 1887, however, when he was sixty-five, he reminded his old friend and fellow Lazzarone Benjamin Gould that from their student days he, Gibbs, had always wanted a laboratory of his own where he could give himself over entirely to research. "The time has at last arrived," Gibbs wrote from Newport, "and this morning I handed [in] my resignation. . . . I have bought [land] here and the plans for my Laboratory are now in the hands of the builder. . . . Congratulate me on being at last a free man."

Benjamin Gould's lot had been more troubled. His quick temper and sharp tongue had not helped. From Harvard in 1875 Gibbs wrote, "He has offended so many people here by his bitter speeches that he has very few friends—*none* in power." Gibbs, Joseph Henry, and others liked and respected him, and they did what little they could, even pushing him for the presidency of the University of California in 1868. His well-to-do wife financed a private observatory for him near Cambridge in the mid-sixties. But setbacks and distractions plagued him. For years he was involved in untangling the business affairs of his late father and other relatives. Failure to get the Harvard Observatory directorship in 1866 embittered him against Peirce (who had opposed his appointment), and he quit his Coast Survey job in anger when Peirce took over the survey in 1867, though closing his service honorably with the first telegraphic determination of transatlantic longitude. He angled pathetically for a Cornell appointment, but got no nibbles. One old friend was saddened in 1867 "to think of the Dr. who has befriended so many, myself included, left as it were without employment and unnoticed." Though elected president of the AAAS in 1868 Gould himself wrote despondently that year, "I have nothing left but my good name." Then his sky brightened at last.

It may have been his work in arranging the Chilean observations of the Gilliss expedition in the fifties that turned Gould's mind toward the stars of the Southern Hemisphere. By the end of the war he had begun sounding out the Argentine government on the subject, and in 1870 it agreed to finance a sky-mapping project under Gould's direction in the Argentine city of Cordoba. This turned out to be the great work of his career. With a staff of five young American astronomers he plunged energetically into building and equipping an observatory, despite formidable deficiencies in local labor and supplies. The Dudley debacle was not repeated this time. While construction went on with painful slowness for two years, Gould resourcefully carried out an elaborate preliminary survey by naked eye and binoculars. Tragedy struck again in 1874 when two of his young daughters and their nurse were swept away and drowned in a swollen stream. But the work went on. Gould became for Argentina what Agassiz had been for the United States, a missionary and organizer of science generally, developing and running a meteorological bureau, determining geographical positions, refining standards of weights and measures, training young Argentines in astronomy, and leaving the Argentine Republic with a permanent national observatory. Now and then he would write his friends at home that the work would surely be done in a year or two. But, as he wrote in 1879, the "mañana" pace of the people, "the unexpectedly wide field opened up to me," and the loss of his wife's expected inheritance during the depression conspired to prolong his stay. "I am very gray & not less obese than when last at home," he wrote his brother in 1880. "I am losing my courage and the years are hurrying on," he wrote in 1882. He was sixty when he came home at last in 1885, his work substantially and nobly done. It had produced fifteen large volumes (for which Gould himself had made more than a million observational judgments) and raised astronomical knowledge of the southern sky to parity with that of the northern.

"Nowadays," Gould had written from Cordoba in 1882, "there is scarcely a mail from home which does not bring the news of the departure of some old friend or acquaintance." Time went on taking its toll of American science generally. Heart trouble had annoyed but not slowed the indefatigable Spencer Baird since the late 1860s. Late in 1885, however, he went so far as to stop working after dinner, and in May 1887, at sixty-four, he gave up work altogether on strict orders from his doctor. That August at Woods Hole, Baird took a final wheelchair tour of the great marine biological laboratory he had called into being, returned to his bed, and died. Asa Gray's turn came next. In 1869, though only fifty-nine, he had come home from a foreign tour with a "venerable white beard" in keeping with current fashion, and in 1873 a small pension endowed by wealthy admirers allowed him at last to give up the teaching he had so grudgingly put up with. By then the passing of John Torrey and William Sullivant had confirmed Gray as the patriarch of American botany. Plant physiology, with laboratories and microscopes, now rivalled taxonomy; but Gray worked away happily on completing his and Torrey's *Flora of North America,* and his mind and collections still served as the clearinghouse of American botanical data. He ranked high in the estimation of Europe, which he toured triumphantly in 1887. Early in 1888, at seventy-seven, he died serenely and without pain.

Gibbs and Gould had corresponded during Gould's long absence, and they saw each other now and then afterwards. For the rest of his days Gould busied himself with analysis of fourteen hundred photographic plates brought back from the Argentine; after his death, friends saw to the publication of his work on them. In 1886 he finally revived his *Astronomical Journal.* This time it lasted. In Gibbs's Newport laboratory, meanwhile, Gibbs extended his work on the complex acids and studied the effect of certain organic compounds on animals. He found time now to pursue gardening with the same enthusiasm and skill, and to entertain occasional visitors on a piazza overlooking his garden and the sea.

At Yale, James Dana also went on working. He still suffered from the obscure illness, perhaps psychosomatic, that incapacitated him for prolonged study and for which the best antidote seemed to be long, geologizing walks and chopping wood. At least it served the purpose (and perhaps was subconsciously meant to) of relieving him from much busy work and routine duties—a disguised blessing like Wolcott Gibbs's affliction, President Eliot. Certainly for an invalid restricted to two or three hours a day, Dana turned out a lot of high-quality work, including revisions of his *Manual of Geology* (the last at eighty-one), a book on corals and coral islands, and another, on characteristics of volcanos, after a trip to Hawaii at seventy-four, as well as a steady stream of articles. In April 1895, two days after his regular walk, with proofs of a new book just corrected, Dana stayed in his bed and died peacefully that evening at eighty-two. So passed another Lazzarone.

Three years later Dana's fellow student of geology and symptoms, that tough old hypochondriac James Hall, died at eighty-six in the midst of his work. Death also came gently for another irascible geologist, Josiah D. Whitney. His last notable geological work, on climate changes in earlier ages, appeared in 1882, but he later spent eight years writing articles for the *Century Dictionary* and meanwhile continued to hold the Sturgis-Hooper Professorship at Harvard. Having finished his year's teaching, and never bedridden, he died peacefully of arteriosclerosis at Lake Sunapee, New Hampshire, in August 1896, at seventy-five. Three weeks later Gould, now seventy-two, wrote Gibbs:

> Do you know, my dear old boy, that it is now a full halfcentury since we met in Berlin, & began to grind out aerial plans for science in America? The world has moved since then, and both of us may claim in our hearts, I think, to have done something in the right direction for our own country; though few of the new tribe are probably aware of it, or ever will be. I have mulled over this a good many times since Whitney was taken away.

On Thanksgiving Day that year, Gould fell down a flight of stairs at his home and died of his injuries before the day was over, leaving his valuable library to the Dudley Observatory. Wolcott Gibbs was now the last of the Lazzaroni.

Perhaps it was Gibbs's growing distinction as a survivor that moved Alexander Agassiz to nominate him for president of the National Academy of Sciences in 1895 to succeed Othniel Marsh. Gibbs had declined the presidency in 1883, but now he accepted. Thus he played a part in the Academy's second most significant nineteenth-century service as a consulting body, after its report on the federal surveys in 1878. The ruthless and accelerating spoliation of forests on public lands had aroused conservationists. At a private meeting on the issue at the home of the Arnold Arboretum director in 1895, Gibbs suggested arranging for the government to request an Academy report on the problem. The conservationist Interior Secretary obliged and even, at Gibbs's urging, provided twenty-five thousand dollars for the Academy committee's expenses. After a Western tour that revealed appalling abuses, the committee persuaded the lame-duck President Cleveland to double the forest reserves by setting aside an added twenty-one million acres. And though the formal report of 1897 came too late to elicit Congressional action that year, its recommendations helped form policy when the Forest Service was established in the Agriculture Department in 1905. By then Gibbs was no longer Academy president, having resigned on grounds of "age and infirmity" in 1900.

Meanwhile another honor had come Gibbs's way. From the secretary of the AAAS, speaking for its council, came a letter in 1896 offering Gibbs the presidency in order "to renew the interest of the leaders of American science in the Association," which, the secretary said, was being squeezed between the Academy and the new specialized societies. He also reported the "unanimous desire of the Association to confer upon you Honorary Fellowship in acknowledgement of what you have done for science." Gibbs accepted, and thus for a year held the distinction of presiding concurrently over the two principal organizations of American scientists.

So ended the astonishing nineteenth century. But Gibbs lived on into the twentieth, for which the proper adjective must wait. Certainly wondrous things were emerging from science in those opening years—x-rays, the electron, radioactivity, radium, quantum theory, and relativity. Gibbs saw them all. His last scientific paper had appeared in 1896, but Rood at Columbia kept him up with some of the new developments, notably radium and x-rays. Wolcott Gibbs had been among the few who had immediately fathomed the work of Willard Gibbs, yet Einstein's special theory of relativity in 1905 may have given him pause, if he knew about it. A new age had clearly arrived. In it, as Albert Michelson's 1907 Nobel Prize in physics signalled, American science would bear a leading part. The dream of the Lazzaroni would be realized. And the new generation would stand on their shoulders.

In 1908, at eighty-six, Gibbs was not only the last of the Lazzaroni but also the last survivor of the fifty original members of the National Academy. That December he died at Newport. Not long before, one of his former students had been consulted by the sculptor of a pair of monumental bronze doors for the Capitol Building in Washington. One panel was to honor "the representative men in American science." It may be seen today at the Capitol, framed by the figures of Joseph Henry and Wolcott Gibbs. No memorial to Gibbs could have been more fitting. He stands now in bronze, as he and his comrades stood in life, at a great door. Thanks in part to him and them, the double doors of science and technology have swung open and the human species has passed through, to an awesome future from which there is no retreat but extinction.

Appendix: A Note on Quantitative Statements

There are a number of published and unpublished statistical studies relevant to this book. I have examined all that I know of, and have used and cited most. Quantitative statements otherwise rest on my analysis of articles in the *Dictionary of American Biography* on 477 scientists and 601 technologists whose working years included any part of the period 1846–76. The analysis covers thirty-one categories of data. Use of that analysis is indicated in the text by quantitative references to "leading scientists" or "*DAB* scientists," and in the source notes by the words "see Appendix."

Other historians' statistical profiles of the nineteenth-century American scientific community usually embrace scientists who published one or more articles each, or belonged to the American Association for the Advancement of Science, or in some other way met a specific but not very exclusive definition of "scientist." Since this book emphasizes the establishment of new patterns and organizations, my analysis concerns itself only with recognized leaders, regardless of whether or not they reflected the characteristics of the rank and file. (As it happens, comparison with other analyses indicates that in most respects they came fairly close to doing so.)

My published essay "A Statistical Profile of American Scientists, 1846–1876" (see bibliography of sources) explains in detail why I consider the *DAB* sound in its criteria of inclusion. That essay also discusses my methodology, prints seventeen key statistical tables for scientists, and comments on my findings. (Note: in line 16 on page 73 of the essay, "231" should be changed to "64.") I have developed similar tables for technologists which have not been published, except in mimeograph for those who heard my paper on the subject at the 1977 convention of the Organization of American Historians. I have, however, deposited a copy of those tables in the Special Collections of the Boston University Library, along with a list of all 1,078 *DAB* articles analyzed, the analysis of each, voluminous printouts of the data, and a number of tables derived from them but not published, though drawn upon for statements in this book.

Source Notes

The following notes are confined exclusively to citation of sources. The sources for each paragraph of text are designated by the last word in the paragraph and the number of the page on which that word falls. All sources except manuscripts are fully cited on first use. Full citations for all manuscript sources are given on pages 421–23, and for other sources cited in more than one chapter on pages 425–31.

The following abbreviations are used in the source notes:

AAAS American Association for the Advancement of Science
ANS Academy of Natural Sciences, Philadelphia
APS American Philosophical Society, Philadelphia
Am. Jour. Sci. *American Journal of Science*
DAB *Dictionary of American Biography*
Dict. Sci. Biog. *Dictionary of Scientific Biography*
Jour. Chem. Educ. *Journal of Chemical Education*
LC Library of Congress

Epigraph

(1844) Alexander Dallas Bache, untitled draft of 1844 National Institute speech beginning "What are the wants of science in the United States?" in Alexander D. Bache Collection, Smithsonian Institution Archives, p. 60.
(1866) Truman Abbe, *Professor Abbe and the Isobars* (New York, 1955), p. 83.

2. *The European Model*

p. 7 [*circumstances*]: *New York Times,* Aug. 25, 1860; R. G. A. Dolby, "The Transmission of Science," *History of Science* 15 (1977): 12.
[*foster*]: Joseph Ben-David, "Scientific Productivity and Academic Organization in Nineteenth Century Medicine," *American Sociological Review* 25 (1960): 828–43.
8 [*&c.*]: J. Morris to S. Haldeman, July 1846, Haldeman MSS.
[*patronage*]: Friedrich Paulsen, *The German Universities: Their Character and Historical Development* (New York, 1895), p. 4; J. Morris to S. Haldeman, July 1846, Haldeman MSS.
[*development*]: Paulsen, *German Universities,* pp. 5–8, 157–59, 161, 177–78, 187–89, 201, 235–36.
[*finding*]: Ibid., pp. 8–10, 12–13.
[*decade*]: John T. Merz, *A History of European Thought in the Nineteenth Century,* 4 vols. (London, 1897), 1: 180; Everett Mendelsohn, "The Emergence of Science as a Profession in Nineteenth-Century Europe," in Karl Hill, ed., *The Management of Scientists* (Boston, 1964), pp. 24–25.
9 [*schools*]: Mendelsohn, "Emergence," pp. 7–11.

[*France*]: Margaret Bradley, "Scientific Education Versus Military Training: The Influence of Napoleon Bonaparte on the *Ecole Polytechnique,*" *Annals of Science* 32 (1975): 425, 427; Mendelsohn, "Emergence," pp. 9, 13–14, 27; René Vallery-Radot, *The Life of Pasteur,* trans. R. L. Devonshire (New York, 1923), pp. 10, 21, 31–32, 42.
[*Germany*]: Merz, *History of European Thought,* 1: 99; Paulsen, *German Universities,* pp. 235–36; Vallery-Radot, *Pasteur,* pp. 81, 100, 103.
[*effort*]: George Haines, *German Influence upon English Education and Science, 1800–1866* (New London, Conn., 1957), p. xi; D. S. L. Cardwell, *The Organisation of Science in England: A Retrospect* (London, 1957), pp. 48–49; Mendelsohn, "Emergence," pp. 6–7.

10 [*entertainment*]: Haines, *German Influence,* pp. 4, 18–19; Cardwell, *Organisation,* pp. 55–56, 81.
[*Faraday*]: George A. Foote, "A Study of Attitudes toward Science in Nineteenth Century England, 1800–1851" (Ph.D. diss., Cornell University, 1950), p. 235; Mendelsohn, "Emergence," p. 27.
[*mid-1840s*]: Charles Lyell, *Travels in North America, in the Years 1841–42,* 2 vols. (New York, 1845), 1: 237–41.
[*government*]: Charles Babbage, *Reflections on the Decline of Science in England* (London, 1830), pp. 9–39; O. J. R. Howarth, *The British Association for the Advancement of Science: A Retrospect, 1831–1931* (London, 1931), pp. 3–7, 12–15; *Report of the First and Second Meetings of the British Association for the Advancement of Science* (London, 1833), p. ix.
[*earlier*]: Cardwell, *Organisation,* pp. 34–36, 67, 74–76; Haines, *German Influence,* pp. 43–44, 53, 57–58.
[*flowering*]: Alexander Vucinich, *Science in Russian Culture: A History to 1860* (Stanford, Calif., 1963), pp. 252, 300, 330, 347, 364, 389.

11 [*consensus*]: Haines, *German Influence,* pp. 16–17, 20–21; Merz, *History of European Thought,* 1: 204–205, 300; Foote, "Attitudes," pp. 15–16; Francis Darwin, *The Life and Letters of Charles Darwin,* 2 vols. (New York, 1887), 1: 82–83.
[*summary*]: B. Silliman, Jr., to A. Gray, Feb. 6, 1846, Gray Herbarium.
[*could*]: Orville A. Roorbach, *Bibliotheca Americana* (New York, 1849), pp. 36, 74, 164, 166, 172; H. Lane to B. Peirce, Aug. 23, 1843, Peirce MSS.; Edwin T. Brewster, *Life and Letters of Josiah Dwight Whitney* (Boston, 1909), pp. 67, 76.

12 [*market*]: Brewster, *Whitney,* p. 47; J. C. Derby, *Fifty Years Among Authors, Books and Publishers* (New York, 1884), pp. 176–77, 180; Hellmut Lehmann-Haupt, *The Book of America* (New York, 1951), pp. 132–35; Allen Johnson and Dumas Malone, eds., *Dictionary of American Biography,* 22 vols. (New York, 1928–58), 15: 280 ("George Putnam"); Earl L. Bradsher, *Mathew Carey* (New York, 1912), pp. 47–48; E. Everett to R. Walsh, Dec. 16, 1846, Everett MSS.; Carl Wittke, *Refugees of Revolution* (Philadelphia, 1942), p. 314.
[*writings*]: J. Cooke to "Dear Sir," Dec. 12, 1850, Brock Collection; A. Hunter Dupree, *Asa Gray* (Cambridge, Mass., 1959), pp. 70–72; Brewster, *Whitney,* p. 73; J. Bachman to V. Audubon, Oct. 23, 1846, Bachman MSS., lists Bachman memberships, including British, Russian, and German societies; H. Lane to B. Peirce, Aug. 23, 1843, Peirce MSS.; *Am. Jour. Sci.* 52 (1846): 146.
[*tardily*]: Bradsher, *Carey,* p. 67n; Clarence Gohdes, *American Literature in Nineteenth-Century England* (New York, 1944), p. 45; B. Gould to B. Peirce, Nov. 28, 1845, Peirce MSS.; Frank Staff, *The Transatlantic Mail* (London, 1956), pp. 81–82; *Proceedings in the City of Washington Respecting Mr. Alexandre Vattemare's System of International Exchange* (Washington, D.C., 1848), p. 14.
[*books*]: Staff, *Transatlantic Mail,* pp. 81–82; H. Schaum to J. LeConte, Apr. 28, 1852, LeConte MSS., APS; H. Rogers to S. Baird, Feb. 28, 1846, Baird MSS.

13 [*evolution*]: W. Struve to Corresponding Secretary of the National Institution at Washington, November 1846, Rhees Collection; Eufrosina Dvoichenko-Markov, "The Pulkovo Observatory and Some American Astronomers of the Mid-19th Century," *Isis* 43 (1952): 244; B. Silliman, Jr., to A. Gray, Feb. 6, 1846, Gray Herbarium; H. Strickland to S. Haldeman, Mar. 30, 1849, Haldeman MSS.; Darwin, *Life and Letters of Darwin,* 1: 477–82.

3. A Procession of Pilgrims

14 [*Silliman*]: William M. and Mabel S. C. Smallwood, *Natural History and the American Mind* (New York, 1941), pp. 57, 59–61, 79, 81, 99–100.
[*official*]: George P. Fisher, *Life of Benjamin Silliman, M.D., LL.D.,* 2 vols. (New York, 1866), 1: 91–96.
15 [*world*]: Ibid., 1: 97, 103–104, 127–30, 136–38, 146, 151, 160, 195–96, 210.
[*supporting*]: John F. Fulton and Elizabeth H. Thomson, *Benjamin Silliman, 1779–1864, Pathfinder in American Science* (New York, 1947), pp. 147–49, 152, 156, 159–61, 273, 276; John F. Fulton, "Science in American Universities, 1636–1946," *Bulletin of the History of Medicine* 20 (1946): 104; Charles Lyell, *A Second Visit to the United States of North America,* 2 vols. (New York, 1849), 1: 235; Charles A. Browne, "The History of Chemical Education in America between the Years 1820 and 1870," *Jour. Chem. Educ.* 9 (1932): 700–701; Allan Nevins and Milton H. Thomas, eds., *The Diary of George Templeton Strong,* 4 vols. (New York, 1952), 1: 16.
[*it*]: *DAB* 7: 245 ("George Gibbs"); Fisher, *Silliman* 1: 256–57, 272–73.
[*scientists*]: Bruce Sinclair, "Americans Abroad: Science and Cultural Nationalism in the Early Nineteenth Century," in Nathan Reingold, ed., *The Sciences in the American Context: New Perspectives* (Washington, D.C., 1979), pp. 36–42; Edgar F. Smith, *James Curtis Booth, Chemist* (Philadelphia, 1922), p. 4; *DAB,* 2: 447–48 ("James Booth"), 9: 536 ("Charles Jackson"), 11: 398 ("Elias Loomis").
16 [*scientist*]: Nathan Reingold, ed., *The Papers of Joseph Henry* (Washington, D.C., 1972–), 1: xix–xxiv, 3, 14; Thomas Coulson, *Joseph Henry: His Life and Work* (Princeton, N.J., 1950), pp. 7–24.
[*inductance*]: Coulson, *Henry,* pp. 334–35.
[*work*]: Ibid., pp. 29, 39–40.
[*experimenter*]: Ibid., pp. 30–38, 144, 332–33.
[*Bache*]: Ibid., pp. 97–98, 106–107.
17 [*humor*]: Ibid., pp. 99–103.
[*England*]: Ibid., pp. 112–14, 335–38.
[*hand*]: Merle M. Odgers, *Alexander Dallas Bache* (Philadelphia, 1947), pp. 4–5, 10–17, 32–33.
[*command*]: Ibid., pp. 99–100, 132, 200–201, 205–207.
18 [*1845*]: Ibid., pp. 104–10, 115–26.
[*me*]: Ibid., pp. 36, 61, 75–76, 79–80; Alexander D. Bache, *Report on Education in Europe* (Philadelphia, 1839), pp. 536–37; A. Bache to J. Henry, Feb. 25, 1846, Henry MSS.
[*science*]: Odgers, *Bache,* pp. 46, 50, 61–63; Coulson, *Henry,* pp. 116–27.
19 [*chemists*]: *DAB* 2: 447–48 ("James Booth"); Smith, *Booth,* p. 4; Mass. College of Agriculture, *Charles Anthony Goessmann* (Cambridge, Mass., 1917), p. 116.
[*1842*]: Brewster, *Whitney,* pp. 5, 14, 26–27, 30, 39, 47, 51–56, 59, 180.
[*masters*]: Ibid., pp. 61, 64, 67–69, 73, 77–78; J. Whitney, Jr., to J. Whitney, Sr., Nov. 26, 1842, July 20, 1843, C. Jackson to J. Whitney, Sr., Feb. 11, 12, 1844, Whitney MSS.
20 [*home*]: Brewster, *Whitney,* pp. 70–71; J. Whitney, Jr., to J. Whitney, Sr., June 22, 1844, Whitney MSS.
[*spring*]: *Amer. Jour. Sci.* 39 (1840): 132–34; *DAB,* 7: 251 ("Oliver Gibbs"), 15: 506 ("James Renwick"); Nevins and Thomas, *Diary of Strong* 1: 3–7, 14, 70; George Gibbs, *The Gibbs Family of Rhode Island* (New York, 1933), p. 121; W. Gibbs to W. Channing, July 24, 1845, Gibbs Family MSS.
[*science*]: W. Gibbs to L. Gibbs, Nov. 30, 1845, Jan. 19, 1846, to W. Channing, July 24, Nov. 4, 11, 1845, Jan. 25, 1846, June 25, 1846, Gibbs Family MSS.; Gibbs, *Gibbs Family,* pp. 110, 136; C. Joy to W. Whitney, Apr. 25, 1851, Whitney MSS.
21 [*much*]: W. Gibbs to W. Channing, July 24, Nov. 4, 1845, Gibbs Family MSS.; J. Whitney, Jr., to J. Whitney, Sr., Jan. 16, 1844, Whitney MSS.
[*studies*]: George C. Comstock, "Benjamin Apthorp Gould," in *National Academy of Sciences Biographical Memoirs* (Washington, D.C., 1924), 17: 155–56.
[*home*]: Ibid., p. 156.
[*science*]: Ibid.; J. Henry to A. Bache, July 16, 1852, Henry MSS.
[*America*]: B. Gould to W. Gibbs, Dec. 6, 1872, Gibbs MSS.

22 [*university*]: Margaret W. Rossiter, *The Emergence of Agricultural Science: Justus Liebig and the Americans, 1840–1880* (New Haven, Conn., 1975), pp. 92–103; *Memorials of John Pitkin Norton* (Albany, N.Y., 1853), pp. 34–43; Richard J. Storr, *The Beginnings of Graduate Education in America* (Chicago, 1953), pp. 54–58; Russell H. Chittenden, *History of the Sheffield Scientific School of Yale University, 1846–1922,* 2 vols. (New Haven, Conn., 1928), 1: 37–42; Gerald T. White, "Benjamin Silliman, Jr., and the Origins of the Sheffield Scientific School," *Ventures* 8 (1968): 23–24.
[*directions*]: Rossiter, *Emergence,* pp. 50–56; Rolf King, "E. N. Horsford's Contribution to the Advancement of Science in America," *New York History* 36 (1955): 307–308; Samuel Rezneck, "Horsford, Early Rensselaer Great," *Rensselaer Review* 3 (1966): 19–20.

23 [*here*]: Henry S. Van Klooster, "Liebig and His American Pupils," *Jour. Chem. Educ.* 33 (1956): 493; King, "Horsford's Contribution," pp. 309–10.
[*histrionics*]: Herbert S. Klickstein, "Charles Caldwell and the Controversy in America over Liebig's 'Animal Chemistry,'" *Chymia* 4 (1953): 311–13; Benjamin Silliman, Sr., *A Visit to Europe in 1851,* 2 vols. (New York, 1853), 2: 293–94.
[*it*]: Silliman, *Visit to Europe* 2: 293–94; Samuel Rezneck, "The European Education of an American Chemist and Its Influence in 19th-Century America: Eben Norton Horsford," *Technology and Culture* 11 (1970): 372; J. Whitney, Jr., to J. Whitney, Sr., Nov. 27, 1846, Jan. 26, 1847, Whitney MSS.; W. Gibbs to W. Channing, Nov. 22, 1846, Gibbs Family MSS.; Justus von Liebig, "An Autobiographical Sketch," in *Annual Report of the Board of Regents of the Smithsonian Institution, 1891* (Washington, D.C., 1891), p. 266.
[*Liebig*]: W. Gibbs to W. Channing, July 24, Nov. 4, 1845, Jan. 25, 1846, to G. Gibbs, Aug. 23, 1846, Gibbs Family MSS.; Brewster, *Whitney,* pp. 79–80.

24 [*nauseam*]: W. Gibbs to G. Gibbs, Oct. 27, 1846, to W. Channing, Nov. 22, 1846, Gibbs Family MSS.; J. Whitney, Jr., to J. Whitney, Sr., Nov. 27, 1846, to W. Whitney, Jan. 26, 1847, Whitney MSS.; Brewster, *Whitney,* pp. 82–83.
[*subjects*]: W. Gibbs to W. Channing, Nov. 22, 1846, to G. Gibbs, Jan. 26, 1847, Gibbs Family MSS.; J. Whitney, Jr., to J. Whitney, Sr., Nov. 27, 1846, Whitney MSS.; Brewster, *Whitney,* pp. 83–85.
[*geologist*]: Comstock, "Gould," pp. 156–57; Brewster, *Whitney,* pp. 86–89.
[*education*]: A. Gray to J. Henry, Jan. 12, 1846, Henry MSS.; *DAB* 16: 94–95 ("Henry Rogers"); B. Peirce to A. Bache, Jan. 29, 1846, Peirce MSS.; J. Hall to J. Webster, Feb. 23, Apr. 2, 1846, J. Webster to J. Hall, May 31, Aug. 25, 1846, E. Horsford to J. Hall, May 1, 1846, Hall MSS.; J. Rogers to W. Rogers, Mar. 26, 1846, Rogers MSS.; E. Everett to J. von Liebig, May 14, 1847, to E. Horsford, Feb. 19, 1847, Everett MSS.; Rossiter, *Emergence,* pp. 56, 66.

25 [*research*]: W. Gibbs to L. Gibbs, July 25, 1846, to G. Gibbs, Feb. 22, 1847, Gibbs MSS.
[*pilgrims*]: Nathan Reingold, "Alexander Dallas Bache: Science and Technology in the American Idiom," *Technology and Culture* 11 (1970): 170–71; Edward S. Holden, *Memorials of William Cranch Bond . . . and . . . George Phillips Bond . . .* (New York, 1897), pp. 229–30; S. C. Chandler, "The Life and Work of Dr. Gould," *Popular Astronomy* 4 (1897): 342.
[*1851*]: John C. Greene, "American Science Comes of Age, 1780–1820," *Journal of American History* 55 (1968): 22, 25, 34–38; Patsy A. Gerstner, "Vertebrate Paleontology, an Early Nineteenth-Century Transatlantic Science," *Journal of the History of Biology* 3 (1970): 137–38, 142–43, 148; Lavinia M. Morehead, *A Few Incidents in the Life of Professor James P. Espy* (Cincinnati, Ohio, 1888), pp. 16–17; P. Browne to S. Haldeman, Oct. 19, 1850, Haldeman MSS.; *DAB* 8: 95 ("Samuel Haldeman"); Edward Lurie, *Louis Agassiz: A Life in Science* (Chicago, 1960), p. 163; C. Dohrn to J. L. LeConte, Dec. 17, 1859, LeConte MSS., APS; Odgers, *Bache,* pp. 152, 198; J. Whitney to W. Whitney, Nov. 4, 1851, Whitney MSS.; *Am. Jour. Sci.* 92 (1866): 137; C. Wetherill to J. Frazer, June 10, 1850, Frazer MSS.; *Am. Jour. Sci.* 55 (1848): 137–38; Silliman, *Visit to Europe* 2: 319.

26 [*patronizing*]: Sinclair, "Americans Abroad," pp. 42–45; Bache, "Wants of Science," pp. 33–34, 67, 69–70, 72.
[*corps*]: Benjamin A. Gould, Jr., *An Address in Commemoration of Sears Cook Walker* (Cambridge, Mass., 1854), p. 7; William J. Rhees, ed., *The Smithsonian Institution: Documents Relative to Its Origin and History, 1835–1899,* 2 vols.

(Washington, D.C., 1901), 1: 359; Bache, "Wants of Science," p. 61; J. Henry to ?, draft, Feb. 27, 1846, Henry MSS.; P. Browne to S. Haldeman, Oct. 19, 1850, Haldeman MSS.; J. Dana to J. Torrey, Jan. 20, 1849, Trent Collection; J. Henry to E. Loomis, draft, Jan. 1849, Henry MSS.

[*people*]: James F. W. Johnston, *Notes on North America,* 2 vols. (Boston, 1851), 2: 284–85; H. Schaum to J. L. LeConte, Jan. 21, 1848, LeConte MSS., APS; S. Baird to J. Leidy, Dec. 5, 1849, Leidy MSS.

[*increasing*]: B. Gould to A. Gould, Apr. 11, 1846, A. Gould MSS.

27 [*tolerate*]: J. Henry to A. Bache, Aug. 9, 1838, Henry MSS., quoted in Nathan Reingold, ed., *Science in Nineteenth Century America: A Documentary History* (New York, 1964), p. 85; W. Gibbs to G. Gibbs, Feb. 22, 1846, Gibbs Family MSS.

[*society*]: *Am. Jour. Sci.* 52 (1846): 153; David E. Smith and Jekuthiel Ginsburg, *A History of Mathematics in America before 1900* (Chicago, 1934), p. 81; Charles A. Browne, "The Role of Refugees in the History of American Science," *Science* 91 (1940): 205–207.

[*science*]: See Appendix.

28 [*that*]: Wesley R. Coe, "A Century of Zoology in America," *Am. Jour. Sci.* 46 (1918): 364; Edward S. Morse, in AAAS *Proceedings* 25 (1876): 139; B. Silliman, Jr., to S. Newcomb, Jan. 31, 1876, Newcomb MSS.

[*Boston*]: J. Lesley to L. Agassiz, Aug. 26, 1871, quoted in Lurie, *Agassiz,* p. 388.

4. Agassiz's Boston

29 [*naturalist*]: Lurie, *Agassiz,* pp. 2–7, 12, 14, 16–17, 38–39, 64–71, 78–79.

30 [*beginning*]: Ibid., pp. 83–87; AAAS *Proceedings* 2: 412, 423.

[*first*]: Lurie, *Agassiz,* pp. 94–100.

31 [*series*]: Ibid., pp. 3, 10, 12, 49, 68, 89–90, 93, 103–105, 108–19; AAAS *Proceedings* 47: 260–65.

[*start*]: Lurie, *Agassiz,* pp. 119–23; Lyell, *Second Visit* 1: 16.

[*bases*]: Donald de B. Beaver, "The American Scientific Community, 1800–1860: A Statistical-Historical Study" (Ph.D. diss., Yale University, 1966), pp. 155–57, 178–80.

32 [*hours*]: J. Henry to J. Torrey, Apr. 11, 1838, Torrey MSS.; J. Henry to C. Wheatstone, Feb. 27, 1846, draft, Henry MSS.; Charles S. Sydnor, *A Gentleman of the Old Natchez Region, Benjamin L. C. Wailes* (Durham, N.C., 1938), p. 171.

[*botany*]: H. Haupt to J. Frazer, Apr. 30, 1850, Frazer MSS.; A. Packard to J. L. LeConte, June 30, 1858, LeConte MSS., APS; J. Woodrow to J. Kimberly, Nov. 13, 1857, University of N.C. Archives; I. Branch to L. Gibbes, Nov. 28, 1859, L. Gibbes MSS., Charleston.

[*instruction*]: J. Kirtland to S. Haldeman, Dec. 28, 1850, Haldeman MSS.; F. Ibbetson to J. L. LeConte, July 19, 1854, LeConte MSS., APS; R. Williamson to A. Bache, Jan. 26, 1860, Rhees Collection.

[*Indiana*]: Beaver, "Scientific Community," pp. 157, 178–79, 355.

33 [*century*]: Brewster, *Whitney,* p. 79; see Appendix.

[*West*]: B. Hedrick to E. Thompson, June 5, 1851, Hedrick MSS.; Lyell, *Second Visit,* vol. 1: 21, 25, vol. 2: 362.

[*astronomers*]: Dorothy G. Wayman, *Edward Sylvester Morse* (Cambridge, Mass., 1942), pp. 6, 36; W. Taylor to B. Wailes, Nov. 22, 1850, Wailes MSS.; see Appendix.

34 [*1840s*]: Bruce W. Stone, "The Role of the Learned Societies in the Growth of Scientific Boston, 1780–1848" (Ph.D. diss., Boston University, 1974), pp. 403, 415–16, 419–25; Elizabeth C. Agassiz, *Louis Agassiz: His Life and Correspondence* (Boston, 1890), pp. 411–12.

[*science*]: Stone, "Learned Societies," pp. 404–12.

[*science*]: Ibid., pp. 403, 412–13.

[*Yankee*]: L. Gibbes to J. Torrey, Oct. 16, 1854, Torrey MSS.

35 [*study*]: Stone, "Learned Societies," p. 519; A. Hunter Dupree, "The National Pattern of American Learned Societies, 1769–1863," in Alexandra Oleson and Sanborn C. Brown, eds., *The Pursuit of Knowledge in the Early American Republic* (Baltimore, 1976), pp. 22–25.

[*existence*]: Stone, "Learned Societies," p. 519; Sally G. Kohlstedt, "The Nineteenth-Century Amateur Tradition," in Gerald Holton and William A. Blanpied, eds., *Science and Its Public: The Changing Relationship* (Dordrecht, Netherlands, 1976), pp. 177, 180–81.

36 [*society*]: William S. W. Ruschenberger, *Report of the Condition of the Academy of Natural Sciences of Philadelphia* (Philadelphia, 1876), p. 40.
[*competition*]: Ralph S. Bates, *Scientific Societies in the United States* (New York, 1958), pp. 39, 51, 69; Brooke Hindle, *The Pursuit of Science in Revolutionary America* (Chapel Hill, N.C., 1956), pp. 263–68; Solon I. Bailey, *The History and Work of Harvard Observatory, 1839 to 1927* (New York, 1931), p. 103.
[*building*]: Thomas T. Bouvé, "Historical Sketch of the Boston Society of Natural History; . . ." in Boston Society of Natural History, *Anniversary Memoirs* (Boston, 1880), pp. 45, 48–52, 56; B. Gould to A. Gould, Apr. 11, 1846, J. Dana to A. Gould, Feb. 16, 1848, A. Gould MSS.

37 [*genius*]: Bouvé, "Historical Sketch," pp. 141, 167–68; *DAB*, 2: 258 ("Jacob Bigelow"), 279–80 ("Amos Binney"), 7: 446–47 ("Augustus Gould"); Kohlstedt, "Amateur Tradition," p. 179.
[*destruction*]: *DAB* 9: 536 ("Charles Jackson").
[*ether*]: *DAB*, 9: 537 ("Charles Jackson"), 7: 269 ("William T. G. Morton"); W. Channing to C. Jackson, May 12, 1847, Oct. 18, 1861, R. Hare to C. Jackson, Feb. 3, 1849, Jackson MSS.

38 [*men*]: *DAB*, 9: 537 ("Charles Jackson"), 13: 269–70 ("William T. G. Morton").
[*investigators*]: J. Torrey to C. Short, Aug. 7, 1851, Short MSS.; *Am. Jour. Sci.* 75 (1858): 155; J. Anthony to S. Haldeman, Dec. 24, 1852, Haldeman MSS.; John C. Clyde, *Life of James H. Coffin* (Easton, Pa., 1881), p. 172; *Am. Jour. Sci.* 76 (1858): 25.
[*works*]: Holden, *Memorials of Bond*, p. 103; Rhees, *Smithsonian Institution: Documents* 1: 256; *Am. Jour. Sci.* 66 (1853): 285; notes on lecture of Sept. 19, 1853, Notebook #1, Kimberly MSS.

39 [*many*]: William Goodwin, "Remarks on the American Colony at Göttingen," Massachusetts Historical Society *Proceedings* 12 (1899): 368–69; James D. B. De Bow, *Statistical View of the United States* (Washington, D.C., 1854), p. 160; William J. Rhees, ed., *Manual of Public Libraries, Institutions, and Societies, in the United States . . .* (Philadelphia, 1859), pp. 16, 122.
[*use*]: Rhees, *Manual of Public Libraries*, pp. 122, 604; Justin Winsor, ed., *The Memorial History of Boston*, 4 vols. (Boston, 1886), 4: 290.
[*go*]: *American Journal of Education* 1 (1856): 369; De Bow, *Statistical View*, p. 161.
[*wildflowers*]: Arthur Gilman, ed., *The Cambridge of Eighteen Hundred and Ninety-Six* (Cambridge, Mass., 1896), pp. 41, 54–57.
[*1850*]: Ibid., pp. 39–40; American Academy of Arts and Sciences, *Memorial of Joseph Lovering* (Cambridge, Mass., 1892), pp. 7–12, 15, 27; *DAB* 19: 593 ("John White Webster").

40 [*Gray*]: Jane L. Gray, ed., *The Letters of Asa Gray*, 2 vols. (Boston, 1893), 1: 343.
[*friendship*]: Dupree, *Gray*, pp. 1–4, 18, 30, 32, 36–39, 55–57, 74, 173, 175, 184.
[*lecturing*]: Ibid., pp. 110–11, 117, 123, 130, 173, 206; J. Torrey to M. Curtis, Nov. 2, 1852, Curtis MSS.
[*students*]: B. Peirce to W. Sprague, Sept. 22, 1855, Gratz Collection.

41 [*States*]: *DAB* 14: 393–94, 396 ("Benjamin Peirce"); *New Orleans Sunday Delta*, Dec. 21, 1856; Edward W. Emerson, *The Early Years of the Saturday Club, 1855–1870* (Boston, 1918), p. 104; Holden, *Memorials of Bond*, p. 226; Dirk J. Struik, *The Origins of American Science* (New York, 1957), pp. 331–32.
[*Peirce*]: Florian Cajori, *The Teaching and History of Mathematics in the United States* (Washington, D.C., 1890), pp. 127–28, 136–37, 140–43; Emerson, *Early Years*, pp. 96–97, 99–101; George F. Hoar, *Autobiography of Seventy Years*, 2 vols. (New York, 1903), 1: 99–100; Henry Cabot Lodge, *Early Memories* (New York, 1913), p. 55.

42 [*day*]: Stone, "Learned Societies," pp. 432–37; Ferris Greenslet, *The Lowells and Their Seven Worlds* (Boston, 1946), pp. 234–35.
[*Agassiz*]: Stone, "Learned Societies," pp. 432–37; Greenslet, *Lowells*, pp. 234–35.
[*ground*]: Agassiz, *Agassiz*, pp. 406–408.

5. Agassiz's America

43 [*thing*]: Agassiz, *Agassiz,* pp. 409–10.
[*21*]: Daniel C. Gilman, *The Life of James Dwight Dana* (New York, 1899), pp. 153–54; Nevins and Thomas, *Diary of Strong* 1: 279.
[*quo*]: B. Silliman, Jr., to J. Bailey, summer of 1846 (dated by internal evidence), Bailey MSS.

44 [*States*]: Gilman, *Dana,* p. 155; Marc Rothenberg, "The Educational and Intellectual Background of American Astronomers, 1825–1875" (Ph.D. diss., Bryn Mawr, 1974), pp. 37–45; Timothy Dwight, *Memories of Yale Life and Men* (New York, 1903), pp. 142–44, 147, 363–64; Agassiz, *Agassiz,* p. 414.
[*Gray*]: Gilman, *Dana,* pp. 280, 282; Dwight, *Memories,* pp. 392–95; Fulton and Thomson, *Silliman,* p. 246; entry for Jan. 4, 1859, Mary Henry Diary.
[*years*]: Dupree, *Gray,* pp. 57–69; Michael L. Prendergast, "James Dwight Dana: The Life and Thought of an American Scientist" (Ph.D. diss., University of California, Los Angeles, 1978), pp. 133, 166, 191.
[*botany*]: Dupree, *Gray,* pp. 55, 63; Prendergast, "Dana," 112–14, 119.
[*it*]: *American Journal of Education* 3 (1857): 148–49.

45 [*survived*]: Beaver, "Scientific Community," p. 230; B. Silliman, Sr., to J. Bailey, Aug. 3, 1843, Bailey MSS.; *Am. Jour. Sci.* 4th ser., 46 (1918): 35.
[*clear*]: Lyell, *Second Visit* 1:238.
[*Boston*]: Ibid., vol. 1: 238, 242, vol. 2: 333.
[*libraries*]: Herman L. Fairchild, *A History of the New York Academy of Sciences* (New York, 1887), pp. 38–45; J. Dana to A. Gould, Feb. 16, 1848, L. Agassiz to A. Gould, no date, but probably late 1847, A. Gould MSS.

46 [*headquarters*]: Beaver, "Scientific Community," pp. 224, 359.
[*endowment*]: *Catalogue of the Officers and Students of Columbia College* (New York, 1848), pp. 4–5, 14.
[*negligible*]: Bayrd Still, *Mirror for Gotham* (New York, 1956), p. 140, illustration opposite p. 107; Frederick Rudolph, *The American College and University* (New York, 1962), pp. 128–30.
[*detail*]: Donald Fleming, *John William Draper and the Religion of Science* (Philadelphia, 1950), pp. 26, 35–41.
[*Yale*]: *DAB* 11: 398–99 ("Elias Loomis"); *Amer. Jour. Sci.* 46 (1918): 354; APS *Proceedings* 86 (1943): 32; Dwight, *Memories,* pp. 384–85; J. Torrey to J. Henry, Mar. 27, 1850, Henry MSS.

47 [*volumes*]: Agassiz, *Agassiz,* pp. 415–16.
[*matter*]: Ellis P. Oberholtzer, *Philadelphia: A History of the City and Its People,* 4 vols. (Philadelphia, 1912), 2: 249–51, 258; George R. Taylor, *The Transportation Revolution, 1815–1860* (New York, 1951), p. 389.
[*it*]: B. Gould to A. Gould, Apr. 11, 1846, A. Gould MSS.; Edward P. Cheyney, *History of the University of Pennsylvania, 1740–1940* (Philadelphia, 1940), pp. 181–82, 288–89, 259, 263; Rhees, *Manual of Public Libraries,* p. 241; Elias Loomis, *The Recent Progress of Astronomy* (New York, 1856), pp. 213–17.

48 [*section*]: Bates, *Scientific Societies,* pp. 24–25; Williams Haynes, *Chemical Pioneers* (New York, 1939), p. 45; Agassiz, *Agassiz,* p. 419.
[*Academy*]: Bates, *Scientific Societies,* pp. 67–68; G. Ord to M. McMichael, Sept. 17, 1845, B. Coates to T. Biddle, Dec. 31, 1845, P. Vaughan to G. Ord, Aug. 18, 1846, financial statement dated Oct. 16, 1846, American Philosophical Society Archives.
[*history*]: Edward J. Nolan, *A Short History of the Academy of Natural Sciences at Philadelphia* (Philadelphia, 1909), pp. 11–13; J. Leidy to L. Gibbes, June 2, 1847, to S. Haldeman, Jan. 2, 1850, F. Hayden to J. Leidy, Jan. 11, 1854, Leidy MSS.; Rhees, *Manual of Public Libraries,* pp. 603–604, 634.
[*Darwinism*]: J. Leidy to E. Burgess, Jan. 26, 1880, Leidy MSS.; *DAB* 11: 150–52 ("Joseph Leidy").

49 [*1859*]: *DAB* 11: 150–52 ("Joseph Leidy").
[*brethren*]: W. Gibbs to J. Leidy, June 16, 1857, Leidy MSS.; "Reminiscences of Edward Nolan," Collection 970, Academy of Natural Sciences; S. Baird to J. Leidy, June 18, 1853, Leidy MSS.
[*specimens*]: William H. Dall, *Spencer Fullerton Baird* (Philadelphia, 1915), pp.

5, 44–46, 141; S. Baird to S. Haldeman, May 22, 1846, Haldeman MSS.; Agassiz, *Agassiz,* p. 424.

50 [*activity*]: William Stanton, *The Leopard's Spots* (Chicago, 1960), pp. 25–41, 102–103; Agassiz, *Agassiz,* pp. 147, 437–38.
[*books*]: Agassiz, *Agassiz,* pp. 420–21; Beaver, "Scientific Community," p. 179; J. Dana to J. Bailey, April 1846, Bailey MSS.; Gilman, *Dana,* p. 145.

51 [*charts*]: A. Hunter Dupree, *Science in the Federal Government* (Cambridge, Mass., 1957), pp. 61–63, 100–105; Agassiz, *Agassiz,* pp. 422–23; A. Bache to J. Frazer, Nov. 9, 1846, Frazer MSS.
[*time*]: Agassiz, *Agassiz,* p. 423; William H. Goetzmann, *Army Exploration in the American West, 1803–1863* (New Haven, Conn., 1959), pp. 4–12.
[*Europe*]: Dupree, *Science in the Federal Government,* pp. 46–47, 70–74; Agassiz, *Agassiz,* pp. 417, 421–22.

52 [*Albany*]: Rhees, *Smithsonian Institution: Documents* 1: 178; Emma Rogers, ed., *Life and Letters of William Barton Rogers,* 2 vols. (Boston, 1896), 1: 229; J. Bailey to J. Hall, Feb. 27, 1846, Hall MSS.; Agassiz, *Agassiz,* p. 426.
[*life*]: John M. Clarke, *James Hall of Albany* (Albany, N.Y., 1923), pp. 12–15, 22, 36, 58–59, 62–63, 67–68, 82; *DAB* 8: 136 ("James Hall"); *Am. Jour. Sci.,* 4th ser. 46 (1918): 69.
[*auspices*]: Clarke, *Hall,* pp. 64, 105, 134–37; *DAB* 8: 136 ("James Hall").

53 [*horseback*]: Clarke, *Hall,* pp. 34, 252–53, 502–503.
[*laboratory*]: Ibid., pp. 69–70; *DAB* 8: 136 ("James Hall").
[*years*]: J. Hall to A. Gould, Dec. 18, 1846, A. Gould MSS.; J. Bailey to J. Torrey, Dec. 29, 1846, Torrey MSS.; Lurie, *Agassiz,* p. 387; Clarke, *Hall,* pp. 85–88, 233.

54 [*America*]: Clarke, *Hall,* pp. 170–71; Lurie, *Agassiz,* pp. 124–25.
[*birds*]: George P. Merrill, *The First One Hundred Years of American Geology* (New Haven, Conn., 1924), p. 435; Albert C. Koch, *Journey Through a Part of the United States of North America in the Years 1844 to 1846* (Carbondale, Ill., 1972), pp. 56–58; Lyell, *Second Visit* 2: 268–69.
[*collecting*]: Angela K. Main, "Thure Kumlien, Koshkonong Naturalist," *Wisconsin Magazine of History* 27 (1933–34): 23, 209, 211, 323–25, 327–28, 330.
[*miles away*]: Arthur Deen, "Frontier Science in Kentucky and the Old Northwest, 1790–1860" (Ph.D. diss., Indiana U., 1938), pp. 379–81, 400–401; Andrew D. Rodgers III, *"Noble Fellow": William Starling Sullivant* (New York, 1940), pp. 107–108, 113, 120.

55 [*moved away*]: Walter B. Hendrickson, "Science and Culture in the American Middle West," in Nathan Reingold, ed., *Science in America Since 1820* (New York, 1976), pp. 34–44; draft of reply, accompanying B. Silliman, Jr., to C. Short, Dec. 5, 1851, Short MSS.; Bates, *Scientific Societies,* p. 49; Max Meisel, *A Bibliography of American Natural History,* 3 vols. (New York, 1924), 3:137–38.
[*life*]: Walter B. Hendrickson, "Natural History and Urban Culture in the Nineteenth Century Middle West," Academy of Science of St. Louis *Transactions* 31 (1958): 238–40; Henry D. Shapiro, "The Western Academy of Natural Sciences of Cincinnati and the Structure of Science in the Ohio Valley, 1810–1850," in Oleson and Brown, *Pursuit of Knowledge,* pp. 239–42; M. J. Klem, "The History of Science in St. Louis," Academy of Science of St. Louis *Transactions* 23, no. 2 (1914): 105, 111, 113–15; B. Shumard to J. Leidy, Jan. 14, 1854, June 18, 1856, Leidy MSS.
[*War*]: Walter B. Hendrickson, *The Arkites and Other Pioneer Natural History Organizations of Cleveland* (Cleveland, 1962), pp. 10–11, 17–20, 29–30.
[*agency*]: R. Kennicott to J. L. LeConte, May 30, 1856, LeConte MSS., APS; Bates, *Scientific Societies,* pp. 48–49; S. A. Forbes, "History of the Former State Natural History Societies of Illinois," *Science* 26 (1907): 892–95.

56 [*scientists*]: W. Chauvenet to A. Bache, Apr. 25, 1860, Bache MSS.; J. Anthony to G. Tryon, Dec. 18, 1862, Tryon-Pilsbry Collection; C. Wetherill to A. Bache, Aug. 22, 1861, Bache MSS.; see Appendix.
[*on*]: De Bow, *Statistical View,* pp. 192–93.
[*associations*]: Joseph Ewan, "San Francisco as a Mecca for Nineteenth Century Naturalists," in California Academy of Sciences, *A Century of Progress in the Natural Sciences, 1853–1953* (San Francisco, 1955), pp. 1–39; J. L. LeConte to J. Bailey, July 1, 1850, Bailey MSS.; T. Logan to J. L. LeConte, Aug. 30, 1854, LeConte MSS., APS; Oscar Lewis, *George Davidson, Pioneer West Coast Scientist* (Berke-

ley, Calif., 1954), p. 12; A. Bache to G. Davidson, May 10, 1850, Davidson MSS.; William F. King, "George Davidson: Pacific Coast Scientist for the U.S. Coast and Geodetic Survey, 1845–1895" (Ph.D. diss., Claremont Graduate School, 1973), p. 54.

57 [*century*]: Ewan, "San Francisco," pp. 9–10; Lewis, *Davidson,* p. 60.
[*entered*]: Lurie, *Agassiz,* pp. 185–86; see Appendix.
[*aristocracy*]: De Bow, *Statistical View,* p. 192.

58 [*science*]: James O. Breeden, "Science in the Old South (1830–1860)," paper delivered at the First Barnard-Millington Symposium on Southern Science and Medicine: Science in the Old South (Oxford, Miss., 1982), p. 6; Thomas C. Johnson, *Scientific Interests in the Old South* (New York, 1936), pp. 126, 132–40, 146–50; Lurie, *Agassiz,* p. 185.
[*South-West*]: J. De Bow to L. Gibbes, Mar. 18, 1846, L. Gibbes MSS., LC; J. Hale to B. Wailes, Dec. 12, 1853, Wailes MSS.; C. Forshey to L. Gibbes, Jan. 31, 1846, L. Gibbes MSS., LC; F. Barnard to E. Hilgard, Apr. 4, 1857, Barnard MSS.
[*fingers*]: Karlem Riess, "Physics and Physical Science in New Orleans, 1800–1860," *American Journal of Physics* 25 (1957): 168–73; Johnson, *Scientific Interests,* p. 155; M. Curtis to A. Gray, Mar. 13, 1860, Gray Herbarium.
[*Boston*]: Johnson, *Scientific Interests,* pp. 161–70; Lester D. Stephens, "Scientific Societies in the Old South" (Paper in Barnard-Millington Symposium, 1982), pp. 5, 8, 19–20.

59 [*name*]: Clement Eaton, *The Growth of Southern Civilization* (New York, 1961), pp. 117–18; Carl Bode, *The American Lyceum* (New York, 1956), pp. 75–78; Johnson, *Scientific Interests,* pp. 50–54, 57.
[*Charleston*]: A. Jaquith to B. Wailes, Apr. 30, 1852, B. Wailes to J. Millington, Aug. 1, 1852, Wailes MSS.; E. Hilgard to J. Hilgard, Oct. 5, 1855, Hilgard MSS.; J. Bachman to V. Audubon, Feb. 18, 1846, Bachman MSS.; H. Ravenel to M. Curtis, June 1, 1847, Curtis MSS.; De Bow, *Statistical View,* p. 159; Clement Eaton, *The Mind of the Old South* (Baton Rouge, La., 1967), p. 225; Claude H. Neuffer, ed., *The Christopher Happoldt Journal* (Charleston, S.C., 1960), pp. 142–43.
[*articles*]: De Bow, *Statistical View,* pp. 157–58; Johnson, *Scientific Interests,* pp. 73–80, 86–93, 143–45.
[*class*]: Johnson, *Scientific Interests,* p. 46; Thomas G. Dyer, "Science in the Antebellum College: The University of Georgia, 1801, 1860" (Paper in Barnard-Millington Symposium, 1982), passim; Breeden, "Science," p. 13; Eaton, *Mind,* p. 243; Ellis M. Coulter, *College Life in the Old South* (New York, 1928), p. 254.

60 [*him*]: Breeden, "Science," pp. 12–15; F. Barnard to J. Hilgard, Feb. 11, 1857, Barnard MSS.
[*own*]: Rogers, *Rogers* 1: 1–2, 181, 255, 333–34.

61 [*prejudices*]: Philip A. Bruce, *History of the University of Virginia, 1819–1919,* 5 vols. (New York, 1920–22), 3: 74–76; Dupree, *Gray,* p. 60; W. Gibbs to G. Gibbs, May 23, 1846, Gibbs Family MSS.; D. Kirkwood to J. Frazer, July 11, 1854, Frazer MSS.; W. Chauvenet to B. Peirce, Sept. 17, 1857, Peirce MSS.
[*him*]: Johnson, *Scientific Interests,* p. 58; W. Powell to B. Wailes, June 29, 1845, Wailes MSS.; John LeConte to A. Bache, June 27, 1854, Rhees Collection; J. Barratt to L. Gibbes, Oct. 18, 1847, L. Gibbes MSS., Charleston; B. Chase to B. Wailes, July 28, 1849, Wailes MSS.
[*it*]: I. Branch to L. Gibbes, Mar. 26, May 6, 1858, L. Gibbes MSS., Charleston; J. Couper to A. Gould, Aug. 7, 1844, A. Gould MSS.
[*master*]: Johnson, *Scientific Interests,* p. 199; Joseph LeConte, *The Autobiography of Joseph LeConte,* ed. William D. Armes (New York, 1903), pp. 175–76.

62 [*science*]: Eaton, *Mind,* pp. 243–44; Breeden, "Science," pp. 23, 28–35; Ronald L. and Janet S. Numbers, "Science in the Old South: A Reappraisal" (Paper in Barnard-Millington Symposium, 1982), pp. 24–25.
[*unreason*]: Stanton, *Leopard's Spots,* pp. 111–12; Numbers, "Science," p. 14; William K. Scarborough, "Science on the Plantation" (Paper in Barnard-Millington Symposium, 1982), pp. 10–27.
[*aristocrats*]: Breeden, "Science," pp. 24, 31; Eaton, *Mind,* p. 244.
[*Mississippi*]: Lyell, *Second Visit* 1: 339–40; J. Anthony to H. Cumings, Sept. 8, 1858, Eyton Collection; Sydnor, *Gentleman,* pp. 192, 202.

63 [*climate*]: M. Curtis to A. Gray, Mar. 9, 1848, June 24, 1847, May 15, 1855, Gray Herbarium; M. Curtis to J. Torrey, Sept. 13, 1849, Torrey MSS.; Rothenberg, "Edu-

cational Background," pp. 164–65; H. Ravenel to M. Curtis, June 25, 1847, Oct. 24, 1856, Curtis MSS.; Prendergast, "Dana," p. 140.
[*effort*]: Breeden, "Science," pp. 4–5; Johnson, *Scientific Interests,* pp. 59–73; see Appendix.

6. Science, American Style

64 [*generally*]: Elmer C. Herber, ed., *Correspondence between Spencer Fullerton Baird and Louis Agassiz* (Washington, D.C., 1963), p. 23.
[*man*]: See Appendix; Rhees, *Smithsonian Institution: Documents* 1: 385.
65 [*scientists*]: Thomas W. Higginson, *Contemporaries* (Boston, 1899), p. 203; M. Curtis to J. Torrey, Sept. 13, 1849, Torrey MSS.; Bache, "Wants of Science," p. 52; Gould, *Address on Walker,* p. 9.
[*it*]: Bache, "Wants of Science," p. 10; *Am. Jour. Sci.* 68 (1854): 201.
[*indoors*]: *Am. Jour. Sci.* 1 (1846): 3; M. Curtis to J. Torrey, Oct. 25, 1852, H. Sartwell to J. Torrey, Oct. 11, 1849, Torrey MSS., ANS.
66 [*prisoner*]: Andrew D. Rodgers III, *John Torrey: A Story of North American Botany* (Princeton, N.J., 1942), p. 244; Rodgers, *"Noble Fellow,"* p. 190; Dall, *Baird,* p. 275; Dupree, *Gray,* pp. 206 and 204–15, passim.
[*data*]: Dupree, *Gray,* pp. 187–88; Dall, *Baird,* pp. 275, 277, 284; Lurie, *Agassiz,* pp. 186–88; Reingold, *Science in Nineteenth Century America,* p. 178.
[*surgeons*]: Dupree, *Gray,* pp. 158, 164, 201, 205, 210–13; Samuel W. Geiser, *Naturalists of the Frontier* (Dallas, Tex., 1948), pp. 16, 121–30, 137–40, 174–75, 187–93; Dall, *Baird,* pp. 237–39; Lurie, *Agassiz,* pp. 188–89; Rodgers, *Torrey,* p. 244; Samuel Rezneck, "The Emergence of a Scientific Community in New York State a Century Ago," *New York History,* July 1962, pp. 222–25.
[*labor*]: Goetzmann, *Army Exploration,* pp. 427–28.
67 [*Darwinism*]: Ibid., pp. 201–204, 322, 325, 388, 425, 431.
[*theory*]: Ibid., pp. 425–26.
[*correlated*]: Ibid., pp. 391, 397, 399, 423.
[*taxonomy*]: Goetzmann, *Army Exploration,* pp. 326–27; William H. Goetzmann, *Exploration and Empire* (New York, 1966), pp. 322–23, 330; Rodgers, *Torrey,* pp. 277–78; Dupree, *Gray,* pp. 207, 213.
[*life*]: Goetzmann, *Army Exploration,* pp. 327, 329–30; Dall, *Baird,* p. 191.
68 [*sight*]: AAAS *Proceedings* 29 (1880): 610; Lurie, *Agassiz,* pp. 206–207, 254–55; Goetzmann, *Exploration and Empire,* p. 330.
[*States*]: Foote, "Study of Attitudes," pp. 15–16; George H. Daniels, *American Science in the Age of Jackson* (New York, 1968), pp. 62–63; John W. Mallet, *Chemistry Applied to the Arts* (Lynchburg, Va., 1868), p. 4.
[*facts*]: Daniels, *American Science,* pp. 70–72, 102–17, 138; *Scientific American* 7 (1852): 229.
69 [*Newberry*]: J. Henry to A. Bache, Dec. 1, 1856, Henry MSS.; Stanley M. Guralnick, *Science and the Ante-Bellum American College* (Philadelphia, 1975), p. 123; Goetzmann, *Army Exploration,* pp. 388, 391–92, 397; AAAS *Proceedings* 16 (1867): 1, 3.
[*miscellany*]: Bliss Perry, ed., *The Heart of Emerson's Journals* (Boston, 1926), p. 223.
[*spot*]: J. Dana to J. Bailey, July 10, 1848, Bailey MSS.; C. Short to B. Silliman, Jr., Dec. 1851, Short MSS.; Samuel S. Sherman, *Increase Allen Lapham* (Milwaukee, 1876), pp. 51, 58; Johnston, *Notes* 2: 286; *Albany Cultivator* 3 (1846): 43; *American Journal of Education* 1 (1856): 319; AAAS *Proceedings* 6 (1851): xlv.
70 [*subject*]: Darwin, *Darwin* 1: 85–86; Agassiz, *Agassiz,* p. 419.
[*it*]: AAAS *Proceedings* 2 (1849): 164; Frederick A. Mitchel, *Ormsby MacKnight Mitchel* (Boston, 1887), p. 121; Johnston, *Notes* 2: 286; H. Schaum to J. L. LeConte, Jan. 21, 1848, LeConte MSS., APS; Herber, *Baird-Agassiz Correspondence,* pp. 77–79. On the rivalry for Bad Lands fossils, see J. Hall to J. Henry, April 24, 1853, F. Meek to J. Hall, May 22, 1853, Hall MSS.; G. Engelmann to S. Baird, June 28, 1853, Baird MSS.; and Reingold, *Science in Nineteenth Century America,* pp. 169–72.
[*bridge*]: Mary L. Ames, ed., *Life and Letters of Peter and Susan Lesley,* 2 vols. (New York, 1909), 1: 144; Simon Newcomb, *The Reminiscences of an Astronomer* (Boston, 1903), p. 70; J. Anthony to S. Roberts, Jan. 23, 1869, Tryon-Pilsbry Collec-

tion; AAAS *Proceedings*, vol. 22 (1873): 3, vol. 34 (1885): 17; *Atlantic Monthly* 82 (1898) 9: 314.

71 [*agenda*]: Walter Houghton, *The Victorian Frame of Mind, 1830–1870* (New Haven, Conn., 1957), pp. 111–12.
[*present*]: Simon Newcomb, "Abstract Science in America, 1776–1876," *North American Review* 122 (1876): 92, 101.
[*Astronomy*]: AAAS *Proceedings*, vol. 18 (1869): 30, vol. 24 (1875): 30, vol. 41 (1892): 18.

72 [*testing*]: Reingold, *Science in Nineteenth Century America*, p. 83; J. Henry to ?, Feb. 26, 1846, Henry MSS.; J. Henry to A. Mayer, Aug. 15, 1868, Hyatt-Mayer Collection. See also B. Silliman to G. Combe, Feb. 20, 1844, MS 7273, f. 134, National Library of Scotland; M. Curtis to A. Gray, June 24, 1847, Gray Herbarium; and AAAS *Proceedings* 6 (1851): xliv.
[*science*]: Nathan Reingold, "American Indifference to Basic Research: A Reappraisal," in George H. Daniels, ed., *Nineteenth-Century American Science* (Evanston, Ill., 1972), pp. 38–62; L. Lesquereux to J. P. Lesley, Jan. 9, 1852, Lesley MSS.; Smithsonian Institution Board of Regents, *Annual Report for 1847* (Washington, D.C., 1848), p. 179; J. Henry to Varnum, June 22, 1847, Henry MSS.; William J. Rhees, ed., *The Smithsonian Institution: Journals of the Board of Regents* (Washington, D.C., 1879), p. 785. See also J. Henry to J. Hammond, June 3, 1849, to C. Babbage, Apr. 27, 1850, and to "My dear M.," July 12, 1869, Henry MSS.

73 [*men*]: W. Brewer to G. Brush, Nov. 22, 1854, Brush Family MSS.; AAAS *Proceedings* 49 (1900): 2; Rutgers University, *Addresses Commemorative of George Hammell Cook* (Newark, N.J., 1891), p. 31.
[*others*]: Alexis de Tocqueville, *Democracy in America*, Vintage paperback ed. in 2 vols. (New York, 1954), 2: 46–47; E. Daniels to J. P. Lesley, Feb. 5, 1865, Lesley MSS.
[*sciences*]: A. Guyot to F. Lieber, Mar. 17, 1860, Lieber MSS.; AAAS *Proceedings* 45 (1896): 29.

7. Becoming a Scientist

75 [*astronomer*]: G. Searle to C. Abbe, Jan. 14, Apr. 6, 1868, Abbe MSS.; *DAB* 16: 534–35 ("Arthur Searle").

76 [*species*]: Anne Rae, "The Psychology of Scientists," in Karl Hill, ed., *The Management of Scientists* (Boston, 1964), p. 69; Francis Galton, *English Men of Science: Their Nature and Nurture* (New York, 1895), pp. 56–175, passim; AAAS *Proceedings* 25 (1876): 139.
[*paleontologist*]: J. Espy to J. Henry, Mar. 20, 1850, Henry MSS.; J. Dana to A. Gould, July 30, 1852, Brock Collection; G. Engelmann to C. Short, Feb. 11, 1851, Short MSS.; C. Hartt to J. Hall, Sept. 7, 1868, Merrill Collection.
[*pallid*]: H. Ravenel to M. Curtis, May 22, 1849, Curtis MSS.; F. Hayden to S. Baird, Feb. 16, 1853, Baird MSS.; Marmaduke B. Wright, *An Address on the Life and Character of the Late Prof. John Locke* (Cincinnati, 1857), pp. 6–7; *Am. Jour. Sci.* 33 (1862): 180.
[*scientists*]: J. Henry to S. Newcomb, May 6, 1857, Newcomb MSS.; Galton, *English Men of Science*, p. 125; AAAS *Proceedings* 16 (1867): 5, 164; Ogden N. Rood, *The Practical Value of Physical Science* (Troy, N.Y., 1859), p. 15.

77 [*it*]: Rae, "Psychology of Scientists," p. 69; A. Hyatt to C. Abbe, Nov. 26, 1864, Abbe MSS.; J. Dana to A. Gould, July 30, 1852, Brock Collection; J. Bachman to J. Haskell, Sept. 9, 1871, Bachman MSS.
[*God*]: Charles E. Rosenberg, *No Other Gods* (Baltimore, 1976), pp. 137–39, 142–43.
[*college*]: See Appendix.

78 [*it*]: Wright, *Address on Locke*, pp. 6–7; LeConte, *Autobiography*, p. 127; Lois W. Burkhalter, *Gideon Lincecum, 1793–1874* (Austin, Tex., 1965), p. 6; Clyde K. Hyder, *Snow of Kansas* (Lawrence, Kans., 1953), pp. 15, 26; Browne, "History of Chemical Education," pp. 700–701; Dall, *Baird*, pp. 36–38; entries for Jan. 14, 16, May 20, 1865, diary of F. Clarke, Clarke MSS.; *DAB* 21: 177–78 ("Frank W. Clarke").
[*impermissible*]: Margaret W. Rossiter, *Women Scientists in America* (Baltimore, 1982), p. xv; Sally G. Kohlstedt, "In from the Periphery: American Women in Science, 1830–1880," *Signs* 4 (1978): 81, 90.

[*lectures*]: Rossiter, *Women Scientists,* pp. 7–9; Deborah Warner, "Science Education for Women in Antebellum America," *Isis* 69 (1978): 58–62.

79 [*entomology*]: Warner, "Science Education," pp. 58, 63–64; Emma L. Bolzau, *Almira Hart Lincoln Phelps* (Philadelphia, 1936), pp. 263, 203–66, passim; Harry B. Weiss, *The Pioneer Century of American Entomology* (New Brunswick, N.J., 1936), p. 186; Leland O. Howard, *A History of Applied Entomology* (Washington, D.C., 1930), pp. 12, 22, 25, 28–29.

[*not*]: Kohlstedt, "In from the Periphery," pp. 84–86, 91; Rossiter, *Women Scientists,* p. 75; Warner, "Science Education," p. 66; Fulton and Thomson, *Silliman,* p. 220; Clarke, *Hall,* p. 180; J. Morris to S. Haldeman, Nov. 21, 1846, Haldeman MSS.; Rodgers, *"Noble Fellow,"* pp. 108, 137; Mitchel, *Mitchel,* p. 168; Susan Walker to A. Bache, Feb. 24, 1853, Rhees Collection. See also Deborah J. Warner, *Graceanna Lewis: Scientist and Humanitarian* (Washington, D.C., 1979), passim.

[*husbands*]: Mary De Lapp, "Pioneer Woman Naturalist [Mrs. James Maxwell]," *Colorado Quarterly,* Summer 1964, pp. 91, 93–94, 96; Ewan, "San Francisco," pp. 24–25.

80 [*status*]: Rossiter, *Women Scientists,* pp. 76–78; Sally G. Kohlstedt, *The Formation of the American Scientific Community: The American Association for the Advancement of Science, 1848–60* (Urbana, Ill., 1976), p. 103; Bouvé, "Historical Sketch," p. 181.

[*recognition*]: Rossiter, *Women Scientists,* pp. xvi, 10.

[*all*]: *DAB* 13: 57–58 ("Maria Mitchell"), 66–67 ("William Mitchell"); Rossiter, *Women Scientists,* p. 12; S. Newcomb to W. Bartlett, May 20, 1862, Newcomb MSS.

[*story*]: Rossiter, *Women Scientists,* pp. xvi–xvii; Kohlstedt, "In from the Periphery," p. 96.

81 [*attainment*]: J. Henry to M. De La Rive, Nov. 12, 1841, quoted in Rexmond C. Cochrane, *The National Academy of Sciences: The First Hundred Years, 1863–1963* (Washington, D.C., 1978), p. 33; AAAS *Proceedings* 18 (1869): 9; Thomas P. Hughes, "Industry through the Crystal Palace" (Ph.D. diss., University of Virginia, 1953), p. 137; Daniels, *American Science,* pp. 13, 34; Howard S. Miller, *Dollars for Research: Science and Its Patrons in Nineteenth-Century America* (Seattle, Wash., 1970), p. 75; J. Henry to J. Mason, June 16, 1848, Henry MSS.

[*passing*]: John C. Greene, "Science and the Public in the Age of Jefferson," *Isis* 49 (1958): 24; Rosenberg, *No Other Gods,* p. 136; Rood, *Practical Value,* pp. 19–20; Johnston, *Notes on North America* 1: 135; G. Engelmann to C. Short, Feb. 11, 1851, Short MSS.; E. Alexander to L. Alexander, Feb. 3, 1856, Alexander MSS.

[*Indiana*]: A. Gray to C. Short, June 24, 1854, Short MSS.; Herbert L. Satterlee, *J. Pierpont Morgan* (New York, 1939), pp. 91–92; F. Hudson to J. Hilgard, Feb. 26, 1855, Hilgard MSS.; R. Gibbes to L. Gibbes, Aug. 4, 1846, L. Gibbes MSS., LC; W. Chauvenet to A. Bache, Apr. 25, 1860, Bache MSS.

[*deal*]: Prendergast, "Dana," p. 105; *DAB* 1: 513 ("Spencer Baird"); F. Hayden to S. Baird, Mar. 5, 1854, Baird MSS.; C. Wagner to A. Hyatt, Jan. 22, 1859, quoted in Francis C. Haber, "Sidelights on American Science as Revealed in the Hyatt Autograph Collection," *Maryland Historical Magazine* 46 (1951): 235.

82 [*book*]: Rosenberg, *No Other Gods,* p. 139; Beaver, "Scientific Community," p. 118.

[*educators*]: See Appendix; De Bow, *Statistical View,* p. 129.

[*category*]: See Appendix.

83 [*treasures*]: Charles O. Paullin, "A Statistical Analysis of the *Dictionary of American Biography*" (unpublished, undated typescript in the *Dictionary of American Biography* MSS., Library of Congress), pp. 244, 254.

[*rare*]: Robert V. Bruce, "A Statistical Profile of American Scientists, 1846–1876," in Daniels, *Nineteenth-Century American Science,* p. 88; Beaver, "Scientific Community," p. 358.

[*motivation*]: Colin B. Burke, *American Collegiate Populations* (New York, 1982), pp. 50, 54; Bruce, *University of Virginia* 3: 182; T. Broun to W. Broun, Oct. 16, 1846, Broun MSS.; Dwight, *Memories of Yale,* p. 56; David Allmendinger, *Paupers and Scholars* (New York, 1975), pp. 4–5, 9, 11–12, 91.

[*80*]: Rudolph, *American College,* p. 219; George A. King, *Theodore Dwight Woolsey* (Chicago, 1956), p. 33; De Bow, *Statistical View,* p. 145; Burke, *American Collegiate Populations,* p. 49.

84 [*volumes*]: Milton H. Thomas, "The Gibbs Affair at Columbia in 1854" (M.A. thesis, Columbia University, 1942), p. 17; *American Journal of Education* 1 (1856): 406; Burke, *American Collegiate Populations,* p. 113; Rudolph, *American College,* pp. 92–93; De Bow, *Statistical View,* p. 141.

[*England*]: Hindle, *Pursuit of Science,* pp. 81, 92–93, 308–17; Rudolph, *American College,* pp. 28–30, 222–25; Guralnick, *Science,* pp. viii, 121, 124; Burke, *American Collegiate Populations,* pp. 56–58, 68.

[*nine*]: Guralnick, *Science,* pp. vii, 116; Burke, *American Collegiate Populations,* pp. 39, 70; Ezekiel P. Belden, *Sketches of Yale College* (New Haven, Conn., 1843), pp. 72–74, 81; Chittenden, *Sheffield School* 1: 22–25; Albert W. Gendebien, "Science the Handmaiden of Religion: The Origins of the Pardee Scientific Course at Lafayette College," *Pennsylvania History* 33 (1966): 137; Frederick A. P. Barnard, *Letters to the Honorable, the Board of Trustees of the University of Mississippi* (Oxford, Miss., 1858), p. 10; Johnson, *Scientific Interests,* p. 26; I. Bernard Cohen, "Harvard and the Scientific Spirit," *Harvard Alumni Bulletin* 50 (1948): 393.

[*backgrounds*]: T. Broun to W. Broun, Oct. 16, 1846, Broun MSS.; AAAS *Proceedings* 33 (1884): 49; American Academy of Arts and Sciences, *Memorial of Joseph Lovering,* p. 12; Cohen, "Harvard," p. 394; J. Torrey to H. Wurtz, Jan. 26, 1858, Wurtz MSS.; Belden, *Sketches,* pp. 75–76; Dupree, *Gray,* p. 200.

85 [*before 1870*]: Rossiter, *Emergence,* pp. 173, 175; Johnson, *Scientific Interests,* pp. 13, 31, 41–43; Jonas Viles, *The University of Missouri: A Centennial History* (Columbia, Mo., 1939), pp. 64–65; Gendebien, "Science," p. 142; Smallwood, *Natural History,* p. 195.

[*the 1870s*]: Guralnick, *Science,* pp. 72–74, 76; Coe, "Century," p. 359; Henry S. Van Klooster, "The Beginnings of Laboratory Instruction in Chemistry in the U.S.A.," *Chymia* 2 (1949): 13–14; Egbert K. Bacon, "A Precursor of the American Chemical Society," *Chymia* 10 (1965): 187–89; D. Barnard to H. Fish, Feb. 14, 1857, Fish MSS., Columbia; Browne, "History of Chemical Education," p. 713.

[*mathematics*]: J. Henry to ?, Feb. 26, 1846, Henry MSS.; Smith and Ginsburg, *Mathematics in America,* pp. 77–78; S. Baird to S. Haldeman, Jan. 15, 1847, Haldeman MSS.; Lurie, *Agassiz,* p. 148; L. Germain to G. Cook, July 8, 1848, W. Ingham to G. Cook, July 27, 1848, Cook MSS.; Browne, "History of Chemical Education," p. 724.

[*texts*]: Elizabeth R. Pennell, *Charles Godfrey Leland,* 2 vols. (Boston, 1906), 1: 40–41; Guralnick, *Science,* p. 54.

86 [*disgusting*]: AAAS *Proceedings* 33 (1884): 38–40; Benjamin H. Hall, *A Collection of College Words and Customs* (Cambridge, Mass., 1851), pp. 27–31; entries for Nov. 15, 1848, Nov. 19, 1849, diary of William H. Brewer, Brewer MSS.; Nevins and Thomas, *Diary of Strong* 2: 327.

[*relief*]: William J. Chute, *Damn Yankee: The First Career of Frederick A. P. Barnard, Educator* (Port Washington, N.Y., 1977), pp. 85–86; B. Hedrick to ?, Oct. 9, 1852, Hedrick MSS.; Smith and Ginsburg, *History of Mathematics,* pp. 67–69.

[*courses*]: Cajori, *Teaching of Mathematics,* pp. 131, 137–38; Bruce, *University of Virginia* 3: 39; Guralnick, *Science,* pp. 56–58.

[*America*]: Guralnick, *Science,* p. 59; B. Hedrick to E. Thompson, Dec. 3, 1851, Hedrick MSS.; S. Tyndale to J. Hilgard, Jan. 28, 1852, Hilgard MSS.; Smith and Ginsburg, *Mathematics in America,* pp. 65, 191, 198.

[*altogether*]: Nevins and Thomas, *Diary of Strong* 2: 326; Raphael Pumpelly, *My Reminiscences,* 2 vols. (New York, 1918), 1: 129; Simon Newcomb, "Exact Science in America," *North American Review* 119 (1874), passim.

87 [*necessary*]: Guralnick, *Science,* pp. 65–70, 76.

[*history*]: Rothenberg, "Educational Background," pp. 58–59, 65, 70–71, 77, 91–92, 104, 125, 142–44; Loomis, *Recent Progress,* pp. 206–208; Truman H. Safford, *The Development of Astronomy in the United States* (Williamstown, Mass., 1888), p. 12; David F. Musto, "Yale Astronomy in the Nineteenth Century," *Ventures* 8 (1968): 9, 11.

[*class*]: Johnson, *Scientific Interests,* pp. 23, 41; S. Haldeman to J. Frazer, Dec. 22, 1851, Frazer MSS.; Loomis, *Recent Progress,* pp. 208–86; see Appendix; William C. Bond, *History and Description of the Astronomical Observatory of Harvard College* (Cambridge, Mass., 1856), p. clxxxi; Musto, "Yale Astronomy," p. 11; AAAS *Proceedings* 49 (1900): 27; Guralnick, *Science,* pp. 89–90, 92.

[*professors*]: Guralnick, *Science,* pp. 95, 109, 112–17.
88 [*settees*]: Guralnick, *Science,* p. 126; Gendebien, "Science," pp. 138–39; *Charleston* (S.C.) *Courier,* Mar. 13, 1850.
[*tried*]: Rudolph, *American College,* pp. 143, 227–28; Guralnick, *Science,* p. 126.
89 [*curtailed*]: Frederick Rudolph, *Curriculum: A History of the American Undergraduate Course of Study Since 1636* (San Francisco, 1977), pp. 78–79, 81–82.
[*general*]: Codman Hislop, *Eliphalet Nott* (Middletown, Conn., 1971), pp. 226–27, 230; Guralnick, *Science,* p. 129.
[*conventionality*]: Rudolph, *American College,* pp. 238–40; Burke, *American Collegiate Populations,* p. 70.
90 [*supposed*]: C. Phillips to D. Swain, Sept. 9, Oct. 8, 1853, University MSS., Southern Historical Collection; Rudolph, *American College,* p. 239.
[*education*]: Johnson, *Scientific Interests,* pp. 28n, 34; Rudolph, *American College,* pp. 231–33.
[*authorities*]: Prendergast, "Dana," p. 78; Storr, *Beginnings,* pp. 1–2, 40–43.
[*midpoint*]: LeConte, *Autobiography,* pp. 104, 124; Fleming, *Draper,* pp. 11–12; William Browning, "The Relation of Physicians to Early American Geology," *Annals of Medical History* 3 (1931): 547–60, 565; see Appendix.
91 [*study*]: See Appendix.
[*price*]: Autobiographical Notes by Eugene Hilgard, p. 10, Hilgard MSS.; Elizabeth A. Osborne, ed., *From the Letter-Files of S. W. Johnson* (New Haven, Conn., 1913), pp. 49, 62, 67; George W. Magee, Jr., ed., *An American Student Abroad: from the Letters of James F. Magee* (Philadelphia, 1932), p. 52; Dall, *Baird,* p. 293.
[*chemists*]: Rossiter, *Emergence,* p. 80; Smithsonian, *Annual Report* 1 (1847): 20; Autobiographical Notes by Eugene Hilgard, p. 10, Hilgard MSS.; Magee, *American Student,* pp. 3–4; D. Barnard to H. Fish, Feb. 14, 1857, Fish MSS., Columbia; see Appendix.
[*fast*]: See Appendix; Rossiter, *Emergence,* p. 80; *American Journal of Education* 2 (1856): 374.
92 [*Europe*]: See Appendix; Lyell, *Second Visit* 2: 280–81; *Am. Jour. Sci.* 72 (1856): 146–48; Charles A. Browne, "European Laboratory Experiences of an Early American Agricultural Chemist—Dr. Evan Pugh (1828–1864)," *Jour. Chem. Educ.* 7 (1930): 510; H. Ward to J. Orton, Dec. 29, 1854, Ward MSS.; Louis Janin to mother, Nov. 3, 1860, Janin MSS.
[*so*]: L. Janin to father, Mar. 21, 1858, Janin MSS.; Browne, "European Laboratory Experiences," pp. 502–3, 508, 510; entry for Mar. 12, 1853, diary of William Kitchell, Kitchell MSS.; Dall, *Baird,* p. 304; C. Joy to W. Whitney, May 17, 1851, Whitney MSS.; Autobiographical Notes of Eugene Hilgard, p. 21, Hilgard MSS.; H. S. Van Klooster, "Liebig," p. 493; Rossiter, *Emergence,* pp. 57, 61; S. Johnson to W. Brewer, Jan. 1854, Brewer MSS.; W. Brewer to G. Brush, Feb. 2, 1857, Brush Family MSS.
[*him*]: Dall, *Baird,* p. 294; Frederick H. Getman, *The Life of Ira Remsen* (Easton, Pa., 1940), p. 31; C. Joy to W. Whitney, Nov. 14, 1851, Whitney MSS.; Magee, *American Student,* pp. 67, 150; Aaron J. Ihde, *The Development of Modern Chemistry* (New York, 1964), p. 264; Browne, "European Laboratory Experiences," pp. 503–506, 510; H. S. Van Klooster, "Friedrich Wöhler and His American Pupils," *Jour. Chem. Educ.* 21 (1944): 160–61; Robert L. Larson, "Charles Frederick Chandler, His Life and Work" (Ph.D. diss., Columbia University, 1950), p. 18; Mass. College of Agriculture, *Goessmann,* pp. 116, 123.
[*first*]: W. Brewer to family, Apr. 8, 1856, Brewer MSS.; C. Joy to W. Whitney, Nov. 14, 1851, Oct. 29, 1852, Whitney MSS.; Magee, *American Student,* pp. 52, 67, 148–49; J. Easter to G. Brush, Dec. 11, 1854, D. Fisher to G. Brush, Jan. 9, 1855, Brush Family MSS.
93 [*war*]: See Appendix.

8. Being a Scientist

94 [*field*]: See Appendix.
[*productive*]: See Appendix; Daniels, *American Science,* pp. 27–28, 35–36; Reingold, *Science,* p. 61; Rodgers, *"Noble Fellow,"* p. 244; Robert Siegfried, "A Study of Chemical Research Publications from the United States before 1880" (Ph.D. diss., University of Wisconsin, 1953), pp. 105–106.

95 [*scientists*]:See Appendix; B. Silliman to R. Hare, May 17, 1847, Smith Collection.
[*should*]: Samuel Soloreichik, "Toxicity: Killer of Great Chemists?", *Journal of Chemical Education* 41 (1964): 282–84; F. Genth to G. Brush, Apr. 27, 1857, Brush Family MSS.; Fisher, *Silliman* 1: 196, 254–55; Nevins and Thomas, *Diary of Strong* 1: 10; entries for Jan. 25, Nov. 14, 1849, diary of W. Brewer, Brewer MSS.; Hoar, *Autobiography* 1: 101.
[*Europe*]: Ernest Child, *The Tools of the Chemist* (New York, 1940), pp. 37, 71–72, 154, 160, 173; *Am. Jour. Sci.* 55 (1848): 154; J. Whitney, Jr., to J. Whitney, Sr., Jan. 26, 1847, Whitney MSS.; W. Gibbs to L. Bradish, June 18, 1851, Bradish MSS.; drafts of unaddressed letters from B. Hedrick, Oct. 1852, Jan. 5, 6, 1854, Hedrick MSS.; draft of unaddressed letter from B. Hedrick, Sept. 15, 1853, Hedrick MSS., UNC Archives; Bacon, "Precursor," p. 187; W. Gibbs to C. Wetherill, Mar. 13, 1855, Wetherill MSS.; W. Gibbs to O. Rood, Oct. 28, 1864, Gibbs MSS.

96 [*apparatus*]: Child, *Tools,* pp. 71, 86–87, 139–40, 196; C. Peirce to A. Bache, June 19, 1861, Bache MSS.; unaddressed draft of letter by B. Hedrick, Jan. 6, 1854, Hedrick MSS.; B. Bakewell, Jr., to G. Cook, Aug. 22, 1862, Cook MSS.
[*analysis*]: Child, *Tools,* pp. 44, 132.
[*strategy*]: Browne, "Chemical Education," p. 696; Gregory P. Baxter, "The Early Days of Chemistry at Harvard," *Harvard Graduate Magazine* 32 (1924): 596; *Am. Jour. Sci.* 46 (1918): 264–66.
[*other*]: See Appendix.

97 [*ninety-one*]: A. Hall to C. Abbe, May 25, 1865, Abbe MSS.; John G. Morris, *Life Reminiscences* (Philadelphia, 1896), pp. 166–67; *DAB* 13: 213 ("John G. Morris").
[*collections*]: F. Hayden to J. Leidy, Feb. 9, 1855, Leidy MSS.; W. Gabb to F. Meek, Jan. 20, 1858, Merrill Collection; Bachman note quoted in Morris, *Life Reminiscences,* p. 169; T. Harris to J. L. LeConte, Sept. 10, 1851, LeConte MSS., APS; J. Barratt to L. Gibbes, Dec. 1, 1849, L. Gibbes MSS., Charleston; H. Ravenel to M. Curtis, July 31, 1847, Oct. 24, 1856, Curtis MSS.; J. Bachman to V. Audubon, Feb. 18, 1846, Bachman MSS.; E. Parrite to S. Haldeman, July 21, 1849, Haldeman MSS.; H. Schaum to J. L. LeConte, Jan. 21, 1848, LeConte MSS., APS; Bouvé, "Historical Sketch," pp. 41, 153.
[*one*]: *Am. Jour. Sci.* 46 (1918): 356, 390; T. Cole to J. Bailey, Apr. 2, 1845, Bailey MSS.; J. Bailey to J. Torrey, Dec. 2, 1847, Torrey MSS.; Herber, *Baird-Agassiz Correspondence,* p. 27; M. Curtis to A. Gray, Feb. 12, 1846, Gray Herbarium.
[*last*]: Smallwood, *Natural History,* pp. 207–209; Child, *Tools,* pp. 165–66; J. Bailey to J. Torrey, May 29, 1847, Oct. 26, 1848, Torrey MSS.; J. Dana to J. Bailey, July 10, 1848, Bailey MSS.; H. Ravenel to M. Curtis, Dec. 29, 1848, Curtis MSS.; *Am. Jour. Sci.,* vol. 55 (1848): 237–40, 285–86, 443, vol. 63 (1852): 290–92, vol. 70 (1855): 142–43; AAAS *Proceedings* 6 (1851): 398; Donald L. Padgitt, *A Short History of the Early American Microscopes* (Chicago, 1975), pp. 6, 8, 11, 13, 21, 31; Rood, *Practical Value,* p. 20.

98 [*specimens*]: See Appendix; *Am. Jour. Sci.* 46 (1918): 204, 218–21, 355–56, 369; Herber, *Baird-Agassiz Correspondence,* p. 5; George G. Simpson, "The Beginnings of Vertebrate Paleontology in North America," APS *Proceedings* 86 (1943): 131, 169, 176.
[*sterility*]: See Appendix; J. Torrey to L. Gibbes, Dec. 6, 1850, Torrey MSS.; Rodgers, *Torrey,* pp. 209, 309; Dupree, *Gray,* pp. 209–10, 214–15; Rodgers, *"Noble Fellow,"* pp. 123–24, 242; J. Torrey to C. Short, Feb. 14, 1852, G. Engelmann to C. Short, Nov. 14, 1853, June 12, 1855, A. Gray to C. Short, June 24, 1854, Short MSS.
[*fields*]: See Appendix; Agassiz, *Agassiz,* p. 437; AAAS *Proceedings* 18 (1869): 33.

99 [*miserable*]: Brewster, *Whitney,* pp. 163, 166, 174–75, 180–81; Ames, *Lesley* 1: 265–66; C. Hartt to L. Hartt, c. 1871, Aug. 29, 1874, Hartt MSS.
[*sea*]: Bache, "Wants of Science," p. 53; Patsy A. Gerstner, "A Dynamic Theory of Mountain Building: Henry Darwin Rogers, 1842," in Reingold, *Science Since 1820,* pp. 104–15; *Am. Jour. Sci.* 46 (1918): 170; Clifford M. Nelson and Fritiof M. Fryxell, "The Ante-Bellum Collaboration of Meek and Hayden in Stratigraphy," in Cecil J. Schneer, ed., *Two Hundred Years of Geology in America* (Hanover, N.H., 1979), p. 196.

100 [*American*]: Prendergast, "Dana," pp. 239, 244–51; R. H. Dott, Jr., "The Geosyncline—First Major Geological Concept 'Made in America'," in Schneer, *Two Hundred Years,* pp. 239–43, 249, 252, 257–58.

[*geology*]: William N. Rice, "The Contributions of America to Geology," *Science* 25 (1907): 162–75; AAAS *Proceedings* 47 (1898): 266.

[*petrology*]: AAAS *Proceedings* 31 (1882): 19; *Am. Jour. Sci.* 46 (1918): 223–24, 246–49.

101 [*currents*]: Reingold, *Science*, p. 60; Bache, "Wants of Science," pp. 52–53; William J. McGee, "Fifty Years of American Science," *Atlantic Monthly* 82 (1898): 313; Daniels, *American Science*, pp. 94–100; *Am. Jour. Sci.* 46 (1918): 352–54.

[*Dept.*]: C. Wetherill to J. Leidy, Aug. 15, 1859, Leidy MSS.; C. Wetherill to A. Bache, Aug. 16, 1859, Rhees Collection; Donald R. Whitnah, *A History of the United States Weather Bureau* (Urbana, Ill., 1961), p. 8; *Scientific American*, Oct. 13, 1849, p. 27; F. Rogers to A. Bache, Sept. 17, 1860, Rhees Collection.

[*per cent*]: See Appendix.

[*forties*]: Deborah J. Warner, "Astronomy in Antebellum America," in Reingold, *Sciences*, p. 55; Samuel F. Bemis, *John Quincy Adams and the Union* (New York, 1956), p. 503.

102 [*imagination*]: Warner, "Astronomy," pp. 61, 65–66; Loomis, *Recent Progress*, pp. 9–53, 121–29, 140–49; Bemis, *Adams*, pp. 503–21.

[*Cambridge*]: Warner, "Astronomy," pp. 61, 63; Bemis, *Adams*, p. 521; Loomis, *Recent Progress*, pp. 206–86; Newcomb, "Abstract Science," p. 108; Holden, *Memorials*, p. 196; Phebe M. Kendall, ed., *Maria Mitchell: Life, Letters, and Journals* (Boston, 1896), p. 96; Safford, *Development*, p. 22; AAAS *Proceedings* 25 (1876), 39–40.

[*astronomy*]: *Am. Jour. Sci.* 73 (1857): 408.

[*1880s*]: Robert P. Multhauf, ed., *Holcomb, Fitz, and Peate: Three 19th Century American Telescope Makers* (Washington, D.C., 1962), pp. 156–70; AAAS *Proceedings* 25 (1876): 45–46; Deborah J. Warner, *Alvan Clark & Sons: Artists in Optics* (Washington, D.C., 1968), pp. 1–37.

[*spectroscopy*]: Bailey, *History*, pp. 38–39; AAAS *Proceedings* 25 (1876): 41; Warner, "Astronomy," p. 67.

103 [*$150,000*]: Bessie Z. Jones and Lyle G. Boyd, *The Harvard College Observatory: The First Four Directorships, 1839–1919* (Cambridge, Mass., 1971), pp. 40–44, 48–50, 54, 96–97; Bond, *History*, pp. iv–vi, xiv–xviii, lxxxv; E. Everett to R. Walsh, Dec. 16, 1846, Everett MSS.; Holden, *Memorials*, pp. 23, 29–30.

[*air*]: Holden, *Memorials*, pp. 31, 45, 58, 78, 266, 269; Jones and Boyd, *Harvard Observatory*, p. 88; Loomis, *Recent Progress*, pp. 97–99, 110–11; Bailey, *History*, pp. 116–17, 196; Elizabeth L. Bond, "The Observatory of Harvard College and Its Early Founders," Cambridge Historical Society *Proceedings* 25 (1939): 82.

[*charts*]: Holden, *Memorials*, pp. 26–27.

104 [*nebulae*]: Warner, "Astronomy," pp. 66–67; AAAS *Proceedings*, vol. 25 (1876): 40–42, vol. 29 (1880): 113.

[*reputation*]: See Appendix; Daniel J. Kevles, *The Physicists* (New York, 1978), pp. 7–8.

[*general*]: Diary of Simon Newcomb, Feb. 9, 1860, to June 8, 1860, passim, and Sept. 20, 1860, Newcomb MSS.; John M. Schofield, *Forty-Six Years in the Army* (New York, 1897), p. 30.

[*century*]: D. Wood to C. Abbe, Aug. 12, 1863, Abbe MSS.; see Appendix.

105 [*explanation*]: Bode, *American Lyceum*, p. 145; G. Emmons to E. Pickering, Aug. 18, 1858, Pickering MSS.; E. Morley to A. Morley, July 21, 1855, Morley MSS.

[*best*]: Clarke, *Hall*, p. 318; Merrill, *First Hundred Years*, p. 264; *Am. Jour. Sci.*, vol. 76 (1858): 326, vol. 73 (1857): 139; J. Henry to A. Mayer, Dec. 7, 1864, Hyatt-Mayer Collection.

106 [*hotly*]: S. Walker to A. Bache, July 30, 1846, National Archives Record Group 23, Private Correspondence of Superintendent's Office; C. Peters to J. Watson, Nov. 9, 1857, Watson MSS.; Holden, *Memorials*, pp. 182–83, 198.

[*sighting*]: Morton Grosser, *The Discovery of Neptune* (Cambridge, Mass., 1962), pp. 110, 139.

[*arrogance*]: Grosser, *Discovery*, pp. 140–41; J. Henry to A. Bache, Mar. 31, 1847, Henry MSS.; E. Everett to J. Bowditch, Mar. 19, 1847, and to H. Holland, June 15, 1847, Everett MSS.; Holden, *Memorials*, pp. 91–92; J. Dana to S. Haldeman, Mar. 14, 1848, Haldeman MSS.; A. Pannekoek, *A History of Astronomy* (London, 1961), pp. 361–62; A. Hall to O. Gibbs., Mar. 11, 1900, Gibbs MSS.

107 [*thought*]: Lurie, *Agassiz*, pp. 193, 207–209.
[*so*]: *Am. Jour. Sci.* 61 (1851): 128; Lurie, *Agassiz*, pp. 205–207.
[*respect*]: Lurie, *Agassiz*, pp. 271–74.
108 [*evidence*]: Dupree, *Gray*, pp. 221, 224–29.
[*uneasy*]: Ibid., pp. 211–17, 224.
109 [*initiates*]: Ibid., pp. 233–46.
[*intervention*]: Ibid., pp. 248–50; Lurie, *Agassiz*, pp. 275–76.
[*this*]: Dupree, *Gray*, pp. 253–61; Lurie, *Agassiz*, pp. 276–81; B. Peirce to A. Bache, Feb. 24, 1859, Rhees Collection.
110 [*1870*]: William F. Sanford, Jr., "Dana and Darwinism," *Journal of the History of Ideas* 26 (1965): 531–36; John A. De Jong, "American Attitudes Toward Evolution Before Darwin" (Ph.D. diss., University of Iowa, 1962), pp. 244–68.
[*won*]: Lurie, *Agassiz*, pp. 255, 266, 300; Rogers, *Rogers* 2: 28; Edward J. Pfeifer, "United States," in Thomas F. Glick, ed., *The Comparative Reception of Darwinism* (Austin, Tex., 1972), pp. 176–81.
[*controversy*]: Dupree, *Gray*, pp. 285–88; Reingold, *Science*, p. 196.
111 [française]: *Am. Jour. Sci.* 80 (1860): 154; Dupree, *Gray*, pp. 272–78.
[*circle*]: Dupree, *Gray*, pp. 295–303; Reingold, *Science*, p. 198.

9. Science and Technology in the Public Mind

115 [*money*]: Kendall, *Mitchell*, pp. 142–43; Robert H. Kargon, *Science in Victorian Manchester* (Baltimore, 1978), pp. 78–79.
[*reports*]: J. Hall to J. Dana, Dec. 12, 1846, Dana MSS.; J. Dana to G. Brush, Sept. 21, 1854, Brush Family MSS.; C. Dewey to J. Torrey, June 18, 1847, Torrey MSS.; C. Wilkes to J. Torrey, Mar. 16, 1851, W. Emory to A. Bache, Mar. 15, 1845, Rhees Collection; G. Warren to J. Leidy, Mar. 8, 1858, Leidy MSS.
116 [*textbooks*]: Rezneck, "Emergence," pp. 218–19, 222–24; E. Loomis to B. Peirce, Apr. 17, 1848, Peirce MSS.
[*facilities*]: Rhees, *Manual*, p. xx; Bode, *American Lyceum*, pp. 48, 85–86, 131–33, 166–68.
[*science*]: Lurie, *Agassiz*, pp. 125, 128–29, 142–44, 161.
[*Observatory*]: *New York Times*, Aug. 22, 1856; Newcomb, "Abstract Science," p. 107; *DAB* 13: 38–39 ("Ormsby Mitchel"); Mitchel, *Mitchel*, pp. 49–52, 158.
117 [*entirely*]: Mitchel, *Mitchel*, pp. 52–53, 95, 157, 171–73; Russell McCormmach, "Ormsby MacKnight Mitchel's *Sidereal Messenger*, 1846–1848," *APS Proceedings*, Feb. 18, 1966, p. 43; Stephen Goldfarb, "Science and Democracy: A History of the Cincinnati Observatory, 1842–1872," *Ohio History* 78 (1969): 176–78.
[*it*]: Mitchel, *Mitchel*, pp. 158, 160–62; O. Mitchel to J. Frazer, Jan. 28, 1851, Frazer MSS.; F. Barnard to E. Hilgard, Aug. 7 (1856?), Barnard MSS.; J. Frazer to A. Bache, Mar. 14, 1859, Bache MSS.; other slighting remarks about Mitchel as a scientist: J. Rogers to W. Rogers, Dec. 19, 1847, Rogers MSS., and B. Gould to A. Bache, June 12, 1860, Bache MSS.; McCormmach, "Mitchel's *Sidereal Messenger*," passim.
[*developed*]: O. Mitchel to B. Peirce, Feb. 15, 1846, Peirce MSS.; AAAS *Proceedings* 25 (1876): 43; Holden, *Memorials*, pp. 220–21; *Atlantic Monthly* 6 (1860): 117–19 (Newcomb's authorship: B. Gould to A. Bache, June 20, 1860, Bache MSS.); Newcomb, "Abstract Science," p. 107.
118 [American]: Fred B. Joyner, *David Ames Wells, Champion of Free Trade* (Cedar Rapids, Iowa, 1939), pp. 21, 224; Donald Zochert, "Science and the Common Man in Ante-Bellum America," in Reingold, *Science Since 1820*, pp. 8–18, 21, 24–25; J. De Bow to L. Gibbes, Oct. 29, 1846, L. Gibbes MSS., LC.
[*materialism*]: Taylor Stoehr, *Hawthorne's Mad Scientists* (Hamden, Conn., 1978), passim; John D. Davies, *Phrenology, Fad and Science* (New Haven, Conn., 1955), pp. 66–67, 75, 142, 171–72.
[*flowering*]: Charles I. Glicksberg, "William Cullen Bryant and Nineteenth-Century Science," *New England Quarterly* 23 (1950): 91–93; Kent K. Kreuter, "The Literary Response to Science, Technology and Industrialism: Studies in the Thought of Hawthorne, Melville, Whitman and Twain" (Ph.D. diss., University of Wisconsin, 1963), pp. 187, 190; Emerson, *Early Years*, pp. 7–8; Lurie, *Agassiz*, pp. 126–27, 147.

[*stars*]: Robert H. Welker, *Birds and Men* (Cambridge, Mass., 1955), pp. 116–18; Walt Whitman, "When I Heard the Learn'd Astronomer," *Leaves of Grass.*

119 [*spiritual*]: Daniels, *American Science,* pp. 36–41; *Springfield* (Mass.) *Republican,* Aug. 3, 1859.

[*Science*]: Rothenberg, "Educational Background," p. 235; J. Cresson to J. Frazer, Jan. 16, 1853, Frazer MSS.; Kendall, *Mitchell,* pp. 29, 239; diary of W. P. Foulke, June 1, 1854, Foulke MSS.

[*professionals*]: John F. McElligott, "Before Darwin: Religion and Science as Presented in American Magazines, 1830–1860" (Ph.D. diss., New York University, 1972), p. 253; Clark A. Elliott, "The American Scientist, 1800–1863: His Origins, Career, and Interests" (Ph.D. diss., Case Western Reserve University, 1970), pp. 97–98.

120 [*papers*]: Rothenberg, "Educational Background," pp. 234–38; Gary Lee Schoepflin, "Denison Olmsted (1791–1859), Scientist, Teacher, Christian" (Ph.D. diss., Oregon State University, 1977), pp. 355–59.

[*membership*]: Schoepflin, "Olmsted," pp. 30–31; McElligott, "Before Darwin," p. 145; Morehead, *Espy,* pp. 18–20; Rutgers, *Addresses,* p. 11; Howard, *History,* p. 45; H. C. Lehman and P. A. Witty, "Scientific Eminence and Church Membership," *Scientific Monthly* 33 (1931): 545. See also Charles E. Rosenberg, "Science and Social Values in 19th Century America," in Arnold Thackray and Everett Mendelsohn, eds., *Science and Values* (New York, 1974), pp. 23–25.

[*Inquisition*]: R. Gibbes to M. Curtis, July 3, 1847, Curtis MSS.; R. Gibbes to B. Silliman, Nov. 1, 1849, Gratz Collection; R. Gibbes to J. Hall, Nov. 5, 1849, Merrill Collection; Eaton, *Mind,* p. 237; AAAS *Proceedings* 18 (1869): 20–21.

[*1850*]: McElligott, "Before Darwin," pp. 9, 23–26, 30, 60; Herbert Hovenkamp, *Science and Religion in America, 1800–1860* (Philadelphia, 1978), p. 98; De Jong, "American Attitudes," pp. 7–8, 76–78.

121 [*past*]: Ronald L. Numbers, *Creation by Natural Law: Laplace's Nebular Hypothesis in American Thought* (Seattle, Wash., 1977), pp. 8–10, 19–26; Charles C. Gillispie, *Genesis and Geology* (Cambridge, Mass., 1951), pp. 121–35.

[*supernatural*]: Numbers, *Creation,* pp. 27, 55, 63–64, 86–87; Hovenkamp, *Science,* pp. 111, 132–34, 206, 208; De Jong, "American Attitudes," pp. 132, 136.

[*Lyell*]: McElligott, "Before Darwin," pp. 131–39, 201–205; De Jong, "American Attitudes," pp. 114–15; Hovenkamp, *Science,* pp. 136–37.

122 [*1852*]: McElligott, "Before Darwin," pp. 139–44; Hovenkamp, *Science,* p. 104; De Jong, "American Attitudes," pp. 10–11; E. Brooks Holifield, "Science and Religion in the Old South: Scientists and Theologians" (Paper in First Barnard-Millington Symposium), pp. 1–2, 5–7, 9–11, 14; Schoepflin, "Olmsted," p. 357.

[*occasions*]: Hovenkamp, *Science,* pp. 41–44; Stanley M. Guralnick, "Geology and Religion before Darwin: The Case of Edward Hitchcock, Theologian and Geologist (1793–1864)," in Reingold, *Science in America,* pp. 116–17, 122–24; Edward Hitchcock, *Reminiscences of Amherst College* (Northampton, Mass., 1863), p. 291; Philip J. Lawrence, "Edward Hitchcock: The Christian Geologist," APS *Proceedings* 116 (1972): 21–22, 28, 33.

[*deposits*]: Hovenkamp, *Science,* pp. 43–49, 145; Guralnick, "Geology," pp. 124–26, 128; Lawrence, "Hitchcock," pp. 28–34.

[*God*]: McElligott, "Before Darwin," pp. 193–98, 329–32; Holifield, "Science," pp. 17, 22–23.

123 [*nature*]: McElligott, "Before Darwin," pp. 258–59, 270–81, 285, 296–304; De Jong, "American Attitudes," pp. 76–81, 141, 147; William H. Longton, "The Carolina Ideal World: Natural Science and Social Thought in Ante Bellum South Carolina," *Civil War History* 20 (1974): 118–34.

[*praise*]: Milton Millhauser, *Just Before Darwin* (Middletown, Conn., 1959), pp. 32–33, 58–118, 162–63; Johnston, *Notes* 2: 440; William E. Barton, *The Soul of Abraham Lincoln* (New York, 1920), p. 169; M. Curtis to A. Gray, Mar. 21, 1846, Gray Herbarium; Daniels, *American Science,* p. 58; LeConte, *Autobiography,* p. 105.

124 [*decade*]: Millhauser, *Just Before Darwin,* pp. 119–26; Gillispie, *Genesis,* p. 163; Dupree, *Gray,* pp. 143–49; J. Dana to A. Gray, June 17, 1845, Feb. 18, 1846, Gray Herbarium; Daniels, *American Science,* pp. 57–59; *Am. Jour. Sci.,* vol. 48 (1845): 395, vol. 61 (1851): 144, vol. 63 (1852): 302, vol. 75 (1858): 203; Prendergast, "Dana,"

pp. 426–32; Edward J. Pfeifer, "The Reception of Darwinism in the United States, 1859–1880" (Ph.D. diss., Brown University, 1957), pp. 2–5.
[Vestiges]: McElligott, "Before Darwin," pp. 329–32.
[*inferior*]: Stanton, *Leopard's Spots,* pp. 100–101; A. Bache to B. Peirce, July 1, 1860, Peirce MSS.; S. Haldeman to J. Sedgwick, Dec. 1859, Haldeman MSS.; Ledger, "Record of Events: 1875," entry for Dec. 1, 1876, Rhees Collection; Edgar J. McManus, *Black Bondage in the North* (Syracuse, N.Y., 1973), p. 211; J. Henry to E. Loomis, Nov. 29, 1841, Loomis MSS.

125 [*on*]: L. Agassiz to R. Agassiz, Dec. 2, 1846, Agassiz MSS.
[*subject*]: Rodgers, *Torrey,* pp. 204, 206; Lurie, *Agassiz,* pp. 143, 256–66; J. Bachman to V. Audubon, Jan. 6, 1848, Bachman MSS.; L. Agassiz to S. Haldeman, May 2, 1850, Haldeman MSS.; J. Torrey to A. Gray, Aug. 27, 1850, Gray Herbarium.
[*species*]: Stanton, *Leopard's Spots,* pp. 65, 69, 75, 77, 97, 125–26, 133, 136.
[*fundamentalism*]: Stanton, *Leopard's Spots,* pp. 138–44, 155, 173, 193–96; A. Gray to J. Dana, Dec. 13, 1857, Dana MSS.; Dupree, *Gray,* pp. 220, 228–29.

126 [*College*]: Hovenkamp, *Science,* pp. 62–65; Guralnick, "Geology," pp. 118–19.
[*1856–57*]: *DAB* 11: 224–25 ("Tayler Lewis"); Morgan B. Sherwood, "Genesis, Evolution, and Geology in America before Darwin: The Dana-Lewis Controversy, 1856–1857," in Cecil J. Schneer, ed., *Toward a History of Geology* (Cambridge, Mass., 1969), pp. 305–308; Prendergast, "Dana," pp. 360–62; J. Dana to A. Gray, Feb. 8, 1856, Gray Herbarium.

127 [*needed it*]: Sherwood, "Genesis," pp. 309–15; Hovenkamp, *Science,* pp. 142–43; De Jong, "American Attitudes," pp. 301–24; Pfeifer, "Reception," pp. 11–14; Prendergast, "Dana," pp. 385–87; *Bibliotheca Sacra* 13 (1856): 89, 92–94, 98–100.
[*supplanting it*]: Gilman, *Dana,* pp. 184–86; C. Joy to J. Dana, Feb. 18, 1856, Dana MSS.; AAAS *Proceedings* 13 (1859): 10; *Springfield* (Mass.) *Republican,* Aug. 20, 1859.
[*agree*]: Numbers, *Creation,* pp. 105–11; De Jong, "American Attitudes," p. 324; Prendergast, "Dana," pp. 512–16; Sherwood, "Genesis," pp. 315–16.

128 [*doubts*]: Greene, "Science and the Public," p. 24; Houghton, *The Victorian Frame of Mind,* pp. 113–14; *Journal of the Franklin Institute* 7 (1844): 42; H. Ravenel to L. Gibbes, July 17, 1855, L. Gibbes MSS., Charleston.
[*science*]: John F. Kasson, *Civilizing the Machine* (New York, 1976), p. 8; Daniel Webster, *The Writings and Speeches of Daniel Webster,* 18 vols. (Boston, 1903), 4: 116–17; Rhees, *Documents,* I, 549; Joel R. Poinsett, *Discourse on the Objects and Importance of the National Institution for the Promotion of Science* (Washington, D.C., 1841), p. 8.
[*illustration*]: Benjamin F. Greene, *The Rensselaer Polytechnic Institute* (Troy, N.Y., 1855), p. 8; William H. Wahl, *Franklin Institute . . . A Sketch of Its Organization and History* (Philadelphia, 1895), pp. 34–35; *Journal of the Franklin Institute,* vol. 7 (1844): 38, 41, vol. 8 (1844): 375, vol. 16 (1848): 439; AAAS *Proceedings* 17 (1868): 60.
[*gin*]: *North American Review* 75 (1852): 364–83; Benjamin Silliman, Jr., *The World of Science, Art and Industry* (New York, 1854), pp. 7–8.

129 [*world*]: *Athenaeum,* Jan. 15, 1848, quoted in Foote, "A Study of Attitudes," pp. 61–62; *Knickerbocker* 31 (1848): 398, quoted in Aurele A. Durocher, "Verbal Opposition to Industrialism in American Magazines, 1830–1860" (Ph.D. diss., University of Minnesota, 1955), p. 42; Worthington C. Ford, ed., *A Cycle of Adams Letters, 1861–1865,* 2 vols. (Boston, 1920), 1: 135.
[*science*]: Bernard Bowron, Leo Marx, and Arnold Rose, "Literature and Covert Culture," *American Quarterly,* Winter 1957; Leo Marx, *The Machine in the Garden* (New York, 1964), pp. 207–8; Marvin Fisher, *Workshops in the Wilderness* (New York, 1967), pp. 149, 154–57; Kasson, *Civilizing the Machine,* pp. 166–68; Agassiz, *Agassiz,* p. 409; *American Journal of Education,* 2 (1856): 356; *Journal of the Franklin Institute* 11 (1845): 23.

130 [*spiritual*]: Roorbach, *Bibliotheca Americana,* pp. 101, 116, 249; Charles A. Fenton, " 'The Bell-Tower': Melville and Technology," *American Literature* 23 (1951): 221; Kreuter, "Literary Response," pp. 101–103, 173.
[*landscape*]: Marx, *The Machine in the Garden,* pp. 11–32; Kreuter, "Literary Response," pp. 54–55, 81–82, 117, 181–83, 191, 287–88.
[*it*]: Kasson, *Civilizing the Machine,* pp. 114–35.

131 [*exalted*]: Marx, *The Machine in the Garden,* pp. 192, 202, 191–226, passim; *United States Magazine and Democratic Review* 18 (1846): 5.
[*annihilation*]: *American Journal of Education* 2 (1856): 363.
[Nature]: Lurie, *Agassiz,* p. 123; Lyell, *Second Visit* 1: 238; J. Bartlett to A. Gould, Oct. 9, 1851, A. Gould MSS.; Dall, *Baird,* pp. 253–54; Hans Huth, *Nature and the American* (Lincoln, Neb., 1972), p. v.

132 [*fire*]: Donald H. Fleming, *Science and Technology in Providence, 1760–1914* (Providence, R.I., 1952), p. 34; Stanley L. Falk, "Soldier-Technologist: Major Alfred Mordecai and the Beginnings of Science in the United States Army" (Ph.D. diss., Georgetown University, 1959), pp. 1–2; Coulson, *Henry,* pp. 151–54; *DAB* 18: 632 ("Daniel Treadwell"); W. Gibbs to G. Gibbs, Jan. 22, 1846, Gibbs Family MSS.; *Am. Jour. Sci.,* vol. 53 (1847): 102–105, vol. 14 (1852): 21–22; *Journal of the Franklin Institute* 12 (1846): 425, 428; J. Rogers to W. and H. Rogers, Dec. 1, 1845 (actually 1846), Rogers MSS.; M. Curtis to Mrs. M. Curtis, Dec. 3, 1846, Curtis MSS.; Hugo Meier, "American Technology in the Nineteenth Century World," *American Quarterly* 10 (1958): 119, 124.
[*Union*]: *Am. Jour. Sci.* 14 (1852): 21–22; Denison Olmsted, "On the Democratic Tendencies of Science," *American Journal of Education* 1 (1856): 164–72; *United States Magazine and Democratic Review* 18 (1846): 4–16; Rhees, *Documents* 1: 357–58.
[*callings*]: W. Paul Strassmann, *Risk and Technological Innovation* (Ithaca, N.Y., 1959), pp. 186–87; *Scientific American* 6 (1851): 221.
[*supposed*]: Bache, "Wants of Science," p. 49; *Southern Literary Messenger* 4 (1848): pp. 374–76.

133 [*fairs*]: Eugene S. Ferguson, "Expositions of Technology, 1851–1900," in Melvin Kranzberg and Carroll W. Pursell, Jr., eds., *Technology in Western Civilization,* 2 vols. (New York, 1967–68), 1: 707–708; Marcus Benjamin, "The Development of Science in New York City," in James G. Wilson, ed., *The Memorial History of the City of New-York,* 4 vols. (New York, 1893), 4: 435; Bode, *American Lyceum,* p. 149; Allan Nevins, ed., *Polk: The Diary of a President* (New York, 1952), pp. 102–103; *American Journal of Education* 2 (1856): 365–66; Arthur A. Ekirch, Jr., *The Idea of Progress in America* (New York, 1951), pp. 119–20; Lewis Mumford, *Technics and Civilization* (New York, 1934), p. 154.
[*charming*]: Thomas P. Hughes, "Industry through the Crystal Palace" (Ph.D. diss., University of Virginia, 1953), pp. 13, 46, 50–51, 125–26, 203, 256–57, 268.
[*chronograph*]: Robert F. Dalzell, Jr., *American Participation in the Great Exhibition of 1851* (Amherst, Mass., 1960), pp. 20, 37–57, 53; J. Henry to A. Bache, June 14, 1850, Henry MSS.; J. Henry to A. Bache, Apr. 11, 1861, Rhees Collection; J. Tyson to J. Kennedy, June 15, 1851, E. Riddle to E. Maynard, Dec. 31, 1851, Maynard MSS.; Zadock Thompson, *Journal of a Trip to London, Paris, and the Great Exhibition* (Burlington, Vt., 1852), p. vi; Hughes, "Industry," pp. 237, 241–42; *North American Review* 75 (1852): 362–63.
[*articles*]: J. Whitney to W. Whitney, June 28, 1851, Whitney MSS.; *North American Review* 75 (1852): 358.
[*showing*]: Ivan D. Steen, "America's First World's Fair," *New-York Historical Society Quarterly* 47 (1963): 258, 260; C. Detmold et al. to President and Directors, August 1852, Crystal Palace MSS.; Ferguson, "Expositions," p. 714; J. Henry to A. Bache, Oct. 14, 1853, Henry MSS.

134 [*description*]: Ferguson, "Expositions," pp. 714–16; Steen, "America's First World's Fair," pp. 274–76; diary of William H. Brewer, September 1853, Brewer MSS.; J. Bachman to J. and K. Bachman, Sept. 13, 1853, Bachman MSS.; Albert B. Paine, *Mark Twain: A Biography,* 4 vols. (New York, 1935), 1: 94–95.
[*Yankees*]: Gene D. Lewis, *Charles Ellet, Jr.* (Urbana, Ill., 1968), p. 131; Abbe, *Abbe,* p. 83.

10. The Wherewithal of Science

135 [*1876*]: Nathan Reingold, "Definitions and Speculations: The Professionalization of Science in America in the Nineteenth Century," in Oleson and Brown, *Pursuit of Knowledge,* pp. 39, 62–63; see Appendix.
[*pillboxes*]: Charles H. Hart, *Memoirs of Samuel Stehman Haldeman*

(Philadelphia, 1881), p. 6; L. Lesquereux to J. Lesley, Apr. 23, 1864, Lesley MSS.; George B. Wood, *Biographical Memoir of Franklin Bache, M.D.* (Philadelphia, 1865), p. 20; T. Harris to J. LeConte, Dec. 21, 1852, LeConte MSS., ANS.

136 [*business*]: A. Blandy to J. LeConte, June 16, 1847, G. Engelmann to J. LeConte, Sept. 13, 1851, LeConte MSS., APS; E. Cresson to G. Belfrage, Jan. 11, 1872, Belfrage MSS.; G. Engelmann to C. Short, Dec. 8, 1849, Short MSS.; J. Bachman to V. Audubon, Nov. 27, 1845, Bachman MSS.; A. Binney to A. Gould, Dec. 14, 1846, A. Gould MSS.; A. Gould to L. Gibbes, Feb. 17, 1847, L. Gibbes MSS., LC; E. Parrite to S. Haldeman, June 29, 1849, Haldeman MSS.; W. Ingham to G. Cook, July 14, 1853, Cook MSS.; I. Lea to E. Ravenel, Apr. 28, 1857, E. Ravenel MSS.

[*Institute*]: Taylor, *Transportation Revolution,* pp. 394–95; Oberholtzer, *Philadelphia* 2: 258; Miller, *Dollars for Research,* p. 139.

[*public*]: A. Sonntag to C. Schott, Feb. 29, 1860, Schott MSS.; I. Hayes to A. Bache, Mar. 24, 1860, Bache MSS.; A. Bache to F. Lieber, Mar. 14, 1860, Lieber MSS.; W. Foulke to A. Bache, May 31, 1860, Bache MSS.; W. Chauvenet to A. Bache, Apr. 25, 1860, Bache MSS.; I. Hayes to A. Bache, Mar. 30, 1860, Bache MSS.; *DAB,* 8: 445 ("Isaac Hayes"), 8: 2 ("Henry Grinnell").

[*towns*]: See Appendix; Smith and Ginsburg, *History of Mathematics,* p. 68; Rudolph, *American College,* p. 193; J. Henry to Dr. Bullions, Aug. 5, 1846, Gratz Collection; J. Wilder to C. Dewey, Aug. 24, 1850, Dewey MSS.; A. Stewart to R. Caruthers, July 16, 1850, Caruthers MSS.; Bruce, *University of Virginia* 3: 98; E. Everett to E. Horsford, Feb. 19, 1847, Everett MSS.; Lurie, *Agassiz,* p. 239; Bond, *History of Harvard Observatory,* pp. lxxx, 1; Lyell, *Second Visit* 1: 159, 191–92.

137 [*mid-fifties*]: Guralnick, *Science,* pp. 135, 143–44; Burke, *American Collegiate Populations,* pp. 47–48; Dupree, *Gray,* p. 199; Lurie, *Agassiz,* p. 239; Nevins, *Diary of Strong* 2: 337; A. Longstreet, Nov. 18, 1858, S. Maupin, Nov. 12, 1858, A. Church, Nov. 12, 1858, F. Barnard, Nov. 15, 1858, L. Garland, Nov. 23, 1858, to D. L. Swain, Univ. of N.C. Archives; Viles, *University of Missouri,* pp. 49, 64; W. Chauvenet to A. Mayer, May 8, 1860, Hyatt-Mayer Collection; D. Kirkwood to J. Warner, July 17, 1858, Warner MSS.; J. Doremus to J. Kimberly Nov. 27, 1857, Kimberly MSS.

[*example*]: C. Gilman to C. Wetherill, July 2, 1856, Wetherill MSS.; Nevins, *Diary of Strong* 2: 337; Bruce, *University of Virginia* 3: 101; H. Tappan to C. Abbe, July 30, 1857, Abbe MSS.; J. Cabell to J. Henry, July 23, 1852, Henry MSS.

[*simultaneously*]: J. Henry to Dr. Bullions, Aug. 5, 1846, Gratz Collection; *DAB* 19: 593 ("John W. Webster"); Lurie, *Agassiz,* p. 243; Jean W. Sidar, *George Hammell Cook* (New Brunswick, N.J., 1976), pp. 47, 51, 56; Rudolph, *American College,* p. 195; J. Torrey to L. Beck, Oct. 24, 1848, Beck MSS.; J. Johnston to J. Torrey, Feb. 28, 1854, Trent Collection; Guralnick, *Science,* pp. 145–46.

[*it*]: A. Stewart to R. Caruthers, Sept. 13, 1850, Caruthers MSS.; Viles, *University of Missouri,* p. 64; Bruce, *University of Virginia* 3: 98–101; E. Horsford to J. Hall, May 1, 1846, Hall MSS.; Henry James, *Charles W. Eliot,* 2 vols. (Boston, 1930), 1: 99–101.

[*upraised*]: Ten-hour loads: F. Wurdemann to L. Gibbes, June 24, 1846, L. Gibbes MSS., LC (at University of Georgia), J. Woodrow to J. Kimberly, Nov. 11, 1857, University of North Carolina Archives (at Oglethorpe College), LeConte, *Autobiography,* p. 166 (at University of S.C., 1857), Dall, *Baird,* p. 183 (at Dickinson College, 1848); fourteen or fifteen-hour loads: A. Caswell to E. Loomis, Dec. 12, 1848, Loomis MSS. (at Brown University), Dall, *Baird,* p. 152 (at Dickinson College, 1846); R. McCulloch to A. Bache, Feb. 23, 1857, Rhees Collection; A. Caswell to E. Loomis, Dec. 20, 1848, Loomis MSS.; petition of W. Gibbs to L. Bradish, 1851, Bradish MSS.; Hal Bridges, *Lee's Maverick General, Daniel Harvey Hill* (New York, 1961), p. 24.

138 [*research*]: J. Henry to B. Peirce, Dec. 30, 1845, Peirce MSS.; *North American Review* 95 (1862): 425; Guralnick, *Science,* p. ix; Rodgers, *Torrey,* p. 202; C. Joy to W. Whitney, Jan. 3, 1854, Whitney MSS.; John H. Van Amringe, *An Historical Sketch of Columbia College* (New York, 1876), p. 73.

[*students*]: Bache, "Wants of Science," pp. 18, 76; AAAS *Proceedings* 6 (1851): xlv; J. Cooke to G. Brush, June 20, 1857, Brush Family MSS.; Denison Olmsted, *The Lecture Delivered before the American Institute of Instruction at Hartford, August, 1845* (Boston, 1846), pp. 87, 91–94; J. Henry to "My dear sir," Aug. 13, 1846, to A. Bache, Jan. 9, 1860, Henry MSS.

[*professors*]: *New York Quarterly* 2 (1853): 448; C. Jackson to J. Whitney, Sr., Feb. 11, 1844, J. Whitney, Jr., to J. Whitney, Sr., Jan. 16, 1844, Whitney MSS.; J. Henry to Dr. Bullions, Aug. 5, 1846, Henry MSS.; R. Gibbes to L. Gibbes, Aug. 9, 1846, L. Gibbes MSS., LC; F. Barnard to E. Herrick, May 17, 1854, Barnard MSS.; E. Everett to E. Horsford, Feb. 19, 1847, Everett MSS.; A. Gray to C. Short, July 5, 1855, Short MSS.; B. Gould to L. Gibbes, Nov. 12, 1849, L. Gibbes MSS., LC.

139 [*DAB*]: Bache, "Wants of Science," p. 56.

[*commence*]: Guralnick, *Science,* pp. 145–46; Lurie, *Agassiz,* p. 129; J. Frazer to S. Haldeman, Nov. 11, 1849, Haldeman MSS.; Fleming, *Draper,* p. 31; Eugene Exman, *The Brothers Harper* (New York, 1965), pp. 274–75; Dupree, *Gray,* p. 203; Chittenden, *Sheffield Scientific School* 1: 46; Anson P. Stokes, *Memorials of Eminent Yale Men,* 2 vols. (New Haven, Conn., 1914), 2: 84; *DAB* 11: 398–99 ("Elias Loomis"); J. Dana to F. Genth, Sept. 15, 1858, Genth MSS.; Edgar F. Smith, *Charles Mayer Wetherill* (New York, 1929), pp. 37–38.

[*benefit*]: see Appendix.

140 [*geology*]: J. Whitney, Jr., to J. Whitney, Sr., Aug. 1, 1845, Whitney MSS.; J. Boynton to J. Frazer, Dec. 14, 1848, Frazer MSS.; W. MacRae to brother, May 10, 1860, MacRae MSS.

[*deposits*]: C. Jackson to J. Whitney, May 7, 1847, Whitney MSS.; Brewster, *Whitney,* pp. 122, 130–35, 145.

[*jobs*]: Harry R. Warfel, ed., *Uncollected Letters of James Gates Percival* (Gainesville, Fla., 1959), pp. xix, 57; Julius H. Ward, *The Life and Letters of James Gates Percival* (Boston, 1866), p. 481; Brewster, *Whitney,* pp. 150–55.

[*mining*]: Rodman W. Paul, *California Gold* (Cambridge, Mass., 1947), pp. 303–307; AAAS *Proceedings* 31 (1882): 17; C. Jackson, "Remarks on Mining Operations," Sept. 2, 1846, American Association of Geologists and Naturalists MSS.; J. Bailey to T. Ringgold, May 30, 1852, Brock Collection.

141 [*life*]: Harold F. Williamson and Arnold R. Daum, *The American Petroleum Industry: The Age of Illumination, 1859–1899* (Evanston, Ill., 1959), pp. 68–72.

[*rods*]: Ibid., pp. 90–91.

[*fever*]: Brewster, *Whitney,* pp. 238–39; J. Henry to Meads, Oct. 11, 1872, Henry MSS.; Clarke, *Hall,* pp. 260–61.

142 [*it*]: Fulton and Thomson, *Silliman,* pp. 196–97; Fisher, *Silliman* 2: 271; B. Silliman, Sr., to daughter, July 4, 1860, Silliman Family MSS.; B. Silliman, Jr., to F. Genth, Dec. 25, 1853, Apr. 9, 1854, Genth MSS.; B. Silliman, Jr., to J. P. Lesley, Aug. 5, 11, 17, 23, 1854, Lesley MSS.

[*science*]: Williamson and Daum, *American Petroleum Industry* 1: 72, 74; J. Dana to G. Brush, Apr. 24, 1855, Brush Family MSS.; B. Silliman, Sr., to "My very dear child," Sept. 12, 1855, Silliman Family MSS.; B. Silliman, Jr., to T. Hunt, Sept. 10, 1855, Dana Family MSS.; J. Dana to G. Brush, Oct. 8, 1855, Brush Family MSS.

[*groping*]: *Scientific American* 7 (1851): 85; Peter C. Welsh, "A Craft That Resisted Change," *Technology and Culture* 4 (1963): 317; B. Silliman, Jr., to H. Wurtz, Mar. 12, 1860, Wurtz MSS.; Siegfried, "Chemical Research Publications," pp. 89, 96.

143 [*ore*]: *Journal of the Franklin Institute* 10 (1845): 67–68; Cyril S. Smith, "The Interaction of Science and Practice in the History of Metallurgy," *Technology and Culture* 2 (1961): 363; *DAB* 9: 392 ("Robert Hunt"); Smith, *Booth,* p. 15.

[*1880*]: Siegfried, "Chemical Research Publications," pp. 81, 84; Deming Jarves, *Reminiscences of Glass-Making* (New York, 1865), p. 4, pp. 52–104 passim; L. Crawford to D. C. Rand & Co., Nov. 28, 1858, Rand MSS.; William C. Geer, *The Reign of Rubber* (New York, 1922), pp. 19, 24; Joe B. Frantz, *Gail Borden* (Norman, Okla., 1951), pp. 224–25; Thomas C. Cochran, *The Pabst Brewing Company* (New York, 1948), pp. 112–13.

[*end*]: *DAB,* 2: 448 ("James Booth"), 13: 162 ("Campbell Morfit"); *Journal of the Franklin Institute* 15 (1848): 123; Williams Haynes, *American Chemical Industry: A History,* 6 vols. (New York, 1954), 1: 148–50.

[*company*]: Haynes, *American Chemical Industry* 1, 158, 298–99; Child, *Tools,* pp. 150–51.

144 [*engineering*]: Charles A. Browne, "Early Philadelphia Sugar Refiners and Technologists," *Journal of Chemical Education* 20 (1943): 522–25; *Journal of the Franklin Institute* 10 (1845): 405.

[*process*]: Haynes, *Chemical Pioneers,* pp. 58, 69; Haynes, *American Chemical*

Industry, 1: 135–36; Edwin T. Freedley, *Leading Pursuits and Leading Men* (Philadelphia, 1856), p. 175.

[*breed*]: Haynes, *Chemical Pioneers,* pp. 26, 30, 89; Haynes, *American Chemical Industry* 1: 184, 216; S. Garrigues to G. Cook, Mar. 3, 1862, Cook MSS.; Collamer M. Abbot, "Isaac Tyson Jr., Pioneer Mining Engineer and Metallurgist," *Maryland Historical Magazine,* March 1965, p. 17.

[*cases*]: Haynes, *American Chemical Industry* 1: 235–36; Haynes, *Chemical Pioneers,* pp. 79–80.

145 [*company*]: Brewster, *Whitney,* pp. 77–79; C. Gilman to C. Wetherill, July 2, 1856, Wetherill MSS.; J. Smith to G. Brush, Sept. 18, 1854, Brush Family MSS.; Osborne, *From the Letter-Files,* p. 22; Larson, "Chandler," pp. 21–22; Smith, *Wetherill,* p. 33; J. Cooke to G. Brush, June 20, 1857, Brush Family MSS.

[*fifties*]: J. Booth to G. Cook, June 20, 1845, Cook MSS.; Smith, *Wetherill,* pp. 22–24; F. Genth to H. Wurtz, Aug. 23, 1855, Wurtz MSS.; F. Genth to G. Brush, Feb. 14, 1859, Brush Family MSS.

[*later*]: Smith, *Booth,* p. 4; C. Jackson to J. Whitney, Sr., Feb. 11, 1844, Whitney MSS.; John J. Beer, "Coal Tar Dye Manufacture and the Origins of the Modern Industrial Research Laboratory," *Isis* 49 (1958): 123–25, 131; Thomas K. Derry and Trevor I. Williams, *A Short History of Technology* (New York, 1960), p. 703; Siegfried, "Chemical Research Publications," p. 108; W. Craw to W. Brewer, Feb. 17, 1854, Brewer MSS.; J. Smith to G. Brush, Jan. 16, 1856, Brush Family MSS.

[*1849–50*]: Percy W. Bidwell and John I. Falconer, *History of Agriculture in the Northern United States, 1620–1860* (Washington, D.C., 1925), p. 320; Rossiter, *Emergence,* pp. 3–26, 29; Margaret W. Rossiter, "The Organization of Agricultural Improvement in the United States, 1785–1865," in Oleson and Brown, *Pursuit of Knowledge,* p. 291; Paul W. Gates, *The Farmer's Age* (New York, 1960), pp. 356–57, 361–62.

146 [*farms*]: Ernest M. Law et al., *State Agricultural Experiment Stations* (Washington, D.C., 1962), pp. 9, 11–12; Gates, *Farmer's Age,* p. 363; John P. Norton, *Address before the Hampshire, Franklin, and Hampden Agricultural Society* (Northampton, Mass., 1849), pp. 3–5, 17; John P. Norton, *Address . . . before the Ontario Co. Agricultural Society* (Canandaigua, N.Y., 1850), pp. 9–15, 18.

[*burst*]: Rossiter, *Emergence,* pp. 45–46, 114–15, 118; Gates, *Farmer's Age,* pp. 316–17, 346.

[*restored*]: Rossiter, *Emergence,* pp. 30–35, 109, 121–23; Rossiter, "Organization," p. 292.

[*footing*]: Rossiter, *Emergence,* pp. 124–25.

147 [*journal*]: *DAB* 15: 96 ("John A. Porter"); Rossiter, *Emergence,* pp. 129–35, 141–42, 150–55; Osborne, *Letter-Files,* pp. 85–86, 97–98, 100; *Am. Jour. Sci.* 82 (1861): 249.

[*significance*]: Rossiter, *Emergence,* p. 141; Hans Jenny, *E. W. Hilgard and the Birth of Modern Soil Science* (Pisa, Italy, 1961), pp. 7, 13–14, 18.

[*research*]: Rossiter, "Organization," pp. 292–93; Rossiter, *Emergence,* p. 174; Johnson, *Scientific Interests,* pp. 17, 41; Lewis C. Gray, *History of Agriculture in the Southern United States to 1860,* 2 vols. (Washington, D.C., 1933), 2: 701; Gates, *Farmer's Age,* p. 317; Osborne, *Letter-Files,* pp. 96, 129; Rosenberg, *No Other Gods,* pp. 146–47.

[*afterward*]: S. Haldeman to E. Evans, Apr. 23, 1849, F. Hough to S. Haldeman, Feb. 21, 1851, Haldeman MSS.; AAAS *Proceedings* 29 (1880): 610; Howard, *History,* pp. 10, 12–13, 30–31; *DAB* 8: 321–22 ("Thaddeus Harris").

148 [*period*]: Howard, *History,* pp. 10, 36–40, 43–48, 180.

[*it*]: Barnard, *Letters,* p. 61; Loomis, *Recent Progress,* pp. 288–89; Jones and Boyd, *Harvard Observatory,* pp. 159–61.

[*electroplating*]: Robert L. Thompson, *Wiring a Continent* (Princeton, N.J., 1947), pp. 248, 440–41; Hughes, "Industry," pp. 58, 82; W. James King, *The Development of Electrical Technology in the Nineteenth Century* (Washington, D.C., 1962), pp. 260, 263–65; Rood, *Practical Value,* pp. 11–12.

[*insignificance*]: Dupree, *Science in the Federal Government,* pp. 103–106.

149 [*existed*]: Agassiz, *Agassiz,* p. 438; Rosenberg, *No Other Gods,* p. 147.

[*strings*]: L. Agassiz to R. Agassiz, Dec. 2, 1846, Agassiz MSS.; Smithsonian, *Annual Report* 1 (1847): 21; Rosenberg, *No Other Gods,* pp. 143–48.

[*NASA*]: *United States Magazine and Democratic Review* 18 (1846): 15; *North*

American Review 75 (1852): 364; *Scientific American* 5 (1850): 173; AAAS *Proceedings* 8 (1855): 292–93, 297; Kendall Birr, "Science in American Industry," in David D. Van Tassel and Michael G. Hall, eds., *Science and Society in the United States* (Homewood, Ill., 1966), p. 44; Nathan Reingold, "Reflections on 200 Years of Science in the United States," in Reingold, *Sciences,* p. 18; Hughes, "Industry," p. 151; Nathan Reingold, "Cleveland Abbe at Pulkowa," *Archives Internationales d'Histoire des Sciences* 17 (1964): 140–47; Alexander D. Bache, *Anniversary Address before the American Institute . . . October 28th, 1856* (New York, 1856), p. 28.

11. The Technological Connection

150 [*gain*]: Edwin T. Layton, Jr., "Mirror-Image Twins: The Communities of Science and Technology," in Daniels, *Nineteenth-Century American Science,* pp. 210–30, passim.

151 [*states*]: See Appendix.

[*experience*]: See Appendix.

[*fraternal*]: Marjorie H. Ciarlante, "A Statistical Profile of Eminent American Inventors, 1700–1860: Social Origins and Role" (Ph.D. diss., Northwestern University, 1978), pp. 193, 259, 267, 330, 333; *Atlantic Monthly,* vol. 2 (1858): 90, vol. 19 (1867): 528, 532; Thomas Ewbank, *The World a Workshop* (New York, 1855), p. 197 and passim; AAAS *Proceedings* 8 (1855): 295–96; Grace R. Cooper, *The Invention of the Sewing Machine* (Washington, D.C., 1968), p. 139; Jacob Schmookler, *Invention and Economic Growth* (Cambridge, Mass., 1966), pp. 199–206 and passim; I. Ingham to G. Cook, Jan. 4, 1856, Cook MSS.; Diary of B. Wailes, Oct. 5, Nov. 2, 23, 1859, Wailes MSS.; Willard I. Toussaint, "Biography of an Iowa Businessman: Charles Mason, 1804–1882" (Ph.D. diss., State University of Iowa, 1963), p. 343.

152 [*briefly*]: AAAS *Proceedings* 8 (1855): 294; Bruce Sinclair, *Philadelphia's Philosopher Mechanics: A History of the Franklin Institute, 1824–1865* (Baltimore, 1975), pp. 195–97, 205–13, 216; *DAB,* 2:80 ("Alfred Beach"), 13:328–29 ("Orson Munn"); *Scientific American,* vol. 1, nos. 18, 25, 50 (1846); vol. 2 (1846): 7, 21–22, 37, vol. 6 (1851): 213; L. Gale to B. Wailes, June 15, 1847, Wailes MSS.; Beaver, "American Scientific Community," p. 24; *Journal of the Franklin Institute,* vol. 10 (1845): Appendix, vol. 13 (1847): 41–43, vol. 14 (1847): 285–86.

[*jobs*]: See Appendix.

[*construction*]: Daniel H. Calhoun, *The American Civil Engineer* (Cambridge, Mass., 1960), pp. 22, 30, 58–61, 182, 194; see Appendix; F. Gardner to "Dear Captain," Dec. 6, 1853, L. Fleming to "My dear Mac," Aug. 7, 1848, H. MacRae to D. MacRae, Aug. 4, 1850, MacRae MSS.

[*engineers*]: W. Gwynn to J. MacRae, Sept. 7, 1850, J. MacRae to D. MacRae, Feb. 12, 1850, H. MacRae to D. MacRae, Aug. 4, 1850, H. MacRae to brother, Sept. 18, 1853, MacRae MSS.; E. Morris to J. Warner, June 19, 1860, N. Jones to J. Warner, Jan. 8, 1860, Warner MSS.; James M. Searles, *Life and Times of a Civil Engineer* (Cincinnati, 1893), pp. 18, 24, 32–33; H. Pomeroy to G. Cook, May 26, Oct. 3, 1853, Cook MSS.; E. Daniels to A. Randall, c. 1858, Daniels MSS.; J. Trautwine to father, Dec. 25, 1843, Trautwine MSS.; Freeman Cleaves, *Meade of Gettysburg* (Norman, Okla., 1960), p. 17; Lewis, *Ellet,* pp. 108–109; James Worrall, *Memoirs of Colonel James Worrall, Civil Engineer* (Harrisburg, Pa., 1887), p. 53.

153 [*to*]: H. Bacon to Mr. West, Sept. 11, 1847, S. Putnam to H. Bacon, Sept. 12, 1854, E. Jennings to H. Bacon, Sept. 12, 1850, Jan. 28, 1853, M. Bacon to H. Bacon, Feb. 22, Apr. 7, 16, June 15, 1853, May 5, Dec. 14, 1854, H. Bacon to J. Hall, May 12, 1851, T. Williams to H. Bacon, July 15, 1850, J. Ilsley to H. Bacon, Feb. 10, 1851, J. Bacon to H. Bacon, June 8, 1854, Bacon MSS.; H. MacRae to brother, May 9, 1852, Jan. 28, 1853, H. MacRae to D. MacRae, Aug. 4, 1850, L. Fleming to J. MacRae, May 16, 1848, J. MacRae to D. MacRae, Mar. 28, 1850, W. Gwynn to J. MacRae, Sept. 19, 1850, F. Gardner to "Dear Captain," Feb. 27, 1854, MacRae MSS.; Danforth H. Ainsworth, *Recollections of a Civil Engineer* (Newton, Iowa, 1893), pp. 7–8, 20; N. Jones to J. Warner, Feb. 14, 1861, Warner MSS.; William H. Wilson, *Reminiscences of a Railroad Engineer* (Philadelphia, 1896), pp. 31, 33, 38; Lewis, *Ellet,* pp. 103, 155; Calhoun, *American Civil Engineer,* pp. 141, 182; W. Gurley to G. Cook, June 12, 1855, Aug. 27, 1858, June 9, 1865, S. Greele to G. Cook, May 14, 1852, Cook MSS.; G. Stengel to J. Hilgard, July 7, 1850, S. Tyndale to J. Hilgard, Sept. 12, 1854, Hilgard MSS.; M. Thayer to G. Dallas, Nov. 11, 1857, Rhees Collection.

[*it*]: E. Morris to J. Warner, June 19, 1860, Warner MSS.; Charles B. Stuart, *Lives and Works of Civil and Military Engineers of America* (New York, 1871), pp. 180, 190; J. MacRae to D. MacRae, Oct. 3, 1848, MacRae MSS.; J. Trautwine to E. Trautwine, Apr. 14, 1858, Trautwine MSS.; Raymond H. Merritt, *Engineering in American Society, 1850–1875* (Lexington, Ky., 1969), p. 119.

[*Pennsylvania*]: Calhoun, *American Civil Engineer,* p. 193; E. Morris to J. Warner, Sept. 15, 1859, Warner MSS.; Worrall, *Memoirs,* p. 67; J. MacRae to D. MacRae, Sept. 14, 1850, MacRae MSS.; Lewis, *Ellet,* p. 108; Merritt, *Engineering,* pp. 63–83.

[*time*]: J. Trautwine to father, Dec. 26, 1843, Trautwine MSS.; Charles W. Hunt, *Historical Sketch of the American Society of Civil Engineers* (New York, 1897), pp. 9–14, 86; Wilson, *Reminiscences,* p. 32; Calhoun, *American Civil Engineer,* pp. 185–86; Merritt, *Engineering,* pp. 99–100; *Scientific American* 5 (1850): 157, 381; John B. Babcock, "The Boston Society of Civil Engineers and Its Founder Members," *Boston Society of Civil Engineers Journal* 23 (1936): 153–54, 157–58.

[*viable*]: Hunt, *Historical Sketch,* pp. 16–17, 20–21, 23–32; Charles W. Hunt, "The First Fifty Years of the American Society of Civil Engineers," American Society of Civil Engineers *Transactions* 48 (1902): 220–24; American Society of Civil Engineers *Transactions* 1 (1872): 3–5.

154 [*credit*]: Poinsett, *Discourse,* p. 31; Brewster, *Whitney,* p. 122; Charles T. Jackson, "Remarks on Mining Operations," Sept. 2, 1846, in American Association of Geologists and Naturalists MSS.; L. Janin to father, May 8, 1859, Janin MSS.; J. Whitney to J. Hall, Dec. 29, 1854, Hall MSS.; J. Hague to G. Hale, Nov. 5, 1863, Jan. 1, 1864, and October 1863 to May 1865, passim, Hague MSS.

[*professionalism*]: L. Janin to parents, Mar. 29, 1863, Janin MSS.; Rodman W. Paul, *Mining Frontiers of the Far West, 1848–1880* (New York, 1963), p. 34; see Appendix.

[*engineering*]: See Appendix.

155 [*culture*]: Monte A. Calvert, *The Mechanical Engineer in America, 1830–1910* (Baltimore, 1967), pp. 3–7, 12–13, 277–78.

[*Engineer*]: Ibid., pp. 14–22.

[*once*]: Ibid., pp. 107–108.

[*engineers*]: E. Blake to Mrs. F. Orcutt, Mar. 5, 1868, Blake Family MSS.; Stokes, *Memorials* 2: 121–22; Charles W. Marsh, *Recollections, 1837–1910* (Chicago, 1910), p. 90; Nehemiah Cleaveland, *A Memoir of Erastus Brigham Bigelow* (Boston, 1860), pp. 13–14; James K. Finch, "Engineering and Science: A Historical Review and Appraisal," *Technology and Culture* 2 (1961), 325–26; John B. Rae, "The 'Know-How' Tradition: Technology in American History," *Technology and Culture* 1 (1960): 141–43; Eugene S. Ferguson, *Kinematics of Mechanisms from the Time of Watt* (Washington, D.C., 1962), pp. 209–15, 219–20, 223, 225; Robert S. Woodbury, *History of the Gear-Cutting Machine* (Cambridge, Mass., 1958), pp. 5, 31, 38–41, 77, 80; Thomas W. Harvey, *Memoir of Hayward Augustus Harvey* (New York, 1900), pp. 3–4, 28; Edwin T. Layton, Jr., "Technology as Knowledge," *Technology and Culture* 15 (1974): 36; Eugene S. Ferguson, "The Mind's Eye: Nonverbal Thought in Technology," *Science* 197 (1977): 827–28, 834; Brooke Hindle, *Emulation and Invention* (New York, 1982), pp. 133–42.

156 [Journal]: Sinclair, *Philadelphia's Philosopher Mechanics,* pp. 29–32, 135–36, 140–49, 176–84, 284–89.

[*century*]: *DAB* 2:529–30 ("Uriah Boyden"); Louis C. Hunter, *Waterpower in the Century of the Steam Engine* (Charlottesville, Va., 1979), pp. 328–29, 333, 335; *DAB* 4:578–79 ("James Francis"); Edwin T. Layton, Jr., "Millwrights and Engineers, Science, Social Roles, and the Evolution of the Turbine in America," in *Sociology of the Sciences: A Yearbook, The Dynamics of Science and Technology* (Dordrecht, Netherlands, 1978), pp. 77, 79; Edwin T. Layton, Jr., "Scientific Technology, 1845–1900: The Hydraulic Turbine and the Origins of American Industrial Research," *Technology and Culture* 20 (1979): pp. 72–73, 76, 78–82.

157 [*efficiency*]: Robert H. Thurston, *History of the Growth of the Steam Engine* (New York, 1878), pp. 381–90, 456; John B. Rae, "Energy Conversion," in Kranzberg and Pursell, *Technology* 1:343; Howard Corning, ed., "A Letter from John Griffen, Ironmaster, 1878," *Journal of Economic and Business History* 3 (1931): 695–97; *DAB* 1:199 ("John F. Allen"); Charles T. Porter, *Engineering Reminiscences* (New York, 1908), pp. 42, 47, 58; *Journal of the Franklin Institute* 47 (1864): 165, 70 (1875): 410.

[*researches*]: Edward W. Sloan III, *Benjamin Franklin Isherwood, Naval Engineer* (Annapolis, Md., 1965), pp. 80–99.

[*armament*]: Falk, "Soldier-Technologist," pp. 57, 83, 228–29, 269, 284, 286–88, 344–55, 386, 410, 482, 594–95.

158 [*Rodman's*]: *DAB* 16:80 ("Thomas Rodman"); Edward G. Williams, "Pittsburgh, Birthplace of a Science," *Western Pennsylvania Historical Magazine,* September 1962, pp. 268–69; B. M. Peirce to B. O. Peirce, Nov. 25, 1859, Peirce MSS.; Theodore F. Rodenbough and William L. Haskins, eds., *The Army of the United States, 1789–1869* (New York, 1896), pp. 130–31; *DAB* 5:29–30 ("John Dahlgren"); Robert V. Bruce, *Lincoln and the Tools of War* (Indianapolis, 1956), pp. 5–6.

[*success*]: E. Appleton to H. Bacon, May 5, 1847, Bacon MSS.; *DAB* 20:70–71 ("Squire Whipple"); Squire Whipple, "On Truss Bridge Building," American Society of Civil Engineers, *Transactions* 1 (1870): 243; Carl W. Condit, *American Building* (Chicago, 1968), pp. 98–100; Layton, "Mirror-Image Twins," p. 223; *DAB* 8:400 ("Herman Haupt"); Herman Haupt, *Reminiscences of General Herman Haupt* (Milwaukee, 1901), pp. xiv–xvi; James A. Ward, *That Man Haupt* (Baton Rouge, La., 1973), pp. 19–20; Richard S. Kirby et al., *Engineering in History* (New York, 1956), p. 231.

[*builders*]: *DAB* 6:87 ("Charles Ellet"); Lewis, *Ellet,* p. 120; David B. Steinman, *The Builders of the Bridge* (New York, 1945), pp. 10–14, 80–81, 97, 170–72; Hamilton Schuyler, *The Roeblings* (Princeton, N.J., 1931), p. 129.

159 [*cannon*]: Carl W. Condit, "Buildings and Construction," in Kranzberg and Pursell, *Technology* 1:374–75; Sinclair, *Philadelphia's Philosopher Mechanics,* pp. 178–85; Layton, "Mirror-Image Twins," pp. 219–21.

[*tons*]: John E. Watkins, "The Development of the American Rail and Track," *Report of the National Museum, 1888–89* (Washington, D.C., 1889), p. 674; Esmond Shaw, *Peter Cooper and the Wrought Iron Beam* (New York, 1960), pp. 19–21, 25; Allan Nevins, *Abram S. Hewitt* (New York, 1935), p. 117; Stephen D. Tucker, "History of R. Hoe & Company, New York" (MSS. in Library of Congress), p. 68.

[*Caltech*]: Treadwell quoted in Stone, "Learned Societies," p. 414.

[*view*]: B. Hedrick to "My dear Sir," May 30, 1853, Hedrick MSS.; Bruce Sinclair, "The Promise of the Future: Technical Education," in Daniels, *Nineteenth-Century American Science,* p. 267; Merritt, *Engineering,* pp. 55–56; Louis How, *James B. Eads* (Boston, 1900), p. 71.

160 [*school*]: *DAB* 18:98 ("Charles Storrow"); Lewis, *Ellet,* pp. 87–92; Falk, "Soldier-Technologist," pp. 255–58, 268, 426–30, 435; B. Silliman to J. Whitworth, Jan. 13, 1858, Silliman Family MSS.; Henry T. Cheever, *Autobiography and Memorials of Ichabod Washburn* (n. p., 1879), pp. 109–10; Chittenden, *Sheffield School* 1:18–21, 69–70; Greene, *Rensselaer,* pp. 9–10, 12; *DAB* 7:299 ("Daniel Gilman"); Daniel C. Gilman, "Scientific Schools of Europe," *American Journal of Education* 1 (1856): 315–25.

[*sixties*]: G. Brush to B. Silliman, Jr., Mar. 5, 1855, A. Rockwell to G. Brush, Nov. 14, 1858, Brush Family MSS.; Pumpelly, *Reminiscences* 1:118, 127–28; L. Janin to father, Mar. 21, Nov. 14, Dec. 5, 1858, Jan. 23, July 13, 1859, Feb. 7, Mar. 6, June 18, 1860, L. Janin to mother, June 11, 1865, Janin MSS.; Greene, *Rensselaer,* p. 47; W. Kitchell to parents, Oct. 3, Dec. 20, 1852, Kitchell MSS.; Clark C. Spence, *Mining Engineers & the American West* (New Haven, Conn., 1970), pp. 26–27, 31–32; *American Journal of Education* 9 (1860): 167–69.

[*professors*]: Rudolph, *American College,* pp. 228–29; Goetzmann, *Army Exploration,* pp. 14–15; Bache, *Anniversary Address,* p. 11; Sinclair, "Promise," p. 257n.

[*defense*]: Calhoun, *American Civil Engineer,* pp. 207–208; Falk, "Soldier-Technologist," pp. 528–30; D. Mahan to J. Phillips, Feb. 5, 1852, University MSS., Southern Historical Collection; Schofield, *Forty-Six Years,* pp. 26–28.

161 [*leaving*]: *DAB* 4:43–44 ("William Chauvenet"); J. Coffin to S. Newcomb, Apr. 13, 1864, Newcomb MSS.; W. Chauvenet to A. Bache, Apr. 25, 1860, Bache MSS.

[*engineering*]: Charles R. Mann, *A Study of Engineering Education* (New York, 1918), pp. 9, 11–12, 37; Samuel Rezneck, *Education for a Technological Society* (Troy, N. Y., 1968), pp. 71, 73, 78–80, 82–83, 93–94, 99–101, 111–15, 121, 123–24; Greene, *Rensselaer,* pp. 3–4, 70; Palmer C. Ricketts, *History of the Rensselaer Polytechnic*

Institute, 1824–1894 (Troy, N.Y., 1895), pp. 94, 96–97, 104; W. Wilson to A. Bache, Mar. 13, 1856, Bache MSS., Smithsonian.

[*century*]: Sinclair, "Promise," pp. 261–63; Spence, *Mining Engineers,* p. 37; John W. Oliver, *History of American Technology* (New York, 1956), p. 244; Edward C. Mack, *Peter Cooper, Citizen of New York* (New York, 1949), pp. 131–32, 247, 264, 266.

162 [*land*]: Rogers, *Rogers* 1:36, 257–59, 421–27.

[*defeat*]: Ibid., vol. 1:274, 288, 333–34, vol. 2:2–77, passim.

[*college*]: Calhoun, *American Civil Engineer,* p. 46.

[*heed*]: Chittenden, *Sheffield School,* pp. 46–47, 50–51.

163 [*desk*]: *Am. Jour. Sci.,* vol. 54 (1847): 305, vol. 57 (1849), 153; Gilman, *Dana,* p. 163; Chittenden, *Sheffield School,* p. 48; O. Rood to H. Wurtz, Feb. 26, 1849, Wurtz MSS.; J. Willet to G. Brush, Aug. 21, 1860, Brush Family MSS.

[*life*]: Chittenden, *Sheffield School,* pp. 50–53, 55–63; Miller, *Dollars,* pp. 92–93; *Memorials of John Pitkin Norton,* p. 64.

[*Sheffield Scientific School*]: Chittenden, *Sheffield School,* pp. 63, 66–74; *DAB* 17:58–59 ("Joseph Sheffield"); G. Brush to F. Genth, May 17, 1860, Genth MSS.

[*Lawrence Scientific School*]: Stone, "Learned Societies," pp. 469–72; B. Peirce to A. Bache, Jan. 29, 1846, Peirce MSS.; E. Everett to J. von Liebig, May 14, 1847, Everett MSS.; Storr, *Beginnings,* pp. 50–52.

164 [*anyway*]: Lurie, *Agassiz,* pp. 138–40, 147, 164–65; *DAB* 6:192 ("Henry Eustis"); C. Phillips to D. Swain, Aug. 27, Sept. 3, 17, 1853, University MSS., Southern Historical Collection; S. Tyndale to J. Hilgard, Nov. 24, 1851, Hilgard MSS.; A. Clapp to H. Bacon, Aug. 5, Sept. 1, 1851, Bacon MSS.; Rossiter, *Emergence,* pp. 75–80; E. Horsford to J. Hall, May 6, 1847, Hall MSS.

[*sharply*]: Arthur Zaidenberg, "From Reforms to Professionalization" (Ph.D. diss., University of California, Los Angeles, 1974), pp. 64, 69, 74–75, 77, 82; Samuel E. Morison, ed., *The Development of Harvard University . . . 1869–1929* (Cambridge, Mass., 1930), p. 426.

[*purpose*]: Merle E. Curti and Roderick Nash, *Philanthropy in the Shaping of American Higher Education* (New Brunswick, N. J., 1965), p. 69; W. Rogers to J. Lesley, Feb. 22, 1852, Lesley MSS.; Bruce, *University of Virginia,* 3:48–49; Johnson, *Scientific Interests,* p. 21; Sinclair, "Promise," pp. 264–66; Rudolph, *American College,* pp. 232–33; W. Martin to W. Graham, Oct. 3, 1866, University of North Carolina Archives; Mann, *Engineering Education,* pp. 9, 15; Gates, *Farmer's Age,* pp. 368–73, 376–80.

165 [*Progress*]: *New York Times,* Aug. 16, 1859.

12. The Public Purse

166 [*research*]: Robert C. Post, *Physics, Patents and Politics* (New York, 1976), pp. 84–101.

[*scientists*]: See Appendix.

[*surveys*]: Bache, "Wants of Science," p. 73; Oscar and Mary F. Handlin, *Commonwealth, A Study of the Role of Government in the American Economy: Massachusetts, 1774–1861* (New York, 1947), p. 250; Lurie, *Agassiz,* pp. 233, 300.

167 [*New York*]: Merrill, *First One Hundred Years,* pp. 127–499.

[*another*]: Ibid., pp. 127, 209, 293; Anne Millbrooke, "Science and Government in the Old South" (Paper in First Barnard-Millington Symposium), p. 1.

[*paleontology*]: Walter B. Hendrickson, "Nineteenth-Century State Geological Surveys," *Isis* 52 (1961): 358–61, 365–68; *Am. Jour. Sci.* 82 (1861): 233; Rogers, *Rogers* 1: 116–18, 190.

[*soils*]: Charles S. Sydnor, "State Geological Surveys in the Old South," in David K. Jackson, ed., *American Studies in Honor of William Kenneth Boyd* (Durham, N.C., 1940), pp. 103–104; Millbrooke, "Science and Government," pp. 2–5.

168 [*European*]: Hendrickson, "State Surveys," pp. 361–64, 368–70; Merrill, *First One Hundred Years,* p. 687; Walter K. Ferguson, *Geology and Politics in Frontier Texas, 1845–1909* (Austin, Tex., 1969), p. 32; G. Engelmann to S. Baird, June 28, 1853, Baird MSS.; J. Dana to J. Whitney, Apr. 6, 1859, Whitney MSS.; Brewster, *Whitney,* p. 176; *Am. Jour. Sci.* 47 (1844): 5.

[*too*]: Clarke, *Hall,* pp. 178–83, 266–69; J. Hall to B. Silliman, Feb. 20, 1849, Gratz Collection.
[*them*]: Clarke, *Hall,* pp. 240–45, 273–80.
[*hands*]: J. Hall to A. Gould, Dec. 18, 1846, Gould MSS.; Clarke, *Hall,* p. 300; see Appendix.

169 [*funds*]: Dupree, *Science in the Federal Government,* pp. 64–65. 91; Bache, "Wants of Science," pp. 59, 73–75; Smithsonian, *Annual Report,* 1849, p. 21.
[*support*]: Dupree, *Science in the Federal Government,* pp. 43–46; Bache, "Wants of Science," p. 46½; Report of Committee on a Communication from the American Academy, Feb. 20, 1846, and J. Frazer to Secretary, Apr. 16, 1846, American Philosophical Society Archives; AAAS *Proceedings* 5 (1851): 243.
[*counsel*]: Rhees, *Smithsonian Institution: Documents* 1: 147–54.
[*navy*]: Bache, "Wants of Science," p. 59; AAAS *Proceedings* 6 (1851); l–li; Bache, *Anniversary Address,* p. 30.

170 [*science*]: Benjamin A. Gould, *An Address in Commemoration of Alexander Dallas Bache . . .* (Salem, Mass., 1868), pp. 37–38.
[*them*]: Rhees, *Smithsonian Institution: Documents* 1: 622, 626–27; J. Henry to A. Gray, Feb. 21, 1849, Historic Letter File, Gray Herbarium.

13. Bache and Maury, Barons of Bureaucracy

171 [*Government*]: Dupree, *Science in the Federal Government,* p. 100; Odgers, *Bache,* p. 140; B. Peirce to A. Bache, Feb. 1, 1845, Rhees Collection; Silliman, *World of Science,* p. 39; *New York Times,* Nov. 17, 1858.
[*stronger*]: Odgers, *Bache,* pp. 140–41, 144.

172 [*Survey*]: A. Bache to J. Hilgard, Jan. 22, 1854, Apr. 22, (1851?), Hilgard MSS.; Gould, *Address on Bache,* pp. 25–27; Odgers, *Bache,* pp. 100, 160; A. Bache to G. Davidson, June 17, 1859, Davidson MSS.; Harold L. Burstyn, "Seafaring and the Emergence of American Science," in Benjamin W. Labaree, ed., *The Atlantic World of Robert G. Albion* (Middletown, Conn., 1975), p. 94.
[*performance*]: George M. Brooke, Jr., *John M. Brooke: Naval Scientist and Educator* (Lexington, Va., 1979), p. 38; Benjamin F. Sands, *From Reefer to Rear Admiral* (New York, 1899), p. 120; Dupree, *Science in the Federal Government,* p. 101; J. Dobbin to A. Bache, June 2, 1857, Bache MSS.; Searles, *Life,* pp. 16–17; Gould, *Address on Bache,* p. 28; Burstyn, "Seafaring," p. 84; Henry H. Humphreys, *Andrew Atkinson Humphreys* (Philadelphia, 1924) pp. 53–55; Hazard Stevens, *The Life of Isaac I. Stevens,* 2 vols. (Boston, 1900), 1: 245.
[*argument*]: Emma M. Maffitt, *The Life and Services of John Newland Maffitt* (New York, 1906), p. 136; Dupree, *Science in the Federal Government,* pp. 101–102; Thomas H. Benton, *Thirty Years' View,* 2 vols. (New York, 1856), 2: 726–29.
[*planning*]: Dupree, *Science in the Federal Government,* pp. 100–102.

173 [*inventors*]: Gould, *Address on Bache,* pp. 26, 39–40; Dupree, *Science in the Federal Government,* pp. 101–102; Post, "Science," pp. 84–85.
[*led*]: Stevens, *Stevens* 1: 253; Dupree, *Science in the Federal Government,* pp. 100–101; A. Bache to Fraley, Oct. 21, 1857, Bache MSS.
[*scientists*]: Post, "Science," pp. 80, 86–89; J. Dana to A. Bache, Oct. 29, 1857, Feb. 2, 1858, Bache MSS.; *Am. Jour. Sci.* 75 (1858): 258.
[*dollars*]: J. Henry to A. Bache, Oct. 16, 1854, July 31, 1855, Bache MSS.; Odgers, *Bache,* pp. 151–52; A. Bache to E. Wharton, Dec. 29, 1846, National Archives Record Group 23, Private Correspondence of Superintendent's Office; *Am. Jour. Sci.,* vol. 55 (1848): 308, vol. 75 (1858): 255.
[*attack*]: Gould, *Address on Bache,* pp. 41–42; J. Dana to A. Bache, Feb. 2, 1858, Bache MSS.; *New York Times,* Nov. 17, 18, Dec. 10, 1858, Mar. 5, 1859; *Am. Jour. Sci.* 77 (1859): 448–49.

174 [*not*]: Reingold, "Alexander Dallas Bache," p. 165; A. Bache to B. Silliman, Sr., Apr. 25, 1854, Silliman Family MSS.; A. Bache to J. Frazer, Apr. 27, 1860, Frazer MSS.
[*America*]: Odgers, *Bache,* pp. 163–64, 186; F. Hudson to J. Hilgard, Mar. 17, 1850, Hilgard MSS.
[*Survey*]: Reingold, "Alexander Dallas Bache," pp. 165, 172–75; Odgers, *Bache,* pp. 181–85, 189.

[*aspect*]: Gould, *Address on Bache,* pp. 34–35; Lurie, *Agassiz,* pp. 131, 178; Dupree, *Science in the Federal Government,* p. 104.

175 [*physics*]: Nathan Reingold, "Research Possibilities in the U.S. Coast and Geodetic Survey Records," *Archives Internationales d'Histoire des Sciences* no. 45 (1958): 344; Odgers, *Bache,* p. 155; Dupree, *Science in the Federal Government,* pp. 102–103.

[*1850s*]: AAAS *Proceedings* 25 (1876): 39–40; Dupree, *Science in the Federal Government,* p. 103.

[*developed*]: Reingold, "Alexander Dallas Bache," pp. 165, 167; Gould, *Address on Bache,* p. 33; Odgers, *Bache,* p. 154.

[*work*]: Gould, *Address on Bache,* p. 27.

176 [*country*]: Ibid., pp. 28–33; Reingold, "Alexander Dallas Bache," p. 165; Dupree, *Science in the Federal Government,* p. 103; Odgers, *Bache,* p. 154; AAAS *Proceedings* 25 (1876): 44; Silliman, *World of Science,* p. xii.

[*history*]: Reingold, "Alexander Dallas Bache," p. 174; Gould, *Address on Bache,* p. 11.

[*operations*]: Reingold, "Alexander Dallas Bache," pp. 168–69.

[*hard*]: Frances L. Williams, *Matthew Fontaine Maury, Scientist of the Sea* (New Brunswick, N.J., 1963), pp. 333–36.

177 [*astronomy*]: Ibid., pp. 1, 13, 20, 24–34, 41, 52, 69–70, 92–94, 106, 109, 111.

[*Washington*]: Ibid., pp. 113–15, 118–19, 121–22, 139, 142–44, 334.

[*establishment*]: *DAB* 7: 292 ("James Gilliss").

178 [*Observatory*]: Williams, *Maury,* pp. 148–51, 155, 157.

[*disunion*]: Ibid., pp. 161, 163–64.

[*establishment*]: M. Maury to W. Rogers, Nov. 23, 1846, Rogers MSS.; Williams, *Maury,* pp. 177, 702; Newcomb, *Reminiscences,* pp. 102, 104; unpublished autobiography of Cleveland Abbe, p. 69, Abbe MSS.

[*office*]: Dupree, *Science in the Federal Government,* p. 107; Charles H. Davis, *Life of Charles Henry Davis, Rear Admiral, 1807–1877* (Boston, 1899), pp. 86–87; Edward L. Towle, "Science, Commerce and the Navy on the Sea-Faring Frontier (1842–1861)" (Ph.D. diss., University of Rochester, 1966), p. 81n; Burstyn, "Seafaring," pp. 99–100.

179 [*Henry*]: Davis, *Davis,* pp. 2–5, 63–64, 75, 79–81, 83–84.

[*science*]: Lodge, *Early Memories,* pp. 195–98, 335; S. Du Pont to T. Sedgwick, Dec. 25, 1852, Crystal Palace MSS.; B. Hedrick to E. Thompson, June 5, 1851, Hedrick MSS.; *Northampton* (Mass.) *Free Press,* Aug. 29, 1865.

[*astronomers*]: *Am. Jour. Sci.* 64 (1852): 319.

[*well*]: B. Hedrick to "My dear Sir," Sept. 15, 1853, University of North Carolina Archives; J. G. de Roulhac Hamilton, *Benjamin Sherwood Hedrick* (Chapel Hill, N.C., 1911), p. 7; James B. Thayer, *Letters of Chauncey Wright* (Cambridge, Mass., 1878), p. 70; J. Winlock to C. Davis, Oct. 9, 1863, Davis MSS.

[*field*]: Williams, *Maury,* p. 237; C. Davis to J. Henry, July 2, 1852, Henry MSS.

180 [*observatories*]: *Am. Jour. Sci.* 64 (1852): 319–23, 326, 334.

[*1849*]: J. Gilliss to A. Bache, Jan. 29, 1858, Bache MSS.; J. Gilliss, circular letter dated June 28, 1859, Frazer MSS.; John P. Harrison, "Science and Politics: Origins and Objectives of Mid-Nineteenth Century Government Expeditions to Latin America," *Hispanic American Historical Review* 35 (1955): 181–84; Wayne D. Rasmussen, "The United States Astronomical Expedition to Chile, 1849–1852," *Hispanic American Historical Review* 34 (1954): 104–105.

[*America*]: A. MacRae to brother, Mar. 21, Aug. 23, 1851, July 19, 1854, and to the father, May 26, 1852, MacRae MSS.; Harrison, "Science and Politics," pp. 182–86; *DAB* 6: 292 ("James Gilliss"); Rasmussen, "U.S. Astronomical Expedition," pp. 106–10.

181 [*honor*]: Holden, *Memorials,* pp. 38, 173–74; Jones and Boyd, *Harvard Observatory,* pp. 89–90, 114–15; J. Gilliss to E. Loomis, Apr. 10, 1848, Loomis MSS.; *DAB* 7: 293 ("James Gilliss"); Rasmussen, "U.S. Astronomical Expedition," pp. 110–12; J. Gilliss, circular letter dated June 28, 1859, Frazer MSS.

[*1858–59*]: Williams, *Maury,* pp. 178, 180, 187–88, 190–92, 195, 258, 696–97; Burstyn, "Seafaring," pp. 95–96, 103, 227, n52.

[*societies*]: Williams, *Maury,* pp. 205–19, 22–23; Burstyn, "Seafaring," 104–106.

182 [*mind*]: Burstyn, "Seafaring," pp. 96–97, 106.

[*direction*]: Williams, *Maury,* pp. 312, 316–17, 320–26; Burstyn, "Seafaring," pp. 98, 103.
[*moot*]: Williams, *Maury,* pp. 230–57; J. Hilgard to A. Bache, Oct. 28, 1858, Bache MSS.
[*savans*]: M. Maury to F. Minor, June 11, 1856, and to J. Minor, Nov. 15, 1858, Maury MSS.; J. Henry to A. Bache, Oct. 3, 1860, W. Chauvenet to A. Bache, Dec. 23, 1858, B. Peirce to A. Bache, July 30, 1860, Bache MSS.; Burstyn, "Seafaring," pp. 108–109.

183 [*all*]: Burstyn, "Seafaring," p. 104.
[*government*]: Towle, "Science, Commerce and the Navy," pp. 106, 477, 497, 499.
[*Medal*]: Reingold, "Bache," p. 170; Williams, *Maury,* pp. 188, 219–20, 223.

184 [*field*]: Towle, "Science, Commerce and the Navy," pp. 337–38, 485–88, 491, 505–506; John Leighly, introduction to reprint of Matthew F. Maury, *The Physical Geography of the Sea* (Cambridge, Mass., 1963), pp. xxviii–xxix.
[*critics*]: Williams, *Maury,* pp. 258–60.
[*languages*]: Leighly, introduction to *Physical Geography,* pp. ix, xiii–xvii, xxvii.

185 [*times*]: Ibid., pp. xxi–xxv, xxviii; Williams, *Maury,* p. 260.
[*review*]: J. Henry to R. Hare, Feb. 3, 1857, Hare MSS.; B. Gould to J. Hilgard, Mar. 28, 1855, Hilgard MSS.; *Am. Jour. Sci.* 69 (1855): 449; J. Dana to A. Bache, May 15, 1858, Bache MSS.; John S. Lupold, "From Physician to Physicist: The Scientific Career of John LeConte, 1818–1891" (Ph.D. diss., University of South Carolina, 1970), pp. 108–109.

186 [*heed*]: Harold L. Burstyn, "William Ferrel," in *Dict. Sci. Biog.* 4:590–91; J. Henry to W. Ferrel, Jan. 2, 1857, Ferrel MSS.; Leighly, introduction to *Physical Geography,* pp. xix–xx.
[*physics*]: Burstyn, "Ferrel," pp. 591–92.
[*bottom*]: Harold L. Burstyn, "Matthew Maury," *Dict. Sci. Biog.* 9: 196; Margaret Deacon, *Scientists and the Sea, 1650–1900* (London, 1971), p. 294.
[*Ferrel*]: Leighly, introduction to *Physical Geography,* p. xx.

14. The Smithsonian, Seedbed of Science

187 [*these*]: Dupree, *Science in the Federal Government,* pp. 66–70.

188 [*body*]: Ibid., pp. 76–79; Paul H. Oehser, *Sons of Science* (New York, 1949), pp. 21–25.
[*Princeton*]: Oehser, *Sons of Science,* p. 25; Dupree, *Science in the Federal Government,* p. 80; Coulson, *Henry,* pp. 178–79.
[*misgivings*]: J. Henry to wife, Dec. 15, 1846, to C. Wheatstone, Feb. 26, 1846, to Dr. Bullions, Oct. 14, 1846, to A. Dean, Oct. 19, 1846, Henry MSS.; J. Henry to J. Torrey, Apr. 11, 1838, Torrey MSS.; J. Rogers to W. Rogers, Apr. 12, 1844, Rogers MSS.
[*Smithsonian*]: J. Henry to wife, Dec. 15, 1846, to A. Bache, Dec. 4, 5, 1846, to ?, Dec. 11, 1846, A. Bache to J. Henry, Dec. 3, 4, 1846, J. Henry to Dr. Ludlow, Dec. 29, 1846, to A. Bache, June 25, July 1, 1847, to R. Hare, June 5, 1847, to W. Seaton, June 28, 1847, Henry MSS.

189 [*Washington*]: W. Ingham to G. Cook, Feb. 20, 1847, Cook MSS.; M. Hope to E. Loomis, July 3, 1848, J. Forsyth to E. Loomis, Nov. 7, 1849, Loomis MSS.; J. Henry to A. Bache, Mar. 31, 1847, to Dr. Foreman, Aug. 9, 1850, Henry MSS.; entry for Nov. 22, 1856, Mary Henry Diary; Coulson, *Henry,* pp. 170–71.
[*night*]: J. Forsyth to E. Loomis, Nov. 7, 1849, Jan. 19, 1853, Loomis MSS.; J. Henry to A. Gray, Jan. 8, 1849, Gray Herbarium; J. Henry to R. Hare, May 11, 1850, Hare MSS.
[*it*]: Typed extract of "Locked Book" entry for Dec. 2, 1854, Henry MSS.
[*publication*]: J. Henry to A. Bache, Sept. 6, 1846, Henry MSS.
[*grand plan*]: Rhees, *Smithsonian Institution: Documents* 1: 432–34.

190 [*expensive plan*]: A. Gray to J. Henry, Jan. 4, 1847, J. Henry to wife, Tuesday evening (Dec. 1846?), Henry MSS.; Geoffrey T. Hellman, *The Smithsonian: Octopus on the Mall* (Philadelphia, 1967), pp. 54–55; Smithsonian Institution, *Annual Report* 1 (1847): 6, 17.
[*1878*]: J. Henry to wife, Dec. 1847?, Jan. 16, 1847, to ?, July 1, 1847, Henry MSS.; *Am. Jour. Sci.* 53 (1847): 286; Rhees, *Smithsonian Institution: Journals,* pp. 724–25; Oehser, *Sons of Science,* p. 39.
[*University*]: J. Henry to A. Bache, Dec. 5, 1846, Henry MSS.

191 [*knowledge*]: J. Henry to wife, Dec. 21, 1846, Jan. 27, 1847, Henry MSS.; Dupree, *Science in the Federal Government*, pp. 76–77; Hellman, *Smithsonian*, p. 65.
[*on*]: Rhees, *Smithsonian Institution: Journals*, pp. 724–25; *DAB* 10: 66 ("Charles Jewett"); Hellman, *Smithsonian*, pp. 65–66; Coulson, *Henry*, p. 212; J. Henry to J. Leidy, May 17, 1854, Leidy MSS.
[*items*]: Hellman, *Smithsonian*, p. 66; Coulson, *Henry*, p. 213; J. Henry to A. Bache, July 27, Sept. 22, 1854, Henry MSS.; Robert W. Johanssen, *Stephen A. Douglas* (New York, 1973), p. 467; J. Dana to J. Henry, Jan. 16, 1855, Dana Family MSS.; J. Henry to J. Dana, May 24, 1855, Dana MSS.; L. Agassiz to C. Upham, Jan. 27, 1855, Henry MSS.; A. Gray to C. Short, May 2, 1855, Short MSS.; *Am. Jour. Sci.*, vol. 69 (1855): 284–87, vol. 70 (1855): 1–21.

192 [*to science*]: H. Herrisse to A. Bache, Mar. 29, 1857, Bache MSS.; P. Carpenter to E. Ravenel, Feb. 5, 1860, E. Ravenel MSS.
[*American science*]: L. Agassiz to C. Upham, Jan. 27, 1855, Henry MSS.; J. Dana to J. Henry, Jan. 16, 1855, Dana Family MSS.
[*electors*]: Hellman, *Smithsonian*, p. 86.
[*Henry*]: J. Henry to L. Beck, Mar. 24, 1848, Henry MSS.; *New York Times*, Aug. 22, 1856; C. Davis to J. Henry, July 2, 1852, Henry MSS.; Coulson, *Henry*, pp. 209–10.

193 [*share*]: George B. Goode, *The Smithsonian Institution, 1846–1896* (Washington, D.C., 1897), p. 485; Smithsonian, *Annual Report*, 1859, pp. 15, 17; J. Henry to Varnum, June 22, 1847, Henry MSS.
[*endowment*]: Rhees, *Smithsonian Institution: Documents* 1: 598.
[*sciences*]: J. Henry to E. Sabine, Aug. 13, 1847, Henry MSS.; Dupree, *Science in the Federal Government*, p. 83; J. Henry to F. Lieber, Oct. 14, 1847, Lieber MSS.; Smithsonian, *Annual Report*, 1847, pp. 173–75.
[*Jewett*]: Smithsonian, *Annual Report*, 1847, pp. 172–77.

194 [*can*]: Ibid., p. 21; J. Henry to R. Hare, Jan. 28, 1848, Henry MSS.; A. Bache to J. Henry, Oct. 19, 1851, Rhees Collection.
[*government*]: J. Henry to wife, Apr. 13, 1847, Henry MSS.; Smithsonian, *Annual Report*, 1852, p. 9; J. Henry to A. Bache, Mar. 31, 1847, Henry MSS.; *Smithsonian Miscellaneous Collections* 18: 788, 793, 796; Smithsonian, *Annual Reports*, 1851, p. 18, 1857, pp. 30–31, 1859, pp. 38, 43.
[*Enterprise*]: Oehser, *Sons of Science*, pp. 48–49; Coulson, *Henry*, pp. 194–96; H. Ravenel to L. Gibbes, June 3, 1857, L. Gibbes MSS., Charleston; Brenda Ball, "Arkansas Weatherman: Dr. Nathan D. Smith," *Arkansas Historical Quarterly* 24 (1965): 79–80.

195 [*Bureau*]: Oehser, *Sons of Science*, pp. 49–51; Coulson, *Henry*, pp. 194, 196–97; Dupree, *Science in the Federal Government*, p. 88; J. Henry to E. Loomis, Aug. 5, 1858, Loomis MSS.
[*sallies*]: J. Henry to E. Loomis, Aug. 5, 1858, Loomis MSS.; Oehser, *Sons of Science*, pp. 48–49; Coulson, *Henry*, pp. 195–98.
[*money*]: Oehser, *Sons of Science*, pp. 54–56.
[*new*]: Ibid., pp. 51–53; Coulson, *Henry*, pp. 193–94; Bouvé, "Historical Sketch," p. 57; Smithsonian Institution, *Annual Report*, 1853, p. 23.

196 [*anthropology*]: Oehser, *Sons of Science*, pp. 41–42; Smithsonian Institution, *Annual Report*, 1847, p. 181; David Lowenthal, *George Perkins Marsh: Versatile Vermonter* (New York, 1958), p. 90; J. Henry to A. Bache, June 25, 1847, Henry MSS.
[*quality*]: Coulson, *Henry*, pp. 181–92, 246–52; Goode, *Smithsonian Institution*, pp. 529–60.
[*it*]: Goode, *Smithsonian Institution*, pp. 486–88, 493–95; Coulson, *Henry*, pp. 190–92; J. Leidy to S. Haldeman, Feb. 1847, Leidy MSS.
[*Baird*]: Smithsonian Institution, *Annual Report*, 1847, p. 84; Rhees, *Smithsonian Institution: Documents* 1: 472; L. Agassiz to J. Henry, Jan. 2, 1847, A. Gray to J. Henry, Jan. 4, 1847, Henry MSS.; Dall, *Baird*, p. 189.

197 [*appointment*]: Dall, *Baird*, pp. 156–60, 165; Lowenthal, *Marsh*, pp. 91–92; J. Henry to wife, July 19, 1850, Henry MSS.
[*collections*]: Dall, *Baird*, pp. 163–65, 189–93, 208; Oehser, *Sons of Science*, p. 68.
[*country*]: J. Henry to A. Gray, Sept. 27, 1851, Gray Herbarium; Dall, *Baird*, pp. 236, 247, 256–57, 272, 274–75.

198 [*Smithsonian*]: J. Morris to S. Haldeman, Feb. 19 (1855?), Haldeman MSS.; J.

Henry to J. Leidy, Mar. 23, 1854, Leidy MSS.; Dall, *Baird,* p. 316; S. Baird to J. Torrey, May 14, 1854, Torrey MSS.
[*scientists*]: Dall, *Baird,* pp. 237–40; Sydnor, *Gentleman,* pp. 193–95; printed circular, "Memoranda in Reference to the Natural History Operations by S. F. Baird," in Suckley MSS.; Oehser, *Sons of Science,* p. 69.
199 [*director*]: Dall, *Baird,* pp. 209–10, 305.
[*history*]: Oehser, *Sons of Science,* pp. 67, 69; Dupree, *Science in the Federal Government,* pp. 83, 85–86; Goode, *Smithsonian Institution,* pp. 315–17, 321–22.
200 [*inhaled*]: Dall, *Baird,* pp. 230–32, 290.

15. Soldiers, Sailors, and Scientists

201 [*willingly*]: J. Whitney, Jr., to J. Whitney, Sr., Dec. 29, 1842, Whitney MSS.
202 [*expenses*]: F. Meek to J. Leidy, Nov. 12, 1856, Leidy MSS.; J. Hall to J. Henry, Apr. 24, 1853, Hall MSS.; Clarke, *Hall,* pp. 245–46.
[*them*]: G. Engelmann to S. Buckley, Nov. 30, 1843, Trent Collection; G. Engelmann to J. Torrey, Mar. 23, 1848, Torrey MSS.
[*cries*]: J. Dana to J. Hall, Aug. 19, 1846, Hall MSS.; Williams, *Maury,* p. 177; Ruschenberger, *Report,* p. 7; B. Silliman, Jr., to J. Bailey, Mar. 14, 1848, Bailey MSS.
[*first*]: Rodgers, *Torrey,* p. 225; J. Peck to J. Bailey, May 25, 1846, Bailey MSS.
203 [*Lobo*]: W. Pease to J. L. LeConte, Jan. 27, 1848, LeConte MSS., APS; J. Bachman to V. Audubon, Dec. 14, 1846, Bachman MSS.
[*plants*]: *DAB* 6: 159–60 ("George Engelmann"); Klem, "History," pp. 113–15; G. Engelmann to J. Torrey, July 19, 1846, Torrey MSS., ANS; Dupree, *Gray,* pp. 162–64.
[*route*]: F. Markoe to S. Haldeman, July 29, 1846, Haldeman MSS.; Goetzmann, *Army Exploration,* pp. 149–51; C. Short to J. Torrey, Aug. 23, 1847, Jan. 26, 1848, Torrey MSS.
[*observations*]: Goetzmann, *Army Exploration,* pp. 127–31, 142–44; W. Emory to J. Frazer, July 28, Aug. 3, 1847, Frazer MSS.
204 [*1859*]: Goetzmann, *Army Exploration,* pp. 154–57, 197–205.
[*environments*]: Edward S. Wallace, *The Great Reconnaissance* (Boston, 1955), passim; Dupree, *Science in the Federal Government,* pp. 93–95.
205 [*one*]: Goetzmann, *Army Exploration,* pp. 305–13.
[*expedition*]: Dall, *Baird,* p. 283; Stevens, *Stevens* 1: 299.
[*subject*]: Dall, *Baird,* p. 298; Dupree, *Science in the Federal Government,* p. 93; Smithsonian, *Annual Report,* 1856, p. 21.
[*species*]: Oehser, *Sons of Science,* pp. 74–76; Dall, *Baird,* pp. 237, 249, 298–99.
206 [*specimens*]: C. Parry to J. Torrey, July 23, 1852, Torrey MSS., ANS; M. Curtis to J. Torrey, Jan. 15, 1851, S. Baird to J. Torrey, May 19, 1853, Torrey MSS.; Rodgers, *Torrey,* pp. 186, 217, 224, 233–34, 265–67, 271, 308; Clarke, *Hall,* p. 300; Goetzmann, *Army Exploration,* pp. 307–308.
[*expeditions*]: John D. Kazar, "The United States Navy and Scientific Exploration, 1837–1860" (Ph.D. diss., University of Massachusetts, 1973), p. 279; Dupree, *Science in the Federal Government,* pp. 61, 95.
[*scientists*]: Kazar, "Navy and Exploration," pp. 267, 274, 281; Burstyn, "Seafaring," p. 82; Vincent Ponko, Jr., *Ships, Seas, and Scientists* (Annapolis, Md., 1973), p. 158.
207 [*Humboldt*]: Dupree, *Science in the Federal Government,* pp. 56–60; Kazar, "Navy and Exploration," pp. 2–5, 21, 36.
[*departments*]: Dupree, *Science in the Federal Government,* pp. 58–59; Kazar, "Navy and Exploration," pp. 43, 47, 52–53; J. Dana to W. Redfield, May 29, 1843, Redfield MSS.
[*physics*]: J. Dana to A. Gray, Feb. 12, Mar. 7, 1846, Gray Herbarium; Kazar, "Navy and Exploration," pp. 67, 69; Dupree, *Science in the Federal Government,* p. 60; Ponko, *Ships,* pp. 30–32; William Stanton, *The Great United States Exploring Expedition of 1838–1842* (Berkeley, Calif., 1975), pp. 362–63.
208 [*labels*]: Stanton, *Exploring Expedition,* pp. 292–303.
[*Europe*]: Ibid., pp. 333–37.
[*published*]: Ibid., pp. 328–33; Rodgers, *Torrey,* pp. 182, 184.
[*journals*]: Kazar, "Navy and Exploration," p. 103; Stanton, *Exploring Expedition,* pp. 349–51, 367.

209 [*geology*]: L. Agassiz to C. Wilkes, Sept. 22, 1854, Wilkes MSS.; Stanton, *Exploring Expedition*, pp. 364–72, 382.
[*it*]: Kazar, "Navy and Exploration," pp. 139–40, 270, 273; Towle, "Science," pp. 79, 81, 83, 493.
[*much*]: Burstyn, "Seafaring," p. 102; Ponko, *Ships*, pp. 33–60; Towle, "Science," pp. 160, 182, 187, 191–92, 215, 217, 222; Williams, *Maury*, p. 44.

210 [*appeared*]: Ponko, *Ships*, pp. 64–65, 91–92, 182–83, 188; Towle, "Science," pp. 136, 348, 351–52, 355, 359, 368–69, 377, 477–79; Kazar, "Navy and Exploration," pp. 236–39, 241–42; Williams, *Maury*, pp. 198–201; Harrison, "Science and Politics," pp. 188–91.
[*science*]: *DAB* 10: 333–34 ("John P. Kennedy"); the standard biography of Kennedy, Charles H. Bohner, *John Pendleton Kennedy* (Baltimore, 1961), says little about his role in navy science; Towle, "Science," pp. 80, 83–85, 481; John P. Kennedy, *Report of the Secretary of the Navy*, Sen. Ex. Doc. 1, 32 Cong., 2d sess., pp. 299, 307–308; Williams, *Maury*, pp. 291–92; Harrison, "Science and Politics," pp. 193–94; Ponko, *Ships*, p. 207.
[*up*]: Kazar, "Navy and Exploration," pp. 144–45, 255–61, 271; Harrison, "Science and Politics," pp. 192–97; Ponko, *Ships*, pp. 108–11, 117, 119, 123; Williams, *Maury*, p. 202.

211 [*another*]: *DAB* 10: 256–57 ("Elisha Kane"); Ponko, *Ships*, pp. 188–98; Kazar, "Navy and Exploration," pp. 242–51.
[*versa*]: Samuel Eliot Morison, *"Old Bruin": Commodore Matthew C. Perry* (Boston, 1967), pp. 134, 278–79, 297, 420–25; Kazar, "Navy and Exploration," pp. 158, 161, 174–75; Ponko, *Ships*, pp. 139–40. Kazar, "Navy and Exploration," p. 161, and Dupree, *Science in the Federal Government*, p. 97, contradict Morison, p. 279, as to Perry's attitude toward civilians. Perry's own statement, which Morison does not mention, is quoted in Ponko, p. 139.

212 [*daguerreotypy*]: Kazar, "Navy and Exploration," pp. 161, 186, 206, 277; Ponko, *Ships*, pp. 206–209; Towle, "Science," pp. 437, 443, 448, 452; Brooke, *Brooke*, p. 80.
[*Wilkes*]: Kazar, "Navy and Exploration," pp. 187–90, 193–95.
[*Academy*]: Ibid., pp. 198, 201–18; Towle, "Science," pp. 455, 461–62.

213 [*forty*]: W. Stimpson to J. Leidy, Dec. 17, 1856, Mar. 14, 1857, Leidy MSS.; Ponko, *Ships*, p. 227; Kazar, "Navy and Exploration," pp. 220–26; *DAB* 18: 32 ("William Stimpson").
[*theory*]: Ponko, *Ships*, pp. 228–29; Kazar, "Navy and Exploration," pp. 228–30; Dupree, *Gray*, pp. 249–51.
[*brothers*]: Kazar, "Navy and Exploration," p. 273; C. Wright to J. Torrey, May 29, 1854, Torrey MSS.; A. MacRae to brother, July 19, 1854, MacRae MSS.; Brooke, *Brooke*, p. 126.
[*tendered*]: J. Newberry to A. Gray, Apr. 24, 1857, Gray Herbarium; W. Hammond to J. L. LeConte, July 25, 1854, Feb. 1, 1855, LeConte MSS., APS.
[*training*]: G. Bond to C. Wilkes, Mar. 28, 1859, Wilkes MSS.; J. Dana to J. Leidy, Jan. 28, 1852, Leidy MSS.; Dupree, *Gray*, pp. 68, 187, 196.

16. Bache and Company, Architects of American Science

217 [*era*]: AAAS *Proceedings*, 1851, p. liii.

218 [*story*]: *Illinois Daily Journal* (Springfield, Ill.), Apr. 12, 1850, quoting *Eco d'Italia* (New York City).
[*group*]: Mark Beach, "Was There a Scientific Lazzaroni?", in Daniels, *Nineteenth-Century American Science*, p. 312 and passim.

219 [*Lazzaroni*]: Walter Cannon, "Scientists and Broad Churchmen: an Early Victorian Network," *Journal of British Studies* 4 (1964): 69–72, 86–87; Roy M. MacLeod, "The X-Club: A Social Network of Science in Late-Victorian England," *Notes and Records of the Royal Society of London* 24 (1969): 307.
[*striking*]: MacLeod, "The X-Club," passim; see also J. Vernon Jensen, "The X Club: Fraternity of Victorian Scientists," *British Journal of the History of Science* 5 (1970): 63–72, and Ruth Barton, "The X-Club: Science, Religion, and Social Change in Victorian England" (Ph.D. diss., University of Pennsylvania, 1976).
[*goals*]: Belver C. Griffith and A. James Miller, "Networks of Informal Communication Among Scientifically Productive Scientists," in Carnot E. Nelson and Don-

ald K. Pollock, eds., *Communication Among Scientists and Engineers* (Lexington, Mass., 1970), pp. 139–40.

220 [*voice*]: Reingold, *Science in Nineteenth-Century America,* pp. 85, 87.

221 [*superintendent*]: C. Davis to B. Peirce, May 23, 1842, Bache MSS.; *DAB* 14: 394–96 ("Benjamin Peirce"); B. Peirce to A. Bache, Dec. 8, 1843, Peirce MSS.
[*business*]: L. Agassiz to A. Bache, Feb. 2, 1847, May 31, 1848, Rhees Collection; Emerson, *Early Years,* p. 101; Gray, *Letters* 1: 349; J. Henry to J. Torrey, Apr. 7, 1849, Torrey MSS.
[*death*]: Chandler, "Gould," p. 342; *DAB* 7: 447–48 ("Benjamin A. Gould"); B. Gould to A. Bache, May 10, 1856, Bache MSS., Smithsonian.

222 [*puller*]: W. Gibbs to G. Gibbs, Feb. 22, 1846, Gibbs Family MSS.; Brewster, *Whitney,* p. 91; recommendation by J. Torrey, Jan. 5, 1848, Gibbs MSS.; J. Torrey to L. Beck, Oct. 24, 1848, Beck MSS.; S. Willis Rudy, *The College of the City of New York: A History, 1847–1947* (New York, 1949), pp. 13, 17; Rodgers, *Torrey,* pp. 212–13; J. Torrey to A. Gray, Nov. 24, 1848, Gray Herbarium; J. Torrey to J. Henry, Jan. 25, 1849, Henry MSS.; J. Stewart to W. Gibbs, Dec. 21, 1848, Gibbs MSS.; Nevins and Halsey, *Diary of Strong* 1: 364.
[*sciences*]: *DAB* 7: 3 ("J. F. Frazer"); A. Bache to J. Frazer, Dec. 11, 1848, Frazer MSS.; J. Dana to A. Bache, Jan. 9, 1857, Jan. 9, 1858, Bache MSS.; J. Dana to A. Bache, Jan. 25, 1852, Rhees Collection; J. Dana to A. Bache, Jan. 9, 1857, Dec. 19, 1857, Feb. 2, 1858, Bache MSS.

223 [*Gray*]: Clarke, *Hall,* p. 229; Brewster, *Whitney,* pp. 123–24, 146.
[*coinage*]: B. Peirce to A. Bache, May 16, 1852, Aug. 30, 1853, Peirce MSS.; F. Rogers to J. L. LeConte, Apr. 20, 1855, J. L. LeConte MSS., APS; Beach, "Lazzaroni," pp. 117–18, dates the earliest use of "Lazzaroni" for the Bache group from a Bache letter ascribed to 1850, but that letter was probably written in 1856 and misfiled because of Bache's ambiguous penmanship.
[*cookery*]: B. Gould to J. Frazer, Nov. 22, 1856, Frazer MSS.; B. Gould to J. Dana, Nov. 22, 1856, Dana MSS.; Lurie, *Agassiz,* p. 182; A. Bache to J. Frazer, Dec. 19, 1857, Frazer MSS.

224 [*result*]: *New York Times,* Nov. 17, 1858; B. Peirce to A. Bache, undated but probably c. Nov. 1, 1858, Apr. 13, 1862, Bache MSS.; A. Bache to J. Frazer, Jan. 3, 1858, Frazer MSS.; W. Gibbs to A. Bache, Nov. 30, 1858, Bache MSS.; B. Peirce to A. Bache, Dec. 23, 1857, Mar. 8, 1861, Rhees Collection; Nathaniel Shaler, *The Autobiography of Nathaniel Southgate Shaler* (Boston, 1909), p. 113; W. Gibbs to A. Bache, Jan. 14, 31, Mar. 13, 1859, Bache MSS.; A. Bache to B. Peirce, Jan. 12, Nov. 22, 1860, Mar. 9, 1861, Dec. 9, 1862, Jan. 26, 1863, Peirce MSS.; A. Bache to J. Frazer, Mar. 5, 1860, Feb. 5, 1863, Frazer MSS.
[*defunct*]: Dupree, *Science in the Federal Government,* pp. 138–40.

17. Support without Strings

225 [*consequences*]: Petition of W. Gibbs to L. Bradish, 1851, Bradish MSS.; J. Henry to Varnum, June 22, 1847, Henry MSS.; L. Agassiz to J. Hall, July 22, 1849, Hall MSS.
[*societies*]: Bache, "Wants of Science," pp. 73–75.
[*research*]: Ibid., pp. 16, 20, 58, 75, 77–79.

226 [*could*]: Ibid., pp. 18, 38–42, 56, 76.
[*faculty*]: Fritz K. Ringer, *The Decline of the German Mandarins* (Cambridge, Mass., 1969), pp. 86–87, 90–95; Reingold, "Reflections," pp. 17–19.
[*limbo*]: Nevins and Thomas, *Diary of Strong* 2: 30, 37; J. Whitney to J. Hall, Jan. 13, 24, 29, 1851, Feb. 5, 1852, Hall MSS.

227 [*Hall*]: J. Whitney to J. Hall, Jan. 29, 1851, Hall MSS.; Storr, *Beginnings,* pp. 61–65; Robert Silverman and Mark Beach, "A National University for Upstate New York," *American Quarterly* 22 (1970): 702.
[*research*]: J. Hall to A. Gould, Dec. 18, 1846, A. Gould MSS.; Clarke, *Hall,* pp. 181, 186–91; Silverman and Beach, "A National University," pp. 703–704.
[*rattle*]: J. Whitney to J. Hall, July 22, 1851, B. Gould to J. Hall, April 1852?, B. Peirce to J. Hall, Oct. 19, 1851, Hall MSS.; Gilman, *Dana,* p. 321; Kohlstedt, *Formation,* pp. 105–107; AAAS *Proceedings* 6 (1851): xlv; circular, "University of Albany," 1851, in Cook MSS.; J. Norton to L. Tucker, Nov. 8, 1851, Gratz Collection.
[*completely fizzled out*]: Rossiter, *Emergence,* p. 124; Silverman and Beach, "A National University," pp. 709–13; Storr, *Beginnings,* pp. 72–74.

228 [*turn "fizzled out"*]: A. Potter to A. Bache, July 16, Dec. 8, 1852, Rhees Collection; A. Bache to J. Frazer, Apr. 22, 1852, Frazer MSS.; Storr, *Beginnings,* pp. 76–81.
[*program*]: *Catalogue of the Officers and Students of Columbia College* (New York, 1853), pp. 4–5, 19; Nevins and Thomas, *Diary of Strong* 1: xxvi–xxvii, xxx; Storr, *Beginnings,* pp. 67–71, 94–96; B. Peirce to A. Bache, May 16, 1852, Peirce MSS.
[*walls*]: W. Gibbs to mother, Nov. 30, 1845, Gibbs Family MSS.; S. Ruggles to A. Potter, July 14, 1852, Gibbs MSS.; J. Norton to J. Dana, July 29, 1852, Dana MSS.; Brewster, *Whitney,* p. 133; Nevins and Thomas, *Diary of Strong* 2: 120; G. Gibbs to W. Gibbs, June 14, 1853, Gibbs Family MSS.

229 [that]: Nevin and Thomas, *Diary of Strong* 2: 139, 143; B. Gould to B. Peirce, Dec. 8, 1853, Peirce MSS.; *Testimonials Presented to the Trustees of Columbia College in Behalf of Dr. Wolcott H.* [sic] *Gibbs* (New York, 1854), pp. 5, 13–15, 20–21, 23–25, and passim; B. Gould to J. Hall, Dec. 8, 1853, Hall MSS.
[*endorsement*]: Nevins and Thomas, *Diary of Strong,* vol. 1: xxx, vol. 2: 146–47; Thomas, "Gibbs Affair," pp. 72–73, 77–79, 89, 104; Daniel G. B. Thompson, *Ruggles of New York* (New York, 1946), pp. 81–82.
[*doings*]: Thomas, "Gibbs Affair," p. 89; E. Jones to H. Fish, Feb. 15, 1854, Fish MSS., Columbia; Thompson, *Ruggles,* pp. 78–79; G. Ogden to H. Fish, Jan. 18, Feb. 12, 1854, Fish MSS., Columbia; S. Ruggles to B. Peirce, Feb. 12, 1854, Peirce MSS.

230 [*age*]: Thompson, *Ruggles,* pp. 84, 86, 88; H. Mann to W. Gibbs, June 13, 1854, W. Gibbs to H. Mann, July 8, 1854, F. Barnard to W. Gibbs, Feb. 26, 1855, Gibbs MSS.; C. Gilman to C. Wetherill, July 2, 1856, Wetherill MSS.; Storr, *Beginnings,* p. 97.
[*case*]: Storr, *Beginnings,* pp. 97–110; Van Amringe, *Historical Sketch,* pp. 69–72, 75; W. Gibbs to A. Bache, Feb. 3, 1857, National Archives Record Group 23, Private Correspondence of the Superintendent's Office; Nevins and Thomas, *Diary of Strong* 4: 220; Richard Hofstadter and Walter P. Metzger, *The Development of Academic Freedom in the United States* (New York, 1955), p. 273.
[*research*]: A. Bache to F. Barnard, Mar. 5, (1864?), Barnard MSS.; Storr, *Beginnings,* pp. 82–84; Alexander D. Bache, "A National University," *American Journal of Education* 1 (1856): 477–78.

231 [*address*]: *American Journal of Education* 2 (1856): 275–77, 281–82; Bache, *Anniversary Address,* pp. 41, 44–45, 48–49.
[*nonscientific*]: Storr, *Beginnings,* pp. 56–58; *American Journal of Education* 2 (1856): 371–74; Fleming, *Science,* pp. 42–43; Chittenden, *History,* pp. 86–89; Brooks M. Kelley, *Yale: A History* (New Haven, Conn., 1974), p. 185.
[*enlargement*]: Ralph Barton Perry, *The Thought and Character of William James,* 2 vols. (Boston, 1935), 1: 204; B. Peirce to A. Bache, Dec. 30, 1855, Peirce MSS.

232 [*duties*]: B. Osgood Peirce, "Biographical Memoir of Joseph Lovering," National Academy of Sciences, *Biographical Memoirs* 6 (1909): 330–33; Rossiter, *Emergence,* pp. 84–87, 197–203.
[*Harvard*]: Emerson, *Early Years,* pp. 421–23; Shaler, *Autobiography,* p. 105; Robert A. McCaughey, "The Transformation of American Academic Life: Harvard University, 1821–1892," *Perspectives in American History* 8 (1974): 263–66.
[*laboratory*]: *DAB* 4: 387 ("Josiah P. Cooke"); J. Whitney to J. Hall, Jan. 13, 1851, Hall MSS.; J. Wyman to A. Gray, Jan. 30, 1851, J. Henry to A. Gray, Feb. 18, 1858, Gray Herbarium; Baxter, "Early Days," pp. 590–91; Miller, *Dollars,* pp. 83–86; Shaler, *Autobiography,* p. 111.

233 [*pride*]: B. Gould to A. Bache, June 20, 1860, Bache MSS.; Lurie, *Agassiz,* pp. 145, 177, 186–88, 190–91, 196–201.
[*series*]: Miller, *Dollars,* pp. 54–55; Lurie, *Agassiz,* pp. 202–207, 248–49.
[*$220,000*]: Lurie, *Agassiz,* pp. 214–20, 227–34.

234 [*them*]: J. Dana to L. Agassiz, Apr. 2, 1859, Agassiz MSS.; Lurie, *Agassiz,* pp. 247–48.
[*triumph*]: Miller, *Dollars,* pp. 57, 67–70; Lurie, *Agassiz,* pp. 245–47, 301–2.
[*Observatory*]: Mary Ann James, "The Dudley Observatory Controversy" (Ph.D. diss., Rice University, 1980), pp. 14, 42–44, 47, 354–55, 359–62.

235 [*minds*]: Ibid., pp. 50–64, 78; Benjamin A. Gould, *Reply to the Statement of the Trustees of the Dudley Observatory* (Albany, N.Y., 1859), pp. 37–40, 44–48; B. Peirce to A. Bache, Aug. 25, 1855, Bache MSS.
[*Observatory*]: Gould, *Reply,* pp. 42–43, 184; James, "Dudley Observatory," p. 76; J. Henry to A. Bache, Sept. 18, 1858, Henry MSS.
[*moon*]: Gould, *Reply,* pp. 95, 98.
[*snapped*]: Ibid., p. 55; James, "Dudley Observatory," pp. 110–13, 125–28, 133.

236 [*employee*]: *DAB* 14: 502–503 ("Christian Peters"); *Defence of Dr. Gould by the Scientific Council of the Dudley Observatory* (Albany, N.Y., 1858), pp. 26–28; Gould, *Reply,* pp. 41, 56–57, 184, 243–44; James, "Dudley Observatory," pp. 139–40.
[*years*]: A. Bache to B. Gould, Nov. 21, 1857, National Archives Record Group 23, Private Correspondence of the Superintendent's Office; Gould, *Reply,* pp. 60–65, 68, 75; J. Henry Diary, Jan. 16, 18, 1858, Henry MSS.; *DAB* 14: 503 ("Christian Peters"); James, "Dudley Observatory," pp. 136, 155–56.
[*member*]: James, "Dudley Observatory," pp. 111–12, 125–27, 154, 160, 202.

237 [*issue*]: Ibid., passim; *Defence of Dr. Gould,* passim; Odgers, *Bache,* p. 197; *Who Withholds Co-operation? Correspondence between the . . . Trustees and the Director* (Albany, N.Y., 1858); *The Dudley Observatory and the Scientific Council: Statement of the Trustees* (Albany, N.Y., 1858); George H. Thacher, *A Key to the Trustees' Statement* (Albany, N.Y., 1858); "Observer," *A Letter to the Majority of the Trustees* (n.p., n.d.); Stephen Van Rensselaer, *The Dudley Observatory: An Address* (Albany, N.Y., 1858); Gould, *Reply,* passim; D. Wood to C. Abbe, Mar. 28, 1864, Abbe MSS.; D. Wells to J. Warner, Feb. 27, 1859, J. Warner to C. Winslow, Sept. 16, 1858, Warner MSS.
[*enlightenment*]: James, "Dudley Observatory," pp. 6–7, 12, 14, 98, 347–50; *DAB* 14: 503 ("Christian Peters").
[*up*]: James, "Dudley Observatory," pp. 1–2, 8, 13–14, 16–18, 111; J. Watson to E. Winslow, Mar. 10, 1865, Watson MSS.

238 [*rights*]: Diary of Joseph Henry, June 29, 1858, Henry MSS.; B. Peirce to A. Bache, Jan. 24, 1858, Rhees Collection; *Boston Daily Advertiser,* Sept. 15, 1858.
[*believed*]: James, "Dudley Observatory," pp. 8, 11, 233; Gould, *Reply,* p. 86; *Defence of Dr. Gould,* p. 52.
[*affair*]: *American Journal of Education* 2 (1856): 282.
[Science]: W. Gibbs to A. Bache, Sept. 7, 24, 1858, Bache MSS.; B. Peirce to A. Bache, Oct. 8, 1858, Jan. 31, 1859, Rhees Collection; J. Dana to J. Hall, Sept. 8, 1858, Hall MSS.

239 [*omitted*]: J. Henry to E. Loomis, Aug. 5, 1858, Loomis MSS.; J. Henry to A. Bache, Sept. 18, 28, Oct. 4, 18, 1858, Henry MSS.; W. Palmer to A. Bache, Sept. 29, 1858, W. Gibbs to A. Bache, Oct. 17, Dec. 12, 1858, Bache MSS.; B. Peirce to A. Bache, Oct. 8, 1858, Rhees Collection; Newcomb *Reminiscences,* p. 82; J. Henry to J. Torrey, Mar. 30, 1859, Torrey MSS.

18. Communication and Conflict

240 [*fifties*]: Thomas T. Bouvé, "Some Reminiscences of Earlier Days in the History of the Society," Boston Society of Natural History *Proceedings* 18 (1876): 246; William Ferguson, *America by River and Rail* (London, 1856), pp. 36–37; Bouvé, "Historical Sketch," pp. 81–84, 92; Boston Society of Natural History, *The Boston Society of Natural History, 1830–1930* (Boston, 1930), p. 13; Bates, *Scientific Societies,* p. 69; John C. Greene, "Science, Learning, and Utility," in Oleson and Brown, *Pursuit of Knowledge,* pp. 14–15; Nolan, *Short History,* p. 15; Hendrickson, *Arkites,* p. 20.

241 [*declined*]: Greene, "Science, Learning, and Utility," pp. 6, 16; R. Stevens to W. Kitchell, July 1860, Kitchell MSS.; Bates, *Scientific Societies,* pp. 48, 51; Shapiro, "Western Academy," p. 242.
[*to*]: Bache, "Wants of Science," pp. 36, 77; Greene, "Science, Learning, and Utility," p. 19; Dupree, "National Pattern," pp. 30–31.
[*them*]: Wayne E. Fuller, *The American Mail* (Chicago, 1972), pp. 61–65; Pliny Miles, *Postal Reform* (New York, 1855), p. 27; Herber, *Correspondence,* p. 60; M. Tuomey to J. Leidy, Nov. 28, 1854, Leidy MSS.
[*everywhere*]: J. Torrey to A. Gray, June 23, 1845, Gray Herbarium; Rodgers, *"Noble Fellow,"* p. 108; Taylor, *Transportation Revolution,* pp. 138–39.

242 [*auxiliaries*]: R. Gibbes to B. Wailes, July 27, 1849, Wailes MSS.; J. Barratt to L. Gibbes, Mar. 27, 1848, L. Gibbes MSS., Charleston; Dupree, *Gray,* p. 201; Lurie, *Agassiz,* p. 188.
[*periodicals*]: AAAS *Proceedings* 2 (1849): 169; A. Gould to J. Phillips, Feb. 21, 1853, Tryon-Pilsbry Collection; Smallwood, *Natural History,* p. 192; Smithsonian, *Annual Report,* 1848, p. 180.

[*this*]: Rodgers, *Torrey*, p. 193; *Am. Jour. Sci.* 61 (1851): 147; Brewster, *Whitney*, p. 121; J. Torrey to C. Short, Jan. 19, 1860, Short MSS.

243 [*1854*]: *Am. Jour. Sci.* 68 (1854): 200–203.
[*fields*]: Allan Nevins, *The Emergence of Lincoln*, 2 vols. (New York, 1950), 2: 448.
[*them*]: Beaver, "Scientific Community," p. 110; Greene, "Science, Learning, and Utility," pp. 8–9, 15.
[*came*]: McCormmach, *"Sidereal Messenger,"* pp. 42, 44–46; *Am. Jour. Sci.* 52 (1846): 442; O. Mitchel to L. Gibbes, Oct. 26, Dec. 14, 1846, July 15, 1847, L. Gibbes MSS., LC.

244 [*1848*]: McCormmach, *"Sidereal Messenger,"* pp. 37, 39, 40–41, 43, 45–46.
[*besides*]: B. Gould to E. Loomis, Sept. 27, 1849, Sept. 24, 1850, Loomis MSS.; AAAS *Proceedings* 4 (1850): 342–43; B. Gould to L. Gibbes, Nov. 12, 1849, L. Gibbes MSS., LC.
[*1886*]: B. Gould to E. Loomis, Sept. 27, 1849, Loomis MSS.; AAAS *Proceedings* 4 (1850): 344; B. Gould to L. Gibbes, Nov. 12, 1849, L. Gibbes MSS., LC; McCormmach, *"Sidereal Messenger,"* p. 47; B. Peirce to A. Bache, Oct. 16, 1853, Peirce MSS.; J. Henry to A. Bache, July 31, 1855, Henry MSS.; Gould, *Reply*, pp. 239–40; Holden, *Memorials*, p. 223.

245 [*editor*]: J. Henry to A. Bache, Sept. 7, 1846, Bache MSS.
[*personalities*]: Prendergast, "Dana," pp. 310–15.
[*others*]: J. Dana to A. Gould, Feb. 16, 1848, A. Gould MSS.; J. Dana to L. Agassiz, Jan. 23, 1853, Agassiz MSS.; J. Henry to J. Coffin, Aug. 21, 1854, Henry MSS.
[*years*]: Bache, "Wants of Science," p. 82; Joyner, p. 20; David A. Wells and George Bliss, Jr., eds., *The Annual of Scientific Discovery* (Boston, 1850), passim; Meisel, *Bibliography* 3: 118; Albert J. Beveridge, *Abraham Lincoln, 1809–1858*, 2 vols. (Boston, 1928), 1: 520; *Am. Jour. Sci.*, vol. 59 (1850): 454, vol. 61 (1851): 447.
[*1861*]: *Am. Jour. Sci.* 33 (1862): 132.

246 [*country*]: L. Lesquereux to J. Lesley, Jan. 9, 1852, Lesley MSS.; Clarke, *Hall*, p. 248; Brewster, *Whitney*, p. 129.
[*it*]: B. Shumard to J. Hall, June 29, 1862, Hall MSS.; J. Dana to S. Baird, Nov. 6, 1865, Baird MSS.; J. Henry to E. Loomis, Aug. 5, 1858, Loomis MSS.; Thomas J. Wertenbaker, *Princeton, 1746–1896* (Princeton, N.J., 1946), p. 222.
[*Canada*]: Cecil J. Schneer, "The Great Taconic Controversy," *Isis* 69 (1978): 174.

247 [*geology*]: Cecil J. Schneer, "Ebenezer Emmons and the Foundations of American Geology," *Isis* 60 (1969): 439–50, passim.
[*unprofessional*]: J. Whitney to J. Hall, Jan. 24, 1851, Hall MSS.; Merrill, *First One Hundred Years*, p. 334; *DAB* 6: 149 ("Ebenezer Emmons").
[*mocked*]: M. Curtis to A. Gray, Oct. 3, 1860, Gray Herbarium; Clarke, *Hall*, pp. 30, 42, 57, 131–35, 315–16.

248 [*1850*]: Schneer, "Taconic Controversy," pp. 174–80; Merrill, *First One Hundred Years*, pp. 597, 599, 601, 606.
[*life*]: Clarke, *Hall*, pp. 204–12, 215.
[*hand*]: *Am. Jour. Sci.* 69 (1855): 406; Schneer, "Taconic Controversy," pp. 182–85, 190; E. Emmons to C. Dewey, Jan. 3, 1861, Dewey MSS.; E. Emmons to H. Ward, Jan. 19, 1861, Ward MSS.; B. Shumard to A. Winchell, Mar. 21, 1861, Winchell MSS.; C. Dewey to J. Torrey, May 15, 1862, Torrey MSS.

249 [*geologists*]: Schneer, "Taconic Controversy," pp. 186–91; Merrill, *First One Hundred Years*, pp. 604–605, 612.
[*personalities*]: Merrill, *First One Hundred Years*, pp. 233–34; J. Whitney to W. Whitney, May 7, 1858, Whitney MSS.; Prendergast, "Dana," pp. 315–24.

250 [*it*]: F. Meek to A. Winchell, Dec. 25, 1861, Winchell MSS.; A. Bache to G. Ord, Jan. 9, 1846, American Philosophical Society Archives; J. Frazer to A. Bache, Mar. 3, 1858, Bache MSS.; B. Peirce to A. Bache, May 8, 1854, Peirce MSS.; Warner, "Astronomy," p. 68; Holden, *Memorials*, pp. 34–36; Jones and Boyd, *Harvard Observatory*, pp. 97–101, 110–15; B. Peirce to A. Bache, May 29, 1858, Rhees Collection; B. Peirce to A. Bache, Dec. 12, 1861, Bache MSS.
[*muster*]: J. Henry to J. Torrey, Mar. 30, 1859, Torrey MSS.

19. Liberty and Union

251 [*him*]: Kohlstedt, *Formation*, pp. 16, 18, 42, 46–50; B. to A. Gould, Apr. 11, 1846, A. Gould MSS.; J. Henry to J. Torrey, c. October 1838, Torrey MSS.

252 [*pressing*]: Kohlstedt, *Formation*, pp. 63–64; Sally Gregory Kohlstedt, "The Formation of the American Scientific Community" (Ph.D. diss., University of Illinois, 1972), p. 102; Hitchcock, *Reminiscences*, pp. 368, 371.
[*1840*]: Hitchcock, *Reminiscences*, pp. 370–71; Clarke, *Hall*, pp. 100–101; Rogers, *Rogers* 1: 154–55; Kohlstedt, *Formation*, pp. 54, 66.
[*BAAS*]: Kohlstedt, *Formation*, pp. 68–70; Hitchcock, *Reminiscences*, pp. 369, 373.

253 [*Markoe*]: Sally Gregory Kohlstedt, "A Step Toward Scientific Self-Identity in the United States: The Failure of the National Institute, 1844," *Isis* 62 (1971): 342–44; Poinsett, *Discourse*, p. 52; Grace E. Heilman and Bernard S. Levin, eds., *Calendar of Joel R. Poinsett Papers in the Henry D. Gilpin Collection* (Philadelphia, 1941), p. 162.
[*role*]: Kohlstedt, "Step," pp. 345–48.
[*Institute*]: Ibid., pp. 339, 353, 358, 360–61; J. Henry to B. Peirce, Nov. 25 (1843?), Peirce MSS.; J. Dana to A. Gould, Apr. 11, 1844, A. Gould MSS.
[*cared*]: Kohlstedt, "Step," p. 344; Madge E. Pickard, "Government and Science in the United States: Historical Backgrounds," *Journal of the History of Medicine* 1 (1946): 282–85, 288; J. Henry to A. Bache, May 27, 1848, Henry MSS.; J. Henry to A. Gray, Dec. 22, 1848, Gray Herbarium; W. Easby to President, National Institute, Feb. 6, 1854, Rhees Collection.
[*Assoc.*]: Kohlstedt, *Formation*, pp. 73–74; Kohlstedt, "Formation," pp. 93, 96, 98; L. Gibbes to A. Gould, Jan. 6, 1848, A. Gould MSS.

254 [*year*]: William H. Hale, "Early Years of the American Association," *Popular Science Monthly* 49 (1896): 502; Taylor, *Transportation Revolution*, pp. 141, 143; AAAS *Proceedings* 5: 247; Dall, *Baird*, p. 264.
[*it*]: Lurie, *Agassiz*, pp. 131–32; Kohlstedt, *Formation*, pp. 79–81; A. Bache to A. Gould, Sept. 18, 1847, A. Gould MSS.; Bates, *Scientific Societies*, p. 74.

255 [*September*]: H. Rogers to S. Haldeman, Feb. 21, 1848, Haldeman MSS.; Rogers, *Rogers* 1: 288; Kohlstedt, *Formation*, pp. 80–85, 87; W. Rogers to W. Redfield, Aug. 24, 1848, Redfield MSS.
[*country*]: J. Torrey to A. Gray, Sept. 28, 1848, Gray Herbarium; AAAS *Proceedings* 1: 82–90, 142; B. Silliman, Jr., to J. Bailey, Dec. 25, 1848, Bailey MSS.; Lurie, *Agassiz*, p. 152.
[*affairs*]: Bache, "Wants of Science," pp. 20–21, 84–87.

256 [*democracy*]: Ibid., p. 84; AAAS *Proceedings* 1: 8.
[*Sanskritist*]: Kohlstedt, "Formation," p. 260; Kohlstedt, *Formation*, pp. 104, 198, 219; *New York Quarterly* 2 (1853): 450 (attribution of authorship is in J. Henry to A. Bache, Oct. 14, 1853, Henry MSS.); *New York Times*, Aug. 25, 1860.
[*vote*]: Kohlstedt, *Formation*, pp. 102, 197; AAAS *Proceedings* vol. 13 (1859): liii, 358, vol. 14 (1860): xlvi, 229.

257 [*use*]: *Baltimore Daily Exchange*, May 4, 5, 1858; *Springfield* (Mass.) *Republican*, Aug. 10, 1859; C. Dewey to J. Torrey, Sept. 21, 1859, Torrey MSS.; C. Dewey to J. Torrey, Dec. 19, 1859, Trent Collection.
[*there*]: *Springfield* (Mass.) *Republican*, Aug. 9, 10, 1859; AAAS *Proceedings* 4 (1850): 163; Kohlstedt, *Formation*, pp. 102, 110.
[*spotlight*]: *Charleston* (S.C.) *Courier*, Mar. 11–19, 1850; *Cincinnati Gazette*, May 6–12, 1851; *Albany Journal*, Aug. 19–25, 1851; *Cleveland Plain Dealer*, July 28–Aug. 3, 1853; *Providence Journal*, Aug. 14–23, 1855; on out-of-town reporters, see *Cincinnati Enquirer*, May 6, 1851, *Montreal Commercial Advertiser*, Aug. 15, 1857, *Springfield* (Mass.) *Republican*, Aug. 4, 1859, *New York Times*, Aug. 4–12, 1859, Aug. 8–11, 1860; B. Gould to E. Loomis, Sept. 24, 1850, Loomis MSS.; J. Henry to C. Dewey, Nov. 7, 1859, Dewey MSS.

258 [*intended*]: Kohlstedt, *Formation*, p. 105; J. Henry to A. Bache, Mar. 29, 1854, Henry MSS.; Kendall, *Mitchell*, p. 23; Dall, *Baird*, p. 263; J. Dana to J. Hall, Sept. 5, 1856, Hall MSS.; W. Whitney to J. Whitney, May 4, 1858, Whitney MSS.; *Springfield* (Mass.) *Republican*, Aug. 10, 1859.
[*society*]: Kendall, *Mitchell*, p. 23; W. Gibbs to A. Bache, May 10, 1858, Bache MSS.; *Washington* (D.C.) *Star*, Apr. 27, 1854; J. Henry to M. Henry, July 31, 1853, Henry MSS.; Brewster, *Whitney*, p. 171; E. Hilgard to J. Hilgard, Aug. 15, 1857, Hilgard MSS.
[*abyss*]: Kohlstedt, *Formation*, pp. 108–109; *New York Times*, Aug. 25, 1860; Meyer Berger, *The Story of the New York Times* (New York, 1951), pp. 250–51.

[*interest*]: Kohlstedt, *Formation,* pp. 112–13.

259 [*objection*]: AAAS *Proceedings* 4 (1850): 164; *Cleveland Plain Dealer,* Aug. 3, 1853.

[*on*]: Brewster, *Whitney,* p. 168; Kohlstedt, *Formation,* pp. 117, 119.

[*program*]: Kohlstedt, *Formation,* p. 109; *New York Times,* Aug. 6, 25, 1860; S. Powel to J. Leidy, Aug. 10, 1860, Leidy MSS.

[*themselves*]: Kohlstedt, *Formation,* pp. 95, 122–24; Miller, *Dollars,* p. 127; AAAS *Proceedings* 6 (1851): liv.

260 [*years*]: Kohlstedt, *Formation,* pp. 89–90, 94, 109–10.

[*specialties*]: AAAS *Proceedings* 2 (1849): 2; J. Barratt to L. Gibbes, June 11, 1850, L. Gibbes MSS., Charleston; I. Lapham to wife, July 31, 1853, Lapham MSS.; *American Journal of Education* 3 (1857): 151; *New York Quarterly* 2 (1853): 458.

261 [*pointless*]: Daniels, *American Science,* pp. 57–60.

[*proceedings*]: Kohlstedt, *Formation,* pp. 97, 139–42; John D. Holmfield, "From Amateurs to Professionals in American Science: The Controversy over the Proceedings of an 1853 Scientific Meeting," APS *Proceedings* 114 (1970): 22–36; *Providence Journal,* Aug. 22, 1855; AAAS *Proceedings* 11 (1857): xxi.

[*democracy*]: *Providence Journal,* Aug. 20, 1855.

262 [*democrats*]: Francis J. Grund, *Aristocracy in America* (London, 1839), passim; Edward Pessen, *Jacksonian America: Society, Personality, and Politics* (Homewood, Ill., 1969), pp. 47–58; see Appendix; Kohlstedt, *Formation,* pp. 155–56.

[*how*]: J. Henry to wife, Jan. 29, 1847; Apr. 11, 1848, Henry MSS.; J. Henry to H. Hagen, May 3, 1861, MCZ Misc. Files; "Extracts from Locked Book" (typed copy), Nov. 9, 1864, Henry MSS.; J. Henry to Dr. Denham, May 25, 1868, to Miss Montague, Apr. 4, 1872, to Mr. Patterson, Dec. 25, 1876, Henry MSS.

[*time*]: Smithsonian, *Annual Report,* 1854, p. 8; Reingold, *Science in Nineteenth-Century America,* pp. 87–88.

263 [*Alexander*]: Elliott, "American Scientist," p. 108.

[*people*]: Rhees, *Smithsonian Institution: Documents* 1: 391; Hindle, *Pursuit,* pp. 190–94, 214–15, 328–29, 352–54, 382, 385; Greene, "Science and the Public," pp. 22–25; Goldfarb, "Science," p. 173; Olmsted, "Democratic Tendencies," pp. 164–72.

264 [*accountability*]: AAAS *Proceedings* 26 (1877): xiv; Kohlstedt, *Formation,* pp. 98, 170, 178, 182.

[*itself*]: Kohlstedt, *Formation,* pp. 97–98, 159–62, 178–79; Diary of W. Brewer, Aug. 14, 1850, Brewer MSS.; J. Henry to A. Bache, Aug. 14, 1850, Henry MSS.

[*Association*]: *Albany Journal,* Aug. 23, 1851; Stanton, *Exploring Expedition,* pp. 321–22; Kohlstedt, *Formation,* pp. 179–81; J. Henry to A. Bache, Sept. 13, 1851, Henry MSS.; A. Bache to J. Henry, Oct. 19, 1851, J. Dana to A. Bache, Sept. 6, 1851, Rhees Collection; J. Dana to J. Hall, May 24, 1854, Hall MSS.

[*ambition*]: Kohlstedt, *Formation,* pp. 171–73, 206.

265 [*plan*]: Ibid., pp. 182–83; J. Henry to wife, Aug. 19, 1855, Henry MSS.; W. Rogers to H. Rogers, August 1855?, Rogers MSS.; *Providence Journal,* Aug. 22, 1855.

[*margin*]: Kohlstedt, *Formation,* p. 184; AAAS *Proceedings* 12 (1858): xv–xviii; W. Rogers to H. Rogers, Sept. 1, 1856, Rogers MSS.

[Times]: W. Rogers to L. Blodget, Sept. 10, 1856, Rogers MSS.; Brewster, *Whitney,* p. 169; Diary of C. Wilkes, Aug. 25, 1856, Wilkes MSS.; *New York Times,* Aug. 22, 1856.

[*place*]: A. Bache to J. Frazer, Sept. 5, 1856, Frazer MSS.; J. Henry to A. Bache, Sept. 18, 1858, Bache MSS., Smithsonian; D. Wells to J. Warner, June 25, 1858, C. Winslow to J. Warner, Aug. 17, 1859, Warner MSS.; AAAS *Proceedings* 13 (1859): lv.

266 [*heretofore*]: E. Agassiz to A. Bache, 1857?, Rhees Collection; A. Bache to B. Peirce, Apr. 3, 22, 1858, Peirce MSS.; A. Bache to J. Frazer, Mar. 22, 1858, Frazer MSS.; W. Gibbs to A. Bache, June 2, 1858, J. Dana to A. Bache, May 15, 1856, Bache MSS.

[*probation*]: W. Rogers to H. Rogers, August 1855, Rogers MSS.; J. Hilgard to E. Hilgard, July 23, 1859, Hilgard MSS.; A. Bache to B. Peirce, Aug. 9, 1860, Peirce MSS.

[*liberty*]: Kohlstedt, *Formation,* p. 189.

[*dodged*]: B. Gould to B. Silliman, Jr., Sept. 21, 1860, Gould MSS.

267 [*storm*]: W. Whitney to wife, Aug. 2, 1860, Whitney MSS.; LeConte, *Autobiography,* p. 178.

[*breather*]: Johannsen, *Douglas,* p. 780.

[*end*]: Chute, *Damn Yankee,* pp. 178–79; A. Bache to B. Peirce, July 1, Aug. 9, 1860,

Peirce MSS.; *New York Times,* Aug. 25, 1860; B. Silliman, Jr., to A. Bache, Aug. 21, 1860, Rhees Collection; *Am. Jour. Sci.* 80 (1860): 298.
[*time*]: *New York Times,* Aug. 25, 1860; Chute, *Damn Yankee,* p. 181; A. Bache to B. Peirce, Aug. 9, 1860, Peirce MSS.; *Springfield* (Mass.) *Republican,* Aug. 10, 1859; *Am. Jour. Sci.* 80 (1860): 301.

268 [*war*]: *New York Times,* Aug. 8, 1860; *Newport* (R.I.) *Advertiser,* Aug. 8, 1860.

20. *Science and the Shock of War*

271 [*bane*]: H. Ravenel to A. Gray, Dec. 11, 1860, M. Curtis to A. Gray, Apr. 17, 1857, Gray Herbarium; J. Torrey to J. Henry, Mar. 27, 1850, Henry MSS.; R. Hare to D. Coxe, Oct. 3, 1850, Hare MSS.; J. Henry to J. Torrey, Nov. 27, 1860, Torrey MSS.; Isabel Anderson, ed., *The Letters and Journals of General Nicholas Longworth Anderson* (New York, 1942), p. 92.

272 [*examiner*]: Emerson, *Early Years,* p. 104; Hamilton, *Hedrick,* passim; C. Manly to B. Hedrick, Oct. 29, 1856, H. Herrisse to B. Hedrick, Feb. 7, 1857, Hedrick MSS.; B. Peirce to B. Hedrick, Apr. 24, 1857, Peirce MSS.
[*abolitionists*]: Fulton and Thomson, *Silliman,* pp. 258–61; B. Silliman, Jr., to J. Hall, Apr. 9, 1847, Merrill MSS.; Dupree, *Gray,* p. 201; H. Rogers to G. Combe, Nov. 14, 1856, Combe MSS.; Rogers, *Rogers* 2: 54–55; Clyde, *Coffin,* pp. 70–73, 76.
[*shells*]: Neil E. Stevens, "Two Southern Botanists and the Civil War," *Scientific Monthly* 9 (1919): 159; Smithsonian, *Annual Report 1848,* p. 142; AAAS *Proceedings* 3 (1850): 215; *American Journal of Education* 3 (1857): 153; J. Torrey to S. Baird, Dec. 24, 1860, Baird MSS.; W. Stimpson to E. Ravenel, Dec. 24, 1860, E. Ravenel MSS.

273 [*overthrown*]: Lupold, "LeConte," pp. 101, 161; LeConte, *Autobiography,* p. 179; M. Curtis to A. Gray, Dec. 5, 12, 1860, Jan. 2, 1861, Gray Herbarium; J. Riddell to G. Welles, July 31, 1862, Rhees Collection; R. Owen to [J. Hall?], Sept. 22, 1863, Merrill MSS.; C. L. Bachman, *John Bachman* (Charleston, S.C., 1888), pp. 282, 352, 354–56, 359; J. Bachman to Dr. Brown, Feb. 14, 1861, Bachman MSS.; D. Mallory to H. Fish, June 9, 1851, Fish MSS., LC; J. Bachman to V. Audubon, Sept. 11, 1851, Jan. 30, 1857, J. Bachman to J. Brown, Feb. 13, 1861, Bachman MSS.
[*move*]: R. Owen to [J. Hall?], Sept. 22, 1863, Merrill MSS.; F. Barnard to R. Gilliss, Feb. 19, 1861, Columbia University Special MSS. Collection; F. Barnard to E. Hilgard, Oct. 3 (1858?), Barnard MSS.
[*sane*]: Walter L. Fleming, *Louisiana State University, 1860–1896* (Baton Rouge, La., 1936), p. 308; J. Henry to wife, Oct. 12, 1863, Henry MSS.; J. Hall to "Friend Kendall," Dec. 9, 1862, Hall MSS.; B. Silliman, Jr., to A. Bache, Mar. 4, 24, 1862, Bache MSS.; W. Gibbs to A. Bache, Feb. 19, 1862, Rhees Collection.
[*country*]: Rodgers, *Torrey,* p. 271; J. Torrey to S. Baird, June 19, 1861, Baird MSS.; Rogers, *Rogers* 2: 54–55, 72; C. Parry to J. Torrey, Feb. 9, 1861, Torrey MSS.

274 [*friendship*]: Gray, *Letters* 2: 469; Dupree, *Gray,* pp. 308, 310; for the Gray-Darwin correspondence on the Civil War, see Gray, *Letters* 2: 472–92, 501–14, 523–33, 536–40, 549, and Ralph Colp, Jr., "Charles Darwin: Slavery and the American Civil War," *Harvard Library Bulletin* 26 (1978): 471–89.
[*last*]: Gray, *Letters* 2: 467; B. Peirce to A. Bache, Jan. 23, 1861, Peirce MSS.; Shaler, *Autobiography,* p. 170; Lurie, *Agassiz,* p. 305; B. Gould to A. Bache, Apr. 13, 1861, J. Hilgard to A. Bache, Jan. 24, 1861, Rhees Collection; W. Gibbs to S. Gay, Aug. 9, 1861, Gay MSS.; Gilman, *Dana,* p. 159.
[*Franklin*]: Fraley to A. Bache, Aug. 24, 1857, J. Hilgard to A. Bache, Sept. 4, 1857, Bache MSS.; A. Bache to J. Henry, Oct. 19, 1851, J. Pearce to A. Bache, Sept. 27, 1857, Rhees Collection; A. Bache to B. Peirce, Feb. 8, 1861, Peirce MSS.
[*praise*]: J. Henry to B. Peirce, Mar. 28, 1866, Henry MSS.

275 [*go*]: George E. Verrill, *The Ancestry, Life and Work of Addison E. Verrill* (Santa Barbara, Calif., 1958), p. 46; J. Henry to A. Bache, Apr. 4, 1862, Rhees Collection; entry for July 3, 1862, Mary Henry Diary; J. Henry to F. Barnard, Feb. 23, 1861, Columbia University Special MSS. Collection; J. Torrey to B. Hedrick, Aug. 27, 1862, Hedrick MSS.
[*roof*]: J. Henry to M. Fillmore, July 15, 1867, to A. Bache, Dec. 5, 1865, and "Locked Book," Nov. 21, 1864, Henry MSS.; J. Henry to A. Bache, June 25, 1861, Rhees Collection; Dall, *Baird,* p. 234; Coulson, *Henry,* p. 243.
[*himself*]: W. Gibbs to A. Bache, Feb. 2, 1862, Rhees Collection.

[*personality*]: entry for Oct. 6, 1864, Mary Henry Diary; J. Henry to ?. Apr. 30, 1866, Henry MSS.

276 [*action*]: entry for June 1, 1861, Mary Henry Diary; Bruce, *Lincoln,* pp. 84–85; Coulson, *Henry,* pp. 238–40; J. Henry to wife, Apr. 21, 1865, Henry MSS.

[*North*]: AAAS *Proceedings* 18 (1869): 33; B. Gould to C. Abbe, Oct. 6, 1864, Abbe MSS.

[*pursuits*]: B. Silliman, Jr., to A. Bache, May 22, 1861, Bache MSS.; J. Hall to J. Lesley, Nov. 4, 1861, Lesley MSS.; F. Hayden to J. Leidy, June 6, 1861, Leidy MSS.; E. Emmons to H. Ward, Jan. 19, 1861, Ward MSS.; A. Murray to J. L. LeConte, Oct. 9, 1861, LeConte MSS., APS.

277 [*1865*]: E. Hilgard to A. Winchell, Oct. 4, 1865, Winchell MSS.; *Am. Jour. Sci.* 91 (1866): 143; *New York Tribune,* Nov. 11, 15, 1862; W. Fry to A. Lincoln, Nov. 14, 1862, Lincoln Collection; U.S., Congress, Senate, Special Committee, *Report on the Fire at the Smithsonian Institution,* Senate Report no. 129, 38th Cong., 2d sess., Feb. 23, 1865, p. 3.

[*lost*]: H. Ravenel to R. Dwight, Apr. 1, 1865, H. Ravenel MSS.; Fleming, *Louisiana State University,* p. 118; J. Millington to A. Bache, Apr. 20, 1863, Rhees Collection; W. Pendleton to E. Alexander, Oct. 9, 1865, F. Smith to E. Alexander, Oct. 11, 1865, Alexander MSS.; W. Hume to L. Gibbes, Aug. 24, 1865, L. Gibbes MSS., LC; S. Elliott to E. Ravenel, Oct. 4, 1861, E. Ravenel MSS.; Bates, *Scientific Societies,* p. 46; Autobiographical Notes of Eugene Hilgard, p. 125, Hilgard MSS.

278 [*acted*]: James G. Gibbes, *Who Burnt Columbia?* (Newberry, S.C., 1902), pp. 35–36; Richard M. Jellison and Phillip S. Swartz, "The Scientific Interests of Robert W. Gibbes," *South Carolina Historical Magazine* 66 (1965): 97; J. McCrady to L. Gibbes, Sept. 25, 1865, J. LeConte to L. Gibbes, Apr. 10, 1865, L. Gibbes MSS., LC; Lupold, "LeConte," pp. 181–82; Lester D. Stephens, *Joseph LeConte: Gentle Prophet of Evolution* (Baton Rouge, La., 1982), pp. 41, 99; Jefferson Davis, *The Rise and Fall of the Confederate Government,* 2 vols. (New York, 1881), 2: 712, 714.

[*forehanded*]: G. Vasey to I. Lapham, Feb. 3, 1862, Lapham MSS.; T. Conrad to J. Hall, Oct. 21, 1862, J. Safford to J. Hall, June 23, 1862, Hall MSS.; G. Engelmann to J. Torrey, Oct. 26, 1861, Torrey MSS.; Lupold, "LeConte," p. 166; E. Ritchie of Boston and J. Chilton of New York to L. Gibbes, various dates in summer and fall of 1860, but especially October and November, L. Gibbes MSS., LC.

279 [*disability*]: Wayne Flynt, "Southern Higher Education and the Civil War," *Civil War History* 14 (1968): passim; Storr, *Beginnings,* pp. 118–28; F. Barnard to ?, Oct. 28, 1863, Barnard MSS.

[*before*]: Ibid.

[*citizens*]: W. Gibbs to A. Bache, Apr. 21, 1861, Rhees Collection; J. Henry to S. Baird, Aug. 16, 1862, W. Rhees to S. Baird, Aug. 16, 1862, Baird MSS.; J. Henry to C. Wetherill, Oct. 17, 1864, Wetherill MSS.; J. Hilgard to E. Hilgard, July 13, 1865, Hilgard MSS.; E. Courtenay to D. Courtenay, July 21, Oct. 2, 1864, Purviance-Courtenay MSS.; F. Hayden to J. Leidy, Nov. 11, 1863, Leidy MSS.

280 [*eminence*]: See Appendix.

[*Carolina*]: Boston Society of Natural History, *Boston Society,* p. 17; L. Agassiz to C. Lütken, September 1864, Agassiz MSS., MCZ; Wayman, *Morse,* pp. 100–101; Bailey, *History,* p. 266; Mass. College of Agriculture, *Goessmann,* p. 135; *Am. Jour. Sci.* 85 (1863): 155.

[*books*]: *DAB* 13: 38–39 ("Ormsby Mitchel"); Robert L. Black, "The Cincinnati Telescope," in *The Centenary of the Cincinnati Observatory* (Cincinnati, 1944), p. 44.

281 [*science*]: Smithsonian, *Annual Report, 1862,* p. 13; M. Rock to W. Darlington, Oct. 20, 1862, MSS. D, New-York Historical Society Miscellaneous MSS.; Main, "Kumlien," p. 334; Angela K. Main, "Life and Letters of Edward Lee Greene," Wisconsin Academy of Sciences *Transactions* 24 (1929): 148–63; F. Hayden to J. Leidy, Apr. 8, Sept. 6, Nov. 11, 1863, Leidy MSS.

[*1863*]: R. Owen to J. Hall, Sept. 22, 1863, Merrill MSS.; E. Andrews to H. Ward, Dec. 12, 1863, Ward MSS.; W. Sherman to F. Barnard, Sept. 4, 1864, Barnard MSS.; G. Horn to J. L. LeConte, Nov. 15, 1864, J. Smith to J. L. LeConte, Jan. 23, 1863, LeConte MSS., APS; A. Bickmore to C. Abbe, Mar. 4, 1863, Abbe MSS.; F. Meek to G. Cook, June 5, 27, 1863, Cook MSS.

[*taste*]: B. Walsh to J. L. LeConte, Sept. 9, 1863, LeConte MSS., APS.

282 [*racism*]: Robert H. Bremner, *The Public Good: Philanthropy and Welfare in the*

Civil War Era (New York, 1980), pp. 39–44; J. Newberry to S. Baird, July 27, 1861, Baird MSS.; J. Newberry to J. Hall, May 14, 1865, Hall MSS.; B. Gould to C. Abbe, Oct. 6, 1864, Mar. 2, 1865, Abbe MSS.; B. Gould to F. Barnard, Feb. 15, 1867, Barnard MSS,; John S. Haller, Jr., *Outcasts from Evolution* (Urbana, Ill., 1971), pp. 22–23, 28–29, 58, 62, 86, 94, 204.

[*it*]: Bachman, *Bachman,* pp. 364–65, 369; J. Bachman to E. Ruffin, Nov. 15, 1864, Bachman MSS.

[*other*]: LeConte, *Autobiography,* p. 240; Lupold, "LeConte," pp. 167, 170; Stevens, "Two Southern Botanists," p. 162; M. Curtis to A. Gray, Sept. 1, 1865, Gray Herbarium.

283 [*West*]: Rogers, *Rogers* 2: 95; F. Hayden to J. Leidy, June 6, 1861, Leidy MSS.; J. Torrey to I. Lapham, Aug. 20, 1861, Lapham MSS.; J. Anthony to G. Tryon, June 27, 1861, Tryon-Pilsbry Collection; A. Fitch to L. Tucker, Aug. 12, 1862, Gratz Collection; J. Henry to J. Coffin, Sept. 19, 1862, Henry MSS.; W. Rogers to J. Dana, Nov. 30, 1862, Dana MSS.; J. Lesley to T. Blackwell, Feb. 10, 1863, Lesley MSS.; J. Kirtland to J. L. LeConte, Jan. 21, 1865, LeConte MSS., APS.

[*up*]: L. Lesquereux to J. Lesley, Aug. 8, 1861, Lesley MSS.; L. Lesquereux to I. Lapham, Mar. 31, 1862, Lapham MSS.; H. Prout to A. Winchell, Dec. 23, 1861, Winchell MSS.; J. Koch to G. Cook, Dec. 22, 1861, Cook MSS.; Sidar, *Cook,* pp. 93–94; Osborne, *Letter-Files,* pp. 136, 158.

[*complaint*]: J. Henry to J. Hall, July 19, 1864, Hall MSS.; J. Henry to A. Bache, Sept. 25, 1864, Bache MSS.; W. Gibbs to F. Genth, May 21, 1864, Genth MSS.; A. Hall to C. Abbe, Nov. 28, 1864, Abbe MSS.; S. Newcomb to father, Feb. 25, 1863, Newcomb MSS.

[*lure*]: E. Horsford to J. Hall, Dec. 17, 1862, Hall MSS.; Allan Nevins, *The War for the Union,* 4 vols. (New York, 1959–71), 3: 257–62.

284 [*prayer*]: J. Henry to C. Dewey, Jan. 16, 1866, Dewey MSS.; F. Hayden to J. Leidy, Jan. 24, 1865, Leidy MSS.; J. Hall to J. Lesley, Mar. 2, 1865, Lesley MSS.; William S. W. Ruschenberger, *A Sketch of the Life of Robert E. Rogers* (Philadelphia, 1885), p. 28; Ames, *Lesley* 1: 393; J. Lesley to J. Hall, Feb. 27, 1865, J. Hall to H. Lee, Apr. 19, 1865, Hall MSS.

[*engineers*]: N. Jones to J. Warner, Feb. 14, 1861, Warner MSS.; Lewis, *Ellet,* p. 170; Harold B. Hancock and Norman B. Wilkinson, "A Manufacturer in Wartime: Du Pont, 1860–1865," *Business History Review* 40 (1966): 214–15; Bureau of the Census, *Historical Statistics of the United States* (1960, Washington, D.C.), pp. 368, 370–71, 428; W. Gurley to G. Cook, July 1, 1861, Apr. 8, Sept. 18, 1863, Apr. 19, 1864, Mar. 14, 1865, Cook MSS.; Paul, *Mining Frontiers,* pp. 65–67; N. Emmons to S. Emmons, Feb. 2, 1863, Emmons MSS.

[*to*]: Gray, *Letters* 2:480, 486.

285 [*news*]: C. Parry to J. Torrey, July 22, 1861, Torrey MSS.; Francis P. Farquhar, ed., *Up and Down California in 1860–1864: The Journal of William H. Brewer* (Berkeley, Calif., 1966), p. 102 and passim; Clarke, *Hall,* pp. 395–96; W. Bartlett to S. Newcomb, Oct. 28, 1861, J. Runkle to S. Newcomb, June 5, 1862, Newcomb MSS.

[*press*]: S. Newcomb diary, entries for May 1861, passim, Sept. 27, 1861, July 12–14, 1864, Apr. 10, 1865, Newcomb MSS.; B. Gould to S. Newcomb, Aug. 19, 1861, H. Eustis to S. Newcomb, Aug. 18, 1862, Newcomb MSS.; B. Peirce to C. Davis, Aug. 20, 1864, S. Newcomb to B. Peirce, Oct. 17, 1864, Peirce MSS.

[*age*]: W. Gibbs to F. Genth, Dec. 12, 1865, Genth MSS.; Ferenc Szabadvary, "Wolcott Gibbs and the Centenary of Electrogravimetry," *Journal of Chemical Education* 41 (1964): 666–67; Prendergast, "Dana," pp. 479–81, 531–35, 549.

286 [*good*]: Clarke, *Hall,* p. 508; Pfeifer, "Darwinism," pp. 50, 53, 55–57; Lurie, *Agassiz,* pp. 306–11.

[*enterprise*]: Wayman, *Morse,* pp. 169–70; L. Agassiz to A. Lincoln, Mar. 21, 1862, Boston University Special Collections; Agassiz, *Agassiz,* p. 578; Lurie, *Agassiz,* p. 336; L. Agassiz to T. Lyman, Apr. 30, 1867, Agassiz MSS., MCZ.

21. War and the Structure of Science

288 [*1862*]: E. Hitchcock to H. Ward, July 19, 1861, Ward MSS.; W. Chauvenet to A. Bache, May 22, 1862, Rhees Collection; Jones and Boyd, *Harvard Observatory,* pp. 117–22.

[*spending*]: Lurie, *Agassiz,* pp. 301, 305, 309, 337.
[*before*]: Bremner, *Public Good,* pp. 80–81; Lurie, *Agassiz,* p. 318; *DAB* 19: 366 ("William J. Walker"); Rogers, *Rogers* 2: 157–59, 238; A. Packard to C. Abbe, May 6, 1865, Abbe MSS.
[*Association*]: Hendrickson, *Arkites,* p. 30; Will E. Edington, "There Were Giants in Those Days," Indiana Academy of Sciences *Proceedings* 44 (1935): 26; Ralph W. Dexter, "History of the Pottsville (Pa.) Scientific Association, 1854–1862," *Science Education* 53 (1969): 32.

289 [*remark*]: Numerous letters to A. Bache, early 1861, in Rhees Collection, particularly from B. Silliman, Jr. (Jan. 4), B. Peirce (Feb. 14, 15), W. Gibbs (Feb. 3), F. Barnard (Mar. 1), and J. Lovering (Mar. 16); AAAS circular, May 27, 1861, Cook MSS.; J. Lovering to H. Ward, Jan. 16, 1862, E. Horsford to H. Ward, Sept. 9, 1862, Ward MSS.; J. Dana to S. Baird, Mar. 10, 1863, Baird MSS.; J. Hilgard to E. Hilgard, July 13, 1865, Hilgard MSS.
[*survived*]: *Am. Jour. Sci.,* 5th ser., 46 (1918): 379; B. Gould to A. Bache, Mar. 6, 1861, Rhees Collection; Cajori, *Teaching,* p. 279; B. Silliman, Jr., to A. Bache, Oct. 22, 1861, Bache MSS.; *Am. Jour. Sci.* 33 (1862): 1; B. Silliman, Jr., to G. Cook, Dec. 1, 1862, Cook MSS.; J. Dana to W. Gibbs, June 6, 1865, Gibbs MSS.
[*reports*]: W. Chauvenet to C. Schott, Aug. 9, 1861, Schott MSS.; J. Dana to G. Cook, Aug. 24, 1861, Cook MSS.; Smithsonian, *Annual Report, 1863,* p. 29; Clyde, *Coffin,* p. 66; Goetzmann, *Army Exploration,* p. 403.
[*Harvard*]: W. Chauvenet to A. Bache, May 22, 1861, Rhees Collection; Viles, *University of Missouri,* pp. 97, 100–101; *DAB* 9: 46 ("Thomas Hill").

290 [*army*]: Daniel J. Pratt, comp., *Statistics of Collegiate Education* (New York, 1865), p. 12; Cheyney, *University of Pennsylvania,* p. 249; A. Bache to C. Wetherill, Aug. 27, 1861, Wetherill MSS.; D. Kirkwood to J. Warner, June 7, 1862, Warner MSS.
[*chemists*]: MS. autobiography of Cleveland Abbe, p. 49, Abbe MSS; Nathan Reingold, "A Good Place to Study Astronomy," *Library of Congress Quarterly Journal* 20 (1963): 211–12; E. Blake to E. Rice, Apr. 10, 1863, Blake Family MSS.; James M. Hart, *German Universities: A Narrative of Personal Experiences* (New York, 1874), p. 60.
[*support*]: Pratt, *Statistics,* p. 7; Gendebien, "Science," pp. 145, 148; *DAB* 14: 200–201 ("Ario Pardee"); Wertenbaker, *Princeton,* pp. 275, 307–308; Sidar, *Cook,* pp. 83, 85.
[*grant*]: Cheyney, *University of Pennsylvania,* p. 242; Rogers, *Rogers* 2:85–86; "Statement for Mr. B. of the Legislature," Apr. 11, 1862, Rogers MSS.; Samuel C. Prescott, *When M.I.T. Was "Boston Tech"* (Cambridge, Mass., 1954), p. 40.

291 [*depression*]: Rezneck, *Education,* pp. 136–43, 147–49, 151; C. Goessmann to G. Cook, Apr. 8, 1862, Cook MSS.
[*work*]: Browne, "History," p. 729; Bacon, "Precursor," pp. 190–93, 196; Osborne, *Letter-Files,* p. 136.
[*1876*]: Verrill, *Verrill,* p. 97; Chittenden, *Sheffield School* 1: 83, 118–19, 121; Rossiter, *Emergence,* p. 140; Osborne, *Letter-Files,* pp. 133–34.
[*it*]: Samuel E. Morison, *Three Centuries of Harvard* (Cambridge, Mass., 1936), p. 303; Henry James, *Charles W. Eliot,* 2 vols. (Boston, 1930), 1: 89.

292 [*rebels*]: Lurie, *Agassiz,* pp. 312–17; A. Verrill to A. Winchell, Feb. 16, 1864, Winchell MSS.
[*it*]: William G. Land, *Thomas Hill, Twentieth President of Harvard* (Cambridge, Mass., 1933), pp. 124–25; B. Peirce to A. Bache, Oct. 26, 1862, Peirce MSS.; L. Agassiz to J. Andrew, Dec. 16, 1862, Agassiz MSS.; Lurie, *Agassiz,* pp. 326–28; Dupree, *Gray,* pp. 314–15.
[*Harvard*]: L. Agassiz to W. Gibbs, Jan. 4, 1863, Agassiz MSS.; W. Gibbs to O. Rood, Apr. 4, 1862, W. Gibbs to "My dear friend," Jan. 6, 1863, Gibbs MSS.

293 [*himself*]: James, *Eliot* 1: 85–113; W. Gibbs to O. Rood, Aug. 24, 1863, Gibbs MSS.
[*15-incher*]: Chute, *Damn Yankee,* pp. 41–42, 72–73, 76, 108–10, 123, 141–42, 146, 160, 173; John Fulton, *Memoirs of Frederick A. P. Barnard* (New York, 1896), pp. 208–12, 235; B. Peirce to A. Bache, Dec. 12, 1861, Bache MSS.
[*1861*]: Chute, *Damn Yankee,* pp. 174–78.

294 [*1862*]: Ibid., pp. 145, 148–49, 158, 168–73, 182, 185–87; F. Barnard to J. Hilgard, Oct. 3, Nov. 11, 1858, Barnard MSS.; F. Barnard to A. Bache, Mar. 1, Apr. 24, 1861, Rhees Collection; F. Barnard to J. Hilgard, Mar. 6, 1861, MSS. B, New-York Historical

Society Miscellaneous MSS.; *New York Times,* Aug. 30, 1864; Jones and Boyd, *Harvard Observatory,* pp. 122–24.
[*successor*]: Chute, *Damn Yankee,* p. 188; A. Bache to B. Peirce, May 26, 28, 1862, Peirce MSS.; A. Bache to C. King, Oct. 8, 1863, Columbia University College Papers; W. Gibbs to O. Rood, Oct. 12, Nov. 9, 1863, O. Rood to W. Gibbs, Dec. 27, 1863, A. Bache to W. Gibbs, Feb. 2, 1864, Gibbs MSS.; J. Barnard to G. Kemble, Oct. 24, 1863, Fish MSS., Columbia.
[*1857*]: Merrill, *First One Hundred Years,* pp. 391, 394, 396; Autobiographical notes of Eugene Hilgard, p. 124, Hilgard MSS.; A. Packard to C. Abbe, Apr. 11, 1863, Abbe MSS.; Sidar, *Cook,* p. 99.

295 [*paleontology*]: Gerald D. Nash, *State Government and Economic Development* (Berkeley, Calif., 1964), pp. 98–103.
[*appropriation*]: F. Hayden to J. Leidy, Feb. 16, 1862, Leidy MSS.; J. Newberry to J. Torrey, Feb. 6, 1862, Torrey MSS.; I. Lapham to J. Potter, Jan. 3, 1862, Lapham MSS.; D. Holloway, Aug. 8, 1861, J. Potter, Apr. 11, 1862, W. Spooner, Sept. 12, 1862, and I. Newton, Sept. 9, 1863, to I. Lapham, Lapham MSS.
[*science*]: George Meade, Jr., *The Life and Letters of George Gordon Meade,* 2 vols. (New York, 1913), 1: 208–12, 215–18.

296 [*standing*]: *DAB* 9: 371–72 ("Andrew Humphreys"), 1: 13 ("Henry Abbot"); Martin Reuss, "Andrew A. Humphreys and the Development of Hydraulic Engineering," *Technology and Culture* 26 (1985): 1–4, 8, 10; A. Humphreys to A. Bache, Apr. 15, 1851, Rhees Collection; *Am. Jour. Sci.* 33 (1862): 181; B. Peirce to A. Bache, Jan. 6, 1862, Bache MSS.; W. Bartlett to S. Newcomb, Jan. 9, 1862, Newcomb MSS.; Goetzmann, *Army Exploration,* p. 431.
[*Engineers*]: Goetzmann, *Army Exploration,* pp. 427–34.
[*him*]: *DAB* 20: 217 ("Charles Wilkes"); Daniel M. Henderson, *The Hidden Coasts* (New York, 1953), p. 218.
[*humbug*]: Williams, *Maury,* pp. 348, 350–55, 363–68; A. Bache to J. L. LeConte, Sept. 24, 1861, LeConte MSS., APS; B. Peirce to A. Bache, Sept. 1, 1861, Bache MSS.

297 [*observatory*]: *DAB* 7: 293 ("James Gilliss"); Sands, *From Reefer,* p. 297; S. Newcomb to father, Sept. 18, 1863, Newcomb MSS.; Newcomb, *Reminiscences,* p. 108; *North American Review* 105 (1867): 386.
[*Davis*]: J. Gilliss to J. Hall, Nov. 27, 1861, Merrill MSS.; *North American Review* 105 (1867): 386–87; S. Newcomb to "Steves," Mar. 3, 1864, to W. Bartlett, May 10, 1864, Newcomb MSS.
[*Spain*]: S. Bent to A. Bache, Apr. 5, 1861, C. Davis to A. Bache, Apr. 8, 1861, Rhees Collection; Davis, *Davis,* pp. 116–17, 156–57, 281–85; Leonard D. White, *The Republican Era: 1869–1901* (New York, 1958), p. 163.

298 [*it*]: Dupree, *Science in the Federal Government,* pp. 100–105, 132; Davis, *Davis,* p. 124; A. Bache to C. Boutelle, Jan. 3, 1861, Rhees Collection.
[*engineers*]: A. Bache to J. Whitney, May 29, 1861, G. McClellan to A. Bache, Jan. 10, 1862, Rhees Collection; Odgers, *Bache,* p. 176; Davis, *Davis,* p. 124.
[*service*]: Reingold, "Bache," p. 165; Davis, *Davis,* pp. 121–22, 124; B. Peirce to A. Bache, Dec. 23, 1861, Jan. 8, 1862, Bache MSS.; G. McClellan to A. Bache, Jan. 10, 1862, R. McKnight to A. Bache, Jan. 13, 1862, Rhees Collection; *New York Times,* Dec. 20, 1861; E. Courtenay, Jr., to D. Courtenay, Jan. 31, 1862, Purviance-Courtenay MSS.

299 [*badly*]: A. Bache to J. Frazer, May 20, 24, Aug. 22, Sept. 24, 1861, Frazer MSS.; A. Bache to W. Rhees, Dec. 16, 1861, Bache MSS.; E. Courtenay, Jr., to D. Courtenay, Jan. 31, 1862, Purviance-Courtenay MSS.
[*physicians*]: A. Bache to J. Frazer, Aug. 25, Oct. 31, 1862, Frazer MSS.; Odgers, *Bache,* pp. 209–12; A. Bache to W. Gibbs, Sept. 26, 1863, Gibbs MSS.
[*live*]: J. Henry to V. Farragut, May 6, 1863, Huntington Misc. MSS.; J. Henry to S. Baird, Aug. 16, 1862, Baird MSS.; J. Henry to N. Bache, Oct. 5, 1867, Henry MSS.; entry for Dec. 6, 1862, Mary Henry Diary; J. Henry to J. Hall, Jan. 2, 1863, Hall MSS.

300 [*specimens*]: Diary of Spencer Baird, Apr. 22–26, 1861, Diary of Joseph Henry, Apr. 9, 11, 1865, Smithsonian Institution Archives; Herber, *Correspondence,* pp. 171–72.
[*consequence*]: Smithsonian, *Annual Reports,* 1861, p. 15, 1863, pp. 13–14, 1864, p. 16, 1865, p. 20; J. Henry to C. Schott, Nov. 28, 1864, Schott MSS.; S. Baird to J. Torrey, Dec. 30, 1861, Torrey MSS.; J. Henry to J. Coffin, Mar. 31, 1864, Henry MSS.
[*business*]: Goode, *Smithsonian,* p. 837; Smithsonian, *Annual Report,* 1861, p. 35.

[*building*]: J. Henry to Dr. Foreman, Aug. 9, 1851, Henry MSS.; U.S., Congress, Senate, Special Committee, *Report on the Fire at the Smithsonian Institution,* Senate Report no. 129, 38th Cong., 2d sess., Feb. 23, 1865, pp. 1–2; Smithsonian, *Annual Report,* 1865, pp. 14–18; Diary of Joseph Henry, Jan. 25, 1865, Smithsonian Institution Archives.

301 [*work*]: *DAB* 17: 514 ("John Stanley"); Hellman, *Smithsonian,* p. 67.
[*Congress*]: Bache, "Wants of Science," pp. 26–31; AAAS *Proceedings,* vol. 6 (1851): xlviii–li, vol. 8 (1854): 16; A. Bache to "My dear Sir," Sept. 25, 1853, Bache MSS.; Bache, *Anniversary Address,* pp. 30–31; Cochrane, *National Academy,* pp. 46–47.
[*research*]: Dupree, *Science in the Federal Government,* pp. 149–50, 152–53; Paul W. Gates, *Agriculture and the Civil War* (New York, 1965), pp. 302, 306–18.

302 [*possibilities*]: Earle D. Ross, *Democracy's College* (Ames, Iowa, 1942), pp. 14–85.
[*Congress*]: Dupree, *Science in the Federal Government,* pp. 136–37; L. Agassiz to A. Bache, Feb. 6, 1863, Rhees Collection; L. Agassiz to H. Wilson, Feb. 5, 1863, Agassiz MSS.; A. Bache to J. Frazer, Apr. 7, 1863, Frazer MSS.
[*21*]: A. Bache to B. Peirce, Dec. 9, 1862, Peirce MSS.; A. Bache to J. Frazer, Feb. 3, Mar. 10, Apr. 7, 1863, Frazer MSS.; Cochrane, *National Academy,* pp. 52–53, 595–96.
[*comment*]: Cochrane, *National Academy,* pp. 53, 56.

303 [*in*]: Ibid., pp. 595–96; Dupree, *Science in the Federal Government,* p. 139; Davis, *Davis,* p. 292; A. Bache to J. Frazer, Mar. 10, 1863, Frazer MSS.
[*community*]: Reingold, *Science in Nineteenth-Century America,* pp. 201, 204; Davis, *Davis,* p. 292; Fleming, *Draper,* p. 110; J. Whitney to G. Brush, Apr. 8, 1863, Whitney MSS.; A. Bache to J. Frazer, Mar. 12, 1863, Frazer MSS.
[*time*]: Cochrane, *National Academy,* pp. 58–62.

304 [*body*]: A. Bache to J. Frazer, Apr. 17, Dec. 26, 1863, Frazer MSS.; Cochrane, *National Academy,* pp. 63, 69–73, 603–604; J. Lesley to "My dear Lyman," May 7, 1863, Lesley MSS.; *Am. Jour. Sci.* 85 (1863): 465 (This quotation is garbled and its meaning thereby changed in Cochrane, *National Academy,* p. 69).
[*government*]: Cochrane, *National Academy,* pp. 80–89, 97, 408; J. Henry to A. Bache, Aug. 21, 1863, to S. Alexander, Jan. 14, 1864, Henry MSS.; G. Welles to Permanent Commission, July 29, 1863, National Archives Record Group 45, Minutes of Permanent Commission, Feb. 11, 1863–Feb. 24, 1864, p. 115; J. Hilgard to W. Gibbs, Mar. 5, 1865, Gibbs MSS.

305 [*skunk*]: A. Gray to S. Baird, Aug. 22, 1864, J. Dana to S. Baird, July 5, 1864, Baird MSS.; Lurie, *Agassiz,* pp. 341–43; J. Dana to J. Lesley, July 18, 1864, J. Lesley to L. Lesquereux, Jan. 15, 1865, Lesley MSS.; J. Henry to A. Bache, Aug. 15, 1864, Henry MSS.; L. Agassiz to J. Henry, Aug. 8, 1864, Peirce MSS.
[*on*]: F. Barnard to J. Hilgard, Dec. 31, 1864, MSS. B, New-York Historical Society Misc. MSS.; Cochrane, *National Academy,* p. 94.
[*sympathy*]: L. Rutherfurd to W. Gibbs, Jan. 12, 1865, Gibbs MSS.; L. Lesquereux to J. Lesley, Aug. 17, 1864, Lesley MSS.

22. *"Small Potatoes"*

306 [*Smithsonian*]: Smithsonian, *Annual Report,* 1862, p. 13.

307 [*industry*]: Nathan Reingold, "Science in the Civil War: The Permanent Commission of the Navy Department," in Reingold, *Science Since 1820,* p. 162; Victor S. Clark, *History of Manufacturing in the United States,* 3 vols. (New York, 1929), 2:12, 18–19; Saul Engelbourg, "The Economic Impact of the Civil War on Manufacturing Enterprise," *Business History* 21 (1979): 150–58; Carl M. Becker, "Entrepreneurial Invention and Innovation in the Miami Valley during the Civil War," *Bulletin of the Cincinnati Historical Society,* January 1964, pp. 7, 19, 23, 25, 27–28; Felicia J. Deyrup, *Arms Makers of the Connecticut Valley . . . 1798–1870* (Northampton, Mass., 1948), p. 96; William S. Dutton, *Du Pont: One Hundred and Forty Years* (New York, 1942). pp. 91–92, 98.
[*navy*]: Bruce, *Lincoln,* pp. 32–33, 69; U.S., Congress, Senate, Committee on the Conduct of the War, *Report on Heavy Ordnance,* Senate Report No. 121, 38th Cong., 2d sess., Feb. 13, 1865, p. 107; Robert U. Johnson and Clarence C. Buel, eds., *Battles and Leaders of the Civil War,* 4 vols. (New York, 1888), 1:616.
[*counterparts*]: Milton F. Perry, *Infernal Machines: The Story of Confederate*

Submarine and Mine Warfare (Baton Rouge, La., 1965), pp. 5–10, 13–17; Brooke, *Brooke*, pp. 51–52, 238, 242–43, 260–71, 281–82; Frank E. Vandiver, *Ploughshares into Swords: Josiah Gorgas and Confederate Ordnance* (Austin, Tex., 1952), pp. 114, 144–45, 154, 161, 183, 192; Emory M. Thomas, *The Confederate Nation: 1861–1865* (New York, 1979), pp. 210–11; *DAB* 15: 329 ("George Rains"); Maurice K. Melton, "Major Military Industries of the Confederate Government" (Ph. D. diss., Emory University, 1978), pp. 99–104, 125, 299–303, 443–45, 496–98, 500, 507–18.

[*projects*]: Bruce, *Lincoln*, pp. 10–14, 34–36, 39–41, 53–54, 62, 70–71, 75–82, 85–87, 89–117, 125–30, 167–68, 172, 190, 192–93, 223–24, 243, 252, 264, 282, 284–88, and passim.

308 [*Smithsonian*]: Reingold, "Science in the Civil War," pp. 162, 164–66, 170, 172; Frederick W. True, ed., *A History of the First Half-Century of the National Academy of Sciences* (Washington, D.C., 1913), pp. 1–2.

[*research*]: *Journal of the Franklin Institute* 45 (1863): 269.

[*successful*]: C. Wetherill to A. Bache, Aug. 22, 1861, Bache MSS.; B. Silliman, Jr., to F. Genth, May 4, 1861, Genth MSS.; E. Horsford to wife, February 1863, July 11, 1862, Jan. 27, 30, Feb. 3, 1863, typescript on "Prof. Horsford's Submarine" by Augustus H. Fiske, Horsford MSS.; Rezneck, "European Education," p. 385.

[*uses*]: Smith, *Wetherill*, p. 38; *DAB* 5:377 ("Robert Doremus"); Johnston & Dow to B. Hedrick, Nov. 22, 1862, Hedrick MSS.; Bruce, *Lincoln*, p. 269; *DAB* 20:571–72 ("Henry Wurtz").

309 [*Ypres*]: Arthur P. Van Gelder and Hugo Schlatter, *History of the Explosives Industry in the United States* (New York, 1927), pp. 771–72; Diary of D. Minor Scales, Mar. 18, Apr. 9, 1863, Duke University Misc. MSS.; Bruce, *Lincoln*, pp. 83, 200–201, 241–45, 247–48, 269–70; *DAB* 5: 377 ("Robert Doremus").

[*along*]: J. Floyd to J. Buchanan, Dec. 5, 1857, National Archives Record Group 107, Letters to the President, 6: 225–26; *Official Records of the Union and Confederate Armies in the War of the Rebellion*, 130 vols. (Washington, D.C., 1880–1901), Series Three, 1: 703; Jeanne McHugh, *Alexander Holley and the Makers of Steel* (Baltimore, 1980), p. 171; Alexander L. Holley, *Treatise on Ordnance and Armor* (New York, 1865), pp. 90, 92–93, 104, 414.

[*cheaper*]: *Scientific American* 4 (1861): 341, 368; T. Prosser to J. Ripley, Aug. 2, 1861, National Archives Record Group 156, Letters Received, 1861, p. 253; J. Ripley to T. Prosser & Sons, Oct. 21, 1861, National Archives Record Group 156, Misc. Letters Sent, 53: 619; *Official Records, Armies*, Series Three, vol. 2: 82, vol. 4: 469.

310 [*battle*]: T. Prosser & Sons to J. Dahlgren, Jan. 13, 1863, National Archives Record Group 74, Letters Received, "Inventions, Rifle & Smooth-bore, 1862–'4," p. 39; Bruce, *Lincoln*, p. 149; McHugh, *Holley*, p. 172.

[*Carolina*]: Ralph W. Donnelly, "Scientists of the Confederate Nitre and Mining Bureau," *Civil War History* 2 (1956): 69–92; Lupold, "LeConte," pp. 170–78.

[*war*]: Richard C. Sheridan, "Alabama Chemists in the Civil War," *Alabama Historical Quarterly* 37 (1975): 265–74.

[*time*]: Williams, *Maury*, p. 376.

311 [*1862*]: Bruce, *Lincoln*, 146–47, 150; Harold B. Hancock and Norman B. Wilkinson, " 'The Devil to Pay!': Saltpeter and the Trent Affair," *Civil War History* 10 (1964): 22–23, 25, 27–28, 31–32.

[*test*]: Bruce, *Lincoln*, pp. 211–14, 225.

312 [*1871*]: Ibid., pp. 225–26, 270, 296; Van Gelder and Schlatter, *History*, p. 118.

23. Many Mansions

313 [*August*]: E. Daniels to J. Lesley, Feb. 5, 1865, A. Gould to J. Lesley, Mar. 1, 1866, Lesley MSS.; B. Gould to F. Barnard, Feb. 7, 1866, Barnard MSS.; J. Dana to H. Ward, Aug. 1, 1866, Ward MSS.; J. Hall to A. Winchell, Aug. 30, 1865, Winchell MSS.; A. Winchell to J. Hall, Sept. 2, 1865, J. Hall to T. Hunt, Mar. 20, 1866, Hall MSS.

314 [*charter*]: AAAS *Proceedings* 15 (1866): 110, 114, 117–18; LeConte, *Autobiography*, pp. 298–99; Ames, *Life of Lesley* 1:507; J. Dana to H. Ward, Aug. 1, 1866, Ward MSS.; W. Whitney to J. Whitney, Aug. 19, 29, 1866, Whitney MSS.; J. Hilgard to E. Hilgard, Apr. 29, 1866, Hilgard MSS.; J. Dana to F. Barnard, Aug. 14, 1866, Barnard MSS.; B. Gould to S. Newcomb, Aug. 24, 1866, Newcomb MSS.; AAAS *Proceedings* 23 (1874): 151 and 1867–77, passim.

[*research*]: A. Gould to J. Lesley, Mar. 1, 1866, Lesley MSS.; AAAS *Proceedings* 22 (1873): 441–42, 23 (1874): 150–51, 26 (1877): xiv–xv, 374; *Hartford Daily Courant*,

Aug. 13, 14, 18, 1874; *New York Times,* Aug. 13, 17, 18, 1874; *Detroit Free Press,* Aug. 18, 1875; *Nashville American,* Sept. 2, 1877.

[*pluralism*]: AAAS, *Summarized Proceedings, 1915–1921* (Washington, D. C., 1921), graph on back cover.

315 [*president*]: Douglas Sloan, "Science in New York City, 1867–1907," *Isis,* 71 (1980): 45–46; Edward Beardsley, *The Rise of the American Chemistry Profession, 1850–1900* (Gainesville, Fla., 1964), pp. 23–25; Herman Skolnik and Kenneth M. Reese, eds., *A Century of Chemistry: The Role of Chemists and the American Chemical Society* (Washington, D. C., 1976), p. 6.

[*dilemma*]: Bates, *Scientific Societies,* pp. 125–26; Beardsley, *American Chemistry Profession,* pp. 26–30.

[*being*]: Cochrane, *National Academy,* pp. 101, 103, 105–106; entries for Jan. 24, 1866, Jan. 23, 1867, Jan. 22, 1868, Aug. 26, 1868, Apr. 16, 1873, Apr. 21, 1874, Diary of Spencer Baird, Smithsonian Institution Archives; entries for Apr. 20, Nov. 2, 1875, Apr. 18, 1876, Diary of Joseph Henry, Smithsonian Institution Archives; *Hartford Courant,* Aug. 14, 17, 1867; J. Hall to J. Torrey, Feb. 15, 1870, Torrey MSS.; G. Hill to S. Newcomb, Nov. 9, 1874, Newcomb MSS.; A. Hunter Dupree, "The National Academy of Sciences and the American Definition of Science," Oleson and Voss, *Organization,* pp. 343, 345.

[*half-century*]: Cochrane, *National Academy,* pp. 100–101, 121–22; J. Henry to wife, Aug. 14, 1866, Henry MSS.; entry for Jan. 23, 1868, Mary Henry Diary; J. Hall to J. Torrey, Feb. 15, 1870, Torrey MSS.; J. Henry to A. Gray, July 8, 1868, A. Gray to J. Henry, Mar. 23, 1870, Henry MSS.; J. Dana to A. Mayer, Mar. 14, 1874, Hyatt-Mayer Collection; J. Henry to J. Frazer, May 11, 1869, Frazer MSS.; J. Henry to "My Dear Sir," May 7, 1869, to S. Alexander, Apr. 22, 1872, Henry MSS.; Dupree, "National Academy," p. 351.

316 [*blame*]: W. Gibbs to F. Genth, Apr. 15, 1874, Genth MSS.; Gerald T. White, *Scientists in Conflict* (San Marino, Calif., 1968), pp. 15–16, 22, 49–62, 71–75, 81–82, 100–101, 140–43.

[*events*]: White, *Scientists,* pp. 109, 152–54, 163, 167–71, 175–76, 180–93, 198–99, 204–205, 215–25.

[*crucial*]: Dupree, "National Academy," pp. 343–45.

[*disbanders*]: Ibid., pp. 352–53; J. Henry to J. Coffin, Jan. (10?), 1866, Henry MSS.; J. Hilgard to E. Hilgard, Jan. 22, 1866, Hilgard MSS.; L. Rutherfurd to W. Gibbs, Nov. 25, 1865, Gibbs MSS.; J. Hall to J. Henry, Jan. 19, 1866, Hall MSS.; A. Gould to J. Lesley, Mar. 1, 1866, Lesley MSS.; A. Caswell to J. Henry, Jan. 13, 1868, Rhees Collection; Cochrane, *National Academy,* pp. 105–109; Fleming, *Draper,* pp. 111–12.

317 [*century*]: Cochrane, *National Academy,* pp. 102, 107, 117–19; Dupree, "National Academy," pp. 345, 349, 351–54.

[*all*]: H. Holt to A. Mayer, May 23, 1874, Hyatt-Mayer Collection.

[*sake*]: Dupree, *Science in the Federal Government,* p. 151; Sidney Fine, *Laissez Faire and the General-Welfare State* (Ann Arbor, Mich., 1956), pp. 352–53; *New York Times,* Dec. 21, 1872.

[*results*]: Fine, *Laissez Faire,* pp. 353–55; Law et al., *State Agricultural Experiment Stations,* pp. 15, 19–23; Rossiter, *Emergence,* pp. 157–71, 176; W. Atwater to S. Baird, Nov. 4, 1875, Baird MSS.; Rosenberg, *No Other Gods,* p. 148; Margaret W. Rossiter, "The Organization of the Agricultural Sciences," in Oleson and Voss, *Organization of Knowledge,* p. 213.

[*work*]: Merrill, *First Hundred Years,* p. 391; John J. Stevenson, "Geological Methods in Earlier Days," *Popular Science Monthly* 86 (1915): 28–31; Ferguson, *Geology and Politics,* pp. 35–36; Hendrickson, "Nineteenth-Century State Surveys," pp. 370–71.

318 [*accordingly*]: Dupree, *Science in the Federal Government,* pp. 152–61.

[*scope*]: Odgers, *Bache,* p. 214; Raymond C. Archibald, *Benjamin Peirce, 1809–1880* (Oberlin, O., 1925), pp. 3–4; *Hartford Courant,* Aug. 16, 1867; Dupree, *Science in the Federal Government,* pp. 202–203.

[*unpredictable*]: Dupree, *Science in the Federal Government,* p. 184; Sands, *From Reefer,* pp. 283–89, 296–97; S. Newcomb to F. Barnard, July 28, 1875, Barnard MSS.; Newcomb, *Reminiscences,* pp. 112–13; C. Abbe, MS. autobiography, pp. 71–72, Abbe MSS.

319 [*appropriations*]: Dupree, *Science in the Federal Government,* p. 186; Arthur L.

Norberg, "Simon Newcomb's Early Astronomical Career," *Isis* 69 (1978): 209–10, 224–25.

[*bureau*]: Whitnah, *United States Weather Bureau,* pp. 15, 17–20.

[*Department*]: Ibid., pp. 22, 33, 38–40, 43, 48–49, 53–56, 60; C. Abbe to G. Barker, Jan. 27, 1876, and to wife, Aug. 25, 1877, Abbe MSS.

[*Wheeler*]: Goetzmann, *Exploration and Empire,* pp. 390–93, 409, 411, 419, 422–24, 429; Dupree, *Science in the Federal Government,* pp. 186–87.

320 [*response*]: Thurman Wilkins, *Clarence King* (New York, 1958), pp. 5–6, 39, 91–92, 94–97; Henry Adams, *The Education of Henry Adams* (Modern Library edition, New York, 1931), pp. 311, 416; Thomas G. Manning, *Government in Science* (Lexington, Ky., 1967), pp. 4–9; Richard A. Bartlett, *Great Surveys of the American West* (Norman, Okla., 1962), pp. 135, 212–13; Goetzmann, *Exploration and Empire,* pp. 430–35, 445.

[*work*]: Bartlett, *Great Surveys,* pp. 206–12; Goetzmann, *Exploration and Empire,* pp. 459–61, 464–66; Wilkins, *King,* p. 208; Manning, *Government,* pp. 13–14.

[*Department*]: Bartlett, *Great Surveys,* pp. 334, 338–39, 350–56; Goetzmann, *Exploration and Empire,* pp. 467–70, 482–83, 485, 487.

321 [*Colorado*]: Goetzmann, *Exploration and Empire,* pp. 490–97; Bartlett, *Great Surveys,* pp. 5–13, 16; Merrill, *First Hundred Years,* pp. 501–506.

[*community*]: Manning, *Government,* pp. 15–18, 20; Bartlett, *Great Surveys,* pp. 17–18, 20, 22, 28, 30, 32, 57, 59, 63, 118–19; Goetzmann, *Exploration and Empire,* pp. 498, 501–502, 514, 527–28; Andrew D. Rodgers III, *John Merle Coulter* (Princeton, N.J., 1944), p. 13.

[*Hayden survey*]: Goetzmann, *Exploration and Empire,* pp. 532–51, 556; Bartlett, *Great Surveys,* pp. 224, 226, 268, 278, 311; William C. Darrah, *Powell of the Colorado* (Princeton, N.J., 1951), pp. 82, 93–94, 152, 181–82; Manning, *Government,* pp. 21–22, 27.

322 [*other survey*]: Goetzmann, *Exploration and Empire,* pp. 554, 562; Manning, *Government,* p. 27; Bartlett, *Great Surveys,* pp. 311, 319–20.

[*King*]: Manning, *Government,* pp. 21–24; Wallace E. Stegner, *Beyond the Hundredth Meridian* (Boston, 1954), pp. 153–54; Goetzmann, *Exploration and Empire,* pp. 554, 563, 565, 569–76; Darrah, *Powell,* pp. 148–51; Bartlett, *Great Surveys,* pp. 313, 319–28.

[*compensation*]: Stegner, *Beyond,* pp. 158–61; Merrill, *First Hundred Years,* pp. 385, 545.

[*years*]: Stegner, *Beyond,* pp. 155–58, 192; Merrill, *First Hundred Years,* pp. 546–47; Manning, *Government in Science,* p. 25.

323 [*science*]: Manning, *Government in Science,* pp. 33–37; Goetzmann, *Exploration and Empire,* pp. 578–82.

[*bureau*]: Manning, *Government in Science,* pp. 38–45, 51, 54–58; Goetzmann, *Exploration and Empire,* pp. 582–85, 588–91, 594–95.

[*everywhere*]: Dupree, *Science in the Federal Government,* p. 202; Stegner, *Beyond,* p. 142.

[*missions*]: Dupree, *Science in the Federal Government,* p. 151.

324 [*required*]: Hellman, *Smithsonian,* p. 94; Oehser, *Sons of Science,* pp. 78–79; Dall, *Baird,* pp. 427, 429.

[*Department*]: Hellman, *Smithsonian,* pp. 94–95, 111; Dupree, *Science in the Federal Government,* p. 237; Oehser, *Sons of Science,* pp. 78–80; Smithsonian, *Annual Report,* 1878, p. 43; Dall, *Baird,* p. 391.

[*Museum*]: Dall, *Baird,* pp. 384–85, 423; Smithsonian, *Annual Reports,* 1876, pp. 8–9, 1877, p. 8; Goode, *Smithsonian,* pp. 837–38.

325 [*in*]: Smithsonian, *Annual Reports,* 1876, pp. 45–46, 1877, pp. 36–37, 46, 1878, p. 41; Hellman, *Smithsonian,* p. 97; Dall, *Baird,* p. 413.

[*quarters*]: Smithsonian, *Annual Reports,* 1876, pp. 8, 11–13, 1877, pp. 7–8.

[*delighted*]: Draft of S. Baird to N. Bishop, June 16, 1868, marked "not sent," Baird MSS.; Dall, *Baird,* pp. 393–95; Smithsonian, *Annual Report,* 1878, pp. 7, 40; Oehser, *Sons of Science,* p. 88.

24. The New Education

326 [*incentive*]: Emerson quoted in Hugh Hawkins, *Between Harvard and America: The Educational Leadership of Charles W. Eliot* (New York, 1972), p. 39.

[*knowledge*]: George E. Peterson, *The New England College in the Age of the University* (Amherst, Mass., 1964), pp. 4–7; Getman, *Remsen*, p. 42.

327 [*it*]: William F. Russell, ed., *The Rise of a University*, 2 vols. (New York, 1937), 1: 72–75; Getman, *Remsen*, pp. 43–44; *DAB*, 11:616 ("James McCosh").

[*extent*]: Hawkins, *Between Harvard*, pp. 82, 85, 91–93, 95–96; Rudolph, *American College*, p. 303; John Higham, "The Matrix of Specialization," in Oleson and Voss, *Organization*, pp. 5–7.

[*system*]: Stanley M. Guralnick, "The American Scientist in Higher Education, 1820–1910," in Reingold, *Sciences*, p. 119; notebook entries, Dec. 6, 20, 1885, Parkman MSS.; Russell, *Rise* 1:346.

[*1871*]: Brooks M. Kelley, *Yale: A History* (New Haven, Conn., 1974), pp. 187, 263; Hawkins, *Between Harvard*, p. 84; Charles W. Eliot, "The New Education," *Atlantic Monthly* 23 (1869): 213; Rudolph, *Curriculum*, pp. 105–107; Sidar, *Cook*, pp. 85, 87, 90, 157.

328 [*eighties*]: AAAS *Proceedings* 22 (1873): 3; Guralnick, "American Scientist," pp. 122–23; D. Gilman to C. Jay, June 11, 1869, Columbia University Special MSS. Collection.

[*caste*]: *Philadelphia Ledger*, Aug. 20, 1868; Eliot, "New Education," pp. 206, 211; Lyman H. Bagg, *Four Years at Yale* (New Haven, Conn., 1871), p. 39; Andrew D. White, "Scientific and Industrial Education in the United States," *Popular Science Monthly* 5 (1874): 173.

[*1873*]: Fulton, "Science," pp. 105, 109; Fulton and Thomson, *Silliman*, p. 217; *Biographical Sketches and Letters of T. Mitchell Prudden, M.D.* (New Haven, Conn., 1927), p. 13; Fabian Franklin, *The Life of Daniel Coit Gilman* (New York, 1910), pp. 90, 97; Chittenden, *Sheffield* 1: 104, 115, 176, 182, 207–208; G. Brush to F. Genth, Oct. 21, 1868, Genth MSS.; Bagg, *Four Years*, p. 37; G. Brush to E. Hilgard, Sept. 6, 1866, Hilgard MSS.; Eliot, "New Education," p. 208; Kelley, *Yale*, pp. 249–50.

[*months*]: Frank W. Clarke, "American Colleges Versus American Science," *Popular Science Monthly* 9 (1876): 476; Eliot, "New Education," p. 210.

329 [*independent position*]: Charles L. Jackson, "Wolcott Gibbs, 1822–1908," *Am. Jour. Sci.*, 4th ser., 27 (1909): pp. 253, 257; *Hartford Courant*, Aug. 16, 1867; J. Henry to wife, Aug. 14, 1866, Henry MSS.; W. Gibbs to O. Rood, May 1, 1866, Gibbs MSS.

[*another position*]: W. Gibbs to O. Rood, Feb. 26, Mar. 21, 1869, Gibbs MSS.; *Atlantic Monthly* 23 (1869): 514.

[*treatment*]: Hawkins, *Between Harvard*, pp. 207–13; Allen D. Bliss, "Men and Machines in Early Harvard Science," *Journal of Chemical Education* 17 (1940): 357; W. Gibbs to O. Rood, Oct. 15, 1871, B. Gould to W. Gibbs, July 8, 1871, Gibbs MSS.; W. Gibbs to J. Lesley, Oct. 29, 1872, Lesley MSS.; Hugh Hawkins, *Pioneer: A History of the Johns Hopkins University, 1874–1889* (Ithaca, N.Y., 1960), p. 45; Jackson, "Gibbs," p. 256; C. Eliot to W. Gibbs, Jan. 21, 1888, Gibbs MSS.

330 [*underdogs*]: Morris Bishop, *Early Cornell, 1865–1900* (Ithaca, N.Y., 1962), p. 74; Rudolph, *Curriculum*, pp. 104–106; *DAB*, 14: 200–201 ("Ario Pardee"), 131–32 ("Asa Packer"); Rudolph, *American College*, p. 246.

[*benefaction*]: Rudolph, *American College*, p. 253; Edward D. Eddy, *Colleges for Our Land and Time* (New York, 1957), pp. 49–50; Cheyney, *University of Pennsylvania*, pp. 254–56, 263–64.

[*training*]: Eddy, *Colleges*, pp. 54–58; Rudolph, *American College*, pp. 255–60; Franklin, *Gilman*, p. 144; F. Nipher to E. Pickering, Mar. 5, 1874, Pickering MSS.

[*both*]: Rudolph, *American College*, pp. 255–57, 279; Hyder, *Snow*, p. 142; Daniel C. Gilman, "On the Growth of American Colleges and Their Present Tendency to the Study of Science," American Institute of Instruction *Proceedings*, 1871, pp. 109–12.

331 [*themselves*]: Rudolph, *American College*, pp. 106–107; Ross, *Democracy's College*, p. 155; Eddy, *Colleges*, p. 59; Mann, *Study*, pp. 6–7; John Rae, "The Application of Science to Industry," in Oleson and Voss, *Organization*, p. 252; Thomas M. Drown, *Technical Training* (Bethlehem, Pa., 1883), p. 13; Clark, *History* 2: 145; American Institute of Mining Engineers, *Memorial of Alexander Lyman Holley* (New York, 1884), p. 28; *DAB* 16: 574–75 ("Coleman Sellers"); Sinclair, "Promise," pp. 267–268.

[*glory*]: American Institute of Mining, Metallurgical, and Petroleum Engineers, *Centennial History . . . , 1871–1970* (New York, 1971), p. 13; Robert H. Thurston,

"The Improvement of the Steam Engine and the Education of Engineers," *Journal of the Franklin Institute* 94 (1872): 23; Spence, *Mining Engineers,* p. 1; Carl Condit, *American Building Art: The Nineteenth Century* (New York, 1960), p. 9; Julius G. Medley, *An Autumn Tour in the United States and Canada* (London, 1873), p. 9.

[*curriculum*]: Chittenden, *Sheffield* 2: 334; Samuel Rezneck, "The Engineering Profession Considers Its Educational Problems," *Association of American Colleges Bulletin* 43 (1957): 410–18; Thomas Egleston, *Technical Education: A Speech* (Philadelphia, 1876), pp. 3–4, 12–14; Alexander Holley, "The Inadequate Union of Engineering Science and Art," in American Institute of Mining Engineers *Transactions* 4 (1876): 200–202, 205–206; William P. Trowbridge, *The Profession of the Mechanical or Dynamical Engineer* (New Haven, Conn., 1870), p. 18.

[*vigor*]: Rezneck, *Education,* pp. 135–37, 142–43, 154; Ricketts, *History,* 114–15.

332 [*school*]: Thomas T. Read, *The Development of Mineral Industry Education in the United States* (New York, 1941), pp. 36–39, 44–46; Spence, *Mining Engineers,* pp. 18, 25; James K. Finch, *A History of the School of Engineering, Columbia University* (New York, 1954), pp. 27–32, 39–40; Fulton, *Barnard,* p. 341; F. Barnard to E. Hilgard, Feb. 19, 1866, Barnard MSS.

[*60*]: Finch, *History,* pp. 35–39, 44–45, 47–50; Van Amringe, *Historical Sketch,* pp. 99–100; Fulton, *Barnard,* p. 423; [Columbia] *School of Mines Quarterly* 35 (1914): 318–19, 321; Read, *Development,* pp. 119–20; Spence, *Mining Engineers,* p. 40.

[*up*]: Trowbridge, *Profession,* p. 3; Calvert, *Mechanical Engineer,* pp. 44–45, 47–48, 53–54.

[*1880s*]: Chittenden, *Sheffield School* vol. 1: 150–51, vol. 2: 335; Herbert F. Taylor, *Seventy Years of the Worcester Polytechnic Institute* (Worcester, Mass., 1937), pp. 3, 12–14, 18, 48, 58, 75, 88; *Scientific American* 135 (1876): 68; *Stillman Williams Robinson: A Memorial* (Columbus, Ohio, 1912), pp. 23–24, 31; Winton U. Solberg, *The University of Illinois, 1867–1894* (Urbana, Ill., 1968), pp. 107, 142–44; Calvert, *Mechanical Engineer,* pp. 48, 56–57.

333 [*1874*]: Curti and Nash, *Philanthropy,* pp. 79–80; *DAB* 13: 254–55 ("Henry Morton"); Calvert, *Mechanical Engineer,* p. 49; Merritt, *Engineering,* pp. 46–50; A. Mayer to H. Morton, Feb. 12, 1870, Hyatt-Mayer Collection; William F. Durand, *Robert Henry Thurston* (New York, 1929), pp. 58–59, 62–63, 233–35; Robert J. Kwik, "The Function of Applied Science and the Mechanical Laboratory during the Period of Formation of the Profession of Mechanical Engineering, as Exemplified in the Career of Robert Henry Thurston, 1839–1903" (Ph.D. diss., University of Pennsylvania, 1974), pp. 93, 128–37, 141, 291; Medley, *Autumn Tour,* p. 122.

[*change*]: Charles A. Browne, "Charles Edward Munroe, 1849–1938," *Journal of the American Chemical Society* 61 (1939): 1305; Daniel J. Kevles, "The Study of Physics in America, 1865–1916" (Ph.D. diss., Princeton University, 1964), pp. 87–88, 165–66, 171–72; Thomas P. Hughes, *Elmer Sperry, Inventor and Engineer* (Baltimore, 1971), p. 15.

[*348*]: Prescott, *M.I.T.,* pp. 51, 55–57, 65–67, 90; Robert H. Richards, *Robert Hallowell Richards, His Mark* (Boston, 1936), p. 37; Harry W. Tyler, *John Daniel Runkle, 1822–1902: A Memorial* (Boston, 1902), p. 14.

[*prospered*]: Prescott, *M.I.T.,* pp. 90, 92, 101–102; Richards, *Richards,* p. 62; J. Runkle to S. Newcomb, Apr. 29, 1878, Newcomb MSS.

334 [*work*]: *Journal of the Society of Arts,* Jan. 4, 1878, pp. 96–97; Horace E. Scudder, "Education by Hand," *Harper's Monthly* 58 (1879): 410–11; Rogers, *Rogers* 2: 303; Tenney L. Davis, "Eliot and Storer, Pioneers in the Laboratory Teaching of Chemistry," *Journal of Chemical Education* 6 (1929): 868–69, 876; Edward C. Pickering, *Progress of the Physical Department of the Mass. Institute of Technology, from 1867 to 1877* (Boston, 1877), pp. 1–10; *DAB,* Supplement Three, p. 631 ("Robert Richards").

[*148*]: Pickering, *Progress,* p. 11; Kevles, "Study," p. 71.

[*America*]: *New Englander* 41 (1882): 435–37; Mann, *Study,* p. 17.

[*only*]: Russell, *Rise* 1: 375.

[*served*]: McCaughey, "Transformation," p. 284; Veysey, *Emergence,* p. 95; Rudolph. *Curriculum,* pp. 121, 128; Burke, *American Collegiate Populations,* p. 71; Bishop, *Early Cornell,* p. 175; Kevles, "Study," p. 143.

335 [*pilgrimages*]: Russell, *Rise* 1: 344; Charles D. Thwing, *The American and the*

German University (New York, 1928), p. 42; Daniel B. Shumway, "The American Students of the University of Göttingen," *German-American Annals* 8 (1910): 187; Getman, *Remsen,* pp. 18, 23, 30–31; Van Klooster, "Friedrich Wöhler," p. 165; *Am. Jour. Sci.,* 4th ser., 46 (1918): 378–79; Sidar, *Cook,* p. 156; J. C. Bartlett, "American Students of Mining in Germany," American Institute of Mining Engineers *Transactions* 5 (1877): 441.

[*portents*]: E. Dana to J. Dana, Feb. 7, 1873, Dana Family MSS.; Bartlett, "American Students," p. 446; Siegfried, "Study," p. 133.

[*year*]: Kelley, *Yale,* p. 257; Hawkins, *Between Harvard,* pp. 55–56; *DAB,* Supplement One, p. 221 ("Edward Dana"); Rudolph, *American College,* p. 335; Higham, "Matrix," pp. 10–11.

[*acceptable*]: H. Rowland to "My Dear Uncle," September 1871, Herter Collection.

336 [*sake*]: Hofstadter and Smith, *American Higher Education* 2: 617; McCaughey, "Transformation," pp. 278–79, 287–90; Hawkins, *Between Harvard,* pp. 53–54, 64; Hugh Hawkins, "University Identity: the Teaching and Research Functions," in Oleson and Voss, *Organization,* p. 286; Bishop, *Early Cornell,* p. 177; Veysey, *Emergence,* p. 95; Rudolph, *Curriculum,* pp. 127–28.

[*fifties*]: Veysey, *Emergence,* pp. 95, 127–29, 158.

[*1875*]: Hawkins, *Pioneer,* pp. 3–4, 9–13, 15.

[*training*]: Ibid., pp. 16–17, 19, 37; Veysey, *Emergence,* p. 160.

337 [*objectives*]: Veysey, *Emergence,* pp. 159, 161–64; Hawkins, "University Identity," p. 286; Hawkins, *Pioneer,* pp. 22, 64–68.

[*zoology*]: Hawkins, *Pioneer,* pp. 22, 27, 29–31, 34–35, 39, 46–48, 65.

[*generally*]: Ibid., pp. 79–82; Rudolph, *American College,* pp. 337–38; abstract of Arthur S. Eichlin, "An Historical Analysis of the Fellowships Program at the Johns Hopkins University, 1876–1889: Daniel Gilman's Unique Contribution" (Ph.D. diss., Loyola University of Chicago, 1976), in *Dissertation Abstracts,* November 1976, p. 2683-A.

[*assistants*]: Rudolph, *Curriculum,* p. 129; Veysey, *Emergence,* p. 174; Eddy, *Colleges,* p. 87; Solberg, *University,* pp. 153–55; AAAS *Proceedings* 32 (1883): 120.

338 [*Year*]: Hawkins, "University Identity," pp. 285–86; Veysey, *Emergence,* pp. 174–76; Rudolph, *Curriculum,* p. 131.

25. Taking Stock

339 [*public*]: Robert C. Post, ed., *1876: A Centennial Exhibition* (Washington, D.C., 1976), pp. 30–31; Ferguson, "Expositions," pp. 718–20; Kasson, *Civilizing,* pp. 162–65.

[*profusion*]: Abram J. Foster, "The Coming of the Electrical Age to the United States" (Ph.D. diss., University of Pittsburgh, 1953), pp. 49–50, 52–53, 56–57; Albert W. Smith, *John Edson Sweet* (New York, 1925), pp. 47–48; Post, *1876,* p. 65.

340 [*it*]: Post, *1876,* p. 63; Robert V. Bruce, *Bell: Alexander Graham Bell and the Conquest of Solitude* (Boston, 1973), pp. 190, 193, 197, 199, 208–209.

[*it*]: Matthew Josephson, *Edison* (New York, 1959), pp. 102n, 131; James D. McCabe, *The Illustrated History of the Centennial Exhibition* (Philadelphia, 1876), p. 859.

[*machines*]: *Journal of the Franklin Institute* 71 (1876): 365–66; Hughes, *Sperry,* pp. 7, 9; J. Smock to G. Cook, May 27, 1876, Cook MSS.; Friedrich Klemm, *A History of Western Technology* (Cambridge, Mass., 1964), p. 328.

341 [Technologist]: Kreuter, "Literary Response," pp. 149–51; *Bibliotheca Sacra* 33 (1876): 585; *DAB* 20: 550–51 ("George Wright"); *Atlantic Monthly* 44 (1879): 138; AAAS *Proceedings* 22 (1873): 5; Houghton, *Victorian Frame,* p. 42n; *Technologist* 1 (1870): 222.

[*societies*]: Robert H. Thurston, "On the Necessity of a Mechanical Laboratory," *Journal of the Franklin Institute* 70 (1875): 412–17; Layton, "Mirror-Image Twins," pp. 222, 225; George Wise, *Willis R. Whitney, General Electric, and the Origins of U.S. Industrial Research* (New York, 1985), pp. 77–79; Haynes, *American Chemical Industry* 1:245; Getman, *Remsen,* pp. 122–23; Howard R. Bartlett, "The Development of Industrial Research in the United States," in National Resources Planning Board, *Research–A National Resource,* Part II, *Industrial Research* (Washington, D.C., 1941), pp. 19, 24–27; Alfred Lief, *"It Floats": The Story of Procter & Gamble* (New York, 1958), pp. 47, 49–50; Andrew Carnegie, *Autobiog-*

raphy of Andrew Carnegie (Boston, 1920), p. 182; *DAB* 5:479 ("Charles Dudley"). [*members*]: Hunt, *Historical Sketch,* pp. 33, 36, 43, 48, 50, 56, 66–68; Hunt, "First Fifty Years," pp. 224–25; Edwin T. Layton, Jr., *The Revolt of the Engineers* (Cleveland, 1971), p. 31.

[*experience*]: American Institute of Mining . . . Engineers, *Centennial History,* pp. 9–11, 15; Layton, *Revolt,* pp. 29–31, 33–34; Merritt, *Engineering,* p. 51.

342 [*flourishes*]: Layton, *Revolt,* p. 35; Bruce Sinclair, *A Centennial History of The American Society of Mechanical Engineers, 1880–1980* (Toronto, 1980), pp. 22–27, 33; Calvert, *Mechanical Engineer,* pp. 109–10.

[*self-criticism*]: H. Middleton to L. Gibbes, Aug. 25, 1876, L. Gibbes MSS., LC; Post, *1876,* pp. 77–79, 137; Dall, *Baird,* pp. 411–13.

[*complied*]: AAAS *Proceedings,* vol. 18 (1869): 6, vol. 22 (1873): 3, vol. 23 (1874): 1–36, vol. 24 (1875): 29ff.; Newcomb, "Exact Science," passim; Clarke, "American Colleges," pp. 467–68; H. Adams to S. Newcomb, Aug. 15, Oct. 25, 1875, Newcomb MSS.; Ernest Samuels, *The Young Henry Adams* (Cambridge, Mass., 1948), p. 276; Newcomb, "Abstract Science," p. 109.

[*America*]: J. L. LeConte to S. Newcomb, Oct. 26, 1874, B. Silliman to S. Newcomb, Oct. 7, 1874, J. Dana to S. Newcomb, Feb. 29, 1876, Newcomb MSS.; J. Anthony to S. Roberts, Jan. 31, 1875, Tryon-Pilsbry Collection; J. Dana to A. Mayer, Apr. 11, 1873, E. Cope to A. Hyatt, June 2, 1870, Hyatt-Mayer Collection; Rodgers, *Sullivant,* p. 243.

343 [*still*]: See Appendix; Dupree, *Gray,* pp. 329–30; Joseph LeConte, *'Ware Sherman* (Berkeley, Calif., 1937), pp. xv–xvi; H. Ravenel to G. Davenport, Nov. 24, 1873, Gray Herbarium; F. Holmes to J. Leidy, Dec. 9, 1865, Mar. 30, 1869, Leidy MSS.; Burkhalter, *Lincecum,* p. 190; Lupold, "LeConte," p. 195; Stephens, *LeConte,* p. 100; J. L. LeConte to P. Uhler, Feb. 8, 1867, MCZ Misc. Files.

[*region*]: Mallet, *Chemistry,* p. 14; Alester G. Holmes and George R. Sherrill, *Thomas Green Clemson: His Life and Work* (Richmond, Va., 1937), pp. 145–55; Bruce, *University of Virginia* 3: 353, 359–60; F. Holmes to J. Leidy, Mar. 30, 1869, Leidy MSS.; F. Barnard to E. Hilgard, Sept. 29, 1870, Barnard MSS.; Rudolph, *American College,* p. 244.

[*Washington*]: J. Henry to H. Wurtz, July 26, 1861, Wurtz MSS.; S. Newcomb to "Dear Brothers," Apr. 29, 1862, and to W. Bartlett, Nov. 9, 1862, J. Henry to S. Newcomb, Nov. 5, 1872, Newcomb MSS.; Diary of Joseph Henry, Jan. 25, 1868, Smithsonian Institution Archives.

[*seven*]: Diary of Joseph Henry, Mar. 8, 1862, Dec. 18, 1864, Henry MSS.; Hugh McCulloch, *Men and Measures of Half a Century* (New York, 1888), p. 262; J. Hilgard to E. Hilgard, Apr. 21, 1867, July 4, 1871, Hilgard MSS.; J. Kirkpatrick Flack, *Desideratum in Washington: The Intellectual Community in the Capital City, 1870–1900* (Cambridge, Mass., 1975), pp. 60–63; AAAS *Proceedings* 25 (1876): xl–xlviii.

344 [*seventeen*]: R. Stearns to G. Tryon, July 27, 1866, Tryon-Pilsbry Collection; LeConte, *Autobiography,* pp. 264–65, 298; J. LeConte to A. Mayer, Jan. 17, 1873, Hyatt-Mayer Collection; AAAS *Proceedings* 25 (1876): xl–xlviii.

[*competitors*]: Daniel J. Kevles, Jeffrey L. Sturchio, and P. Thomas Carroll, "The Sciences in America, Circa 1880," *Science* 209 (1980): 27; see Appendix; Bruce, "Statistical Profile," p. 72.

[*divinity*]: AAAS *Proceedings* 47 (1898): 267; *Am. Jour. Sci.,* 4th ser., 46 (1918): 119–21, 141, 158, 162; Merrill, *First Hundred Years,* p. 488; Lurie, *Agassiz,* pp. 346–49, 353–57.

345 [*scientist*]: Charles Schuchert and Clara Mae LeVene, *O. C. Marsh, Pioneer in Paleontology* (New Haven, Conn., 1940), pp. 169, 176, 262–64, 334; Henry F. Osborn, *Cope: Master Naturalist* (Princeton, N.J., 1931), pp. 180–81; James Penick, "Professor Cope Vs. Professor Marsh," *American Heritage* 22, no. 5 (August 1971): 5, 9–10, 91–95; "Reminiscences," Edward Nolan MSS.

[*world*]: Osborn, *Cope,* pp. 159–60, 216; J. Leidy to J. Corson, June 20, 1872, Leidy MSS.; G. Horn to J. L. LeConte, Mar. 24, 1870, LeConte MSS., APS; Schuchert, *Marsh,* pp. 169–70; Alpheus S. Packard, "A Century's Progress in American Zoology," *American Naturalist* 10 (1876): 595.

346 [*truth*]: Schuchert, *Marsh,* pp. 230–40; Pfeifer, "Reception," pp. 59–61, 87–90, 94–95, 97–98.

[*opportunity*]: Lurie, *Agassiz*, pp. 372–73; L. Agassiz to B. Wilder, Mar. 27, 1871, J. Wyman to B. Wilder, May 1871, Wilder MSS.

347 [*Dana*]: Edward J. Pfeifer, "The Genesis of American Neo-Lamarckism," *Isis* 56 (1965): 156–58, 160, 162; Peter J. Bowler, "Edward Drinker Cope and the Changing Structure of Evolutionary Theory," *Isis* 68 (1977): 259–61, 264–65; Prendergast, "Dana," 494–521.

[*alone*]: AAAS *Proceedings* 25 (1876): 147–48, 30 (1881): 263; Packard, "Century's Progress," pp. 597–98; Joel A. Allen, "Progress of Ornithology in the United States during the Last Century," *American Naturalist* 10 (1876): 549; Howard, *History*, p. 73; T. Bland to G. Tryon, Sept. 19, 1865, Tryon-Pilsbry Collection.

[*Germans*]: See Appendix; *Am. Jour. Sci.*, 4th ser. 46 (1918): 373; Packard, "Century's Progress," p. 598.

[*rich*]: See Appendix; Siegfried, "Study," pp. 85, 110, 118, 136, 142; AAAS *Proceedings* 28 (1879): 214–15; Daniel J. Kevles, "The Physics, Mathematics, and Chemistry Communities: A Comparative Analysis," in Oleson and Voss, *Organization*, p. 148; Hawkins, *Pioneer*, pp. 140–41.

348 [*physics*]: *Am. Jour. Sci.*, 4th ser., 46 (1918): 264, 266–67; *DAB*, Supplement One, pp. 177–78 ("Frank Clarke"); Siegfried, "Study," pp. 112, 116.

[*1880s*]: AAAS *Proceedings*, vol. 25 (1876): 46–48, vol. 29 (1880): 111, vol. 32 (1883): 117; Jones and Boyd, *Harvard Observatory*, pp. 176–78; L. Boss to S. Newcomb, July 10, 17, 1876, Newcomb MSS.; Safford, *Development*, pp. 26–27.

[*rotation*]: Jones and Boyd, *Harvard Observatory*, pp. 177, 184; Howard N. Plotkin, "Edward C. Pickering, the Henry Draper Memorial, and the Beginnings of Astrophysics in America," *Annals of Science* 35 (1978): 365–66; *DAB* 20: 624 ("Charles Young"), 10: 594 ("Samuel Langley").

[*pull*]: Bruce, "Statistical Profile," pp. 71–72.

349 [*downtown*]: AAAS *Proceedings* 24 (1875): 29–34; Smith and Ginsburg, *History*, pp. 82–83, 102; Bagg, *Four Years*, pp. 320–21, 323, 325–26; Taylor, *Seventy Years*, p. 110.

[*mechanics*]: Smith and Ginsburg, *History*, pp. 112–13, 119–20, 130, 198–99; Helena M. Pycior, "Benjamin Peirce's Linear Associative Algebra," *Isis* 70 (1979): 537, 547–51.

[*physicists*]: Kevles, "Study," pp. 20, 72, 117, 135, 137; Kevles, *Physicists*, pp. 7–8; Kevles, "Communities," p. 140; AAAS *Proceedings*, vol. 23 (1874): 35, vol. 32 (1883): 121.

350 [*century*]: Bruce, "Statistical Profile," pp. 71–72; *DAB* 16: 198–99 ("Henry Rowland"); Hawkins, *Pioneer*, pp. 45–47.

[*career*]: Lynde P. Wheeler, *Josiah Willard Gibbs: The History of a Great Mind* (New Haven, Conn., 1952), pp. 7, 11, 16, 22, 24–28, 32, 35, 39–44, 57–59, 63; Gibbs, *Gibbs Family*, pp. 3, 138.

351 [*salary*]: Wheeler, *Gibbs*, pp. 63, 71–74, 84–85, 87–88, 91, 94, 96, 99.

[*pages*]: Ibid., pp. 71, 75, 77–82, 96.

352 [*blue*]: Ibid., p. 83; A. Einstein to Mrs. J. Whitney, July 15, 1926, Newcomb MSS.

[*abroad*]: Kevles et al., "Sciences," pp. 28–29; Kevles, "Communities," pp. 140, 146; J. Henry to "My Dear Sir," May 7, 1869, Henry MSS.; Newcomb, "Abstract Science," p. 110.

[*threat*]: George B. Goode, "The Beginnings of American Science: The Third Century," Smithsonian, *Annual Report*, 1897, 2: 462; Newcomb, "Abstract Science," p. 110; J. Dana to C. Dewey, Jan. 22, 1866, Dewey MSS.; J. Dana to F. Hayden, Feb. 1, 1875, Merrill MSS.; B. Silliman to H. Draper, Jan. 6, 1876, Draper MSS.

353 [*year*]: Nolan, *Short History*, pp. 17–18; Howard, *History*, pp. 51, 66–68; B. Gould to S. Newcomb, Sept. 15, 1869, Newcomb MSS.; Comstock, "Gould," p. 168; *Am. Jour. Sci.*, 4th ser., 46 (1918): 380; Cajori, *Teaching*, pp. 279–80; Smith and Ginsburg, *History*, p. 116; Beardsley, *Rise*, pp. 36–40; Hawkins, *Pioneer*, pp. 73–76.

[*themes*]: Newcomb, "Abstract Science," p. 118; John Tyndall, *Six Lectures on Light* (New York, 1873), pp. 169–72, 174; AAAS *Proceedings*, vol. 22 (1873): 3, vol. 29 (1880): 610–11.

[*arts*]: AAAS *Proceedings* vol. 18 (1869): 18, vol. 26 (1877): 64, 67–70, vol. 27 (1878): 136–37.

[*America*]: Ibid. 32 (1883): 116–17; *New York Times*, Sept. 19, 1873.

354 [*appeal*]: Frank W. Clarke, "Scientific Dabblers," *Popular Science Monthly* 1 (1872): 600; *Nation* 4 (1867): 32–34; J. Powell to J. Leidy, June 19, 1874, Leidy MSS.; J. LeConte to L. Gibbes, Sept. 11, 1867, L. Gibbes MSS., LC; Miller, *Dollars,* pp. 119–23.
[*year*]: *New York Times,* July 22, 1866, Aug. 24, 1873; *Buffalo Express,* Aug. 23, 1876; *Nashville American,* Sept. 2, 1877; AAAS *Proceedings* 24 (1875): 341; J. Powell to J. Leidy, June 19, 1874, Leidy MSS.; J. Hilgard to E. Hilgard, June 7, 1874, Hilgard MSS.; Don D. Walker, "The *Popular Science Monthly,* 1872–1878" (Ph.D. diss., University of Minnesota, 1956), pp. 6, 37; Joseph A. Boromé, "The Evolution Controversy," in Donald Sheehan and Harold C. Syrett, eds., *Essays in American Historiography* (New York, 1960), pp. 178–79; William E. Leverette, Jr., "E. L. Youmans' Crusade for Scientific Autonomy and Respectability," *American Quarterly* 17 (1965): 12–14.
[*date*]: *Smithsonian Institution Miscellaneous Collections* 18: 783; Gilman, "Growth," pp. 106–107; Franklin, *Gilman,* p. 132; *Scientific American* 34 (1876): 208; *DAB* 13: 455 ("Simon Newcomb"); Houghton, *Victorian Frame,* pp. 35–36.

355 [*Tyndall*]: Kevles, *Physicists,* pp. 14, 17; *Buffalo Commercial Advertiser,* Aug. 23, 1876; Ames, *Lesley* 1: 514; Donald Fleming, "American Science and the World Scientific Community," *Cahiers d'Histoire Mondiale* 8 (1965): 675.
[*assumed*]: Abbe, *Abbe,* p. 65; AAAS *Proceedings* 18 (1869): 31.
[*public*]: Pfeifer, "Reception," pp. 66–69, 78–79, 118–20; Boromé, "Evolution," pp. 179–85; John Dillenberger, *Protestant Thought and Natural Science* (New York, 1960), pp. 234, 240–41.
[*cosmogony*]: Hofstadter and Metzger, *Development,* pp. 330–35, 342; Gilman, *Dana,* p. 191.

356 [*example*]: Goode, "Beginnings," p. 465; A. Rice to S. Newcomb, May 24, 1879, Newcomb MSS.; *Atlantic Monthly* 18 (1866): 128; AAAS *Proceedings* 18 (1869): 19; Haber, "Sidelights," p. 248.
[*older*]: Goode, "Beginnings," p. 465; Ralph W. Dexter, "The Salem Meeting of the American Association for the Advancement of Science," *Essex Institute Historical Collections,* 93 (1957): 263; *Hartford Daily Courant,* Aug. 20, 1874; *Detroit Evening News,* Aug. 14, 1875; *New York Times,* Aug. 20, 1875.

26. Epilogue

357 [*race*]: Lurie, *Agassiz,* pp. 344, 352; A. Hyatt to C. Abbe, June 17, 1865, Abbe MSS.; Thayer, *Letters of Wright,* p. 88; *Am. Jour. Sci.* 91 (1866): 407–408; C. Hartt to Lucy, Nov. 24, 1868, Hartt MSS.; AAAS *Proceedings* 22 (1873): 20.

358 [*sixty-six*]: L. Agassiz to T. Lyman, Sept. 15, 1866, Agassiz MSS., MCZ; Lurie, *Agassiz,* pp. 350, 352, 360–61, 366–71, 382, 380–81, 388; Ralph W. Dexter, "The 'Salem Secession' of Agassiz Zoologists," *Essex Institute Historical Collections* 101 (1965), 37–38; L. Agassiz to B. Wilder, Nov. 25, 1873, Wilder MSS.
[*triumph*]: Williams, *Maury,* pp. 465–66, 469, 475–77; Davis, *Davis,* pp. 332–34; *DAB* 14: 394–95 ("Benjamin Peirce"); Prescott, *M.I.T.,* pp. 103–104.

359 [*man*]: Jackson, "Gibbs," p. 256; O. Rood to A. Mayer, Apr. 11, 1877, Hyatt-Mayer Collection; W. Gibbs to B. Gould, Sept. 27, 1887, to O. Rood, May 31, 1891, Gibbs MSS.
[*last*]: O. Rood to W. Gibbs, Nov. 2, 1873, W. Gibbs to O. Rood, June 22, 1875, Gibbs MSS.; J. Henry to B. Peirce, Mar. 28, 1866, to A. Gray, Oct. 24, 1868, Henry MSS.; Comstock, "Gould," pp. 158, 161–62; B. Gould to B. Wilder, Sept. 30, 1867, Wilder MSS.; G. Searle to C. Abbe, Oct. 20, 1867, Abbe MSS.; B. Gould to G. Ellis, Mar. 12, 1868, Ellis MSS.

360 [*northern*]: Comstock, "Gould," pp. 162–67; B. Gould to E. Holden, Feb. 1, 1873, Dwight MSS.; B. Gould to F. Parkman, Dec. 5, 1874, July 1, 1876, Oct. 8, 1882, Parkman MSS.; B. Gould to J. Watson, May 29, 1879, Watson MSS.; B. Gould to brother, Oct. 28, 1879, Feb. 1, 1880, Ellis MSS.; B. Gould to J. L. LeConte, Apr. 5, 1882, LeConte MSS., ANS.
[*pain*]: B. Gould to F. Parkman, Oct. 8, 1882, Parkman MSS.; Dall, *Baird,* pp. 385, 402, 405–408; Dupree, *Gray,* pp. 341, 349–50, 353, 390–91, 393, 411, 419.
[*sea*]: B. Gould to W. Gibbs, Dec. 6, 1872, Sept. 8, 1896, Gibbs MSS.; B. Gould to brother, Oct. 21, 1881, Ellis MSS.; W. Gibbs to E. Atkinson, Jan. 22, 1888, Atkinson MSS.; Comstock, "Gould," p. 168; Jackson, "Gibbs," pp. 257–59.

361 [*another Lazzarone*]: Prendergast, "Dana," pp. 298–304; Gilman, *Dana,* pp. 266–67, 390–94.
[*Lazzaroni*]: Brewster, *Whitney,* pp. 377, 382; B. Gould to W. Gibbs, Sept. 8, 1896, Gibbs MSS.; Comstock, "Gould," p. 169; Richard G. Olson, "The Gould Controversy at Dudley Observatory," *Annals of Science* 27 (1971): 275.
362 [*1900*]: Cochrane, *National Academy,* pp. 157, 159, 161–64.
[*scientists*]: F. Putnam to W. Gibbs, Sept. 14, 1896, Gibbs MSS.
[*shoulders*]: W. Gibbs to O. Rood, May 18, 1876, Feb. 9, 1896, Nov. 10, 1901, Gibbs MSS.
[*extinction*]: *Am. Jour. Sci.* 27 (1909): 258; C. Munroe to Miss Bettens, May 4, 1908, Gibbs MSS.; *Art in the United States Capitol* (Washington, D.C., n.d.), p. 350.

Manuscript Sources Cited

Cleveland Abbe MSS., Library of Congress
Henry L. Abbot MSS., Houghton Library, Harvard University
Louis Agassiz MSS., Houghton Library, Harvard University (Agassiz MSS.)
Louis Agassiz MSS., Museum of Comparative Zoology, Harvard University (Agassiz MSS., MCZ)
Edward P. Alexander MSS., Southern Historical Collection, University of North Carolina
American Association of Geologists and Naturalists MSS., Academy of Natural Sciences, Philadelphia
American Philosophical Society Archives, Philadelphia
Edward Atkinson MSS., Massachusetts Historical Society
Alexander D. Bache MSS., Smithsonian Institution (Bache MSS., Smithsonian)
Alexander D. Bache MSS., Library of Congress (Bache MSS.)
John Bachman MSS., Charleston (S.C.) Museum
Henry D. Bacon MSS., Huntington Library (Bacon MSS., Huntington)
Henry D. Bacon MSS., Southern Historical Collection, University of North Carolina (Bacon MSS.)
Jacob W. Bailey MSS., Boston Museum of Science
Spencer F. Baird MSS., Smithsonian Institution
Frederick A. P. Barnard MSS., Columbia University
Lewis C. Beck MSS., Rutgers University
George F. Becker MSS., Library of Congress
G. W. Belfrage MSS., Academy of Natural Sciences, Philadelphia
Blake Family MSS., Yale University
Boston University Special Collections
Luther B. Bradish MSS., New-York Historical Society
William H. Brewer MSS., Yale University
Brock Collection, Huntington Library
William L. Broun MSS., Southern Historical Collection, University of North Carolina
Brush Family MSS., Yale University
Robert L. Caruthers MSS., Southern Historical Collection, University of North Carolina
Frank W. Clarke MSS., Library of Congress
Columbia University College Papers
Columbia University Special MSS. Collection
George Combe MSS., National Library of Scotland, Edinburgh
George H. Cook MSS., Rutgers University
Crystal Palace MSS., New-York Historical Society
Moses A. Curtis MSS., Southern Historical Collection, University of North Carolina
Dana Family MSS., Yale University
James D. Dana MSS., Beinecke Library, Yale University
Edward Daniels MSS., Wisconsin Historical Society

George Davidson MSS., Bancroft Library, University of California, Berkeley
Charles H. Davis MSS., Henry Ford Museum, Dearborn, Michigan
Chester Dewey MSS., University of Rochester
Henry Draper MSS., New York Public Library
Duke University Miscellaneous MSS.
Theodore F. Dwight MSS., Massachusetts Historical Society
George E. Ellis MSS., Massachusetts Historical Society
Samuel F. Emmons MSS., Library of Congress
Edward Everett MSS., Massachusetts Historical Society
Eyton Collection, American Philosophical Society, Philadelphia
William Ferrel MSS., Duke University
Hamilton Fish MSS., Columbia University (Fish MSS., Columbia)
Hamilton Fish MSS., Library of Congress (Fish MSS., LC)
William P. Foulke MSS., Academy of Natural Sciences, Philadelphia
John F. Frazer MSS., American Philosophical Society, Philadelphia
Sydney H. Gay MSS., Columbia University
Frederick A. Genth MSS., Pennsylvania State University
Lewis R. Gibbes MSS., Charleston (S.C.) Museum (L. Gibbes MSS., Charleston)
Lewis R. Gibbes MSS., Library of Congress (L. Gibbes MSS., LC)
Oliver Wolcott Gibbs MSS., Franklin Institute, Philadelphia
Gibbs Family MSS., Wisconsin Historical Society
Augustus A. Gould MSS., Houghton Library, Harvard University (A. Gould MSS.)
Benjamin A. Gould MSS., Beinecke Library, Yale University (Gould MSS.)
Gratz Collection, Pennsylvania Historical Society
Gray Herbarium, Harvard University
James D. Hague MSS., Huntington Library
Samuel S. Haldeman MSS., Academy of Natural Sciences, Philadelphia
James Hall MSS., New York State Library
Robert Hare MSS., American Philosophical Society, Philadelphia
Charles F. Hartt MSS., Duke University
Benjamin S. Hedrick MSS., Duke University (Hedrick MSS.)
Benjamin S. Hedrick MSS., University of North Carolina Archives (Hedrick MSS., UNC Archives)
Joseph Henry MSS., Smithsonian Institution
Mary Henry Diary, Smithsonian Institution
Christian A. Herter Collection, Columbia University
Eugene W. Hilgard MSS., Bancroft Library, University of California, Berkeley
Eben N. Horsford MSS., Rensselaer Polytechnic Institute
Huntington Library Miscellaneous MSS., San Marino, California (Huntington Misc. MSS.)
Hyatt-Mayer Collection, Princeton University
Charles T. Jackson MSS., Massachusetts Historical Society
Louis Janin MSS., Huntington Library
John Kimberly MSS., Southern Historical Collection, University of North Carolina
William Kitchell MSS., Rutgers University
Increase A. Lapham MSS., Wisconsin Historical Society
John L. LeConte MSS., Academy of Natural Sciences, Philadelphia (LeConte MSS., ANS)
John L. LeConte MSS., American Philosophical Society, Philadelphia (LeConte MSS., APS)
Joseph Leidy MSS., Academy of Natural Sciences, Philadelphia
J. Peter Lesley MSS., American Philosophical Society, Philadelphia
Francis Lieber MSS., Huntington Library
Robert T. Lincoln Collection, Library of Congress
Elias Loomis MSS., Yale University
Henry MacRae MSS., Duke University
Matthew F. Maury MSS., Library of Congress
Edward Maynard MSS., Library of Congress
George P. Merrill Collection, Library of Congress
William P. Miles MSS., Southern Historical Collection, University of North Carolina
Edward W. Morley MSS., Library of Congress

Museum of Comparative Zoology Miscellaneous Files, Harvard University (MCZ Misc. Files)
National Archives Record Group 23 (Coast and Geodetic Survey)
National Archives Record Group 45 (Naval Records Collection)
National Archives Record Group 74 (Navy Bureau of Ordnance)
National Archives Record Group 107 (Office of the Secretary of War)
National Archives Record Group 156 (Army Ordnance Department)
New-York Historical Society Miscellaneous MSS.
Simon Newcomb MSS., Library of Congress
Edward Nolan MSS., Academy of Natural Sciences, Philadelphia
Francis Parkman MSS., Massachusetts Historical Society
Benjamin Peirce MSS., Houghton Library, Harvard University
Edward C. Pickering MSS., Harvard University Archives
Purviance-Courtenay MSS., Duke University
Daniel C. Rand MSS., Duke University
Edmund Ravenel MSS., Charleston (S.C.) Museum (E. Ravenel MSS.)
Henry W. Ravenel MSS., Charleston (S.C.) Museum (H. Ravenel MSS.)
William C. Redfield MSS., Yale University
William P. Rhees Collection, Huntington Library
William B. Rogers MSS., Massachusetts Institute of Technology
Charles A. Schott MSS., Library of Congress
William Sharswood MSS., American Philosophical Society, Philadelphia
Charles W. Short MSS., Southern Historical Collection, University of North Carolina
Silliman Family MSS., Yale University
Edgar F. Smith Collection, University of Pennsylvania
Smithsonian Institution Archives
South Carolina Historical Society Miscellaneous MSS.
Southern Historical Collection, University of North Carolina
George Suckley MSS., Huntington Library
John Torrey MSS., Academy of Natural Sciences, Philadelphia (Torrey MSS., ANS)
John Torrey MSS., New York Botanical Garden (Torrey MSS.)
John C. Trautwine MSS., Cornell University
Trent Collection, Duke University Hospital Library
Tryon-Pilsbry Collection, Academy of Natural Sciences, Philadelphia
Benjamin L. C. Wailes MSS., Duke University
Henry A. Ward MSS., University of Rochester
John Warner MSS., American Philosophical Society, Philadelphia
James C. Watson MSS., Michigan Historical Collections, University of Michigan, Ann Arbor
Charles M. Wetherill MSS., Edgar F. Smith Library, University of Pennsylvania
William D. Whitney MSS., Yale University
Burton G. Wilder MSS., Cornell University
Charles Wilkes MSS., Library of Congress
Alexander Winchell MSS., Michigan Historical Collections, University of Michigan, Ann Arbor
Henry Wurtz MSS., New York Public Library

Other Sources Cited in More than One Chapter

All sources except manuscripts are fully cited on first use.

Abbe, Truman. *Professor Abbe and the Isobars: The Story of Cleveland Abbe, America's First Weatherman.* New York, 1955.

Agassiz, Elizabeth C. *Louis Agassiz: His Life and Correspondence.* Boston, 1890.

American Academy of Arts and Sciences. *Memorial of Joseph Lovering, Late President of the Academy.* Cambridge, Mass., 1892.

American Institute of Mining, Metallurgical, and Petroleum Engineers. *Centennial History of the American Institute of Mining, Metallurgical, and Petroleum Engineers, 1871–1970.* New York, 1971.

Ames, Mary L., ed. *Life and Letters of Peter and Susan Lesley.* 2 vols. New York, 1909.

Bache, Alexander Dallas. *Anniversary Address before the American Institute of the City of New York, at the Tabernacle, October 28th, 1856.* New York, 1856.

———. Untitled draft of 1844 National Institute speech beginning "What are the wants of science in the United States?" in Alexander D. Bache Collection, Smithsonian Institution Archives. Washington, D.C. (Cited as Bache, "Wants of Science.")

Bacon, Egbert K. "A Precursor of the American Chemical Society—Chandler and the Society of Union College." *Chymia* 10 (1965): 183–97.

Bagg, Lyman H. *Four Years at Yale.* New Haven, Conn. 1871.

Bailey, Solon I. *The History and Work of Harvard Observatory, 1839 to 1927.* New York, 1931.

Barnard, Frederick A. P. *Letters to the Honorable, the Board of Trustees of the University of Mississippi.* Oxford, Miss., 1858.

Bates, Ralph S. *Scientific Societies in the United States.* New York, 1958.

Baxter, Gregory P. "The Early Days of Chemistry at Harvard." *Harvard Graduate Magazine,* 32 (1924): 589–97.

Beardsley, Edward. *The Rise of the American Chemistry Profession, 1850–1900.* Gainesville, Florida, 1964.

Beaver, Donald de B. "The American Scientific Community, 1800–1860: A Statistical-Historical Study." Ph.D. dissertation, Yale University, 1966.

Bode, Carl. *The American Lyceum.* New York, 1956.

Bond, William C. *History and Description of the Astronomical Observatory of Harvard College.* Cambridge, Mass., 1856.

Boston Society of Natural History. *The Boston Society of Natural History, 1830–1930.* Boston, 1930.

Bouvé, Thomas T. "Historical Sketch of the Boston Society of Natural History; . . ." In *Anniversary Memoirs,* Boston Society of Natural History, pp. 14–250. Boston, 1880.

Bremner, Robert H. *The Public Good: Philanthropy and Welfare in the Civil War Era.* New York, 1980.

Brewster, Edwin T. *Life and Letters of Josiah Dwight Whitney.* Boston, 1909.

Brooke, George M., Jr. *John M. Brooke: Naval Scientist and Educator.* Lexington, Va., 1979.

Browne, Charles A. "The History of Chemical Education in America between the Years 1820 and 1870." *Journal of Chemical Education* 9 (1932): 696–743.

Bruce, Philip A. *History of the University of Virginia, 1819–1919.* 5 vols. New York, 1920–1922.

Bruce, Robert V. *Lincoln and the Tools of War.* Indianapolis, 1956.

———. "A Statistical Profile of American Scientists, 1846–1876." In Daniels, *Nineteenth-Century American Science.*

Burke, Colin B. *American Collegiate Populations: A Test of the Traditional View.* New York, 1982.

Burkhalter, Lois W. *Gideon Lincecum, 1793–1874.* Austin, Tex., 1965.

Burstyn, Harold L. "Seafaring and the Emergence of American Science." In Labaree, Benjamin W., ed., *The Atlantic World of Robert G. Albion,* pp. 76–109. Middletown, Conn., 1975.

Cajori, Florian. *The Teaching and History of Mathematics in the United States.* Washington, D.C., 1890.

Calvert, Monte A. *The Mechanical Engineer in America, 1830–1910.* Baltimore, 1967.

Chandler, S. C. "The Life and Work of Dr. [B. A.] Gould." *Popular Astronomy* 4 (1897): 341–47.

Massachusetts College of Agriculture. *Charles Anthony Goessmann.* Cambridge, Mass., 1917.

Cheyney, Edward P. *History of the University of Pennsylvania, 1740–1940.* Philadelphia, 1940.

Child, Ernest. *The Tools of the Chemist.* New York, 1940.

Chittenden, Russell H. *History of the Sheffield Scientific School of Yale University, 1846–1922.* 2 vols. New Haven, Conn., 1928.

Chute, William J. *Damn Yankee; The First Career of Frederick A. P. Barnard, Educator.* Port Washington, N.Y., 1977.

Clark, Victor S. *History of Manufacturing in the United States.* 3 vols. New York, 1929.

Clarke, Frank W. "American Colleges Versus American Science." *Popular Science Monthly* 9 (1876): 467.

Clarke, John M. *James Hall of Albany.* Albany, N.Y., 1923.

Clyde, John C. *Life of James H. Coffin.* Easton, Pa., 1881.

Cochrane, Rexmond C. *The National Academy of Sciences: The First Hundred Years, 1863–1963.* Washington, D.C., 1978.

Coe, Wesley R. "A Century of Zoology in America." *American Journal of Science* 46 (1918).

Cohen, I. Bernard. "Harvard and the Scientific Spirit." *Harvard Alumni Bulletin* 50 (1948): 393–98.

Comstock, George C. "Benjamin Apthorp Gould." *National Academy of Sciences Biographical Memoirs* 17 (1924): 155–80.

Coulson, Thomas. *Joseph Henry: His Life and Work.* Princeton, N.J., 1950.

Curti, Merle E., and Nash, Roderick. *Philanthropy in the Shaping of American Higher Education.* New Brunswick, N.J., 1965.

Dall, William H. *Spencer Fullerton Baird.* Philadelphia, 1915.

Daniels, George H. *American Science in the Age of Jackson.* New York, 1968.

———, ed. *Nineteenth-Century American Science: A Reappraisal.* Evanston, Ill., 1972.

Darwin, Francis. *The Life and Letters of Charles Darwin.* 2 vols. New York, 1887.

Davis, Charles H. *Life of Charles Henry Davis, Rear Admiral, 1807–1877.* Boston, 1899.

De Bow, James D. B. *Statistical View of the United States.* Washington, D.C., 1854.

De Jong, John A. "American Attitudes toward Evolution before Darwin." Ph.D. dissertation, University of Iowa, 1962.

Dupree, A. Hunter. *Asa Gray.* Cambridge, Mass., 1959.

———. "The National Pattern of American Learned Societies, 1769–1863." In Oleson and Brown, *Pursuit of Knowledge.*

———. *Science in the Federal Government.* Cambridge, Mass., 1957.

Dwight, Timothy. *Memories of Yale Life and Men, 1845–1899.* New York, 1903.

Eaton, Clement. *The Mind of the Old South.* Baton Rouge, La., 1967.

Elliott, Clark A. "The American Scientist, 1800–1863: His Origins, Career, and Interests." Ph.D. dissertation, Case Western Reserve University, 1970.

Emerson, Edward W. *The Early Years of the Saturday Club, 1855–1870.* Boston, 1918.
Ewan, Joseph. "San Francisco as a Mecca for Nineteenth Century Naturalists." In California Academy of Sciences, *A Century of Progress in the Natural Sciences, 1853–1953.* San Francisco, 1955.
Falk, Stanley L. "Soldier-Technologist: Major Alfred Mordecai and the Beginnings of Science in the United States Army." Ph.D. dissertation, Georgetown University, 1959.
Ferguson, Eugene S. "Expositions of Technology, 1851–1900." In Kranzberg and Pursell, *Technology in Western Civilization* 1.
Ferguson, Walter K. *Geology and Politics in Frontier Texas, 1845–1909.* Austin, Texas, 1969.
First Barnard-Millington Symposium on Southern Science and Medicine: Science in the Old South. Oxford, Miss., 1982.
Fisher, George P. *Life of Benjamin Silliman, M.D., LL.D.* 2 vols. Philadelphia, 1866.
Fleming, Donald. *John William Draper and the Religion of Science.* Philadelphia, 1950.
———. *Science and Technology in Providence, 1760–1914.* Providence, R.I., 1952.
Foote, George A. "A Study of Attitudes toward Science in Nineteenth Century England, 1800–1851." Ph.D. dissertation, Cornell University, 1950.
Franklin, Fabian. *The Life of Daniel Coit Gilman.* New York, 1910.
Fulton, John. *Memoirs of Frederick A. P. Barnard.* New York, 1896.
Fulton, John F., and Thomson, Elizabeth H. *Benjamin Silliman, 1779–1864, Pathfinder in American Science.* New York, 1947.
Fulton, John F. "Science in American Universities, 1636–1946, with Particular Reference to Harvard and Yale." *Bulletin of the History of Medicine* 20 (1946): 97–111.
Gates, Paul W. *The Farmer's Age.* New York, 1960.
Gendebien, Albert W. "Science the Handmaiden of Religion: The Origins of the Pardee Scientific Course at Lafayette College." *Pennsylvania History* 33 (1966).
Getman, Frederick H. *The Life of Ira Remsen.* Easton, Pa., 1940.
Gibbs, George. *The Gibbs Family of Rhode Island and Some Related Families.* New York, 1933.
Gilman, Daniel C. *The Life of James Dwight Dana.* New York, 1899.
———. "On the Growth of American Colleges and Their Present Tendency to the Study of Science." American Institute of Instruction, *Proceedings,* 1871.
Goetzmann, William H. *Army Exploration in the American West, 1803–1863.* New Haven, Conn., 1959.
———. *Exploration and Empire.* New York, 1967.
Goldfarb, Stephen. "Science and Democracy: A History of the Cincinnati Observatory, 1842–1872." *Ohio History* 78 (1969): 172–223.
Goode, George B. *The Smithsonian Institution, 1846–1896.* Washington, D.C., 1897.
Goessman, Charles Anthony: *See* Massachusetts College of Agriculture
Gould, Benjamin A. *An address in commemoration of Alexander Dallas Bache, delivered August 6, 1868, before the American Association for the Advancement of Science.* Essex Institute, Salem, Mass., 1868.
———. *An Address in Commemoration of Sears Cook Walker.* Cambridge, Mass., 1854.
Gray, Jane L., ed. *The Letters of Asa Gray.* 2 vols. Boston, 1893.
Greene, Benjamin F. *The Rensselaer Polytechnic Institute.* Troy, New York, 1855.
Greene, John C. "Science and the Public in the Age of Jefferson." *Isis,* 49 (1958): 13–25.
Guralnick, Stanley M. *Science and the Ante-Bellum American College.* Philadelphia, 1975.
Haber, Francis C. "Sidelights on American Science as Revealed in the Hyatt Autograph Collection." *Maryland Historical Magazine,* 46 (1951): 233–56.
Hamilton, J. G. de Roulhac. *Benjamin Sherwood Hedrick.* Chapel Hill, N.C., 1911.
Harrison, John P., "Science and Politics: Origins and Objectives of Mid-Nineteenth Century [U.S.] Government Expeditions to Latin America." *Hispanic American Historical Review* 35 (1955): 175–202.
Hawkins, Hugh. *Pioneer; A History of the Johns Hopkins University, 1874–1899.* Ithaca, N.Y., 1960.
Haynes, Williams. *American Chemical Industry: A History.* 6 vols. New York, 1954.
———. *Chemical Pioneers: The Founders of the American Chemical Industry.* New York, 1939.
Hellman, Geoffrey T. *The Smithsonian: Octopus on the Mall.* Philadelphia, 1967.

Hendrickson, Walter B. *The Arkites and Other Pioneer Natural History Organizations of Cleveland.* Cleveland, 1962.

———. "Nineteenth-Century State Geological Surveys, Early Government Support of Science." *Isis* 52 (1961): 357–71.

Herber, Elmer Charles, ed. *Correspondence between Spencer Fullerton Baird and Louis Agassiz, Two Pioneer American Naturalists.* Washington, D.C., 1963.

Hindle, Brooke. *The Pursuit of Science in Revolutionary America, 1735–1789.* Chapel Hill, N.C., 1956.

Hitchcock, Edward. *Reminiscences of Amherst College.* Northampton, Mass., 1863.

Hoar, George F. *Autobiography of Seventy Years.* 2 vols. New York, 1903.

Hofstadter, Richard, and Metzger, Walter P. *The Development of Academic Freedom in the United States.* New York, 1955.

Holden, Edward S. *Memorials of William Cranch Bond, . . . and . . . George Phillips Bond, . . .* New York, 1897.

Houghton, Walter. *The Victorian Frame of Mind, 1830–1870.* New Haven, Conn., 1957.

Howard, Leland O. *A History of Applied Entomology (Somewhat Anecdotal).* Washington, D.C., 1930.

Hughes, Thomas P. *Elmer Sperry, Inventor and Engineer.* Baltimore, 1971.

———. "Industry through the Crystal Palace: a Study of the Great Exhibition Held in London, 1851." Ph.D. dissertation, University of Virginia, 1953.

Hunt, Charles W. "The First Fifty Years of the American Society of Civil Engineers, 1852–1902." American Society of Civil Engineers, *Transactions* 48 (1902): 220–226.

———. *Historical Sketch of the American Society of Civil Engineers.* New York, 1897.

Hyder, Clyde K. *Snow of Kansas.* Lawrence, Kansas, 1953.

Jackson, Charles L. "Wolcott Gibbs, 1822–1908." *American Journal of Science,* Fourth Series, 27 (1909): 253–59.

James, Henry. *Charles W. Eliot.* 2 vols. Boston, 1930.

Johanssen, Robert W. *Stephen A. Douglas.* New York, 1973.

Johnson, Allen, and Malone, Dumas, eds. *Dictionary of American Biography.* 22 vols. New York, 1928–58.

Johnson, Thomas C. *Scientific Interests in the Old South.* New York, 1936.

Johnston, James F. W. *Notes on North America,* 2 vols. Boston, 1851.

Jones, Bessie Z., and Boyd, Lyle G. *The Harvard College Observatory: The First Four Directorships, 1839–1919.* Cambridge, Mass., 1971.

Joyner, Fred B. *David Ames Wells, Champion of Free Trade.* Cedar Rapids, Iowa, 1939.

Kasson, John F. *Civilizing the Machine: Technology and Republican Values in America, 1776–1900.* New York, 1976.

Kendall, Phebe M., ed. *Maria Mitchell: Life, Letters, and Journals.* Boston, 1896.

Kevles, Daniel J. *The Physicists: The History of a Scientific Community in Modern America.* New York, 1978.

———. "The Study of Physics in America, 1865–1916." Ph.D. dissertation, Princeton University, 1964.

Klem, M. J. "The History of Science in St. Louis." Academy of Science of St. Louis *Transactions* 23, no. 2, 1914.

Kohlstedt, Sally G. *The Formation of the American Scientific Community: The American Association for the Advancement of Science, 1848–60.* Urbana, Ill., 1976.

Kranzberg, Melvin, and Pursell, Carroll W., Jr. *Technology in Western Civilization.* 2 vols. New York, 1967.

Kreuter, Kent K. "The Literary Response to Science, Technology and Industrialism: Studies in the Thought of Hawthorne, Melville, Whitman and Twain." Ph.D. dissertation, University of Wisconsin, 1963.

Larson, Robert L. "Charles Frederick Chandler, His Life and Work." Ph.D. dissertation, Columbia University, 1950.

Law, Ernest M., et al. *State Agricultural Experiment Stations.* Washington, D.C., 1962.

Layton, Edwin T., Jr. "Mirror-Image Twins: The Communities of Science and Technology." In Daniels, *Nineteenth-Century American Science.*

LeConte, Joseph. *The Autobiography of Joseph LeConte.* Edited by William D. Armes. New York, 1903.

Lewis, Gene D. *Charles Ellet, Jr.: The Engineer as Individualist, 1810–1862.* Urbana, Ill., 1968.

Lodge, Henry Cabot. *Early Memories.* New York, 1913.
Loomis, Elias. *The Recent Progress of Astronomy.* New York, 1856.
Lupold, John S. "From Physician to Physicist: The Scientific Career of John LeConte, 1818–1891." Ph.D. dissertation, University of South Carolina, 1970.
Lurie, Edward. *Louis Agassiz: A Life in Science.* Chicago, 1960.
Lyell, Charles. *A Second Visit to the United States of North America,* 2 vols. New York, 1849.
Mack, Edward C. *Peter Cooper, Citizen of New York.* New York, 1949.
Main, Angela K. "Thure Kumlien, Koshkonong Naturalist." *Wisconsin Magazine of History* 27 (1933–34): 17–39, 194–220, 321–343.
Mallet, John W. *Chemistry Applied to the Arts.* Lynchburg, Va., 1868.
Mann, Charles R. *A Study of Engineering Education.* New York, 1918.
Massachusetts College of Agriculture. *Charles Anthony Goessmann.* Cambridge, Mass., 1917.
McCaughey, Robert A. "The Transformation of American Academic Life: Harvard University 1821–1892." *Perspectives in American History* 8 (1974): 239–332.
McCormmach, Russell. "Ormsby MacKnight Mitchel's *Sidereal Messenger,* 1846–1848." American Philosophical Society *Proceedings,* Feb. 18, 1966.
Meisel, Max. *A Bibliography of American Natural History.* 3 vols. New York, 1924.
Memorials of John Pitkin Norton. Albany, N.Y., 1853.
Merrill, George P. *The First One Hundred Years of American Geology.* New Haven, Conn., 1924.
Merritt, Raymond H. *Engineering in American Society, 1850–1875.* Lexington, Ky., 1969.
Miller, Howard S. *Dollars for Research: Science and Its Patrons in Nineteenth-Century America.* Seattle, Wash., 1970.
Mitchel, Frederick A. *Ormsby MacKnight Mitchel.* Boston, 1887.
Morehead, Lavinia M. *A Few Incidents in the Life of Professor James P. Espy.* Cincinnati, 1888.
Nevins, Allan, and Thomas, Milton H., eds. *The Diary of George Templeton Strong.* 4 vols. New York, 1952.
Newcomb, Simon. "Abstract Science in America, 1776–1876." *North American Review* 122 (1876).
———. "Exact Science in America." *North American Review* 119 (1874).
———. *The Reminiscences of an Astronomer.* Boston, 1903.
Nolan, Edward J. *A Short History of the Academy of Natural Sciences at Philadelphia.* Philadelphia, 1909.
Oberholtzer, Ellis P. *Philadelphia: A History of the City and Its People.* 4 vols. Philadelphia, 1912.
Odgers, Merle M. *Alexander Dallas Bache, Scientist and Educator, 1806–1867.* Philadelphia, 1949.
Oehser, Paul H. *Sons of Science.* New York, 1949.
Oleson, Alexandra, and Brown, Sanborn C., eds. *The Pursuit of Knowledge in the Early American Republic: American Scientific and Learned Societies from Colonial Times to the Civil War.* Baltimore, 1976.
Oleson, Alexandra, and Voss, John, eds. *The Organization of Knowledge in Modern America, 1860–1920.* Baltimore, 1979.
Olmsted, Denison. "On the Democratic Tendencies of Science." *American Journal of Education* 1 (1856).
Osborne, Elizabeth A., ed. *From the Letter-Files of S. W. Johnson.* New Haven, Conn., 1913.
Pfeifer, Edward J. "The Reception of Darwinism in the United States, 1859–1880." Ph.D. dissertation, Brown University, 1957.
Poinsett, Joel R. *Discourse on the Objects and Importance of the National Institution for the Promotion of Science.* Washington, D.C., 1841.
Post, Robert C. "Science, Public Policy, and Popular Precepts: Alexander Dallas Bache and Alfred Beach as Symbolic Adversaries." In Reingold, *Sciences.*
Prendergast, Michael L. "James Dwight Dana: The Life and Thought of an American Scientist." Ph.D. dissertation, University of California, Los Angeles, 1978.
Prescott, Samuel C. *When M.I.T. Was "Boston Tech."* Cambridge, Mass., 1954.
Pumpelly, Raphael. *My Reminiscences.* New York, 1918.
Reingold, Nathan. "Alexander Dallas Bache: Science and Technology in the American

Idiom." *Technology and Culture* 11 (1970): 163–177.
———. "Reflections on 200 Years of Science in the United States." In Reingold, *Sciences.*
———, ed. *Science in America Since 1820.* New York, 1976.
———, ed. *Science in Nineteenth Century America: A Documentary History.* New York, 1964.
———, ed. *The Sciences in the American Context: New Perspectives.* Washington, D.C. 1979.
Rezneck, Samuel. *Education for a Technological Society: A Sesquicentennial History of Rensselaer Polytechnic Institute.* Troy, N.Y., 1968.
———, "The Emergence of a Scientific Community in New York State a Century Ago." *New York History,* July 1962, pp. 211–38.
———, "The European Education of an American Chemist and Its Influence in 19th-Century America: Eben Norton Horsford." *Technology and Culture* 11 (1970): 366–88.
Rhees, William J. *Manual of Public Libraries, Institutions, and Societies, in the United States . . .* Philadelphia, 1859.
———, ed. *The Smithsonian Institution: Documents Relative to Its Origin and History, 1835–1899.* 2 vols. Washington, D.C., 1901.
———, ed. *The Smithsonian Institution: Journals of the Board of Regents.* Washington, D.C., 1879.
Ricketts, Palmer C. *History of the Rensselaer Polytechnic Institute, 1824–1894.* Troy, N.Y., 1895.
Rodgers, Andrew D., III. *John Torrey: A Story of North American Botany.* Princeton, N.J., 1942.
———. *"Noble Fellow": William Starling Sullivant.* New York, 1940.
Rogers, Emma, ed. *Life and Letters of William Barton Rogers.* 2 vols. Boston, 1896.
Rood, Ogden N. *The Practical Value of Physical Science.* Troy, N.Y., 1859.
Roorbach, Orville A. *Bibliotheca Americana. American Publications . . . 1820 to 1848, . . .* New York, 1849.
Rosenberg, Charles E. *No Other Gods.* Baltimore, 1976.
Ross, Earle D. *Democracy's College.* Ames, Iowa, 1942.
Rossiter, Margaret W. *The Emergence of Agricultural Science: Justus Liebig and the Americans, 1840–1880.* New Haven, Conn., 1975.
Rothenberg, Marc. "The Educational and Intellectual Background of American Astronomers, 1825–1875." Ph.D. dissertation, Bryn Mawr, 1974.
Rudolph, Frederick. *The American College and University, A History.* New York, 1962.
———. *Curriculum.* San Francisco, 1977.
Ruschenberger, William S. W. *Report of the Condition of the Academy of Natural Sciences of Philadelphia.* Philadelphia, 1876.
Rutgers University. *Addresses Commemorative of George Hammell Cook.* Newark, N.J., 1891.
Safford, Truman H. *The Development of Astronomy in the United States.* Williamstown, Mass., 1888.
Sands, Benjamin F. *From Reefer to Rear Admiral.* New York, 1899.
Schofield, John M. *Forty-Six Years in the Army.* New York, 1897.
Searles, James M. *Life and Times of a Civil Engineer.* Cincinnati, 1893.
Shaler, Nathaniel S. *The Autobiography of Nathaniel Southgate Shaler.* Boston, 1909.
Shapiro, Henry D. "The Western Academy of Natural Sciences of Cincinnati and the Structure of Science in the Ohio Valley, 1810–1850." In Oleson and Brown, *Pursuit of Knowledge.*
Sidar, Jean W. *George Hammell Cook: A Life in Agriculture and Geology.* New Brunswick, N.J., 1976.
Siegfried, Robert. "A Study of Chemical Research Publications from the United States before 1880." Ph.D. dissertation, University of Wisconsin, 1953.
Silliman, Benjamin, Jr. *The World of Science, Art and Industry.* New York, 1854.
Sinclair, Bruce. *Philadelphia's Philosopher Mechanics: A History of the Franklin Institute, 1824–1865.* Baltimore, 1975.
———. "The Promise of the Future: Technical Education." In Daniels, *Nineteenth-Century Science.*
Smallwood, William M. and Mabel S. C. *Natural History and the American Mind.* New York, 1941.
Smith, David E., and Ginsburg, Jekuthiel. *A History of Mathematics in America before 1900.* Chicago, 1934.

Smith, Edgar F. *Charles Mayer Wetherill.* New York, 1929.
———. *James Curtis Booth, Chemist.* Philadelphia, 1922.
Smithsonian Institution Board of Regents. *Annual Report.* Washington, D.C., 1847–78.
Spence, Clark C. *Mining Engineers and the American West.* New Haven, Conn., 1970.
Stanton, William. *The Leopard's Spots.* Chicago, 1960.
———. *The Great United States Exploring Expedition of 1838–1842.* Berkeley, Calif., 1975.
Stephens, Lester D. *Joseph LeConte: Gentle Prophet of Evolution.* Baton Rouge, La., 1982.
Stevens, Hazard. *The Life of Isaac I. Stevens.* 2 vols. Boston, 1900.
Stokes, Anson P. *Memorials of Eminent Yale Men.* 2 vols. New Haven, Conn., 1914.
Stone, Bruce W. "The Role of the Learned Societies in the Growth of Scientific Boston, 1780–1848." Ph.D. dissertation, Boston University, 1974.
Storr, Richard J. *The Beginnings of Graduate Education in America.* Chicago, 1953.
Sydnor, Charles S. *A Gentleman of the Old Natchez Region, Benjamin L. C. Wailes.* Durham, N.C., 1938.
Taylor, George R. *The Transportation Revolution, 1815–1860.* New York, 1951.
Taylor, Herbert F. *Seventy Years of the Worcester Polytechnic Institute.* Worcester, Mass., 1937.
Thayer, James B. *Letters of Chauncey Wright.* Cambridge, Mass., 1878.
Thomas, Milton H. "The Gibbs Affair at Columbia in 1854." M.A. thesis, Columbia University, 1942.
Tocqueville, Alexis de. *Democracy in America.* Vintage paperback edition in 2 vols. New York, 1954.
Towle, Edward L. "Science, Commerce and the Navy on the Sea-Faring Frontier (1842–1861)." Ph.D. dissertation, University of Rochester, 1966.
Van Amringe, John H. *An Historical Sketch of Columbia College.* New York, 1876.
Van Klooster, Henry S. "Friedrich Wöhler and His American Pupils." *Journal of Chemical Education* 21 (1944): 158–170.
———. "Liebig and His American Pupils." *Journal of Chemical Education* 33 (1956): 493–97.
Verrill, George E. *The Ancestry, Life and Work of Addison E. Verrill.* Santa Barbara, Calif., 1958.
Viles, Jonas. *The University of Missouri: A Centennial History.* Columbia, Mo., 1939.
Warner, Deborah J. "Astronomy in Antebellum America." In Reingold, *Sciences.*
Wayman, Dorothy G. *Edward Sylvester Morse.* Cambridge, Mass., 1942.
Wertenbaker, Thomas J. *Princeton 1746–1896.* Princeton, N.J., 1946.
Whitnah, Donald R. *A History of the United States Weather Bureau.* Urbana, Ill., 1961.
Williams, Frances L. *Matthew Fontaine Maury, Scientist of the Sea.* New Brunswick, N.J., 1963.
Wright, Marmaduke B. *An Address on the Life and Character of the Late Prof. John Locke.* Cincinnati, 1857.

Acknowledgments

Once upon a time, when John F. Kennedy was president and the world was young, the late Allan Nevins persuaded me to embark on a study that grew over the years into something more than either of us had expected. However the result in hand may be received by others, I shall remain grateful to him for propelling me into the exhilaration of this long (though intermittent) chase. I am also grateful to Nathan Reingold and Edwin T. Layton, who read the early chapters of my first draft on science and technology, respectively, and sent me detailed, constructive, written critiques. Charles Rosenberg, Daniel J. Kevles, and my Boston University colleague David D. Hall each read several chapters of that early effort and gave me suggestions and encouragement. My former Wisconsin student Edward H. Beardsley confirmed and augmented my evaluation of the Civil War's impact on the chemistry profession. At a later stage my colleague Saul Engelbourg read and concurred in my view of the war's impact on civilian technology. And Sally Gregory Kohlstedt read and generally endorsed my treatment of the American Association for the Advancement of Science, on which her book remains definitive for the antebellum years. All these reassurances helped keep the long interruptions in this enterprise from becoming terminal.

At Boston University I have improvidently neglected to steer my doctoral candidates into the subjects I was working on, though I have made extensive use of dissertations written elsewhere. Among my many graduate students, therefore, only three happened to contribute directly to this particular undertaking: John B. Cusack, who, as my research assistant in 1964, expertly analyzed and coded more than a thousand *Dictionary of American Biography* articles, as well as looking up hundreds of call numbers; George Wise, who read several chapters on technology and made helpful suggestions; and Bruce W. Stone, whose outstanding dissertation on Boston scientific societies to 1848 (see Other Sources Cited) barely touched my time period but furnished valuable insights into the Boston background of my story.

In the summer of 1966 the Henry E. Huntington Library granted me a fellowship that helped me make the most of its rich resources. Boston University not only gave me regular sabbaticals but also agreed to an arrangement of time and courses in late years that has eased and expedited what had begun to seem like an endless journey. The staffs of several score libraries and archives have been uniformly helpful, as evidenced by my manuscript citations. My source notes further attest my indebtedness to hundreds of my fellow historians, past and present. I wish I had space also to salute by name the dozens of unsung but indispensable bibliographers on whose work I and many others have drawn without specific acknowledgment.

From first to last my sister Marilyn and her husband Wendell J. Greene have given me essential aid and comfort, as have my nieces Constance Fenner and Deborah Talaba and their families. For this long book, as for my previous ones, Marilyn has typed many hundreds of clean pages from palimpsests of scrawls and interlineations. Once again, thanks.

Madbury, New Hampshire
July 4, 1985

Index

A NOTE ABOUT THE AUTHOR

Robert V. Bruce graduated from the University of New Hampshire with a degree in mechanical engineering and received his Ph.D. in history from Boston University. In 1984 he was named Professor Emeritus of Boston University where he has taught since 1955, and still teaches each fall. His other honors include being named a Guggenheim Fellow, 1957–58, Fellow of the Henry E. Huntington Library, 1966, and Fellow of the Society of American Historians. Bruce resides in Madbury, New Hampshire, and is the author of three previous books: *Lincoln and the Tools of War* (1956), *1877*: Year of Violence (1959) and *Bell: Alexander Graham Bell and the Conquest of Solitude* (1973).